IET ENERGY ENGINEERING SERIES 96

Industrial Power Systems with Distributed and Embedded Generation

Other volumes in this series:

Volume 1	**Power Circuit Breaker Theory and Design** C.H. Flurscheim (Editor)
Volume 4	**Industrial Microwave Heating** A.C. Metaxas and R.J. Meredith
Volume 7	**Insulators for High Voltages** J.S.T. Looms
Volume 8	**Variable Frequency AC Motor Drive Systems** D. Finney
Volume 10	**SF_6 Switchgear** H.M. Ryan and G.R. Jones
Volume 11	**Conduction and Induction Heating** E.J. Davies
Volume 13	**Statistical Techniques for High Voltage Engineering** W. Hauschild and W. Mosch
Volume 14	**Uninterruptible Power Supplies** J. Platts and J.D. St Aubyn (Editors)
Volume 15	**Digital Protection for Power Systems** A.T. Johns and S.K. Salman
Volume 16	**Electricity Economics and Planning** T.W. Berrie
Volume 18	**Vacuum Switchgear** A. Greenwood
Volume 19	**Electrical Safety: A guide to causes and prevention of hazards** J. Maxwell Adams
Volume 21	**Electricity Distribution Network Design, 2nd Edition** E. Lakervi and E.J. Holmes
Volume 22	**Artificial Intelligence Techniques in Power Systems** K. Warwick, A.O. Ekwue and R. Aggarwal (Editors)
Volume 24	**Power System Commissioning and Maintenance Practice** K. Harker
Volume 25	**Engineers' Handbook of Industrial Microwave Heating** R.J. Meredith
Volume 26	**Small Electric Motors** H. Moczala *et al.*
Volume 27	**AC-DC Power System Analysis** J. Arrillaga and B.C. Smith
Volume 29	**High Voltage Direct Current Transmission, 2nd Edition** J. Arrillaga
Volume 30	**Flexible AC Transmission Systems (FACTS)** Y.-H. Song (Editor)
Volume 31	**Embedded Generation** N. Jenkins *et al.*
Volume 32	**High Voltage Engineering and Testing, 2nd Edition** H.M. Ryan (Editor)
Volume 33	**Overvoltage Protection of Low-Voltage Systems, Revised Edition** P. Hasse
Volume 36	**Voltage Quality in Electrical Power Systems** J. Schlabbach *et al.*
Volume 37	**Electrical Steels for Rotating Machines** P. Beckley
Volume 38	**The Electric Car: Development and future of battery, hybrid and fuel-cell cars** M. Westbrook
Volume 39	**Power Systems Electromagnetic Transients Simulation** J. Arrillaga and N. Watson
Volume 40	**Advances in High Voltage Engineering** M. Haddad and D. Warne
Volume 41	**Electrical Operation of Electrostatic Precipitators** K. Parker
Volume 43	**Thermal Power Plant Simulation and Control** D. Flynn
Volume 44	**Economic Evaluation of Projects in the Electricity Supply Industry** H. Khatib
Volume 45	**Propulsion Systems for Hybrid Vehicles** J. Miller
Volume 46	**Distribution Switchgear** S. Stewart
Volume 47	**Protection of Electricity Distribution Networks, 2nd Edition** J. Gers and E. Holmes
Volume 48	**Wood Pole Overhead Lines** B. Wareing
Volume 49	**Electric Fuses, 3rd Edition** A. Wright and G. Newbery
Volume 50	**Wind Power Integration: Connection and system operational aspects** B. Fox *et al.*
Volume 51	**Short Circuit Currents** J. Schlabbach
Volume 52	**Nuclear Power** J. Wood
Volume 53	**Condition Assessment of High Voltage Insulation in Power System Equipment** R.E. James and Q. Su
Volume 55	**Local Energy: Distributed generation of heat and power** J. Wood
Volume 56	**Condition Monitoring of Rotating Electrical Machines** P. Tavner, L. Ran, J. Penman and H. Sedding
Volume 57	**The Control Techniques Drives and Controls Handbook, 2nd Edition** B. Drury
Volume 58	**Lightning Protection** V. Cooray (Editor)
Volume 59	**Ultracapacitor Applications** J.M. Miller
Volume 62	**Lightning Electromagnetics** V. Cooray
Volume 63	**Energy Storage for Power Systems, 2nd Edition** A. Ter-Gazarian
Volume 65	**Protection of Electricity Distribution Networks, 3rd Edition** J. Gers
Volume 66	**High Voltage Engineering Testing, 3rd Edition** H. Ryan (Editor)
Volume 67	**Multicore Simulation of Power System Transients** F.M. Uriate
Volume 68	**Distribution System Analysis and Automation** J. Gers
Volume 69	**The Lightening Flash, 2nd Edition** V. Cooray (Editor)
Volume 70	**Economic Evaluation of Projects in the Electricity Supply Industry, 3rd Edition** H. Khatib
Volume 72	**Control Circuits in Power Electronics: Practical issues in design and implementation** M. Castilla (Editor)
Volume 73	**Wide Area Monitoring, Protection and Control Systems: The enabler for Smarter Grids** A. Vaccaro and A. Zobaa (Editors)
Volume 74	**Power Electronic Converters and Systems: Frontiers and applications** A.M. Trzynadlowski (Editor)

Volume 75 **Power Distribution Automation** B. Das (Editor)
Volume 76 **Power System Stability: Modelling, analysis and control** B. Om P. Malik
Volume 78 **Numerical Analysis of Power System Transients and Dynamics** A. Ametani (Editor)
Volume 79 **Vehicle-to-Grid: Linking electric vehicles to the smart grid** J. Lu and J. Hossain (Editors)
Volume 81 **Cyber-Physical-Social Systems and Constructs in Electric Power Engineering** S. Suryanarayanan, R. Roche and T.M. Hansen (Editors)
Volume 82 **Periodic Control of Power Electronic Converters** F. Blaabjerg, K. Zhou, D. Wang and Y. Yang
Volume 86 **Advances in Power System Modelling, Control and Stability Analysis** F. Milano (Editor)
Volume 87 **Cogeneration: Technologies, optimisation and implementation** C.A. Frangopoulos (Editor)
Volume 88 **Smarter Energy: From smart metering to the smart grid** H. Sun, N. Hatziargyriou, H.V. Poor, L. Carpanini and M.A. Sánchez Fornié (Editors)
Volume 89 **Hydrogen Production, Separation and Purification for Energy** A. Basile, F. Dalena, J. Tong and T.N. Veziroğlu (Editors)
Volume 90 **Clean Energy Microgrids** S. Obara and J. Morel (Editors)
Volume 91 **Fuzzy Logic Control in Energy Systems with Design Applications in MATLAB®/Simulink®** İ.H. Altaş
Volume 92 **Power Quality in Future Electrical Power Systems** A.F. Zobaa and S.H.E.A. Aleem (Editors)
Volume 93 **Cogeneration and District Energy Systems: Modelling, analysis and optimization** M.A. Rosen and S. Koohi-Fayegh
Volume 94 **Introduction to the Smart Grid: Concepts, technologies and evolution** S.K. Salman
Volume 95 **Communication, Control and Security Challenges for the Smart Grid** S.M. Muyeen and S. Rahman (Editors)
Volume 97 **Synchronized Phasor Measurements for Smart Grids** M.J.B. Reddy and D.K. Mohanta (Editors)
Volume 98 **Large Scale Grid Integration of Renewable Energy Sources** A. Moreno-Munoz (Editor)
Volume 100 **Modeling and Dynamic Behaviour of Hydropower Plants** N. Kishor and J. Fraile-Ardanuy (Editors)
Volume 101 **Methane and Hydrogen for Energy Storage** R. Carriveau and D.S.-K. Ting
Volume 104 **Power Transformer Condition Monitoring and Diagnosis** A. Abu-Siada (Editor)
Volume 106 **Surface Passivation of Industrial Crystalline Silicon Solar Cells** J. John (Editor)
Volume 107 **Bifacial Photovoltaics: Technology, applications and economics** J. Libal and R. Kopecek (Editors)
Volume 108 **Fault Diagnosis of Induction Motors** J. Faiz, V. Ghorbanian and G. Joksimović
Volume 110 **High Voltage Power Network Construction** K. Harker
Volume 111 **Energy Storage at Different Voltage Levels: Technology, integration, and market aspects** A.F. Zobaa, P.F. Ribeiro, S.H.A. Aleem and S.N. Afifi (Editors)
Volume 112 **Wireless Power Transfer: Theory, technology and application** N. Shinohara
Volume 115 **DC Distribution Systems and Microgrids** T. Dragičević, F. Blaabjerg and P. Wheeler
Volume 117 **Structural Control and Fault Detection of Wind Turbine Systems** H.R. Karimi
Volume 119 **Thermal Power Plant Control and Instrumentation: The control of boilers and HRSGs, 2nd Edition** D. Lindsley, J. Grist and D. Parker
Volume 120 **Fault Diagnosis for Robust Inverter Power Drives** A. Ginart (Editor)
Volume 123 **Power Systems Electromagnetic Transients Simulation, 2nd Edition** N. Watson and J. Arrillaga
Volume 124 **Power Market Transformation** B. Murray
Volume 126 **Diagnosis and Fault Tolerance of Electrical Machines, Power Electronics and Drives** A. J.M. Cardoso
Volume 128 **Characterization of Wide Bandgap Power Semiconductor Devices** F. Wang, Z. Zhang and E.A. Jones
Volume 130 **Wind and Solar Based Energy Systems for Communities** R. Carriveau and D.S.-K. Ting (Editors)
Volume 131 **Metaheuristic Optimization in Power Engineering** J. Radosavljević
Volume 132 **Power Line Communication Systems for Smart Grids** I.R.S. Casella and A. Anpalagan
Volume 905 **Power System Protection, 4 volumes**

Industrial Power Systems with Distributed and Embedded Generation

Radian Belu

The Institution of Engineering and Technology

Published by The Institution of Engineering and Technology, London, United Kingdom

The Institution of Engineering and Technology is registered as a Charity in England & Wales (no. 211014) and Scotland (no. SC038698).

First published 2018

The Institution of Engineering and Technology
Michael Faraday House
Six Hills Way, Stevenage
Herts, SG1 2AY, United Kingdom

www.theiet.org

British Library Cataloguing in Publication Data
A catalogue record for this product is available from the British Library

ISBN 978-1-78561-152-0 (hardback)
ISBN 978-1-78561-153-7 (PDF)

Typeset in India by MPS Limited

To my wife Paulina Belu, my best friend and partner in life,
for her patience and support, and to the memory
of my parents Grigore and Gheorghita Belu.

Contents

Preface

In the last decades, our society and every country are facing with energy challenges, such as energy conservation, the use of increased energy-efficient equipment, devices or systems, security of the energy supply, energy portfolio diversification, sustainability, or pollution reduction and control. The book goal and objective is to provide the engineering students, as well as the engineers and technicians interested in industrial power distribution and renewable energy systems with essential knowledge of the major technologies, their fundamental principles, characteristics, and how they work and how they are evaluate in order to properly select the optimum system or equipment. The book covers major disciplines: basic and fundamental knowledge in power systems, such as power engineering basic, motors and transformers, and in building and industrial power distribution, such as load characteristics and calculations, load and motor centers, building electrical systems and lighting, or motor protection and control. Four chapters of the book are giving comprehensive knowledge of the major renewable energy topics, such as solar and wind energy, photovoltaics, geothermal energy and small hydro-power systems, major energy storage technologies, and a brief description of microgrids, distributed generation issues and energy management. This book originates from courses that the author taught in the areas of energy and power engineering, renewable energy, industrial power distribution, or building electrical systems. Technical content and presentations are independent of any specific technology, being neutral to any specific technology, while these are adhering to several premises of the renewable energy or industrial energy systems. The book is intended both as a textbook and as reference book for students, instructors, engineers, and professionals interested in industrial energy systems, renewable energy, and electrical systems design. The author is indebted to the students and colleagues and co-professionals for their feedback and suggestions over the years, and last but not the least to the editor technical staff for support and help.

Chapter 1
Introduction, review of electric circuits

Objectives and abstract

Power systems are undergoing significant changes in terms of how they are operated, how electricity is generated and transferred to the users, and how the consumers interact and participate with the power systems. The main focus of this book is to provide the engineers, students, or interested readers with the essential knowledge of the power and energy systems, as well as main energy technologies including how they work and operate, and how they are evaluated and selected for specific applications. The purpose of this chapter is to introduce the engineers, students, or interested readers to the contemporary energy system issues and challenges, and brief the historical perspective of the power system evolution. The sections of this chapter are giving a quite comprehensive description of electric circuit theorems and solutions methods, direct current (DC) and alternative current (AC) circuits, power in AC circuits, and other important issues, terms, and definitions. The last section of the chapter gives a brief summary of the unit system and measurements. Several examples are included in sections to help in better understanding of the chapter. The reader must be fully aware that good understanding of DC and AC circuit theorems and solving methods, power in AC circuits, and the measurements and units are vital for the understanding of power and energy systems, analysis, design, operation, and management of these systems. These are the chapter objectives, goals, and aims. The chapter may be useful and recommended even for the readers who are fully familiar with the topics of the chapter.

1.1 Introduction, power system structure, brief history of power system

Power engineering, power system analysis, and theory are very broad subjects to be covered, even at the most basic level in a single textbook. Power and electrical engineers are concerned with each and every step and aspect of the process of electricity generation, transmission, distribution, and utilization. Power industry is probably the largest and most complex industry in the history of mankind. Electrical and power engineers who work in this industry encounter challenging problems in designing and shaping future power systems, to deliver increasing amounts of electrical energy with highest quality standards, in safe, clean, efficient, and economical manner. Power system industry and utilities significantly contribute to the welfare, life standard, progress, and technological advances of humanity. The growth and demand of electrical energy in the world, as a whole and in each country, in the past half century was phenomenal, with over 50 times the growth and demand rates in all other energy forms used during the same period. Today, the installed electrical power capacity per capita in the US is estimated at about 3 kW, with similar values in the EU and developed countries. There is also significant increase in the installed capacity per capita in China, Brazil, or India and in most of the developing countries. In 1878, Thomas Edison began to work on the electric light and formulated the concept of centrally generated power with distributed lighting serving the surrounding area. Historically, Edison Electric Illuminating Company of New York inaugurated the first commercial power plant at the Pearl Street station in 1882, with six engine-dynamo (DC generator) sets, with a capacity of four 250-HP boilers. It uses a 110-V DC underground distribution with copper cables insulated with jute wrapping. In 1882, the first water wheel-driven generator was installed in Appleton, Wisconsin. The introduction of the DC motor by Sprague Electric, the growth of incandescent lighting, and the development of three-wire 220-V DC systems allowed load to increase somewhat and promoted the expansion of Edison's DC systems. However, the voltage problems and transmission over long distances continued in the DC systems. These limitations of maximum distance and load were overcome by the development of William Stanley of a practical transformer. With the transformer, the ability to transmit power at high voltage and lower line voltage drops made AC more attractive than DC. The first practical AC distribution was installed by W. Stanley at Great Barrington, Massachusetts, in 1885 for Westinghouse, who acquired the American rights to the transformer from its British inventors. The first single-phase AC line of 4 kV was operated in United States in 1889, between Oregon City and Portland, over 20 km distance. The first three-phase line in Germany became operational in 1891, transmitting power 179 km at 12 kV. Southern California Edison Company established the first three-phase 2.3 kV system in 1893. The invention of induction machines by Nikola Tesla in 1887 helped to replace DC motors in many applications and hastened the advances in the use of AC electrical systems. In 1884, Timisoara, Romania was the first city of Europe with electric street lighting installed.

From 1882 through 1972, the electric utility industry grew at a remarkable pace, growth was based on the outstanding technological and scientific advances and engineering creativity. In the United States, only the electric energy sales have grown over 400 times during this period. A growth rate was 50 times the growth rate in all energy forms used during the same period. During the early period, the electricity was generated in steam-powered and water-powered turbine plants. Today, steam turbine accounts for about 85% of the US electricity generation, while hydro-turbine for about 7%. Gas turbines are often used to meet the peak load demands. Renewable energy generation, mainly wind and solar energy, is also expected to grow considerably in the near future. Steam turbines are fueled primarily by coal, gas, oil, and uranium. In 1957, nuclear units of 90-MW steam turbines were installed at the Shipping-port Atomic Power Station, near Pittsburg. The United States is the world's largest supplier of commercial nuclear power, and generated 33% of the world's nuclear electricity in 2013. After 1990s, the fuel of choice for new power plant in the United States was natural gas due to its availability, low cost, higher efficiency, lower pollutant emissions, shorter construction time, safety, and lack of controversy associated with this type of power plants. In the past three decades, there was an increasing trend to utilize the renewable energy sources for electricity generation, due to their abundance, almost zero emissions during the operation, and technological advances, making them economically viable options. Renewable energy sources include other conventional hydro-power plants, solar-thermal, wind energy, photovoltaics, biomass, or ocean energy. In 2012, approximately 12% of the US electricity was generated by renewable energy sources, while in Germany about 31% of electricity generated from renewables was achieved in 2014.

1.2 Electric circuit review

Electrical circuits connect power supplies to loads, such as: *resistors*, *capacitors*, *inductors*, *electric motors*, *heaters*, or *electrical lamps*. The connection between the supply and the load is made by soldering with wires or conductors that are often called as leads, or with many kinds of connectors and terminals. Energy is delivered from the source to the user on demand by turning a switch to ON. Sometimes several circuit elements are connected to the same lead, which is called a common lead for those elements. Various parts of the circuits are called circuit elements, which can be in series or in parallel. Electric network theory deals with two primitive quantities, which we will refer to as: potential (or voltage) and current. Current is the actual flow of charged carriers, while difference in potential is the force that causes that flow. As we will study, potential is a single-valued function that may be uniquely defined over the nodes of a network. Current, on the other hand, flows through the branches of the network. Figure 1.1 shows the basic notion of a branch, in which a voltage is defined across the branch and a current is defined to flow through the branch. A network is a collection of such elements, connected together by wires.

Network topology is the interconnection of its elements. That, plus the constraints on voltage and current imposed by the elements themselves, determines the performance of the network, described by the distribution of voltages and currents throughout the network. Two important concepts must be described initially. These

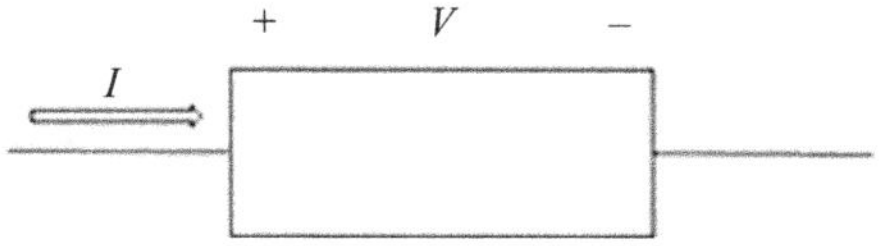

Figure 1.1 Basic circuit element

are of "loop" and "node". A *node* is a point in a circuit where three or more elements are interconnected. A *branch* is a current path between two nodes. Each branch in a circuit can have only one current in it although a branch may have no current. A *loop* is a closed path consisting of different branches with different currents in each branch. So loops in the network are any closed path through two or more elements of the network. Any nontrivial network will have at least one such loop. The two fundamental laws of network theory are known as Kirchhoff's Voltage Law (KVL), and Kirchhoff's Current Law (KCL). These laws describe the topology of the network, and arise directly from the fundamental laws of electromagnetics. They are simply stated as: (a) KVL states that around any loop of a network, the sum of all voltages, taken in the same direction, is zero:

$$\sum_{loop} v_k = 0 \tag{1.1}$$

And (b) KCL states that at any node of a network, the sum of all currents entering the node is zero:

$$\sum_{node} i_k = 0 \tag{1.2}$$

It is worth to note that KVL is a discrete version of Faraday's law, valid to the extent that no time-varying flux links the loop. At any point, where there is a node formed by the junction of various current carrying branches, by current conservation, the sum of the currents into the node must be equal the sum of the currents out of the node (otherwise charge would build up at the junction). KCL is just the current conservation, allowing for no accumulation of the electrical charge at the node. The rules for determining ΔV across a resistor and a battery with a designated travel direction are shown in Figure 1.2. Note that the choice of travel direction is arbitrary. The same equation is obtained whether the closed loop is traversed clockwise or counterclockwise.

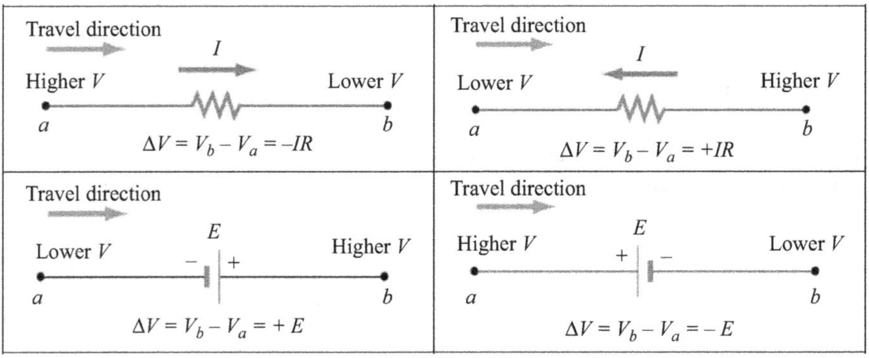

Figure 1.2 Rules for determining potential difference across resistors and batteries

Example 1.1: Consider a source of emf, V_{in} that is connected in series to two resistors, R_1 and R_2, as shown in Figure 1.3. Calculate the voltage across the resistor R_2.

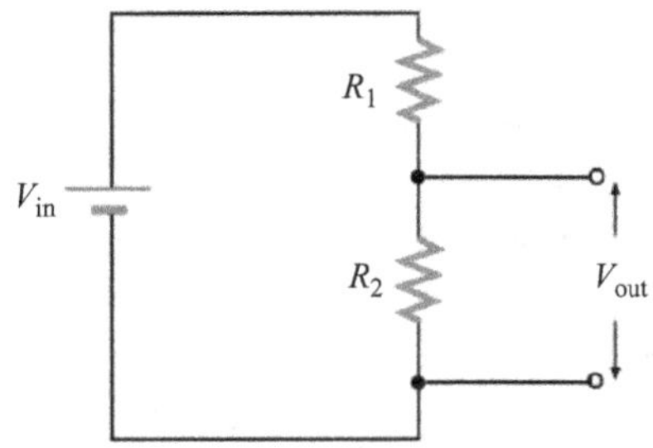

Figure 1.3 Voltage divider

Solution: From the KVL (loop rule), the potential difference, V_{out}, across resistor R_2 is less than V_{in}, and is given by:

$$V_{in} - IR_1 - IR_2 = 0 \Rightarrow I = \frac{V_{in}}{R_1 + R_2}$$

Thus, the potential difference, V_{out}, across resistor R_2 is given by:

$$V_{out} = R_2 I = \frac{R_2}{R_1 + R_2} V_{in}$$

This circuit is called a *voltage divider*, having many applications in electronics.

Network elements affect voltages and currents in one of three ways: (1) voltage sources constrain the potential difference across their terminals to be of some fixed value (the value of the source); (2) current sources constrain the current through the branch to be of some fixed value; and (3) all other elements impose some sort of relationship, either linear or nonlinear, between voltage across and current through the branch. Voltage and current sources can be either independent or dependent. Independent sources have values which are, as the name implies, independent of other variables in a circuit. Dependent sources have values which depend on some other variable in a circuit. A common example of a dependent source is the equivalent current source used for modeling the collector junction in a transistor. Typically, this is modeled as a current dependent current source, in which collector current is taken to be directly dependent on the emitter current. Such dependent sources must be handled with some care, for certain tricks we will be discussing later do not work with them. For the present time, we will consider, in addition to voltage and current sources, only impedance elements, which impose a linear relationship between voltage and current. The most common of these is the resistance, which imposes the relationship which is often referred to as Ohm's law:

$$v(t) = Ri(t) \tag{1.3}$$

Example 1.2: The voltage across a 33 kΩ resistance, R is 660 mV. What is the current through that resistance?

Solution: From (1.3), the current through the resistance is given by:

$$I = \frac{V}{R} = \frac{660 \times 10^{-3}}{33 \times 10^{3}} = 220 \times 10^{-6} \text{ A, or } 220\ \mu\text{A}$$

1.2.1 Linearity and superposition

An extraordinarily powerful notion of the network theory is ***linearity***. This property has two essential elements, stated as follows:

- For any single input x yielding output y, the response to an input $k{\cdot}x$ is $k{\cdot}y$ for any value of k.
- If, in a multi-input network the input x_1 by itself yields output y_1 and a second input x_2 by itself yields y_2, then the combination of inputs x_1 and x_2 yields the output $y = y_1 + y_2$.

  This is important for two reasons:

1. It tells us that the solution to certain problems involving networks with multiple inputs is actually easier than we might expect: if a network is linear, we may solve for the output with each separate input, and then add the outputs. This is called ***superposition***.
2. It also tells us that, for networks that are linear, it is not necessary to actually consider the value of the inputs in calculating response. What is important is a system function, or a ratio of output to input.

Superposition is an important principle when dealing with linear networks, and can be used to make analysis easier. If a network has multiple independent sources, it is possible to find the response to each source separately and then add up all of the responses to find the total response. A particularly important ramification of the property of linearity is expressed in the notion of equivalent circuits. To wit: if we are considering the response of a network at any given terminal pair, that is, a pair of nodes that have been brought out to the outside world, it follows from the properties of linearity that, if the network is linear, the output at a single terminal pair (either voltage or current) is the sum of two components:

1. The response that would exist if the excitation at the terminal pair were zero.
2. The response forced at the terminal pair by the exciting voltage or current.

This notion may be expressed with either voltage or current as the response. These yield the Thevenin and Norton equivalent networks, which are exactly equivalent. At any terminal pair, the properties of a linear network may be expressed in terms of either Thevenin or Norton equivalents. The Thevenin equivalent circuit is

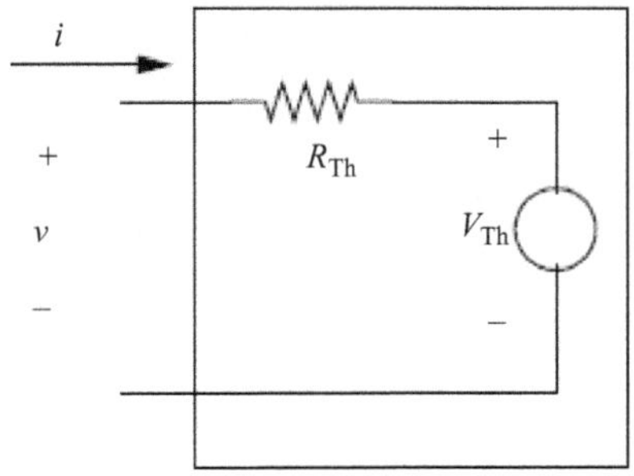

Figure 1.4 *Thevenin equivalent network*

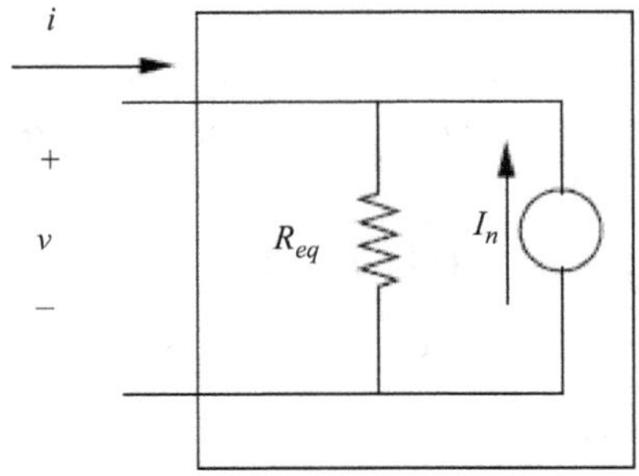

Figure 1.5 *Norton equivalent network*

shown in Figure 1.4, while the Norton equivalent circuit is shown in Figure 1.5. The Thevenin and Norton equivalent networks have the same impedance. Further, the equivalent sources are related by the simple relationship:

$$V_{Th} = R_{eq} I_N \tag{1.4}$$

Thevenin equivalent voltage is the open-circuit voltage, which is the internal source voltage that would appear at the terminals of the equivalent circuit in open-circuit. Similarly, the Norton equivalent current is the same as minus the short-circuit current.

1.2.2 DC vs AC power systems

The current of a DC circuit, as shown in Figure 1.6, consisting of a battery and a pure resistive load can be calculated, by using Ohm's law as ration between the source voltage V_{DC} and the load resistance, R as:

$$I_{DC} = \frac{V_{DC}}{R} \tag{1.5}$$

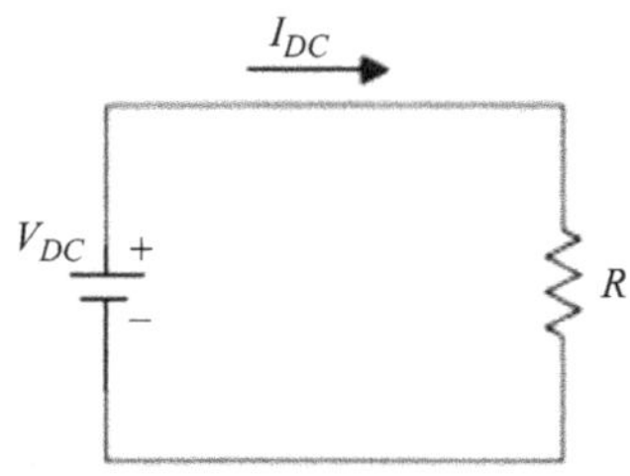

Figure 1.6 A simple DC circuit

The power, P, provided by the voltage source is given by:

$$P_{DC} = V_{DC} \cdot I_{DC} = \frac{V_{DC}^2}{R} = RI_{DC}^2 \tag{1.6}$$

Example 1.3: What is the power dissipated in a 1.8 kΩ resistance, if a current of 45 mA flows through it?

Solution: Using (1.6), the dissipated power is:

$$P_{DC} = RI_{DC}^2 = 1.8 \times 10^2 \cdot \left(45 \times 10^{-3}\right)^2 = 3.6 \text{ W}$$

The DC voltage and current waveforms are shown in Figure 1.7. It is also worth mentioning that these DC quantities are real numbers and not complex numbers.

There is another category of circuits, the AC circuits. Electric power systems usually involve sinusoidally varying (or nearly so) voltages and currents. That is, voltage and current are functions of time that are nearly pure sine waves at a fixed frequency. In North America, Canada, Brazil, and the Eastern Japan, the frequency is 60 Hz. In rest of the world, the electrical frequency is 50 Hz. Normal power system operates at this fixed frequency, so we will study how these systems operate in this mode. Accordingly, this section opens with a review of complex numbers

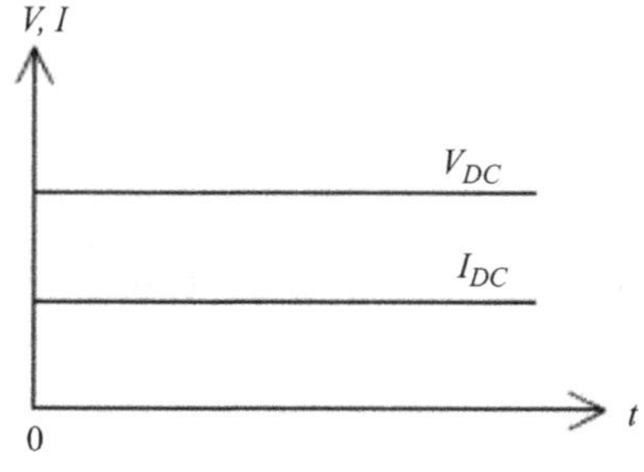

Figure 1.7 Voltage and current waveforms of the simple DC circuit

and with the representation of voltage and current as complex amplitudes with complex exponential time dependence. The discussion proceeds, through impedance, to describe a pictorial representation of complex amplitudes, called phasors. Power is then defined and in sinusoidal steady state, reduced to complex form. Finally, flow of power through impedances and a conservation law are discussed.

This section deals with the alternating voltages and currents with associated energy flows. The focus is on sinusoidal steady-state conditions, in which virtually all quantities of interest may be represented by single and complex numbers. As in power systems, the sinusoidal voltages are generated and consequently, most likely sinusoidal currents are flowed in the generation, transmission, and distribution systems. Sinusoidal quantities are assumed throughout this material, unless otherwise specified. In general, a set of typical steady-state voltage and current waveforms of an AC circuit can be drawn as shown in Figure 1.8, and their mathematical expressions can be written as follows:

$$v(t) = V_m \cos(\omega t + \phi) \tag{1.7}$$

and

$$i(t) = I_m \cos(\omega t + \theta) \tag{1.8}$$

Here, V_m and I_m are the maximum (peak) values or the amplitudes of the voltage and current waveforms, ω is the angular frequency in radians/s, ϕ and θ are the voltage and current phase angle, respectively, with respect to the reference in degrees or in radians. Usually in power engineering, the voltage is taken as reference. The period in Figure 1.8 can be 360° or 2π. In some cases, the period can be in time, for instance, 0.016667 s for 60 Hz. A sinusoidal function is specified by three parameters: amplitude, frequency, and phase. The amplitude gives the maximum value or height of the curve, as measured from the neutral position (the total distance from crest to trough is thus twice the amplitude). The frequency gives the number of complete oscillations per unit time. Alternatively, one can specify the rate of oscillation in terms of the inverse of frequency, the period. The period is simply the duration of one complete cycle. The phase indicates the starting point of

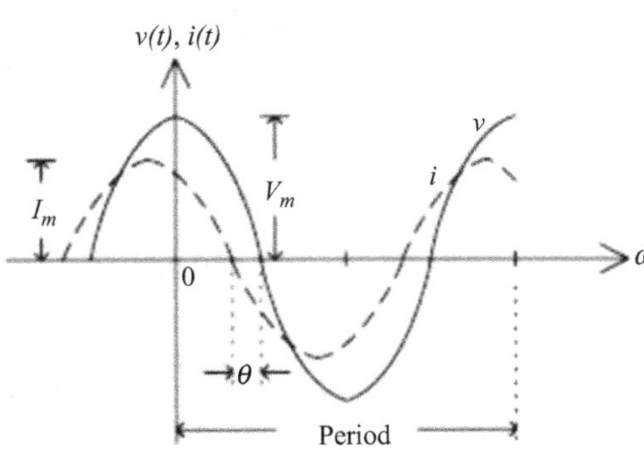

Figure 1.8 Typical voltage and current waveforms

the sinusoid. In other words, the phase angle specifies an angle by which the curve is ahead or behind of where it would be, had it started at time zero. The period (measured in seconds), frequency (measured in Hz or cycles/s), and the angular frequency (velocity) (measured in radians/s) are related through:

$$T = \frac{1}{f} = \frac{2\pi}{\omega}$$

1.2.3 Resistive, inductive, and capacitive circuit elements

Consider a purely resistive circuit with a resistor connected to an AC generator, as shown in Figure 1.9. Applying Kirchhoff's voltage law yields to the instantaneous voltage drop across the resistor:

$$v(t) - V_R(t) = v(t) - Ri(t) = 0 \tag{1.9}$$

The power dissipated by a resistor in AC is given by:

$$p(t) = Ri^2(t) \tag{1.10}$$

So far, we have dealt with circuit elements which have no memory and which, therefore, are characterized by instantaneous behavior. The expressions, which are used to calculate what these elements are doing, are algebraic (and for most of such elements are linear too). As it turns out, much of the circuitry we will be studying can be characterized, with complex parameters. We will be briefly discussing the behavior of the inductors and capacitors.

Symbols for capacitive and inductive circuit elements are shown in Figure 1.10. They are characterized by the first-order derivative relationships between voltage and current:

$$i_C(t) = C\frac{dv_C(t)}{dt} \qquad v_L(t) = L\frac{di_L(t)}{dt} \tag{1.11}$$

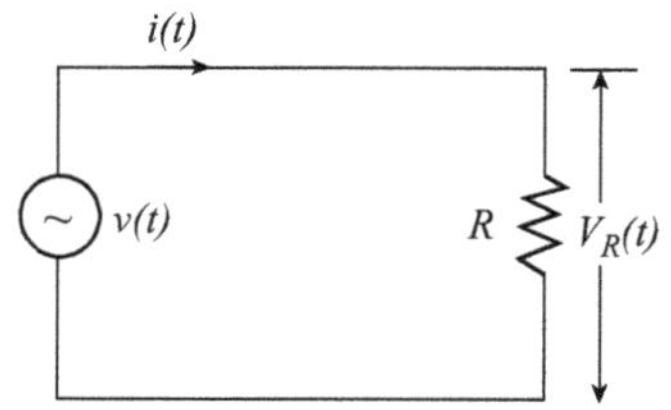

Figure 1.9 A purely resistive circuit

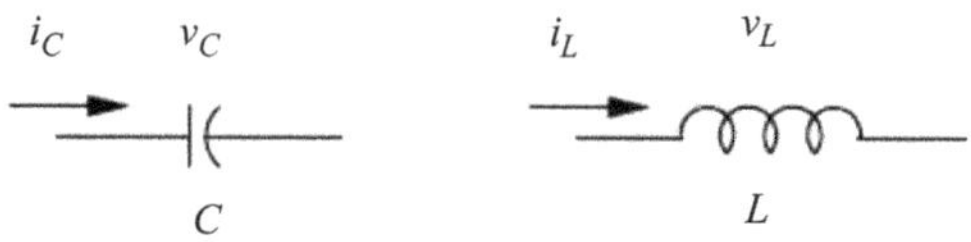

Figure 1.10 Capacitance and inductance symbols

The expressions of (1.11) are representing Ohm's law for a capacitor and an inductor, respectively, and are *nonlinear* as the time derivatives are involved in their characterization However, the expressions describing their behavior in networks will become ordinary differential equations. If the $i(t)$ in (1.11) are the sinusoidal values as in (1.7), then the voltage across a resistor is given by:

$$v(t) = RI_m \sin(\omega t + \phi) = V_m \sin(\omega t + \phi)\ V \tag{1.12}$$

Using sinusoidal current and voltage in (1.11), we are getting for the inductor:

$$\begin{aligned} v(t) &= L\frac{dI_m \sin(\omega t + \phi)}{dt} = \omega LI_m \cos(\omega t + \phi) \\ &= \omega LI_m \sin(\omega t + \phi + \pi/2) = V_m \sin(\omega t + \phi + \pi/2)\ V \end{aligned} \tag{1.13}$$

In (1.12), the voltage leads the current by 90°. For capacitor from (1.11), we will get the following relationship:

$$\begin{aligned} i(t) &= C\frac{dV_m \sin(\omega t + \theta)}{dt} = \omega CV_m \cos(\omega t + \theta) \\ &= \omega CV_m \sin(\omega t + \theta + \pi/2)\ \text{A} \end{aligned} \tag{1.14}$$

In this case, the current leads the voltage by 90°. Equations (1.12–1.14) can be written in a standard form using the impedance concept as:

$$v(t) = Zi(t) \tag{1.15}$$

The values of the impedance corresponding to the values of resistance, inductance, and capacitance are, respectively

$$Z = R + j0$$

$$Z = 0 + j\omega L$$

$$Z = 0 - \frac{j}{\omega C}$$

The impedances of a resistor in series with an inductor (R–L circuit), a resistor in series with a capacitor (R–C circuit), and a resistor in series with an inductor and a capacitor (R–L–C circuit) are expressed as:

$$\begin{aligned} Z_{R-L} &= R + jX_L = R + j\omega L \\ Z_{R-C} &= R + jX_C = R - j\omega C \\ Z &= R + j(X_L - X_C) \end{aligned} \tag{1.16}$$

Example 1.4: A 120 V adjustable frequency AC voltage source is connected across an RLC series circuit. At frequency of 60 Hz, the resistance is 5 Ω, the inductive reactance is 4 Ω, and the capacitive reactance is 3 Ω. Compute the following quantities:

1. The load impedance at 60 Hz.
2. Frequency at which the load impedance is equal to the resistance only.

Solution:

1. The load impedance, as given in (1.16) is given by:

$$Z_L = R + j(X_L - X_C) = 5 + j(4 - 3) = 5 + j = 5.1 < 11.31^\circ\ \Omega$$

2. The load inductance and capacitance are:

$$X_L = 2\pi f L => L = \frac{X_L}{2\pi f} = \frac{4}{2 \times 3.14 \times 60} = 10.616 \times 10^{-3}\ \text{H} = 10.616\ \text{mH}$$

$$X_C = \frac{1}{2\pi f C} => C = \frac{1}{2\pi f X_C} = \frac{1}{2 \times 3.14 \times 60 \times 3} = 884.64\ \mu\text{F}$$

The frequency at which the load is purely resistive (resonance frequency) occurs when the inductive and capacitive reactances are equal:

$$X_L = X_C => f_0 = \frac{1}{2\pi\sqrt{LC}} = \frac{1}{2\pi\sqrt{10.616 \times 10^{-3} \cdot 884.6 \times 10^{-6}}} = 51.96\ \text{Hz}$$

1.2.4 Effective or RMS value

The effective or root mean square (RMS) value of a periodic signal is equal to the magnitude of a DC signal which produces the same heating effect as the periodic signal when applied across a load resistance. Consider a periodic signal, $v(t)$, as in (1.15), then its mean value over one period is defined by the:

$$V_{av} = \frac{1}{T}\int_0^T v(t)dt \tag{1.17}$$

The average value of sinusoidal current or voltage over one period is zero. Integrating (1.10) over one period, we can estimate the heat produced by the AC circuit in one cycle as:

$$P_{AC} = \frac{1}{T}\int_0^T Ri^2(t)dt \tag{1.18}$$

The average heat produced by AC and DC waveforms, (1.18) and (1.6), are the same, and

$$RI^2 = \frac{1}{T}\int_0^T R\,i^2(t)dt$$

The above relationship yields to the definition of the RMS current:

$$I_{RMS} = \sqrt{\frac{1}{T}\int_0^T i^2(t)dt} \tag{1.19}$$

Similarly for the voltage waveform, the relationship is:

$$V_{RMS} = \sqrt{\frac{1}{T}\int_0^T v^2(t)dt} \tag{1.20}$$

For an alternating current and voltage, the above expressions are:

$$I_{RMS} = \frac{I_m}{\sqrt{2}} = 0.707\ I_m \text{ and } V_{RMS} = \frac{V_m}{\sqrt{2}} = 0.707\ V_m \tag{1.21}$$

The average power absorbed by a resistor can be written as:

$$P = RI_{RMS}^2 = \frac{V_{RMS}^2}{R} \tag{1.22}$$

Example 1.5: An inductor has a 54 Ω reactance at 60 Hz. What is the peak value of the current if the inductor is connected to a 50 Hz and 100 V RMS source?

Solution: The inductance at 50 Hz is calculated as:

$$X_L = 2\pi f L$$
$$X_{L(50\text{ Hz})} = 2\pi 50 \cdot L = 2\pi 50 \cdot \frac{X_{L(60\text{ Hz})}}{2\pi 60} = \frac{5}{6} \times 54 = 45\ \Omega$$

Solving for the peak current:

$$I_{peak} = \frac{\sqrt{2}I_{RMS}}{X_{L(50\text{ Hz})}} = \frac{\sqrt{2}\cdot 100}{45} = 3.13\text{ A}$$

1.3 Phasor representation

In analyzing AC circuits or the real reactive power situation for a specific type of load, electrical engineers make use of concepts, such as *impedance*, *phasor*, and *complex power*. In physics and engineering, a phasor (a portmanteau of phase vector, is a complex number representing a sinusoidal function whose amplitude (A), angular frequency (ω), and initial phase (θ) are time-invariant. It is related to a more general concept called analytic representation, which decomposes a sinusoid into the product of a complex constant and a factor that encapsulates the frequency and time dependence. These topics belong to the electrical circuit courses. Before reviewing these concepts, it is useful to have a brief presentation of some important mathematical relationships. Euler's formula relates complex exponentials and trigonometric functions, through the expression:

$$e^{j\theta} = \cos\theta + j\sin\theta \tag{1.23}$$

Here $j = \sqrt{-1}$ is the complex unit. In electrical engineering, we are using j, not i, which is used traditionally for current. By adding and subtracting Euler's formula and its complex conjugate, we get two important relationships:

$$\cos\theta = \frac{1}{2}\left(e^{j\theta} + e^{-j\theta}\right) \text{ and } \sin\theta = \frac{1}{2j}\left(e^{j\theta} - e^{-j\theta}\right) \tag{1.24}$$

These relationships are called *the inverse Euler formulae*, called by some mathematicians, as *definitions* of cos(θ) and sin(θ), very important formulae, which are extensively used in the electrical and power engineering. Time-domain voltage and current expressions in (1.7) and (1.8) can be expressed in so-called *phasor form* as:

$$\vec{V} = V_m\langle\theta_v \tag{1.25}$$

and

$$\vec{I} = I_m\langle\theta_i \tag{1.26}$$

However, we usually take RMS values of voltage and current, rather than the amplitude or peak values as magnitudes of the phasors.

$$\begin{aligned}\vec{V} &= V\langle\theta_v \\ \vec{I} &= I\langle\theta_i\end{aligned} \tag{1.27}$$

In other words, an AC current as one of (1.8) is represented by a vector of length I rotating at an angular velocity (frequency) ω rad/s as shown in Figure 1.11. Since the actual value of $i(t)$ depends on the phase angle of the rotating vector, the rotating vector is named *phasor*. A phasor represents a time-varying sinusoidal waveform by a fixed complex number. Depending on the sign of ($\theta_v - \theta_i$), we have either the current is lagging ($\theta_v > \theta_i$) the voltage or leading ($\theta_v < \theta_i$) the voltage. The voltage and current phasors can have phase difference between their maximum values. However, the phasors and complex impedances are only relevant to the sinusoidal sources. In the electrical power industry, since the voltage is given by the generator or the utility, the power engineers always take the voltage as reference phasor and then designate the current as leading or lagging the voltage.

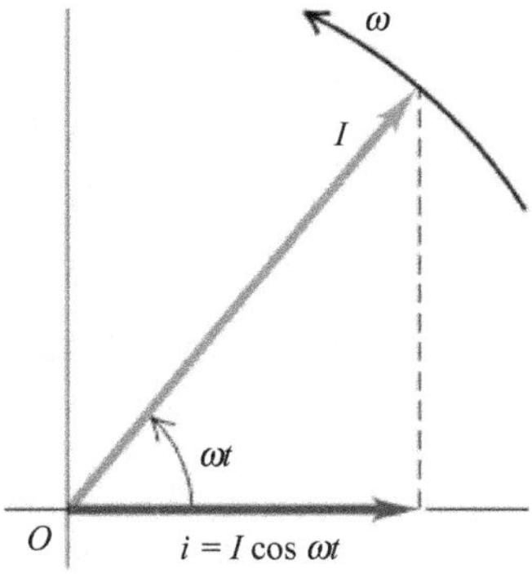

Figure 1.11 Rotating phasor I representing an AC current

Example 1.6: If the voltage and the current in an electric circuit are:

$$v(t) = \sqrt{2}(30)\cos(\omega t + 30^\circ)$$
$$i(t) = \sqrt{2}(5)\cos(\omega t - 20^\circ)$$

Write the phasor forms of $v(t)$ and $i(t)$, and then find the average power P into the network.

Solution:

$$\vec{V} = 30\langle 30^\circ \text{ V},\ \vec{I} = 5\langle -20^\circ \text{ A}$$
$$\theta_v - \theta_i = 30^\circ - (-20^\circ) = 50^\circ$$
$$P = 30 \times 5\cos(50^\circ) = 96.5 \text{ W}$$

The Ohm's and Kirchhoff's laws and other network theorems can be expressed straightforward in phasor notation, applying in the same manner in AC as in DC, except that all numbers in AC are phasors (complex numbers), each with magnitude and phase angle, making the AC algebra complex. The nodal analysis, mesh analysis, superposition, Thevenin, and Norton equivalent source models are valid in phasor notation as they are in DC. For a general form of AC power delivered from a source to a load, we consider the voltage and current phasors:

$$\vec{V} = Z\vec{I} \tag{1.28}$$

Here $Z = R + jX$ is the circuit impedance. Kirchhoff's circuit laws work with phasors in the complex form:

$$\sum_{k=1}^{N} \vec{I}_k = 0, \text{ and } \sum_{k=1}^{M} \vec{V}_k = 0 \tag{1.29}$$

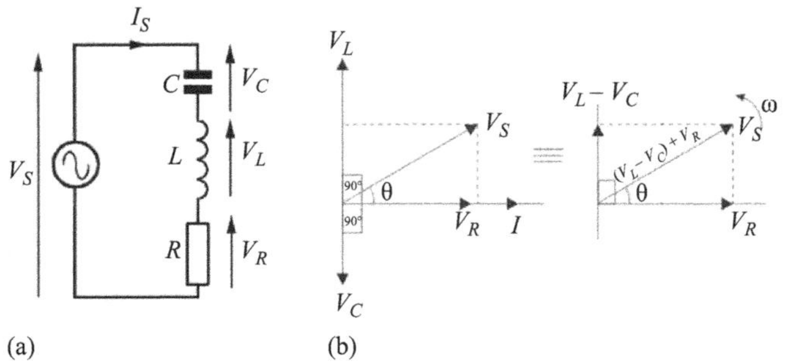

Figure 1.12 R–L–C circuit connection (a) and phasor diagram (b)

An R–L–C circuit as one shown in Figure 1.12(a) is composed by a resistor, an inductor, and a capacitor connected in series. Its phasor diagram is shown in Figure 1.12(b).

The load impedance can be expressed as:

$$\begin{aligned} \vec{Z} &= \vec{R} + \vec{X}_L + \vec{X}_C \\ Z &= R + j(X_L - X_C) = R + j\left(\omega L - \frac{1}{\omega C}\right) \end{aligned} \tag{1.30}$$

The magnitude of the impedance and the phase angle applied voltage and other voltages can be expressed as:

$$\begin{aligned} |Z| &= \sqrt{R^2 + (X_L - X_C)^2} \\ \tan(\theta) &= \frac{X_L - X_C}{R} \end{aligned} \tag{1.31}$$

Depending on the inductive reactance magnitude X_L with respect to the capacitive reactance X_C, the phase angle can be positive, negative, or zero. When $X_L = X_C$, the impedance in the circuit in Figure 1.12(a) is the equivalent to the resistance only, and the phase angle is zero. At the resonance, we have, $\omega L = 1/\omega C$ and:

$$\omega_0 = \frac{1}{\sqrt{LC}} \text{ and } f_0 = \frac{1}{2\pi\sqrt{LC}} \tag{1.32}$$

Example 1.7: A 120 V adjustable frequency AC source is connected across an R–L–C circuit, as shown in Figure 1.12(a). At 60 Hz, the resistance is 5 Ω, the inductive reactance is 3 Ω, and the capacitive reactance is 4 Ω. Calculate the resonant frequency of this circuit, and the load current at 60 Hz and at resonance frequency.

Solution: First we calculate the inductance and the capacitance and then the resonant frequency:

$$L = \frac{X_L}{2\pi f} = \frac{3}{2\pi 60} = 7.96 \text{ mH}$$

$$C = \frac{1}{2\pi f \times X_C} = \frac{1}{2\pi 60 \cdot 4} = 663.15 \text{ μF}$$

And

$$f_0 = \frac{1}{2\pi\sqrt{LC}} = 69.3 \text{ Hz}$$

The impedance at 60 Hz is:

$$Z_L = R + j(X_L - X_C) = 5 + j(3-4) = 5 - j = 5.1 < -11.31^\circ \ \Omega$$

The currents at 60 Hz and the resonant frequency are, respectively:

$$I_L = \frac{120 \text{ V}}{5.1 < -11.31^\circ \ \Omega} = 23.53 < 11.32^\circ \text{ A}$$

$$I_{res} = \frac{120}{5} = 24.0 \text{ A}$$

1.4 Power in single-phase AC circuits

Electrical power is of essential importance in any power system analysis, while the value of instantaneous power is the product of instantaneous voltage times current and is given by:

$$p(t) = v(t)i(t) = V_m I_m[\cos(\omega t)\cos(\omega t - \phi)] \tag{1.33}$$

where

$$\begin{aligned} v(t) &= V_m \cos(\omega t) \\ i(t) &= I_m \cos(\omega t - \phi) \end{aligned} \tag{1.34}$$

By trigonometric manipulating (1.1) can be rewritten as:

$$p(t) = \frac{1}{2} V_m I_m[\cos(\phi) - \cos(2\omega t - \phi)] \tag{1.35}$$

If we introduce the RMS (or effective) values of voltage and current, (1.35) can be written as:

$$p(t) = V_{RMS} I_{RMS} \cos(\phi) - V_{RMS} I_{RMS} \cos(2\omega t - \phi) \tag{1.36}$$

The power pulsates around and average first term in (1.36) at double frequency. Equation (1.36) can be transformed into the alternate form:

$$p(t) = V_{RMS} I_{RMS} \cos(\phi)(1 - \cos(2\omega t)) - V_{RMS} I_{RMS} \sin(\phi)\sin(2\omega t) \quad (1.37)$$

The power has been decomposed in two components; the first one pulsates around average value but is never negative, while the second one has a zero average value. At this juncture, we can introduce the following two quantities:

$$\begin{aligned} P &= V_{RMS} I_{RMS} \cos(\phi) \quad \text{(real or active power)} \\ Q &= V_{RMS} I_{RMS} \sin(\phi) \quad \text{(reactive power)} \end{aligned} \quad (1.38)$$

We can write (1.37) more compactly as:

$$p(t) = P(1 - \cos(2\omega t)) - Q \sin(2\omega t) \text{ (W)} \quad (1.39)$$

These concepts are of fundamental importance. The *real* or *active* power defined as the average value of $p(t)$ means the useful power transmitted, the *reactive* power is by definition equal to peak value of that component that travels back and forth on the line, has a zero average so is capable of no useful work. Both P and Q have dimension watts (W). However, to emphasize the fact that the latter represents reactive (or nonactive) power is measured in *volt-amperes reactive* (VAR). Larger and more practical units are kilovar (kVAR) and megavar (MVAR).

In any electric circuit, all the real (active) powers supplied to various circuit components sum the total real (active) power supplied:

$$\text{Total active power supplied} = \sum_k P_k = \sum_k R_k I_k^2 \quad (1.40)$$

Similarly, all the reactive powers supplied to the various circuit components sum the total reactive power supplied:

$$\text{Total reactive power supplied} = \sum_k Q_k = \sum_k X_k I_k^2 \quad (1.41)$$

where X_k and Q_k are negative, if the component is capacitive and positive, if the component is inductive. These are the expression of power conservation in AC circuits. The AC power is given by the product of voltage and complex conjugate of the current phasor. Therefore, the AC power is also a phasor, a complex quantity; hence the name of *complex power* is expressed by:

$$\vec{S} = V_{RMS} \langle \theta_v \cdot I_{RMS} \langle -\theta_i = V_{RMS} I_{RMS} \langle \theta_v - \theta_i = V_{RMS} I_{RMS} \langle \theta \quad (1.42)$$

Complex power is measured in VA (volt-amps), kVA, or MVA. By using (1.38) and (1.42), the complex power becomes:

$$\vec{S} = P + jQ \quad (1.43)$$

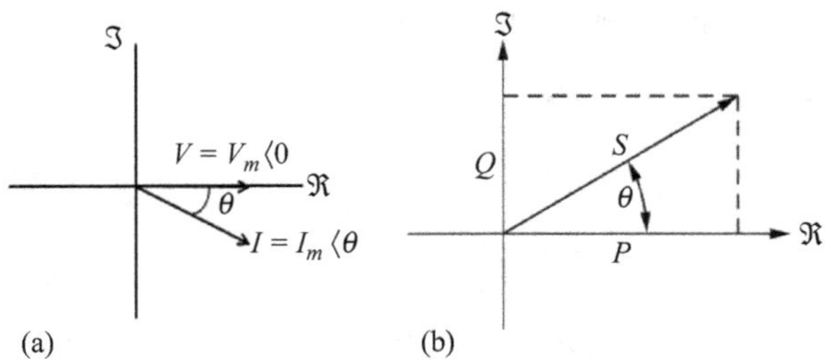

Figure 1.13 (a) Voltage and current phasor diagram and (b) power triangle

The ratio of the active power to the complex power is the power factor, expressed as:

$$PF = \frac{P}{S} = \cos(\theta) \tag{1.44}$$

The AC power absorbed by a load of impedance Z can be computed as product of the voltage across the Z times the complex conjugate of current through Z, that is:

$$\vec{S} = \vec{V} \times \vec{I}^* = ZI^2 = (R + jX)I^2 = P + jQ \tag{1.45}$$

where

$$\begin{aligned} P &= VI\cos(\theta) \\ Q &= VI\sin(\theta) \end{aligned} \tag{1.46}$$

In (1.46), $I{\cdot}\cos(\theta)$ is the current component in phase with the voltage phasor in Figure 1.13(a), and results in real power transfer P. Whereas, $I{\cdot}\sin(\theta)$ is the current component, that is, at 90° angle to the voltage phasor, Figure 1.13(a) and results in reactive power Q but no average real power. The power triangle corresponding to the phasors in Figure 1.13(a) is shown in Figure 1.13(b). From (1.45), the magnitude of S, also called the *apparent power* is expressed as:

$$|S| = \sqrt{P^2 + Q^2} \tag{1.47}$$

And

$$\theta = \tan^{-1}\left(\frac{P}{Q}\right) \tag{1.48}$$

Example 1.8: Suppose that the voltage and the current in Figure 1.13, in sinusoidal form are:

$$\begin{aligned} v(t) &= \sqrt{2}(10)\cos(\omega t + 50^\circ) \\ i(t) &= \sqrt{2}(5)\cos(\omega t + 20^\circ) \end{aligned}$$

Find the complex power.

Solution: First we compute the voltage and current phasors as:

$$\vec{V} = 10\langle 50^\circ \text{ V}$$
$$\vec{I} = 5\langle 20^\circ \text{ A}$$

The complex power is computed as:

$$\bar{S} = \vec{V}\vec{I}^* = 10\langle 50^\circ 5\langle -20^\circ = 150\langle 30^\circ = 129.9 + j75.0 \text{ VA}$$

And

$$P = 129.9 \text{ W}$$
$$Q = 75.0 \text{ VAR}$$

Notice that in impedance $Z = R + jX$, the real power P is absorbed only in the resistance R, and the active power Q is absorbed only in the reactance, X.

Example 1.9: A 120-V, 60 Hz is delivering power to a 0.20 mF capacitor in parallel with 20 mH coil that has a 10 Ω winding resistance. Determine the current, PF, and apparent, active and reactive powers delivered by the source. (Notice: practical capacitors have a negligible resistance; however, the inductors may have significant resistance due to long wires needed for the coil.)

Solution: We are determining first the circuit impedances. The capacitor impedance is:

$$Z_{Cap} = -\frac{j}{\omega C} = -\frac{j}{2\pi 60 \times 0.20 \times 10^{-3}} = -j13.26\ \Omega = 13.26\langle -90^\circ\ \Omega$$

The coil impedance is computed as:

$$Z_{Coil} = R + j\omega L = 10 + j2\pi 60 \times 20 \times 10^{-3} = 10 + j7.54\ \Omega$$
$$= 12.52\langle 37^\circ\ \Omega$$

Total impedance is estimated using the parallel combination of capacitor and coil impedances:

$$Z_{Total} = \frac{Z_{Cap}Z_{Coil}}{Z_{Cap} + Z_{Coil}} = \frac{13.26\langle -90^\circ \times 12.52\langle 37^\circ}{13.26\langle -90^\circ + 12.52\langle 37^\circ} = \frac{166.02\langle -53^\circ}{11.53\langle -29.8^\circ}$$

$$Z_{Total} = 10.06\langle -23.2^\circ\ \Omega$$

The capacitor current is computed as:

$$I_{Cap} = \frac{120}{13.26\langle -90^\circ} = 9.05\langle 90^\circ \text{ A}$$

While the inductor/coil current is computed as:

$$I_{Coil} = \frac{120}{12.52\angle 37^\circ} = 9.58\angle -37^\circ \text{ A}$$

The input current is computed using Kirchhoff's current law as:

$$I_{line} = I_{Cap} + I_{Coil} = j9.05 + 7.653 - j5.673 = 7.653 + j3.287$$

$$I_{line} = 8.330\angle 23.3^\circ$$

Overall circuit power factor is:

$$PF = \cos^{-1}(23.3^\circ) = 0.92 \text{ or } 92.2\%$$

Apparent power delivered by the source is computed as:

$$S = VI^*_{line} = 120 \times 8.330\angle -23.3^\circ = 999.6\angle -23.3^\circ = 918.2 - j395.2 \text{ VA}$$

While the active (real) and reactive powers are:

$$P = 918.2 \text{ W}$$

$$Q = -395.2 \text{ VAR}$$

Example 1.10: Consider a circuit composes of a series R–L branch in parallel with a capacitance, C with the following parameters:

$$R = 0.5\ \Omega,\ X_L = 0.8\ \Omega,\ \text{and } X_C = 1.25\ \Omega$$

Assume that: $V = 220\angle 0^\circ$ V

Calculate the input current, the active, reactive, and apparent power into the circuit.

Solution: The current into the R–L branch is given by:

$$Z = R + jX_L = 0.5 + j0.8 = \sqrt{0.5^2 + 0.8^2}\angle \tan^{-1}\left(\frac{0.8}{0.5}\right) = 0.943\angle 58^\circ\ \Omega$$

$$I_Z = \frac{220\angle 0^\circ}{0.934\angle 53^\circ} = 233.2\angle -58^\circ \text{ A}$$

The power factor of R–L branch is:

$$PF_Z = \cos(53^\circ) = 0.53 \text{ or } 53\%$$

The current into the capacitor branch is:

$$I_Z = \frac{220\angle 0^\circ}{1.25\angle -90^\circ} = 176\angle 90^\circ \text{ A}$$

The input current, I_t is given by the Kirchhoff' current law as:

$$I_t = I_Z + I_C = 233.2\angle{-58^\circ} + 176\angle{90^\circ} = 123.7 - j197.7 + j176.0$$
$$I_t = 123.7 - j21.7 = 125.6\angle{-10^\circ}\ \text{A}$$

The circuit power factor is:

$$PF = \cos(-10^\circ) = 0.985 \text{ or } 98.5\%$$

Notice that magnitude to the input current is lesser than that of R–L branch and the overall power factor is higher than that of the R–L branch circuit. This is the capacitor effect and is an example of the power factor correction. The apparent power into the circuit is:

$$S_t = VI_t^* = (220\angle{0^\circ})(125.6\angle{10^\circ}) = 27{,}632\angle{10^\circ}\ \text{VA}$$

Or in algebraic form:

$$S_t = 27.213 + j4.796\ \text{kVA}$$

Thus, the active and reactive powers are:

$$P = 27.213\ \text{kW}$$
$$Q = 4.796\ \text{kVAR}$$

The physical significance of the apparent, active (real), and reactive power should be understood in the context of power engineering. The cost of most of the electrical equipment, such as generators, transformers, and transmission lines is proportional to $|S|(=VI)$, since their electrical insulation level and the magnetic core size for a give power system frequency depends on the voltage V, while the conductor size depends on the RMS current I. The real power P has a physical significance since it represents the useful work being performed plus losses. Under most operating conditions, it is desirable to have the reactive power zero, or it results in increased apparent power.

1.5 Power factor correction—brief introduction

In simple words, *power factor* is simply a name given to the ratio of "actual" power (active power) being used in a circuit, expressed in watts or more commonly kilowatts (kW), to the power which is "apparently" being drawn from the mains, expressed in volt-ampere or more commonly kilovolt-ampere (kVA). A power factor of less than one means that the voltage and current waveforms are not in phase, reducing the instantaneous product of the two waveforms (V·I). Real power is the capacity of the circuit for performing work in a particular time. Apparent power is the product of the current and voltage of the circuit. Due to energy stored in the load and returned to the source, or due to a nonlinear load that distorts the wave shape of the current drawn from the source, the apparent power will be greater than the real power. In an electric power system, a load with a low power factor draws more current than a load with a high power factor for the same amount of useful power transferred. The higher currents increase the energy lost in the distribution system and require larger wires and other equipment. Because of the costs of larger equipment and wasted energy, electrical utilities will usually charge a higher cost to industrial or commercial customers where there is a low power factor. All modern industries utilize electrical energy in some form or other. Two basic categories of load encountered in AC networks are resistive and inductive loads. The inductive loads tend to degrade or lower the power factor. Among the disadvantages of low power factor are:

1. Increased authorities cost since more current has to be transmitted and this cost is directly billed to the consumers on maximum demand kVA systems.
2. Causes overloaded generators, transformers, and distribution lines within a plant, resulting in greater voltage drops and power losses, all representing waste, inefficiency, and needless wear and tear on the industrial electrical equipment.
3. Reduces load handling capability of the electrical system of the plant.

Most electrical supply authorities have changed to kVA demand systems from the inefficient kW demand system. Consumers are now billed and penalized for their inefficient systems according to the apparent power being used or even penalized for plants with power factor below a predetermined value. The most practical and economic power factor improvement devices are the capacitor, capacitor banks on synchronous condensators.

1.6 Electrical energy

The energy that is consumed by a load is the power delivered to the load, P, over a certain period of time. The SI unit of energy is Joule (J). However, the practical units used in power engineering are: watt-hours (Wh), kWh, MWh, etc. If the power delivered to the load is constant during a given time period τ, the energy is expressed as:

$$E = P \cdot \tau \tag{1.49}$$

For time-dependent power, the energy over a given period of time, τ is computed by the integral over that period of time:

$$E = \int_0^{\tau} P \cdot dt \tag{1.50}$$

For discrete power, the energy is the summation of the power for each time interval, expressed as:

$$E = \sum_k P_k \cdot t_k \tag{1.51}$$

Example 1.11: An electric load is connected across a 120 V source. The load impedance is changing over a 24-h period as follows:

Time interval	Load impedance (Ω)	Power factor angle (Degrees)
8:00–10:00 AM	10	30
10:00 AM–2:00 PM	15	0
2:00 PM–6:00 PM	20	10
6:00 PM–10:00 PM	30	45
10:00 PM–12:00 AM	15	30
12:00–8:00 AM	5	50

Compute the energy consumed by the load during this 24-h period.

Solution:

Time interval	Load current (A)	Load power $V_sI\cos(\theta)$ W	Time period (h)	Energy (Wh)
8:00–10:00 AM	4.0	415.8	2.0	831.6
10:00 AM–2:00 PM	8.0	960.0	4.0	3840.0
2:00 PM–6:00 PM	6.0	709.1	4.0	2836.4
6:00 PM–10:00 PM	4.0	339.6	4.0	1358.4
10:00 PM–12:00 AM	8.0	831.5	2.0	1663.0
12:00–8:00 AM	20.0	1201.1	8.0	9608.8

Total energy consumed over a 24-h period is: 20138.2 Wh or 20.138 kWh.

Example 1.12: A load is connected to a 120 V source. The load power is expressed as:

$$P(t) = 30 + 120\sin(24t) \text{ kW}$$

Here t is the time in hours. Estimate the energy consumed by this load after 1 h and after 24 h.

Solution: By using (1.44), the time-dependent energy is expressed as:

$$E(t) = \int_0^{\tau} P(t)\cdot dt = \left[30t - 120\frac{\cos(24t)}{24}\right]_0^{\tau} = 30\tau - 5\cos(24\tau) + 5$$

After 1 h, $\tau = 1$

$$E(1) = 30(1) - 5\cos(24) + 5 = 30.432 \text{ kWh}$$

After 24 h, $\tau = 24$

$$E(1) = 30(24) - 5\cos(24 * 24) + 5 = 729.060 \text{ kWh}$$

1.7 Measurement units used in energy systems

To make comparisons of various quantities and to quantify the magnitude of physical quantities, we need a good understanding of units, their definitions, and techniques of dimensional analysis. Most of the readers are already familiar with units used in energy system measurements, such as: watt, joule, British Thermal Unit (BTU), and multiples of them. The most common units used in this book are defined here, while some units that are unique to specific technologies or power system subsections are defined later where they arise. We are also assuming that the reader already understands basic metric units, such as meter, kilogram, Kelvin and Celsius degrees, and seconds. The system of measure in use in most of the world is the SI (an abbreviation of the International System) or metric system. Most of the energy-related units, The United States does not use the metric system, however, the amperes, volt, watt, and kilowatt-hour, used in electricity and electrical engineering in the US are SI units. There are several units from various systems of units employed in energy and power fields, but whatever possible the use of the SI units is used. SI is founded on seven SI base units for seven base quantities assumed to be mutually independent, as given in Table 1.1. Other quantities, called derived quantities, are defined in terms of the seven base quantities via a system of a system of quantity equations.

In engineering and science, the dimensional analysis is the analysis of the relationships between different physical quantities by identifying their fundamental dimensions (e.g., length, mass, time, and electric charge) and units of measure (e.g., miles vs kilometers, or pounds vs kilograms vs grams) and tracking these dimensions as calculations or comparisons are performed. Converting from one dimensional unit to another is often somewhat complex. Dimensional analysis, or more specifically the factor-label method, also known as the unit-factor method, is a widely used technique for performing such conversions using the rules of algebra. Any physically meaningful equation (and any inequality and in-equation) must have the same dimensions on the left and right sides. Checking this is a common application of performing dimensional analysis. Dimensional analysis is also routinely used as a check on the plausibility of derived equations and computations. It is generally used to categorize the types of physical quantities and units based on their relationship or dependence on other units. The SI units for these derived

Table 1.1 SI fundamental units

Base quantity	Name	Symbol
Length	Meter	m
Mass	Kilogram	kg
Time	Second	s
Electric current	Ampere	A
Temperature	Kelvin	K
Amount of substance	Mole	Mol
Luminous intensity	Candela	Cd

Table 1.2 Most common energy units

Unit name	Definition
Joule (J)	Work done by ace of 1 N acting through 1 m (also W-s)
Erg	Work done by 1 dyne force acting through 1 cm
Calorie (Cal)	Heat needed to rise the temperature of 1 g of water by 1 °C
BTU	Heat needed to rise the temperature of 1 lb of water by 1 °F
kWh	Energy of 1 kW of power flowing for 1 hour, 3600 J
VA	Volt-ampere may be used to measure apparent power
Quad	10^{15} BTU
Electron-Volt	Energy gained by an electron through 1 V potential difference
Foot-pound	Work done by 1 lb force acting through 1 ft
Megaton	Energy released when a million tons of TNT explodes

quantities are obtained from the principles and equations of physics and the seven SI base units. The fact that energy exists in many forms was one of the reasons that we have several units for this physical quantity. For example, for heat, we have calories, BTUs, Joules, ergs, and foot-pound for mechanical energy, kilowatt hours (kWh) for electrical energy, and electron-Volts (eV) for nuclear and atomic energy. However, since all describe the same fundamental quantity, there are conversion relationships or factors relating them. Table 1.2 lists some of the most common energy units.

Metric units are adapted to specific applications by adding a prefix that denotes a multiple of 10, applied to the unit in question. The symbols for submultiplier prefixes are all in lowercase, while the symbols for multipliers are lower case (ex. kilo) and upper case letters. A list of these prefixes is given in the appendix section of this book. The prefix names are derived mostly for Greek and Latin with some severe corruptions. A familiarity with the practical meaning of each order of magnitude can help engineers and practitioners to avoid calculation errors from incorrect manipulation of scientific notation and prefixes.

1.8 Chapter summary

In this chapter, we discussed the basic electric circuit theorems and some of the techniques used in analyzing electric circuits. A historical note on the power system evolution is included in this chapter. We reviewed the concepts of phasors and explained how the current and voltage can be represented in phasor forms. We discussed the concept of impedance and the resonance frequency. The concepts of complex power, real (active), and reactive power are also reviewed. The notion of RMS values of current and voltage, phasor diagram, power factor, and power triangle are introduced and discussed. We also presented a short summary of the most common measurement units used in electrical and power engineering. The use of power triangle and improvements of power factor are also briefly discussed. Energy concept is introduced and presented in the subsection of the chapter, followed by the discussion of the fundamental measurement units of SI system, and the most common units used in power and energy engineering and their definitions.

Further readings

[1] A.R. Daniels, Introduction to Electrical Machines, The Macmillan Press, London, 1976.

[2] O.I. Elgerd, Basic Electric Power Engineering, Addison-Wesley, Reading Massachusetts, 1977.

[3] V. Del Toro, Basic Electric Machines, Prentice Hall, Englewood Cliffs, New Jersey, 1985.

[4] R.D. Schultz and R.A. Smith, Introduction to Electric Power Engineering, John Wiley & Sons, New York, 1988.

[5] A.C. Williamson, Introduction to Electrical Energy Systems, Longman Scientific & Technical, England, 1988.

[6] J.R. Cogdell, Foundations of Electric Power, Prentice-Hall, Upper-Saddle River, New Jersey, 1999.

[7] S.J. Chapman, Electric Machinery and Power System Fundamentals, McGraw Hill, Singapore, 2002 (2nd ed.).

[8] C.I. Huber, Electric Machines, Prentice Hall, Upper Saddle River, New Jersey, 2002

[9] M.A. Pai, Power Circuits and Electro-mechanics, Stipes Publishing L.L.C., Champaign, Illinois, 2003.

[10] J. Andrews and N. Jelley, Energy Science, Oxford University Press, Oxford, UK, 2007.

[11] P. Schavemaker and L. van der Sluis, Electrical Power System Essentials, John Wiley and Sons, Chichester, UK, 2008.

[12] M.E. El-Hawary, Introduction to Power Systems, Wiley—IEEE Press, New York, 2008.

[13] Y. Goswami and F. Keith (eds.), Energy Conversion, CRC Press, Boca Raton, Florida, 2008.

[14] M.A. El-Sharkawi, Electric Energy: An Introduction (2nd ed.), CRC Press, Boca Raton, Florida, 2009.

[15] J.L. Kirtley, Electric Power Principles: Sources, Conversion, Distribution and Use, Wiley, Chichester, UK, 2010.

[16] B.K. Hodge, Alternative Energy Systems and Applications, Wiley, Hoboken, New Jersey, 2010.

[17] J.F. Kreider and S. Krumdieck, Principles of Sustainable Energy, CRC Press, Boca Raton, Florida, 2011.

[18] N. Mohan, Electric Power Systems—A First Course, Wiley, Hoboken, New Jersey, 2012.

[19] F.M. Vanek, L.D. Albright, and L.T. Angenent, Energy Systems Engineering, McGraw Hill, New York, 2012.

[20] S.W. Blume, Electric Power System Basics for the Nonelectrical Professional (2nd ed.), Wiley and Sons, New York, 2016.
[21] N. Jenkins and J. Ekanayake, Renewable Energy Engineering, Cambridge University Press, Cambridge, UK, 2017.

Questions and problems

1. Electric power systems provide energy in a variety of commercial and industrial settings. Make a list of the most common systems and devices that receive electric power in: (a) large office building, (b) industrial facility, and (c) in a construction site.
2. What is the significance of the apparent power?
3. What is the advantage of using phasors in AC electric circuits?
4. What is power factor?
5. What is the real, reactive, and apparent power? What units are they measured?
6. Two parallel resistors carry current of 3 A and 5 A. If the resistance of one is 90 Ω, what is the resistance of the other?
7. A resistor of 4.7 kΩ has a voltage of 23.5 mV across it. Determine the current and express it in a convenient unit.
8. A DC generator transmits power over a transmission lines having a resistance of 0.35 Ω to a 10 KW load. The load voltage is 135 V. What is the generator voltage? How much power is lost by the transmission line?
9. An incandescent light bulb rated at 60 W will dissipate 60 W, as heat and light when connected across a 120-V ideal voltage source. If three of these bulbs are connected in series across the same source, determine the power each bulb will dissipated.
10. If 280 mA flows through 1.8 MΩ resistor, what is the power dissipated in it and the voltage across the resistance in kilovolts?
11. Voltage and current of an electric load can be expressed by the following expressions:

$$v(t) = 340\sin(377t + 30^\circ)\ \text{V}$$
$$i(t) = 50\sin(377t + 60^\circ)\ \text{A}$$

Calculate the following: (a) RMS voltage and current, (b) current frequency, (c) phase shit between current and voltage, (d) average voltage, and (e) load impedance.

12. A 100 V, 60 Hz AC circuit with a 26.5 μF capacitor. What is the current in the circuit? What is the relationship between the voltage and the current in the circuit?
13. A 50 Ω resistor, a 0.10 H inductor, and a 10.0 μF capacitor are connected in series to a 60 Hz power source. The RMS circuit current is 3.50 A. Find the RMS voltage across: (a) resistor, (b) inductor, (c) capacitor, (d) the RLC combination, and (d) sketch the phasor diagram of this circuit.

14. A series consists of a resistance of 15 Ω, and inductance of 0.8 mH, and capacitance of 100 μ. What is the impedance of the circuit at 60 Hz? Is the total impedance capacitive or inductive?
15. A sinusoidal voltage 280·cos(377*t*) V is applied across a circuit element, drawing a current of 100·cos(377*t*–30°) A. Determine the average power absorbed in the element.
16. Find the average real power into the terminals of the two-terminal network shown in

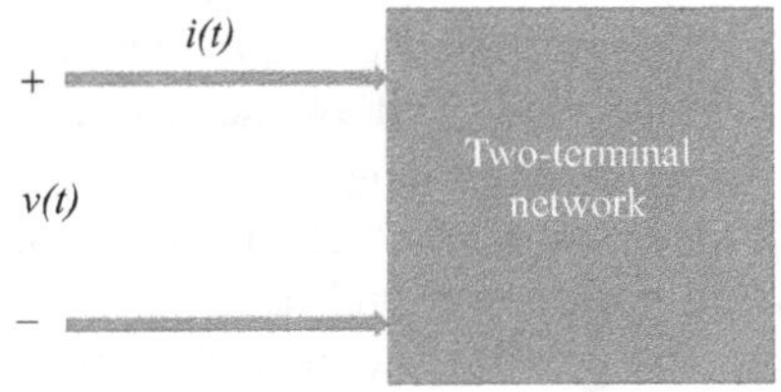

Figure P1.1 for each of the following cases

(a) $$v(t) = 100\cos(377t + 15°)\text{ V}$$ $$i(t) = -10\cos(377t + 45°)\text{ A}$$

(b) $$v(t) = 100\cos(377t - 75°)\text{ V}$$ $$i(t) = -10\cos(377t + 15°)\text{ A}$$

(c) $$v(t) = \sqrt{2}100\cos(377t - 30°)\text{ V}$$ $$i(t) = \sqrt{2}10\cos(377t - 45°)\text{ A}$$

17. A single-phase source supplies a load of $R = 30\ \Omega$ connected in parallel with capacitive reactance $X_C = 20\ \Omega$. The voltage across the load is:

$$v(t) = \sqrt{2}120\cos(\omega t - 45°)\text{ V}$$

(a) Find the phasors of voltage, total current, instantaneous current, and complex power.
(b) Repeat (a) if instead of capacitive reactance, we have an inductive reactance $X_L = 20\ \Omega$.

18. An AC voltage of 120 V is connected to a 45 Ω resistor. What is the power in the circuit?
19. An electric load consists of a 6 Ω resistance, a 6 Ω inductive reactance, and an 8 Ω capacitive reactance connected in series. The series impedance is connected across a voltage source of 120 V. Compute the following: (a) power factor of the load; (b) source current; (c) the apparent, real, and reactive powers in this circuit.
20. For a circuit consisting of a 0.25 H inductor in series with a parallel combination of 4.5 Ω resistor and a 0.125 F capacitor, find the circuit impedance if the angular frequency is $\omega = 100$ rad/s.

21. A 120 V (RMS) source voltage is applied to a $24\angle 30^\circ\ \Omega$ load impedance. Calculate the average, RMS, peak load currents, and the average power delivered to the load.
22. A parallel circuit consisting of two impedances

$$Z_1 = 15\angle 45^\circ\ \Omega$$

$$Z_2 = 30\angle -30^\circ\ \Omega$$

is connected across of voltage source: $\vec{V} = 120\angle 0^\circ$ V.

Find: (a) the current in each circuit branch and the total current, (b) the complex power in each branch and the total complex power, (c) the overall power factor of the circuit, and (d) draw the power triangles for each branch circuit and the combined power triangle.
23. A 10 HP, 60 Hz, 220 V single-phase electric motor operates at 90% efficiency while running at rated horsepower and has a power factor of 0.707 lagging. Find: (a) the current drawn by the motor, and (b) the real and reactive power absorbed by the motor.
24. A transformer rated at 1200 kVA is operated at full load (i.e., carrying rated kVA) at a PF = 0.70 lagging. A capacitor bank is added to improve the overall power factor to become 0.9 leading. Find the capacitor bank reactive power required.
25. A parallel circuit consists of two impedances:

$$Z_1 = 30\angle 45^\circ\ \Omega$$

$$Z_2 = 45\angle -30^\circ\ \Omega$$

Find: (a) total impedance, the current in each branch and the total current; (b) the total complex power; and (c) the overall power factor of the circuit.
26. Three loads in parallel are:
- Load 1: 300 kVA at PF = 0.67 lagging
- Load 2: 180 KW at PF = 0.85 leading
- Load 3: $283 + j100$ kVA

Find the total complex power and the overall power factor of the combination. What is the capacitor reactive power to make the overall power factor PF = 0.90 leading. The capacitor is connected across the parallel combination.
27. Determine the average and the RMS values of the square wave current:

$$i(t) = \begin{cases} 10\text{ A}, & \text{for } 0 \le t \le 2.5 \text{ ms} \\ -10\text{ A}, & \text{for } 2.5 \le t \le 5 \text{ ms} \end{cases}$$

28. The RMS voltage and current of a capacitive load is 120 V and 10 A in a 60 Hz power system. The instantaneous power consumed by the load has no average component. Compute: (a) real (active) and reactive powers computed by the load, (b) power factor, and (c) frequency of the reactive power.

29. Find the RMS value of $v(t)$ if $v(t)$ is a sinusoid that is offset by a DC value:

$$v(t) = 2\sin(\omega t) + 2.5$$

30. Powered by a 220 V RMS voltage source, a load consisting of three impedances, $Z_1 = 20 + j20\ \Omega$, $Z_2 = 10 + j20\ \Omega$, and $Z_3 = 10 - j20\ \Omega$ are connected in parallel. Determine the real power drawn by each of the impedance and by the load.
31. A single-phase circuit is placed across a 120-V RMS, 60-Hz source, with an ammeter, a voltmeter, and a wattmeter connected. The instruments indicate 12 A, 120 V, and 800 W, respectively. Find the power factor, the phase angle, the circuit impedance, and the circuit resistance.
32. If the voltage and current given below are supplied by a source to a circuit or load, determine:
    (a) The power supplied by the source which is dissipated as heat or work in the circuit (load).
    (b) The power stored in reactive components in the circuit (load).
    (c) The power factor angle and the power factor.

$$\bar{V}_s = 45 < 0.366(rad)\ \text{V}, \quad \text{and } \bar{I}_s = 13.5 < -0.500(rad)$$

33. A load is connected to a 120 V source. The load power is expressed as:

$$P(t) = 45 + 150\sin(24t)\ \text{kW}$$

Here t is the time in hours. Estimate the energy consumed by this load after 1 h and after 24 h

34. Consider an RLC series circuit with $R = 25\ \Omega$, $L = 6.00$ H, and $C = 25\ \mu$F. The circuit is connected to a 10 V RMS, 600 Hz power supply.
    (a) What is the voltage across the RLC combination?
    (b) Which is the largest average power delivered to the resistor, to the capacitor, or the inductor?
    (c) Find the average power delivered to the circuit.
35. Three parallel loads draw from a 460 V source. The active power drawn by each of the sources are 30 kW, 12 kW, and 18 kW, respectively, while the corresponding reactive powers are 18 kVAR, 9 kVAR, and 15 kVAR, respectively. Determine the combined power factor and total apparent power drowns from the source.
36. The heating element in an electric heater has a resistance of 250 Ω. Find the power dissipated in the heater when it is connected to a voltage source of 240 V rms.
37. A 10 kW electric motor is connected to a source of 120 V (RMS), 60 Hz, and the result is a lagging PF of 0.8. To correct the PF to 0.95 lagging, a capacitor is placed in parallel with the motor. Calculate the current drawn from the source with and without the capacitor connected. Determine the value of the capacitor required to make the correction.

38. The power consumed by an electric load is expressed as:

$$P(t) = 150\left(1 - \exp\left(-\frac{t}{12}\right)\right) \text{ kW}$$

where t is the time expressed in hours. Compute the energy consumed by the load in 12 h and in 24 h.

Chapter 2
Power system basics

Objectives and abstract

The generation, transmission, and distribution of the electricity is the business of the large utility companies, being performed through complex networks of interconnected generators, transformers, transmission lines, control, monitoring, measurement, and protection equipment, developed over a century. The chapter starts with a brief power system description and presentation, some details on major power system components, and functions. Various fuels (e.g., coal, natural gas, oil, nuclear energy, or water power) or energy sources (e.g., wind energy, solar energy, ocean energy, or geothermal energy) are used to generate electricity in modern power systems. Most of the modern power systems are three-phase, as it enabled more efficient and economical energy generation and transmission to the users. Power is changed from three-phase to single-phase only for consumers living very near. However, the power distribution is usually changed from three-phase to single-phase networks near consumers (end-users). In modern power systems, the generator voltages are transformed into high voltages for the efficient and economic transmission at long distances, then the voltages are lowered to the levels required by the consumers near the industrial and residential locations. The chapter also contains a brief introduction on the per-unit system, a very useful tool to simplify calculations and analysis of the power system. A brief discussion of the frequency characteristics and issues is the focus of the last section of the chapter. The chapter topics are the basics to understand the electric motor, transformer, and generator operations and characteristics, so it is vital that the readers must have good understanding of three-phase and per-unit systems and must be able to apply them.

2.1 Introduction, power system basics

The electric power system is a complex technical system with main function is to deal with generation, consumption, and storage of electric power. A power system is an interconnected network with components converting nonelectrical energy continuously into the electrical form and transporting the electrical energy from the generating sources to the loads/users. This chapter will give a conceptual introduction to the electric power system as a main part of the energy system. It will also

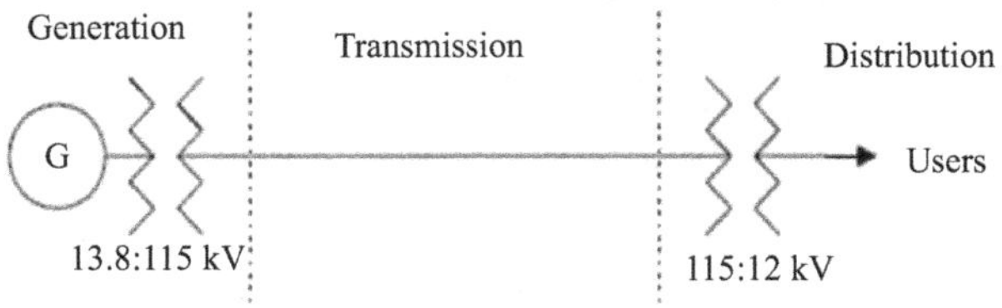

Figure 2.1 A basic structure of a simplified power system

go through a brief description on fundamental modeling of the AC system and the power flow, main components of a power system, its operation and planning, including aspects of reliability and market. The chapter conceptually describes the technical and fundamental characteristics and performances of the electric power system with main function to deliver and transfer electricity between generation, consumption and storage of electric power. A sustainable energy system involves key components: increasing use of renewable energy resources, increased energy efficiency, and use of electricity in the transportation sector. As an effect of this, some key questions for the electric power system are integration of intermittent electricity generation, e.g., from wind power, and connections of electrical vehicles to the electrical distribution system and different solutions for energy storage.

Modern power systems are made up of three major distinct components or subsystems: generation, transmission, and distribution, as shown in Figure 2.1. The generation subsystem includes power plant generation units, for example, turbines and generators. The energy resources used to generate electricity in most of the power plants are fossil fuels (e.g., coal, natural gas, or oil), nuclear, or hydropower. In the last three decades, there was an increasing trend to include renewable (alternative) energy sources for electricity energy generation, especially in the distribution section. Nuclear power and hydroelectric power are nonpolluting energy sources, while the last one is also renewable energy source. At present, in the United States, hydropower account for about 8% and nuclear power 20% of the electricity generation. The generated electricity is transmitted by a complex network composed of transmission lines, transformers, control, and protective equipment. Transmission lines are used to transfer electrical energy from power plants to load centers. Transformers are used to step-up the voltage at power plants to very high values (200–1,200 kV), in order to reduce currents and losses, and reducing the size of transmission wires and implicitly reducing the overall cost of the transmission system. The transmission lines carry power over lines to load centers, where appropriate, step it down to lower voltages up to 35 kV at bulk power substations. At load centers, the transmission line voltages are reduced by step-down transformers to lower values (4.5–35 kV) for power distribution networks. Some large industrial customers are supplied from these substations. This is known as the transmission–subtransmission systems or sections. Distribution of power to commercial and residential users takes place through a distribution system consisting of substations where step-down transformers lower the voltages to a range of 2.4–69 kV. Power is carried by main feeders to the specific areas where there are lateral feeders to step it down to customer levels. At customer sites, the voltage is further reduced to values, such as 120V, 208 V, 280/277 V, etc. as required by the users. Power systems are extensively monitored, controlled, and protected. These complex transmission and distribution networks encompass larger areas or regions. Each power system has several levels of protection to minimize or avoid the effects of any damaged or nonoperational system component, on the system's ability to provide safe and reliable electricity to all customers. Any power system serves the basic function to supply customers with electricity as economically and reliably as possible. In summary, the main functions of three subsystems of the power systems are:

1. Generation subsystem—Generating and/or sources of electrical energy.
2. Transmission and subtransmission subsystems—Transporting electrical energy from its sources to load centers with high voltages (115 kV and above) to reduce losses.
3. Distribution—Distributing electrical energy from substations (in the range 44–12 kV) to end users/customers.
4. Consumers, users, or utilization subsystem.

This basic structure of a power system is shown in Figure 2.1. In this complex power system structure, electro-mechanical systems play a key role. An essential component of power systems is three-phase AC synchronous generators or alternators. The electric generator converts nonelectrical energy provided by the prime mover, usually steam or hydro turbines to electrical energy. The source of mechanical power, commonly known as prime mover, may be hydraulic turbines or steam turbines. The function of turbine is to rotate electrical generators by converting the thermal energy of the steam or the kinetic energy of the water into rotating mechanical energy. In thermal power plants, fossil fuels or nuclear reactions are used to produce high temperature steam, eventually passed through the turbine blades causing the turbine to rotate. Typical hydroelectric plants consist of dam, holding water upstream at high elevations with respect to the turbine. When electricity is needed, the water flows through the hydro turbine blades through penstocks, rotating the generator. Since the generator is mounted on the turbine shaft, the generator rotates with the turbine generating electricity. At load level, the bulk of the energy is consumed through electrical motors, mostly induction type.

The generators used in a power plant are synchronous machines (alternators). Synchronous generators have two synchronized rotating electromagnetic fields, one produced by the rotor driven at synchronous speed and excited by a DC circuit, and the other one is produced in the stator windings by the three-phase armature currents. The excitation system maintains the generator voltage and control the reactive power flow. Due their structure and construction, AC generators can generate high power and voltage, typically of 30 kV. However, induction or DC generators can be found in standalone and low-power distributed generation. Synchronous machines have magnetic field circuits mounted on the rotor and is firmly connected to the turbine shaft. The alternator stator has windings wrapped around its core in a three-phase configuration. Insulation requirements and other practical design issues limit the generated voltage to some low values, up to 30 kV. The generator voltages (5–kV) are not high enough for efficient power transmission, being stepped-up by transmission transformers. The devices connecting generators to transmission subsystem and from transmission subsystem to distribution subsystem are the transformers. The transformers transfer the electricity with very high efficiency from one level of the voltage to another one, suitable for specific applications. Their main functions are stepping up the lower generation voltage to the higher transmission voltage and stepping down the higher transmission voltage to the lower distribution voltage. The main advantage of having higher voltage in transmission system is to reduce the losses in the grid. Since transformers operate at constant power, when the voltage is higher, then the current has a lower value. Therefore, the losses, a function of the current square, will be lower at a higher voltage. However, when the electrical energy is delivered to the load centers, the voltage is stepped down for safer distribution and usage requirements. When the electric power reaches customers' facilities, it is further stepped down to the required levels depending on the various standards worldwide. The electricity in an electric power system may undergo four or five transformations between its generation and consumers.

A power system is predominantly in steady state operation or in a state that could with sufficient accuracy be regarded as steady state. In a power system, there are always small load changes, switching actions, and other transients occurring so that in a strict mathematical sense, most of the variables are varying with the time. However, these variations are most of the time so small that an algebraic, i.e., not time varying model of the power system is justified.

The electricity is delivered from the generation ends (power plants) to the loads (consumers) transmission lines and transformers. The bulk of the electricity, produced into the power plants are transmitted to the load centers over long distance high-voltage transmission lines, operating at very high voltages, 220–1,200 kV. The lines that distribute the electrical power within an area are called medium-voltage distribution lines. There are several other categories, such as subtransmission and high-voltage distribution line or power distribution networks, which will be discussed later in this book. The transmission lines are high-voltage conductors (wires) mounted on tall towers to prevent them to be in contact with humans, trees, animals, buildings, equipment, or ground. High-voltage towers are normally made up of galvanized steel, 25–45 m in height, to achieve strength and durability needed in harsh environments, as shown in Figure 2.2. The higher is the voltage of the wire, the taller is the tower. Since steel is electrical conductive material, high-voltage wires are not mounted directly on the towers. Instead, the insulators made up of nonconductive materials mounted on the towers are used to hold the conductors away from the tower structure. Insulators, of various shapes and designs, are strong enough to withstand the static and dynamic forces exerted by the conductors during windstorm, freezing rains, earthquakes, or animal impacts.

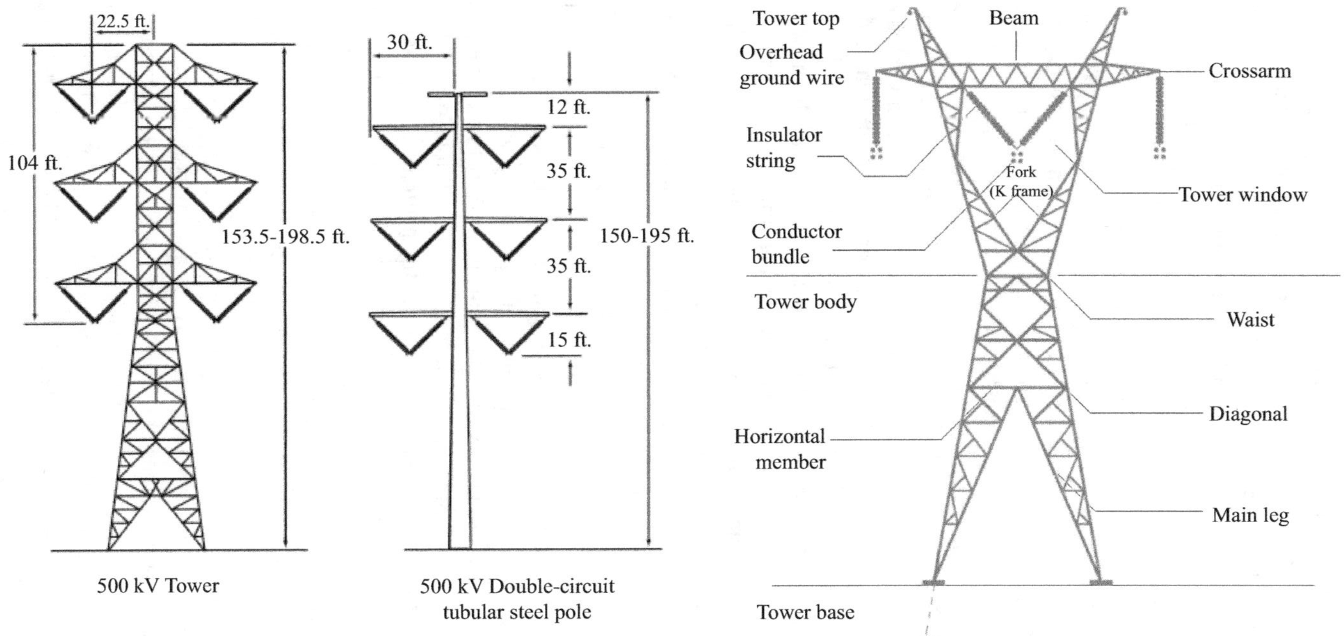

Figure 2.2 High-voltage transmission towers

2.2 Sources of energy

Today, the vast majority of energy, including electricity consumed around the world is provided by fossil fuels, such as coal, oil, and natural gas. At the same time, all fossil fuels are not the same in terms of their availability and their environmental impacts. Fossil fuels are coming from layers of prehistoric carbonaceous materials, compressed over millions of years to form solid, liquid, or gas in high energy-dense concentrations. These can be easily extracted, transported, and combusted to meet human energy needs. Fossil fuels are varying in terms of energy density, the energy content per unit of mass or volume, current consumption rates in different countries or world regions, remaining availability of resources, and emissions per unit of energy released during combustion. Since the time when coal surpassed wood as a leading energy source used by humans for energy needs, fossil fuels are becoming the dominant primary energy source worldwide, counting for over 85% of all energy production. Because fossil fuels are so important for meeting our energy needs, sustainable energy consumption must consider high-efficient fuel combustion technologies, energy efficiency and conservation, distributed generation, and the move toward the use of renewable and alternative energy sources.

From the three major fossil fuels used today, coal cannot be used directly in internal combustion engines, so its applications are limited to the stationary ones, mostly in industrial applications, but also on limited scale, in some of the domestic applications. Natural gas and oil are more flexible in a way they are combusted, being used extensively for transportation, and also for stationary applications. Natural gas requires less preprocessing before combustion and has less adverse environmental impacts than the other two. One important characteristic of the fossil fuels is the energy density. The energy density is a measure of the amount of energy per unit of mass or volume available in that resource. It is an important parameter as greater quantities of fossil fuels with lower energy density will be required in the energy conversion process, thereby increasing the overall cost and the environmental impact. Coal energy density is expressed in GJ/ton or millions of BTU/ton, the oil energy density in GJ/barrel or BTU/barrel, while the natural gas density is expressed in kJ/m^3 or BTU/ft^3, when the gas is at constant atmospheric pressure. However, in practice, natural gas is compressed in order to save the space during transportation and storage. Typical values of coal energy density are between 15 and 30 GJ/ton, about 5.2 GJ/barrel for crude oil, and 36.5 MJ/m^3 for natural gas. Among all three, coal is the most difficult to extract, transport, and use. The cost of these fossil fuels varies largely from country to country, year to year, and region to region with price variations and fluctuations not dictated only by the market. However, when the energy prices rise disproportionately, they tend to choke off the economic growth, leading to the reduction in demands, so the energy prices tend to fall again. On long run, high energy costs motivated research and applications of high energy-efficient technologies, easing the pressure on energy demand, and allowing the cost growth to slow down or reverse.

Nuclear fuels are heavy materials, used for energy generation, mostly uranium (U), but plutonium (Pu) is also used. The term, nuclear energy, has different meanings to different people. For an energy engineer, nuclear energy is the controlled release of the nuclear fuel energy for electricity generation. Nuclear energy can be obtained either through *fission* (heavy atom splitting resulting in two or more smaller atoms), or fusion (joining together of two light atoms into a larger one). Energy release during such processes is transferred to a working fluid, and then the thermal energy is converted into electricity in a similar conversion process as one in a conventional thermal power plant. The energy releases during such processes is much larger than the chemical energy releases during the combustion process. There are no or very little pollutant emissions during construction and operation of a nuclear power plant, as well as in all other related processes. However, managing the radioactive materials requires unique techniques, including reactor shielding, careful moving and transportation of by-products and waste, power plant control and monitoring, etc. Data for uranium production are not freely available for political and security reasons. The largest uranium producers are Canada, Australia, Niger, Namibia, Russia, United States, and former USSR states. About 80% of the consumption of nuclear energy is mostly distributed in North America and Europe, and the rest mostly in Asia.

Hydroelectric power plants harness the energy of the Earth hydrologic cycle, while converting it into the electrical energy. Water from ocean, lakes, rivers, plants, etc. absorbs the solar energy, evaporates into the atmosphere forming clouds, eventually return back to ground as precipitations, and through the ocean and lakes through water streams. The motion of water through ocean and lakes is due to the kinetic energy, which can be harnessed by hydro-power plants and converted into electrical energy. If water is stored at high elevations, usually in a reservoir, it poses potential energy proportional to that elevation. When this water is allowed to flow from high to low elevation, the potential energy is converted into kinetic energy, and converted into electricity through a hydro-power turbine. The common types of hydroelectric power plants are impoundment hydroelectric (involving a large water reservoir formed by a dam), diversion hydroelectric (some water of a river with strong current is diverted through hydropower turbines), and pumped storage hydroelectric systems, used as a form of energy storage. Amount of electricity generation depends on the water head behind the dam, reservoir capacity, flow rate, topography, and efficiency of the hydroelectric power plant components. The electricity generation system description is beyond the scope of this book, interested readers are directed to the book references or elsewhere in the literature.

It is commonly accepted that the earth's fossil energy resources are limited, and sometimes in future, their production will come beyond their peaks. At the same time, there is strong opposition against strengthening the nuclear power in many parts of the world. In this scenario, renewable energy resources will have to contribute more and more to the world's ever rising needs of the energy in the future. The major renewable energy resources are the sun, with some forms also attributed to the moon and the earth. Notable for their contribution to the current energy demand are water, wind, solar energy, and biomass. Renewable energy is becoming increasingly used in electricity generation due to the significant technological advances in wind turbine, photovoltaic systems, energy storage, power electronics, and control technologies. For the most part, renewable energy sources also provide clean energy, or energy that emits few greenhouse gases or pollutants. For this reason, many policy experts and scientists advocate renewable energy sources over traditional fossil fuels. The difficulty is in achieving the technology, infrastructure, and political support to make this transition. According to NREL reports, electricity supply and demand can be balanced in every hour of the year in each region with nearly 80% of electricity from renewable resources, nearly 50% from variable renewable generation, according to simulations of 2,050 power system operations. However, as renewable electricity generation increases, additional transmission infrastructure is required to deliver electricity generation from cost-effective remote renewable resources to load centers, enable reserve sharing over greater distances, and smooth output profiles of variable resources by enabling greater geospatial diversity. Many of the system flexibility resources and options described above can benefit from transmission infrastructure enhancements to enable the transfer of power and sharing of reserves over large areas to accommodate the variability of wind and solar electricity generation in combination with variability in electricity demand.

2.3 Power system structure and components

The device converting mechanical energy to electrical energy is called a generator. Synchronous machines can produce high power reliably with high efficiency, and therefore, are widely used as generators in power systems. A generator serves two basic functions. The first one is to produce active power (MW), and the second function, frequently forgotten, is to produce reactive power (MVAR). The discussion on generators will be limited to the fundamentals related to these two functions. More details related to the dynamic performance of the synchronous generators can be found in the references at the end of this chapter or elsewhere in the literature. The mechanical structure of generators is out of the scope of this material. A simplified turbine-generator-exciter system is shown in Figure 2.3. The turbine, or the prime mover, controls the active power generation. For instance, by increasing the valve

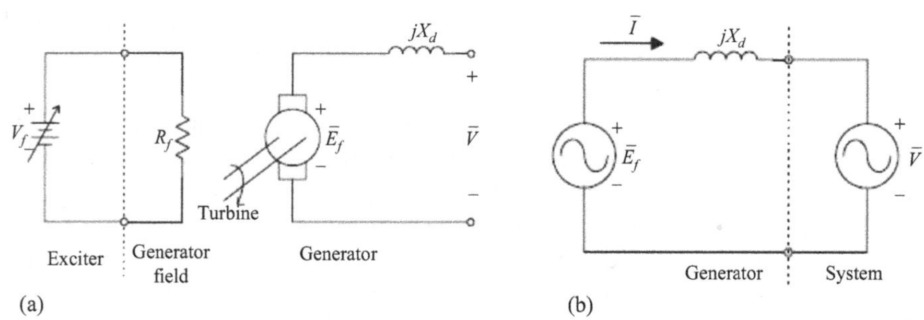

Figure 2.3 (a) Electrical representation of a simplified turbine-generator-exciter system and (b) the per-phase steady-state equivalent circuit of a synchronous generator

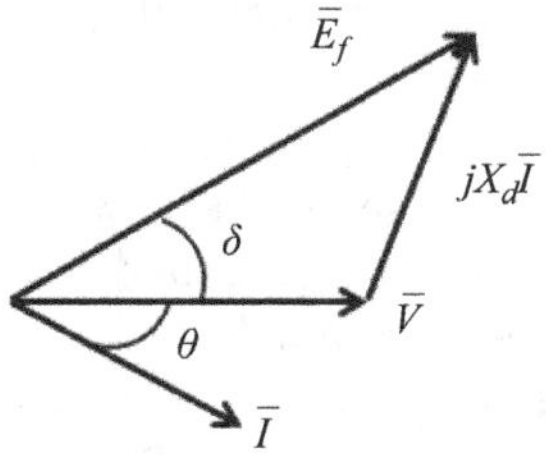

Figure 2.4 Phasor representation of the induced generator voltage

opening of a steam turbine, more active power can be generated and vice versa. The exciter, represented as an adjustable DC voltage source, controls the filed current that controls the internal generated voltage source, the so-called excitation voltage, E_f. In this way, the generator terminal voltage, V, is controlled.

The steady-state equivalent circuit of a synchronous generator can be drawn as an internal voltage source and its (direct-axis) synchronous reactance in series, as shown in Figure 2.3. The system is represented with an infinite bus, which holds a constant voltage. The generator terminal voltage or system voltage is usually chosen as the reference, therefore, a zero degree phase angle. Then, through a phasor representation, the generator internal generated voltage can be obtained as:

$$\overline{E}_f = \overline{I}(jX_d) + \overline{V} = E_f < \delta \tag{2.1}$$

Here the angle δ is called the generator power (torque) angle, and X_d is the synchronous reactance, as shown in the Figure 2.4. Figure 2.4 is showing the graphical representation of these quantities, together with the current and voltage, in the phasor and per-phase notation of the internal generated voltage. Such representations can be very useful in the generator power and characteristics calculations. The per-phase analysis of the complex power injected into the power system, by the synchronous generator can be calculated by:

$$\overline{S} = \overline{V} \cdot \overline{I}^* = \frac{VE_f}{X_d}\sin\delta + j\left[\frac{VE_f}{X_d}\cos\delta - \frac{V^2}{X_d}\right] = P + jQ$$

The maximum value of the active power, P_{max}, is referred to as the steady-state stability limit and can be calculated as the limit for a power (torque) angle of 90° (sin90° is equal to 1). It is worth mentioning that when the generator active power increases, the power angle also increases. However, at P_{max}, the power angle is 90°, the angle cannot be increased any further, since the generator cannot maintain synchronism with the rest of the power system.

2.4 Three-phase systems

The generation, transmission, and distribution of electric power is accomplished by means of three-phase circuits. An AC generator designed to develop a single sinusoidal voltage for each rotation of the shaft (rotor) is referred to as a single-phase AC generator. If the number of coils on the rotor is increased in a specified manner, the result is a poly-phase AC generator, which develops more than one AC phase voltage per rotation of the rotor. As it was mentioned in the generating station, three sinusoidal voltages are generated having the same amplitude but displaced in phase by 120°, so-called a balanced source. If the generated voltages reach their peak values in the sequential order *abc*, the generator is said to have a positive phase sequence, as shown in Figure 2.6(a). If the phase order is *acb*, the generator is said to have a negative phase sequence, as shown in Figure 2.6(b). In a three-phase system, the instantaneous power delivered to the external loads is constant rather than pulsating as it is in a single-phase circuit. Also, three-phase motors, having constant torque tend to start and run much better than single-phase motors. This feature of three-phase power, coupled with the inherent efficiency of its transmission compared to single-phase (less wire for the same delivered power), accounts for its universal use. In general, three-phase systems are preferred over single-phase systems for the transmission of power for many reasons, such as:

1. Thinner conductors can be used to transmit the same kVA at the same voltage, which reduces the amount of copper required (typically about 25% less).
2. The lighter lines are easier to install, and the supporting structures can be less massive and farther apart.
3. Three-phase equipment and motors are running smoothly and have preferred running and starting characteristics compared to single-phase systems because of a more even flow of power to the transducer than can be delivered with a single-phase supply.
4. In general, the larger motors are three-phase because they are essentially self-starting and do not require a special design or additional starting circuitry.

Three-phase sinusoidal voltages and currents are generated with the same magnitude but are displaced in phase by 120°, by what is called a balanced source or generator, as shown in Figure 2.5. In a three-phase generator, three identical coils a, b, and c are separated by an angle of 120° from each other, and the generator is turned by prime

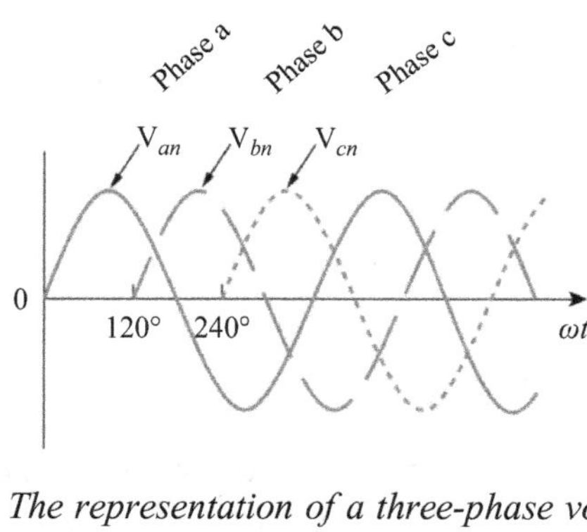

Figure 2.5 The representation of a three-phase voltage system

mover, a system of three voltages, V_{an}, V_{bn}, and V_{cn} with the same magnitude and separated by 120° phase angles are produced:

$$
\begin{aligned}
V_{an} &= V_m \sin(\omega t) = V_m \angle 0^\circ \\
V_{bn} &= V_m \sin(\omega t - 120^\circ) = V_m \angle -120^\circ \\
V_{cn} &= V_m \sin(\omega t - 240^\circ) = V_m \sin(\omega t + 120^\circ) = V_m \angle -240^\circ
\end{aligned}
\tag{2.2}
$$

where V_m is the peak value or the magnitude of the generated voltage. The sum of the three waveform voltages, by using trigonometric identities is:

$$
\begin{aligned}
V &= V_{an} + V_{bn} + V_{cn} \\
&= V_m[\sin(\omega t) + \sin(\omega t - 120^\circ) + \sin(\omega t - 240^\circ)] = 0
\end{aligned}
\tag{2.3}
$$

Three-phase systems may be labeled either by 1, 2, 3 or a, b, c, or sometimes, by using the three natural color, red, yellow, and blue to represent them. The phase sequence is quite important for the transmission, distribution, and use of electrical power. If the generated voltages reach their peak values in the sequential order *abc*, the generator is said to have a positive phase sequence, shown in Figure 2.6(a). If the phase order is *acb*, the generator is said to have a negative phase sequence, as shown in Figure 2.6(b).

The three single-phase voltages can be connected to form practical three-phase systems in two ways: (1) star or wye (Y) connections (circuits), or (2) delta (Δ) connections (circuits), as are shown in Figure 2.7. In the Y-connection, one terminal of each generator coil is connected to a common point or neutral *n* and the other three terminals represent the three-phase supply. In a balanced three-phase system, knowledge of one of the phases gives the other two phases directly. However, this is not the case for an unbalanced supply. In a star connected supply, it can be seen that the line current (current in the line) is equal to the phase current (current in a phase). However, the line voltage is not equal to the phase voltage. In a three-phase system the instantaneous power delivered to the external loads is constant rather than pulsating, as it is in a single-phase circuit. Also, three-phase motors, having constant torque, start and run much better than single-phase motors. This feature of three-phase power, coupled with the inherent efficiency of its

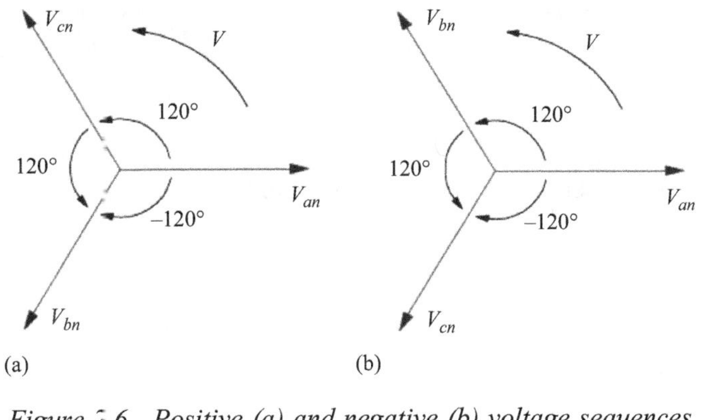

Figure 2.6 Positive (a) and negative (b) voltage sequences

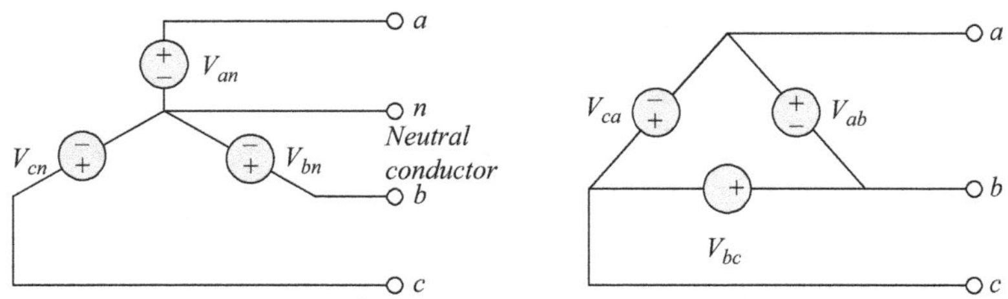

(a) Wye connected system (b) Delta connected system

Figure 2.7 Wye and Delta connected sources

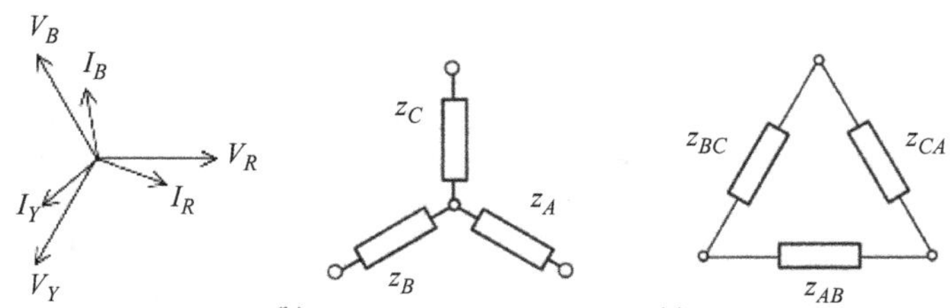

(a) (b) (c)

Figure 2.8 (a) Phasor diagram, (b) Y-connected load, and (c) delta connection

transmission compared to single-phase (less wire for the same delivered power), accounts for its universal use. A power system has Y-connected generators and usually includes both Δ- and Y-connected loads. Generators are rarely Δ-connected because if the voltages are not perfectly balanced, there will be a net voltage and consequently a circulating current around the Δ loop. Also, the phase voltages are lower in the Y-connected generator and thus less insulation is required. Figure 2.8 shows a Y-connected generator supplying balanced Y-connected loads through a three-phase line. In the Y-connected circuits, the phase voltage is the voltage between any line (phase) and the neutral point, represented by V_{an}, V_{bn}, and V_{cn}, while the voltage between any two lines is called the line or line-to-line voltage, represented by V_{ab}, V_{bc}, and V_{ca}, respectively. For a balanced system, each phase voltage has the same magnitude and we define:

$$|V_{an}| = |V_{bn}| = |V_{cn}| = V_P \tag{2.4}$$

Here V_P denotes the effective magnitude of the phase voltage. We can show that:

$$V_{ab} = V_{an} - V_{bn} = V_P(1 - 1\langle -120^\circ) = \sqrt{3}V_P\langle 30^\circ \tag{2.5}$$

Similarly relationships can be obtained as:

$$\begin{aligned} V_{bc} &= \sqrt{3}V_P\langle -90^\circ \\ V_{ca} &= \sqrt{3}V_P\langle 150^\circ \end{aligned} \tag{2.6}$$

In a balanced three-phase Y-connected voltage system, the line voltage V_L whose magnitude is related to the phase voltage magnitude through:

$$V_L = \sqrt{3}V_P \tag{2.7}$$

Example 2.1: A three-phase generator is Y-connected, as shown in Figure 2.7. The magnitude of each phase voltage is 220 V RMS. For *abc* phase sequence, write the three-phase voltage equations, and calculate the line voltage magnitude.

Solution: The expressions of the phase voltages are:

$$V_{an} = 220\angle 0^\circ \text{ V}$$
$$V_{bn} = 220\angle -120^\circ \text{ V}$$
$$V_{cn} = 220\angle 120^\circ \text{ V}$$

While the magnitude for the

$$V_{LL} = \sqrt{3} \times V_P = \sqrt{3} \times 220 = 380.6 \text{ V}$$

For a balanced system, the angles between the phases are 120° and the magnitudes are all equal. Thus the line voltages would be 30° leading the nearest phase voltage. Calculation will easily show that the magnitude of the line voltage is $\sqrt{3}$ times the phase voltage. A current flowing out of line terminal, I_L (the effective value of the line current) is the same the phase current I_P (the effective value of the phase current) for the Y-connected circuits, thus:

$$I_L = I_P \tag{2.8}$$

In delta connection, as shown in Figure 2.7(b), the line and the phase voltages have the same magnitude:

$$|V_L| = |V_P| \tag{2.9}$$

Similarly in the case of a delta connected supply, the current in the line is $\sqrt{3}$ times the current in the delta. In a manner similar as for the Y-connected sources, we can easily prove:

$$\begin{aligned} I_{ab} &= \sqrt{3}I_P\angle 0^\circ \\ I_{bc} &= \sqrt{3}I_P\angle -90^\circ \\ I_{ca} &= \sqrt{3}I_P\angle 150^\circ \end{aligned} \tag{2.10}$$

A balanced three-phase current system in a delta connection yields into a corresponding set of balanced line currents related as:

$$I_L = \sqrt{3}I_P \tag{2.11}$$

where the I_L is denoting the magnitude of the three equally line currents of the system.

2.4.1 Balanced loads

A load on a three-phase supply usually consists of three impedances and one of the ways in which these can be connected, wye (Y) or star connection or delta (Δ), as shown in Figure 2.8. A balanced load would have the impedances of the three phase equal in magnitude and in phase. Although the three phases would have the phase angles differing by 120° in a balanced supply, the current in each phase would also have phase angles differing by 120° with balanced currents. Thus, if the current is lagging (or leading) the corresponding voltage by a particular angle in one phase, then it would lag (or lead) by the same angle in the other two phases as well (Figure 2.8(a)). For the same load, star connected impedance and the delta connected impedance will not have the same value. However, in both cases, each of the three phases will have the same impedance, as shown in Figure 2.8(b) and (c).

In Y-connected loads, the line currents are taken from the supply bare, therefore, equal to the phase currents of the load. In Figure 2.9, the star point of load is connected to the load star point and the neutral current, in phasor notation is expressed as:

$$I_N = I_A + I_B + I_C \tag{2.12}$$

With the neutral connected to the three-phase loads, the phase voltages across the corresponding phase of the load are:

$$V_A = Z_A I_A; \quad V_B = Z_B I_B; \quad V_C = Z_C I_C \tag{2.13}$$

For a balanced set of supply voltages and loads, the phasor sum of the load voltages and currents is always zero. Since for balanced loads the neutral current is zero, the neutral conductor between the load and the source is not strictly needed, and is frequently omitted. If the load impedances are not balanced, with the neutral conductor connected, a neutral current is flowing. However, if the neutral is not connected, then the star point neutral departs from the supply neutral (so-called ***floating neutral***). The following equations then apply:

$$\begin{aligned} V_{AB} &= Z_A I_A - Z_B I_B \\ V_{BC} &= Z_B I_B - Z_C I_C \\ I_A + I_B + I_C &= 0 \end{aligned} \tag{2.14}$$

These equations are sufficient to solve for the load phase currents (equal to the source line currents) and hence the load voltages.

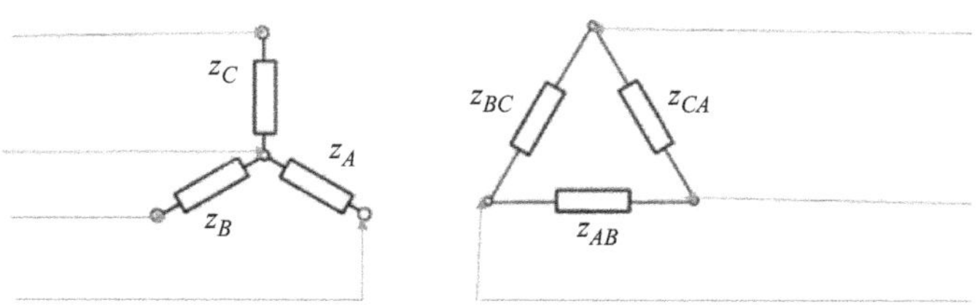

Figure 2.9 Star wins the neutral and delta connected loads

In the delta connected loads, the load phase voltages are equal to the source line voltages for a balanced supply and loads. The relationships between supply line currents and the load phase currents in a phasor form are expressed as:

$$\begin{aligned} I_A &= I_{AB} - I_{CA} \\ I_B &= I_{BC} - I_{AB} \\ I_C &= I_{CA} - I_{BC} \end{aligned} \tag{2.15}$$

A distinct advantage of a consistent set of notation adopted in a three-phase circuit analysis is the symmetry of the expressions, resulting in an additional way to check their consistency and correctitude. By using Ohm's law, the phase currents are given by:

$$I_{AB} = \frac{V_{AB}}{Z_{AB}}; \quad I_{BC} = \frac{V_{BC}}{Z_{BC}}; \quad I_{CA} = \frac{V_{CA}}{Z_{CA}} \tag{2.16}$$

Example 2.2: A Y-connected balanced three-phase load consisting of three impedances of

$$Z_L = 44\angle 30^\circ\ \Omega$$

The loads are supplied with the balanced phase voltages:

$$\begin{aligned} V_{an} &= 220\angle 0^\circ\ \text{V} \\ V_{bn} &= 220\angle -120^\circ\ \text{V} \\ V_{cn} &= 220\angle 120^\circ\ \text{V} \end{aligned}$$

Calculate: (a) the phase currents; and (b) the line-to-line phasor voltages

Solution:

(a) The phase currents are computed as:

$$\begin{aligned} I_{an} &= \frac{220\angle 0^\circ}{44\angle 30^\circ} = 5\angle -30^\circ\ \text{A} \\ I_{bn} &= \frac{220\angle -120^\circ}{44\angle 30^\circ} = 5\angle -150^\circ\ \text{A} \\ I_{cn} &= \frac{220\angle 120^\circ}{44\angle 30^\circ} = 5\angle 90^\circ\ \text{A} \end{aligned}$$

(b) Applying (2.5) and (2.6) the line-to-line voltages are obtained as:

$$\begin{aligned} V_{ab} &= V_{an} - V_{bn} = 220\angle 0^\circ - 220\angle -120^\circ = 220\sqrt{3}\angle 30^\circ\ \text{V} \\ V_{bc} &= V_{bn} - V_{cn} = 220\angle -120^\circ - 220\angle 120^\circ = 220\sqrt{3}\angle -90^\circ\ \text{V} \\ V_{ca} &= V_{cn} - V_{cn} = 220\angle 120^\circ - 220\angle 0^\circ = 220\sqrt{3}\angle 150^\circ\ \text{V} \end{aligned}$$

Example 2.3: Repeat example 2.2 if the three impedances are Δ-connected.

Solution:

(a) From previous example, we have:

$$V_{ab} = 220\sqrt{3}\angle 30^\circ \text{ V}$$
$$V_{bc} = 220\sqrt{3}\angle -90^\circ \text{ V}$$
$$V_{ca} = 220\sqrt{3}\angle 150^\circ \text{ V}$$

The currents in each of the load impedances are:

$$I_{ab} = \frac{220\sqrt{3}\angle 30^\circ}{44\angle 30^\circ} = 5\sqrt{3}\angle 0^\circ \text{ A}$$

$$I_{bc} = \frac{220\sqrt{3}\angle -90^\circ}{44\angle 30^\circ} = 5\sqrt{3}\angle -120^\circ \text{ A}$$

$$I_{ca} = \frac{220\sqrt{3}\angle 150^\circ}{44\angle 30^\circ} = 5\sqrt{3}\angle 120^\circ \text{ A}$$

(b) The line currents are the computed using (2.15) as:

$$I_a = I_{ab} - I_{ca} = 5\sqrt{3}\angle 0^\circ - 5\sqrt{3}\angle -120^\circ = 15\angle 30^\circ \text{ A}$$
$$I_b = I_{bc} - I_{ab} = 5\sqrt{3}\angle -120^\circ - 5\sqrt{3}\angle 120^\circ = 15\angle -90^\circ \text{ A}$$
$$I_c = I_{ca} - I_{bc} = 5\sqrt{3}\angle 120^\circ - 5\sqrt{3}\angle -120^\circ = 15\angle 150^\circ \text{ A}$$

2.4.2 Mixed connection circuits, wye–delta connection

The source and the load are not always connected in the same manner. For example, the load can be Δ-connected and the source Y-connected, or vice versa. In either case, attention must be paid to the line and phase quantity calculations. The phase and line voltage and current relationships established in the subsections of the previous chapter apply here. We can infer this by examining the following example:

Example 2.4: Figure 2.10 is showing a balanced Δ-connected load supplied by balanced 120 V Y-connected source. The line current is $I_3 = 15\angle 75^\circ$ A. Compute the load impedance.

Solution: First we compute the voltage across the load:

$$\overline{V}_{12} = \sqrt{3}V_1\angle 30^\circ = \sqrt{3}(120\angle 0^\circ)\angle 30^\circ = 208\angle 30^\circ \text{ V}$$

The load current is:

$$I_{12} = \frac{I_1}{\sqrt{3}\angle -30^\circ} = \frac{I_3\angle -120^\circ}{\sqrt{3}\angle -30^\circ} = \frac{15\angle 75^\circ - 120^\circ}{\sqrt{3}\angle -30^\circ} = 5.77\angle -15^\circ \text{ A}$$

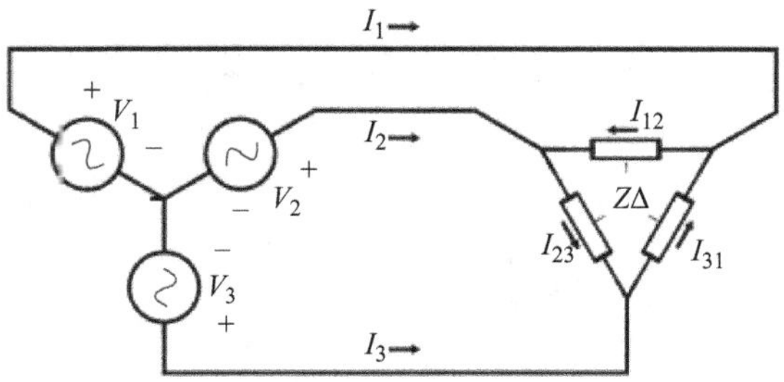

Figure 2.10 Wye-connected source supplying a delta connected load

The load impedance is computed by Ohm's law as:

$$Z_\Delta = \frac{V_{12}}{I_{12}} = \frac{208\langle 30^\circ}{5\ 77\langle -15^\circ} = 36\langle 45^\circ\ \Omega$$

2.5 Power relationships in three-phase circuits

Assuming a three-phase is supplying a three-phase balanced load, as in Figure 2.8 with three-phase sinusoidal phase voltages:

$$\begin{aligned} v_a(t) &= \sqrt{2}V_P \sin(\omega t) \\ v_b(t) &= \sqrt{2}V_P \sin(\omega t - 120^\circ) \\ v_c(t) &= \sqrt{2}V_P \sin(\omega t + 120^\circ) \end{aligned} \tag{2.17}$$

With current flowing through the load given by:

$$\begin{aligned} i_a(t) &= \sqrt{2}I_P \sin(\omega t - \phi) \\ i_b(t) &= \sqrt{2}I_P \sin(\omega t - 120^\circ - \phi) \\ i_c(t) &= \sqrt{2}I_P \sin(\omega t + 120^\circ - \phi) \end{aligned} \tag{2.18}$$

where ϕ is the phase angle between the voltage and current in each phase.

The instantaneous power supplied to one phase of the load (Equation (1.3) of the previous chapter), as in Figure 2.11 is:

$$p(t) = v(t) \cdot i(t)$$

Therefore, the instantaneous power in each of the three phases of the load is:

$$\begin{aligned} p_a(t) &= v_a(t) \cdot i_a(t) = 2VI \sin(\omega t)\sin(\omega t - \theta) \\ p_b(t) &= v_b(t) \cdot i_b(t) = 2VI \sin(\omega t - 120^\circ)\sin(\omega t - 120^\circ - \theta) \\ p_c(t) &= v_c(t) \cdot i_c(t) = 2VI \sin\left(\omega t - 240^\circ\right)\sin(\omega t - 240^\circ - \theta) \end{aligned} \tag{2.19}$$

The total instantaneous power flowing into the load is the expresses as:

$$p_{3\phi}(t) = v_a(t)i_a(t) + v_b(t)i_b(t) + v_c(t)i_c(t)\ \text{W} \tag{2.20}$$

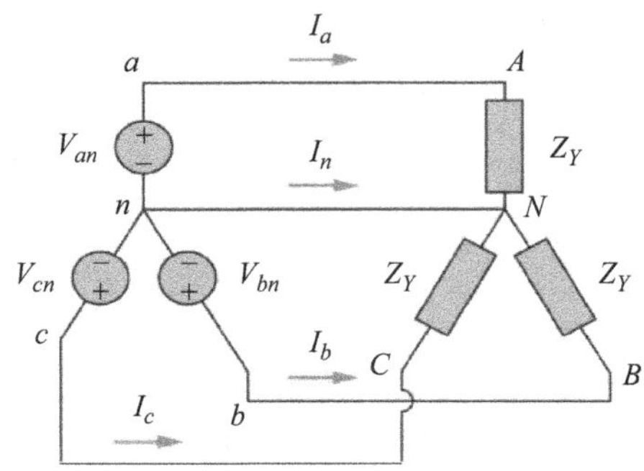

Figure 2.11 A Y-connected generator supplying a Y-connected load

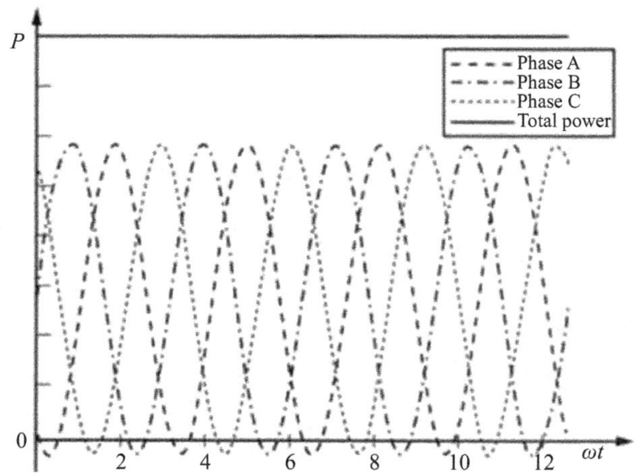

Figure 2.12 Power in a three-phase balanced system

By substituting expressions for phase voltages and currents, (2.17) and (2.18), respectively in (2.13) and using additional trigonometric identity:

$$\cos(\alpha) + \cos(\alpha - 120°) + \cos(\alpha - 240°) = 0$$

Equation (2.19) can be re-written as the equation below:

$$p_{3\phi}(t) = 3|V||I|\cos(\phi) = 3\text{P W} \tag{2.21}$$

Here $|V| = \sqrt{2}V_P$ and $|I| = \sqrt{2}I_P$ are the peak magnitude (amplitude) of the phase voltage and current. Equation (2.21) represents a very important result. In other words: *In a balanced three-phase system, the sum of the three individually pulsating phase powers adds to a constant, nonpulsating total active power of magnitude three times the real (active) power in each phase.* A graphical representation of this important consequence of three-phase balanced system is shown in Figure 2.12. However, one has to keep in mind that (2.21) is valid only for balanced conditions.

The single phase power equations (Chapter 1) are applied to each phase of Y- or Δ-connected three-phase loads. The real, active, and apparent powers supplied to a balanced three-phase load are:

$$
\begin{aligned}
P &= 3V_\phi I_\phi \cos(\theta) = 3ZI_\phi^2 \cos(\theta) \\
Q &= 3V_\phi I_\phi \sin(\theta) = 3ZI_\phi^2 \sin(\theta) \\
S &= 3V_\phi I_\phi = 3ZI_\phi^2
\end{aligned}
\tag{2.22}
$$

Angle θ is again the angle between the voltage and the current in any of the load phase, and the power factor of the load is the cosine of this angle. We can express the powers of (2.22) in terms of line quantities, regardless the connection type (wye or delta) as:

$$
\begin{aligned}
P &= \sqrt{3}V_{LL}I_L \cos(\theta) \\
Q &= \sqrt{3}V_{LL}I_L \sin(\theta) \\
S &= \sqrt{3}V_{LL}I_L
\end{aligned}
\tag{2.23}
$$

We have to keep in mind that angle θ in (2.23) is the angle between the *phase voltage* and *the phase current*, not the angle between the line-to-line voltage and the line current.

Example 2.5: The terminal line-to-line voltage of three-phase generator equals 13.2 kV. It is symmetrically loaded and delivers an RMS current of 1.350 kA per phase at a phase angle of 24° lagging. Compute the power delivered by this generator.

Solution:

The RMS value of the phase voltage is:

$$|V| = \frac{13.2}{\sqrt{3}} = 7.621 \text{ kV/phase}$$

The per-phase active (real) and reactive power are given by:

$$
\begin{aligned}
P &= 7.621 \times 1.350 \cdot \cos(24^\circ) = 9.399 \text{ MW/phase} \\
Q &= 7.621 \times 1.350 \cdot \cos(24^\circ) = 4.185 \text{ MVAR/phase}
\end{aligned}
$$

The instantaneous powers in phase *a*, *b*, and *c* are pulsating and are given by:

$$
\begin{aligned}
p_a(t) &= 9.399(1 - \cos(2\omega t)) - 4.185 \sin(2\omega t) \\
p_b(t) &= 9.399(1 - \cos(2\omega t - 120^\circ)) - 4.185 \sin(2\omega t - 120^\circ) \\
p_a(t) &= 9.399(1 - \cos(2\omega t - 240^\circ)) - 4.185 \sin(2\omega t - 240^\circ)
\end{aligned}
\tag{2.24}
$$

The total (constant) three-phase power is:

$$P_{3\phi} = 3 \times 9.399 = 28.197 \text{ MW}$$

The fact that three-phase *active (real) power* is constant tempts us to believe that the *reactive power* in a three-phase is zero (as in a DC circuit). However, the reactive power is very much present in *each phase* as shown in (2.22). The reactive power per phase is 4.185 MVAR.

Example 2.6: A three-phase load draws 120 kW at a power factor of 0.85 lagging from a 440 V bus. In parallel with this load, a three-phase capacitor bank that is rated 50 kVAR is inserted, find:

1. The line current without the capacitor bank.
2. The line current with the capacitor bank.
3. The PF without the capacitor bank.
4. The PF with the capacitor bank.

Solution:

1. From the three-phase active power formula, the magnitude of the load current is:

$$I_{Load} = \frac{P}{\sqrt{3}V_L \times PF} = \frac{120 \times 10^3}{\sqrt{3}440(0.85)} = 185.25 \text{ A}$$

$$I_{Load} = 185.25\langle -\cos^{-1}(0.85) = 185.25\langle -31.8^\circ \text{ A}$$

2. The line current of the capacitor bank (a pure reactive load) is:

$$I_{Cap} = \frac{50 \times 10^3}{\sqrt{3}440} = 65.6\langle 90^\circ \text{ A}$$

The line current is:

$$I_L = I_{Load} + I_{Cap} = 160.6\langle -11.5^\circ \text{ A}$$

3. The PF without capacitor bank is PF = 0.85
4. The PF with capacitor bank is:

$$PF = \cos(11.5^\circ) = 0.98$$

2.6 Per-unit system

The per-unit (p.u.) value representation of electrical variables in power system and electric machine computation is a common and useful practice. In the power systems analysis, field of electrical engineering, a per-unit system is the expression of system quantities as fractions of a defined base unit quantity. An interconnected power system typically consists of many different voltage levels given in a system containing several transformers and/or rotating machines. The *per-unit system* simplifies the analysis of complex power systems by choosing a common set of base parameters in terms of which, all systems quantities are defined. The different voltage levels disappear and the overall system reduces to a set of impedances. Calculations are simplified because quantities expressed as per-unit do not change when they are referred from one side of a transformer to the other. This can be a pronounced advantage in power system analysis where large numbers of transformers may be encountered. Moreover, similar types of apparatus will have the impedances lying within a narrow numerical range when expressed as a per-unit fraction of the equipment rating, even if the unit size varies widely. Conversion of per-unit quantities to volts, ohms, or amperes requires knowledge of the base that the per-unit quantities were referenced to. The main idea of per-unit system is to absorb large difference in absolute values into base relationships. Representations of elements in the system with per-unit values become more uniform. The per-unit numerical value of any quality is the ratio of its value to a chosen base quantity of the same dimension. A per-unit quantity is a normalized quantity with respect to the chosen base value. There are several reasons for using a per-unit system:

- Similar apparatus (generators, transformers, lines) will have similar per-unit impedances and losses expressed on their own rating, regardless of their absolute size. Because of this, per-unit data can be checked rapidly for gross errors. A per-unit value out of normal range is worth looking into for potential errors.
- The per-unit values for various components lie within a narrow range regardless of the equipment rating.
- Manufacturers usually specify the apparatus impedance in per-unit values.
- Use of the constant $\sqrt{3}$ is reduced in three-phase calculations.
- Per-unit quantities are the same on either side of a transformer, independent of voltage level.
- By normalizing quantities to a common base, both hand and automatic calculations are simplified.
- It improves numerical stability of automatic calculation methods.
- Per-unit data representation yields important information about relative magnitudes.
- Ideal for computer simulations.

The definition of the per-unit value of a quantity is:

$$p.u.\text{ value} = \frac{\text{Actual value}}{\text{Base (reference) value of the same dimension}} \tag{2.25}$$

The complete characterization of a per-unit system requires that all four base values be defined. Given the four base values, the per-unit quantities are defined as:

$$V_{p.u.} = \frac{V}{V_{base}}; \quad I_{p.u.} = \frac{I}{I_{base}}; \quad S_{p.u.} = \frac{S}{S_{base}}; \quad Z_{p.u.} = \frac{Z}{Z_{base}} \tag{2.26}$$

The per-unit system was developed to make manual analysis of power systems easier. Although power-system analysis is now done by computer, results are often expressed as per-unit values on a convenient system-wide base. The base value is always a real number and the per-unit value is dimensionless. Five quantities are involved in this calculation: the current, the voltage, the complex power, the impedance, and the phase angle. Phase angles are dimensionless; the other four quantities are completely described by knowing only two of them. Usually, the nominal line or equipment voltage is known as well as the apparent (complex) power, so these two quantities are often selected for base value calculation. For example, considering a single-phase system, then the expression of base current is:

$$I_{base} = \frac{S_{base(1-\phi)}}{V_{base(LN)}} \tag{2.27}$$

The expression of the base impedance is:

$$Z_{base} = \frac{V^2_{base(LN)}}{S_{base(1-\phi)}} = \frac{V_{base(LN)}}{I_{base}} \tag{2.28}$$

The magnitude of the base current in a three-phase system can be calculated as:

$$I_{base} = \frac{S_{base(3-\phi)}}{\sqrt{3}V_{base(LL)}} \tag{2.29}$$

The base impedance can be calculated as:

$$Z_{base} = \frac{V^2_{base(LL)}}{S_{base(3-\phi)}} = \frac{V_{base(LL)}}{\sqrt{3}I_{base}} \tag{2.30}$$

Per-unit quantities obey the circuit laws, thus:

$$\begin{aligned} S_{p.u.} &= V_{p.u.}I^*_{p.u.} \\ V_{p.u.} &= Z_{p.u.}I_{p.u.} \end{aligned} \tag{2.31}$$

For a three-phase system, the phase impedance in per-unit is given by:

$$Z_{p.u.} = \frac{V^2_{p.u.}}{S^*_{L(p.u.)}} \tag{2.32}$$

Example 2.6: Assuming that line voltage of 735 KV for 120 MVA transmission line with the impedance:

$$Z = 4.50 + j75.30\ \Omega$$

Calculate the per-unit transmission line resistance, reactance, and impedance.

Solution:

$$Z = \sqrt{4.5^2 + 75.3^2}\angle \tan^{-1}\left(\frac{75.3}{4.5}\right) = 75.4\angle 86.6^\circ\ \Omega$$

The base impedance is:

$$Z_{base} = \frac{V_{base}^2}{S_{base}} = \frac{\left(735 \times 10^3\right)^2}{120 \times 10^6} = 4502\ \Omega$$

The per-unit transmission line resistance, reactance, and impedance are computed as:

$$R_{p.u.} = \frac{4.5}{4{,}502} = 9.996 \times 10^{-4}\ \text{p.u.}$$

$$X_{p.u.} = \frac{75.3}{4{,}502} = 0.01673\ \text{p.u.}$$

$$Z_{p.u.} = \frac{75.4}{4{,}502} = 0.01675\ \text{p.u.}$$

Usually, if none are specified, the p.u. values given are on the nameplate ratings as base. There are situations when the base for the system is different from the base for each particular generator or transformer, hence it is important to be able to express the p.u. value in terms of different bases. The rule for the impedances is:

$$Z_{p.u.(new)} = Z_{p.u.(old)} \frac{S_{base(new)}}{S_{base(old)}} \cdot \frac{V_{base(new)}^2}{V_{base(old)}^2} \tag{2.33}$$

Example 2.7: Convert the impedance value of the Example 2.5 to the new base 240 MVA and 345 kV.

Solution: We have

$$Z_{p.u.(old)} = 9.996 \times 10^{-4} + j0.01673$$

for a 120 MVA, and 735 kV base. With a new base 240 MVA and 345 kV, by using the impedance conversion relationship (3.30):

$$Z_{p.u.(new)} = Z_{p.u.(old)} \left(\frac{240}{120}\right) \cdot \left(\frac{735}{345}\right)^2 = 9.0775 \cdot Z_{(p.u.)(old)}$$

And

$$Z_{p.u.(new)} = 0.0091 + j0.1519$$

Example 2.8: A three-phase generator rated at 350 MVA and 21 kV has per-phase reactance of 0.35 p.u. on its own base. The generator is placed in a system where the bases are 100 MVA and 13.8 kV. Find the reactance of the generator in per-unit on the new base.

Solution: Applying (3.30) we have:

$$X_{p.u.(new)} = 0.35\left(\frac{100}{350}\right)\left(\frac{21}{13.8}\right)^2 = 0.232 \text{ p.u.}$$

2.7 Voltage and frequency characteristics

The utility frequency or (power) line frequency, the frequency of the alternating current in an electric power grid transmitted from a power plant to the end-user. In large parts of the world, this is 50 Hz, although it is 60 Hz in the United States and parts of Asia. During the development of commercial electric power systems in the late nineteenth and early twentieth centuries, several different frequencies (and voltages) had been used. Large investment in equipment at one frequency made standardization a slow process. However, as of the turn of the twenty-first century, places that now use the 50 Hz frequency tend to use 220–240 V, and those that now use 60 Hz tend to use 100–127 V. Both frequencies coexist today (Japan uses both) with no great technical reason to prefer one over the other and no apparent desire for complete worldwide standardization. Unless specified by the manufacturer to operate on both 50 and 60 Hz, appliances may not operate efficiently or even safely if used on anything other than the intended frequency. Several factors influence the choice of frequency in an AC system. Lighting, electrical motors, transformers, generators, and transmission lines all have characteristics which depend on the power frequency. All of these factors interact and make selection of a power frequency, a matter of considerable importance. The best frequency is a compromise between contradictory requirements. An accurate model for an electrical load is very important for a power system. Loads, such as electrical motors, lighting, and heating, show different characteristics with the change of voltage and frequency. In a power system, voltage and frequency at load bus always change due to the disturbances, which cause the fluctuation in the load power. This load power variation is not same for all the loads, and depends on the characteristics of the load connected to the load bus. For the purpose of stability analysis of a system, it is very essential to determine the effects of the changes of the loads due to the voltage and frequency changes.

In the late nineteenth century, designers would pick a relatively high frequency for systems featuring transformers and arc lights, so as to economize on transformer materials, but would pick a lower frequency for systems with long transmission lines or feeding primarily motor loads or rotary converters for producing direct current. When large central generating stations became practical, the choice of frequency was made based on the nature of the intended load. Eventually, improvements in machine design allowed a single frequency to be used both for lighting and motor loads. A unified system improved the economics of electricity production since system load was more uniform during the course of a day. The induction motor was found to work well on frequencies around 50–60 Hz, but with the materials available in the 1890s would not work well at a frequency of, say,

133 Hz. There is a fixed relationship between the magnetic pole number in the induction motor field, the alternating current frequency, and the rotation speed; so, a given standard speed limits the choice of frequency (and the reverse). Once AC electric motors became common, it was important to standardize the frequency for compatibility with the customer's equipment. Generators operated by slow-speed reciprocating engines will produce lower frequencies, for a given number of poles, than those operated by a high-speed steam turbine. For very slow prime mover speeds, it would be costly to build a generator with enough poles to provide a high AC frequency. Also, synchronizing two generators to the same speed was found to be easier at lower speeds. Very early isolated AC generating schemes used arbitrary frequencies based on convenience for steam engine, water turbine, and electrical generator design. Frequencies between 16⅔ and 133⅓ Hz were used on different systems. For example, the city of Coventry, England, in 1895 had a unique 87 Hz single-phase distribution system that was in use until 1906. The proliferation of frequencies grew out of the rapid development of electrical machines in the period 1880 through 1900. In the early incandescent lighting period, single-phase AC was common and typical generators were 8-pole machines operated at 2,000 RPM, giving a frequency of 133 Hz. Though many theories exist and quite a few entertaining urban legends, there is little certitude in the details of the history of 60 Hz versus 50 Hz. The German company AEG (descended from a company founded by Edison in Germany) built the first German generating facility to run at 50 Hz. At the time, AEG had a virtual monopoly and their standards spread to the rest of Europe. Westinghouse Electric decided to standardize on a higher frequency to permit operation of both electric lighting and induction motors on the same generating system. Although 50 Hz was suitable for both, in 1890 Westinghouse considered that existing arc-lighting equipment operated slightly better on 60 Hz, and so that frequency was chosen. The operation of Tesla's induction motor, licensed by Westinghouse in 1888, required a lower frequency than the 133 Hz common for lighting systems at that time. In 1893, General Electric Corporation, which was affiliated with AEG in Germany, built a generating project at Mill Creek, California using 50 Hz, but changed to 60 Hz a year later to maintain market share with the Westinghouse standard. In the early days of electrification, so many frequencies were used that no one value prevailed (London in 1918 had 10 different frequencies). As the 20th century continued, more power was produced at 60 Hz (North America) or 50 Hz (Europe and most of Asia). Standardization allowed international trade in electrical equipment and later, the interconnection of power grids. It wasn't until after World War II with the advent of affordable electrical consumer goods that more uniform standards were enacted. In Britain, a standard frequency of 50 Hz was declared as early as 1904, but significant development continued at other frequencies. The National Grid implementation starting in 1926 compelled the standardization of frequencies among the interconnected electrical service providers. The 50 Hz standard was completely established only after World War II. By about 1900, European manufacturers had mostly standardized on 50 Hz for new installations. Remnant installations at other frequencies persisted well after the Second World War. Because of the cost of conversion, some parts of the

distribution system may continue to operate on original frequencies even after a new frequency is chosen. A 25 Hz power was used in Ontario, Quebec, the northern United States, and for railway electrification, until 1950. The 15 kV AC rail networks, used in Germany, Austria, Switzerland, Sweden, and Norway, still operate at 16⅔ Hz. However, other AC railway systems are energized at the local commercial power frequency, 50 Hz or 60 Hz. Power frequencies as high as 400 Hz are used in aircrafts, space-crafts, submarines, computer power server rooms, and military equipment. Transformers, for example, can be made smaller because the magnetic core can be much smaller for the same voltage level. Induction motors turn at a speed proportional to frequency, so a high frequency power supply allows more power to be obtained for the same motor volume and mass. Transformers and motors for 400 Hz are much smaller and lighter than at 50 or 60 Hz, which is an advantage in aircraft and ships. United States military standard MIL-STD-704 exists for aircraft use of 400 Hz power.

Regulation of power system frequency for timekeeping accuracy was not commonplace until after 1926 and Laurens Hammond's invention of the electric clock driven by a synchronous motor. During the 1920s, Hammond gave away hundreds of such clocks to power station owners in the US and Canada as incentive to maintain a steady 60-cycle frequency, thus rendering his inexpensive clock uniquely practical in any business or home in North America. Developed in 1933, The Hammond Organ used a synchronous AC clock motor to maintain perfect pitch, based on power-line frequency stability. Today, AC-power network operators regulate the daily average frequency so that clocks stay within a few seconds of correct time. In practice, the nominal frequency is raised or lowered by a specific percentage to maintain synchronization. Over the course of a day, the average frequency is maintained at the nominal value within a few hundred parts per million. In the synchronous grid of Continental Europe, the deviation between network phase time and UTC (based on International Atomic Time) is calculated at 08:00 each day in a control center in Switzerland. The target frequency is then adjusted by up to ± 0.01 Hz ($\pm 0.02\%$) from 50 Hz as needed, to ensure a long-term frequency average of exactly 50 Hz $\times$ 60 s $\times$ 60 min $\times$ 24 h $=$ 4,320,000 cycles per day. In North America, whenever the error exceeds 10 s for the east, 3 s for Texas, or 2 s for the west, a correction of ± 0.02 Hz (0.033%) is applied. Time error corrections start and end either on the hour or on the half hour. Real-time frequency meters for power generation in the United Kingdom are available online—an official National Grid one, and an unofficial one maintained by Dynamic Demand. Real-time frequency data of the synchronous grid of Continental Europe is available on websites, such as mainsfrequency.com and gridfrequency.eu. The Frequency Monitoring Network (FNET) at the University of Tennessee measures the frequency of the interconnections within the North American power grid, as well as in several other parts of the world. These measurements are displayed on the FNET website. Smaller power systems may not maintain frequency with the same degree of accuracy. In 2011, The North American Electric Reliability Corporation (NERC) discussed a proposed experiment that would relax frequency regulation requirements for electrical grids which would reduce the long-term accuracy of clocks and other devices that use the 60 Hz grid frequency as a time base.

The primary reason for accurate frequency control is to allow the flow of alternating current power from multiple generators through the network to be controlled. The trend in system frequency is a measure of mismatch between demand and generation, and so is a necessary parameter for load control in interconnected systems. Frequency of the system will vary as load and generation change. Increasing the mechanical input power to a synchronous generator will not greatly affect the system frequency but will produce more electric power from that unit. During a severe overload caused by tripping or failure of generators or transmission lines, the power system frequency will decline, due to an imbalance of load versus generation. Loss of an interconnection, while exporting power (relative to system total generation) will cause system frequency to rise. Automatic generation control is used to maintain scheduled frequency and interchange power flows. Control systems in power plants detect changes in the network-wide frequency and adjust mechanical power input to generators back to their target frequency. This counteracting usually takes a few tens of seconds due to the large rotating masses involved. Temporary frequency changes are an unavoidable consequence of changing demand. Exceptional or rapidly changing mains frequency is often a sign that an electricity distribution network is operating near its capacity limits, dramatic examples of which can sometimes be observed shortly before major outages. Frequency protective relays on the power system network sense the decline of frequency and automatically initiate load shedding or tripping of interconnection lines, to preserve the operation of at least part of the network. Small frequency deviations (i.e., 0.5 Hz on a 50 Hz or 60 Hz network) will result in automatic load shedding or other control actions to restore system frequency. Smaller power systems, not extensively interconnected with many generators and loads, will not maintain frequency with the same degree of accuracy. Where system frequency is not tightly regulated during heavy load periods, the system operators may allow system frequency to rise during periods of light load, to maintain a daily average frequency of acceptable accuracy. Portable generators, not connected to a utility system, need not tightly regulate their frequency, because typical loads are insensitive to small frequency deviations. Power system stability is classified as rotor angle stability and voltage stability. Voltage stability is in power systems which are heavily loaded, disturbance, or have a shortage of reactive power. Nowadays, the demand of electricity has dramatically increased and a modern power system becomes a complex network of transmission lines interconnecting the generating stations to the major load points in the overall power system in order to support the high demand of consumers.

2.8 Chapter summary

The conception, by Nikola Tesla of a poly-phase, alternating current system was one of the most important innovations in the history of the electrical engineering. Most of the power system today from energy generation, transmission, and large part of the distribution, is three-phase system. In the earlier days of the power generation, Tesla not only led the battle if the power system uses DC or AC but also proved that the three-phase electric power was the most efficient way to generate, transfer, and use electricity. The essential feature of a three-phase system is although all currents are sinusoidal alternating waveforms if the system is balanced the total instantaneous power of the system is constant. A poly-phase electric generator or motor converts power from one form to another without fluctuations or pulsations, with constant energy stored in the electromagnetic field. The *per-unit system* simplifies the analysis of complex power systems by choosing a common set of base parameters in terms of which, all systems quantities are defined. The different voltage levels disappear and the overall system reduces to a set of impedances. Standard frequency used in North America is 60 Hz, while the Europe and the rest of the word are using 50 Hz frequency.

Further readings

[1] N. Cohn, Control of Generation and Power Flow on Interconnected Systems. Wiley, New York, 1971.

[2] O. Elgerd, Basic Electric Power Engineering, Addison-Wesley, Massachusetts, 1977.

[3] D.J. Glover, T. Overbye, and M.S. Sarma, Power System Analysis and Design (5th ed.), Cengage Learning, USA, 2012.

[4] F.A. Furfari, The Evolution of Power-Line Frequencies 133⅓ to 25 Hz, Industry Applications Magazine, IEEE, Sep/Oct 2000, Volume 6(5), pp. 12–14.

[5] E.L. Owen, The Origins of 60-Hz as a Power Frequency, Industry Applications Magazine, IEEE, Volume 3(6), Nov/Dec 1997, p. 8, 10, 12–14.

[6] G. Neidhofer, 50-Hz Frequency: How the Standard Emerged from a European Jungle, IEEE Power and Energy Magazine, July/August 2011, pp. 66–81.

[7] A. Daniels, Introduction to Electrical Machines, The Macmillan Press, UK, 1976.

[8] R.D. Schultz and R.A. Smith, Introduction to Electric Power Engineering, John Wiley & Sons, New York, 1988.

[9] A.C. Williamson, Introduction to Electrical Energy Systems, Longman Scientific & Technical, England, 1988.

[10] J.R. Cogdell, Foundations of Electric Power, Upper-Saddle River, Prentice-Hall, NJ, 1999.

[11] S.J. Chapman, Electrical Machinery and Power System Fundamentals (2nd ed.), McGraw Hill, New York, 2002.

[12] P. Schavemaker and L. van der Sluis, Electrical Power System Essentials, John Wiley and Sons, Chichester, UK, 2008.

[13] M.E. El-Hawary, Introduction to Power Systems, Wiley—IEEE Press, New York, 2008.

[14] M.A. El-Sharkawi, Electric Energy: An Introduction (2nd ed.), CRC Press, Boca Raton, FL, 2009.

[15] J.L. Kirtley, Electric Power Principles: Sources, Conversion, Distribution and Use, Wiley, Chichester, UK, 2010.

[16] F.M. Vanek, L.D. Albright, and L.T. Angenent, Energy Systems Engineering, McGraw Hill, New York, 2012.
[17] N. Mohan, Electric Power Systems—A First Course, Wiley, USA, 2012.
[18] F.M. Vanek, L.D. Albright, and L.T. Angenent, Energy Systems Engineering, McGraw Hill, New York, 2012.
[19] S.W. Blume, Electric Power System Basics for the Nonelectrical Professional (2nd ed.), Wiley and Sons, USA, 2016.
[20] J.E. Fleckenstein, Three-Phase Electrical Power, CRC Press, Boca Raton, FL, 2016.
[21] S.L. Herman, Electrical Transformers and Rotating Machines, Cengage Learning, UK, 2017.

Questions and problems

1. What types of connections are possible for three-phase sources and loads?
2. A Y-connected load has a 460 V voltage applied to it. What is its phase voltage?
3. If the load in previous question is Δ-connected, what will be its phase voltage?
4. What are the advantages of the three-phase systems?
5. List the advantages and disadvantages of high-voltage and low-voltage transmission lines.
6. What is function of power plant generators?
7. What is meant by the term "balanced" in a three-phase system?
8. What is the phase sequence? What is its importance?
9. In a balanced Δ-connected circuit with all resistive loads, what is the led or lag of the line currents with respect to line voltages?
10. What are the relationships between the phase and the line voltages and current for a Y-connection?
11. What are the relationships between the phase and line voltages and current for a Δ-connection?
12. Write the relationships for active, reactive, and apparent powers in three-phase circuits in terms of both line and phase voltages and currents.
13. A Y-connected balanced three-phase source is supplying power to a balanced three-phase load. The source phase voltage and current are given by:

$$v(t) = 420 \sin(377t + 30^\circ)\ \text{V}$$
$$i(t) = 90 \sin(377t + 15^\circ)\ \text{V}$$

Calculate (a) the RMS and line-to-line voltages and currents; (b) supply frequency; (c) power factor at the source side; (d) three-phase active, reactive, and apparent powers supplied to the load; and (e) the load impedance if the load is balanced and Y-connected.

14. A 120-HP, three-phase, 480-V induction motor operates at 0.88 PF lagging. Find (a) active, reactive, and apparent powers consumed per-phase; and (b) Suppose that the motor is supplied from a 460-V source through a feeder

whose impedance is $0.3 + j0.5\ \Omega$ per-phase. In this case, calculate the motor side voltage, the source power factor, and the transmission efficiency.

15. Repeat problem 7 if the motor's efficiency is 85%.
16. A three-phase wye-connected generator has the phase voltage 230 V. Calculate the phase voltages with angles and the line voltages.
17. The magnitude of each phase voltage of an unbalanced load is 220 V RMS. The load impedances are:

$$Z_A = 6 + j8\ \Omega$$
$$Z_B = 4 + j6\ \Omega$$
$$Z_A = 3 + j4\ \Omega$$

Calculate (a) line currents; and (b) the neutral current.

18. The per-phase reactance, 10 kVA, 120 V, Y-connected synchronous generator is 12 Ω. Determine the per-unit reactance, considering the base values are 10 kVA and 120 V.
19. The per-phase load impedance of a three-phase delta-connected load is $4 + j6\ \Omega$. If a 480 V, three-phase supply is connected to this load, find the magnitude of (a) phase current; and (b) line current.
20. Calculate the RMS value, supply frequency, and the phase shift in degrees for the AC voltage given by:

$$v(t) = 180\sin(300t + 0.866)\ V$$

21. For the balanced circuit in Figure P2.1 the magnitude of voltage is 270 V, calculate the power delivered to each of the resistors of the unbalanced load.

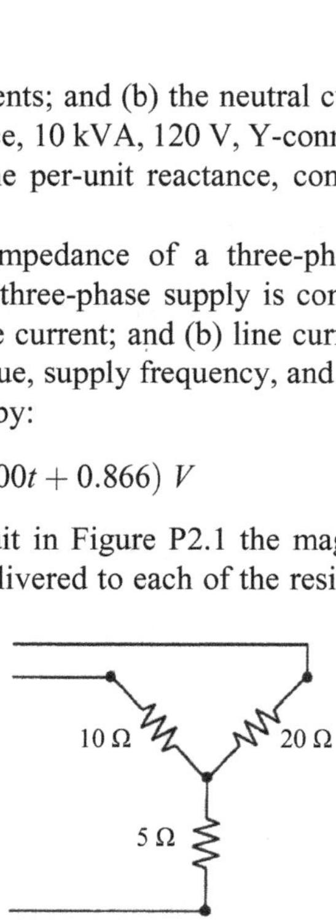

Figure P2.1

22. A single-phase load absorbs 200 kVA power from 35 kV busbar. Determine the base current and base impedance by considering those quantities as base values.
23. A three-phase, 460 V, wye-connected source is supplying power to a wye-connected balanced load. Calculate the load impedance, if the load current I_a is 10 A and is in phase with the line-to-line voltage, V_{bc}.
24. The base quantities of a system are 450 kVA and 35 kV. The per-unit impedance of this system is 0.035. Determine the actual impedance value.

25. The voltage and current measured for a wye-connected load are:

$$\overline{V}_{AB} = 230\langle 45^\circ \text{ V}$$
$$\overline{I}_C = 6\langle 130^\circ \text{ A}$$

1. Calculate the power factor angle.
2. Calculate the real power consumed by the load.

26. A balanced three-phase load is connected to a 480 V feeder. The line current I_A, in phase with the line-to-line voltage V_{BC} is equal to 12.5 A. Compute the load impedance if (a) the load is wye-connected; and (b) the load is delta-connected.
27. Three loads are connected in parallel across a 12.47 kV power supply. One is resistive 63 kW load, the second one is an induction motor of 72 kW and 63 kVAR, and the last one is a capacitive load drawing 180 kW at 0.85 PF. Find the total apparent power, power factor, and power supply current.
28. A single-phase transformer is rated 220/4,400 V and 5.0 kVA. The reactance of the transformer is 0.1 Ω measured from the low-voltage side. Determine the reactance of the transformer on its own base.
29. Repeat problem 19 using a new base: apparent power of 10 kVA and a voltage of 440 V.
30. A balanced star-connected load is fed from a 460 V, 60 Hz, three-phase supply. The resistance in each phase of the load is 30 Ω and the load draws a total power of 15 kW. Calculate (a) the line current drawn, (b) the load power factor, and (c) the load inductance.
31. A three-phase load is connected in star to a 460 V, 60 Hz supply. Each phase of the load consists of a coil having inductance 0.2 H and resistance 45 Ω. Calculate the line current.
32. If the load specified in problem 28 is connected in delta, determine the values for phase and line currents.
33. A balanced Y-connected load have a phase impedance of $8 + j6\ \Omega$ is connected to a three-phase supply of 460 V. Determine (a) the phase voltage; (b) the phase and line currents; (c) the power factor at the load; and (d) the power consumed by the load.

Chapter 3
Transformers and electrical motors

Objectives and abstract

Electric motors and transformers can be found in applications from computer disk drives, appliances, cars, power supplies, laptop and cell phone chargers, to industrial process lines, pumping systems, conveyers, escalators, and many more. For example, modern building services are heavily reliant upon electrical motors and drives, found in air conditioning, ventilation, and heating systems, or elevators. All electric motors have a stationary part (component), the stator, a moving part (component), the rotor, and interface with energy supply and eventually control and protection subsystems. Transformers are electric energy convertors, operating on Faraday's law of induction, designed to step up or step down the voltages as required by the loads. Motors and transformers are introducing a great deal of flexibility in power transmission, distribution, and use. The precision of control, introduced by electric motors, through the application of torque, speed, and position control has been critical and central to all aspects of industrial processes, automation, or transportation. This chapter focuses on the analysis and description of electric transformers and main types of electric motors used in commercial and industrial applications. The reader is advised that the chapter content and materials are critical for the understanding of several book topics. Motor and transformer construction, components, characteristics, and performances are included and discussed in details, together with short discussion of the common transformer and motor applications. A short discussion of energy conversion is also included in this chapter. This chapter also discusses on equivalent circuits, power, torque, and losses of transformers, induction and synchronous motors, as well as the most common type of DC motors. Several examples are also included in all chapter sections to facilitate the chapter topics understanding.

3.1 Introduction

The objectives of this chapter are to introduce the fundamental notions and concepts of the electromechanical energy conversion, leading to a good understanding of the operation of various electromechanical energy converters, such as transformers and electric motors. The subject of electromechanical energy conversion is one that should be of particular interest to both, the electrical and nonelectrical engineer, because it forms one of the important points of contact between electrical engineering and other engineering disciplines. The emphasis of this chapter is on explaining the properties and characteristics of each type of electrical machine, with its advantages and disadvantages with regard to other types, and on classifying these machines in terms of their performances, functionalities, characteristics, and the most common field of applications. At the end of this chapter, the readers are able to describe the principles of operation of transformers, DC and AC motors, and generators, understand and interpret the nameplate data of an electric machine, understand and interpret the voltage-current or torque-speed characteristic of an electric generator or electric motor, respectively. Finally, the reader learns, understands, and specifies the requirements of an electric machine given for an application.

3.1.1 Transformers in electrical systems

Transformers are commonly used in applications which require the conversion of AC voltage from one voltage level to another. They are essential part of power systems, having the role to convert electrical energy at one voltage to some other voltage. In order to effectively transmit power over long distances without prohibitive line losses, the voltage from the generator (a maximum of output voltage of approximately 25–30 kV) must be increased to a significantly higher level (from approximately 150 kV up to 750 kV). Transformers must also be utilized on the distribution end of the line to step the voltage down (in stages) to the voltage levels required by the consumer. Transformers also have a very wide range of applications outside the power area. Transformers are essential components in the design of DC power supplies. They can provide DC isolation between two parts of a circuit. Transformers can be used for impedance matching between sources and loads or sources and transmission lines. They can also be used to physically insulate one circuit from another for safety. The transformer is a valuable apparatus in electrical power systems, for it enables us to utilize different voltage levels across the system for most economical value. There are two broad categories of transformers: electronic transformers, which operate at very low power levels, and power transformers, which process thousands of watts of power. The basic principle of operation of both types of transformers is the same. Power transformers are used in power generation, transmission, and distribution systems to raise or lower the level of voltage to the desired levels. In power systems, transmission lines are typically operated on voltages that are substantially higher than either generation or utilization voltages, for two reasons:

1. Transmission line losses are proportional to the square of the current and higher voltages generally produce lower transmission line losses.
2. Transmission line capacity to carry electrical power is roughly proportional to the square of the voltage, so higher voltages allow intensive use of the transmission line right-of-way. It means that to enable transmission lines to carry high power levels, high transmission voltages are required. This is true in both cases of underground cables and overhead transmission lines.

A power distribution transformer or service transformer is a transformer that provides the final voltage transformation in the electric power distribution system, stepping down the voltages used in the distribution lines to the level required by the customers. An electric power distribution system is the final stage in the delivery of electric power, carrying the electricity to individual consumers. Distribution substations connect to the transmission system and lower the transmission voltage to medium voltage ranging between 2 kV and 35 kV with the use of transformers. Primary distribution lines carry this medium voltage power to the distribution transformers located near the customer's premises. Distribution transformers again lower the voltage to the utilization voltage of household appliances and typically feed several customers through secondary distribution lines at this voltage. Commercial and residential customers are connected to the secondary distribution lines through the electrical networks, the so-called service drops. Customers demanding larger amount of power are connected directly to the primary power distribution networks or to the subtransmission level. If mounted on a utility pole, they are called pole-mount transformers. If the distribution lines are located at ground level or underground, distribution transformers are mounted on concrete pads and locked in steel cases, the pad-mount transformers. Distribution transformers normally have ratings less than 200 kVA, although some national standards describe units up to 5,000 kVA. Since distribution transformers are energized for 24 h a day (even when they don't carry any load), reducing iron losses has an important role in their design. As they usually don't operate at full load, they are designed to have maximum efficiency at lower loads.

3.1.2 Electromechanical energy conversion systems

Electric machines are a means of converting energy. Motors take electrical energy and produce mechanical energy. Electric motors are used to power hundreds of devices we use in everyday life. Motors come in various sizes. Huge motors that can take loads of 1,000's of horsepower (HP) are typically used in the industry. Some examples of large motor applications include elevators, electric trains, hoists, conveyor belts, and heavy metal rolling mills. Examples of small motor applications include motors used in automobiles, robots, hand power tools, and food blenders. Micro-machines are electric machines with parts as small as the size of red blood cells, and find many applications in medicine, instrumentation, or control. Electric motors are broadly classified into two different categories: direct current and alternating current motors. Within these categories, there are numerous types, each offering unique abilities and characteristics, making each type suitable for a set of specific applications. In most cases, regardless of type, electric motors consist of a stator (stationary field) and a rotor (the rotating field or armature) and operate through the interaction of magnetic flux and electric current to produce rotational speed and torque. DC motors are distinguished by their ability to operate from direct current. There are different kinds of DC and AC motors, but they all work on the similar principles. In this chapter, we will study their basic principle of operation and their characteristics. It is important to understand the motor characteristics so we can choose the right one for the required application. An electromechanical energy conversion device is essentially a medium of transfer between an input side and an output side. Three electrical machines (DC, induction, and synchronous) are extensively used for electromechanical energy conversion.

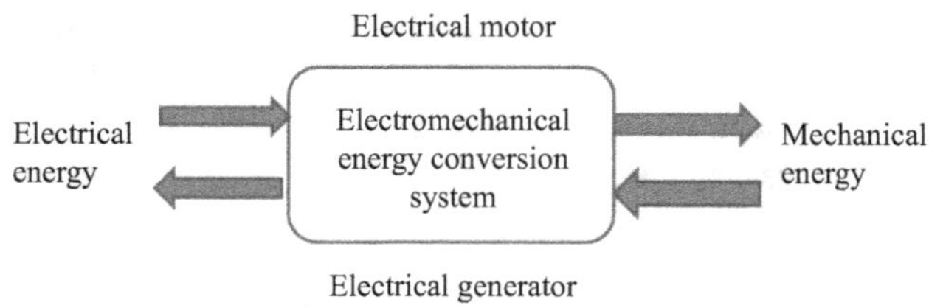

Figure 3.1 Electromechanical energy conversion systems, electric motor, and electric generator

Electromechanical energy conversion occurs when there is a change in magnetic flux linking a coil, associated with mechanical motion. Generators and motors are primarily rotating machines. The rotating machines are called motors when they consume electrical energy, and are referred to as generators when they produce electrical energy, as shown in Figure 3.1. In practical applications, while DC machines are almost always single phase, AC machines can be single-phase or three-phase types.

3.2 Transformer theory, construction, and design

A transformer consists of two or more windings that are magnetically coupled using a ferromagnetic core. For a two-winding transformer, the winding connected to the AC supply is referred to as the ***primary*** while the one connected to the load is referred to as the ***secondary***. A time-varying current passing through the primary coil produces a time-varying magnetic flux density within the core. According to Faraday law, the variable magnetic flux induces a voltage into the secondary terminals. In order to understand how a transformer operates, two inductors placed in close proximity to one another can be magnetically coupled, while such magnetic coupled circuits are extended to the transformers. After understating the relationships between voltages and currents, the practical considerations regarding the use of transformers are examined. The purpose of a power distribution transformer is to reduce the primary voltage of the electric distribution system to the utilization voltage serving the customer. Distribution transformers are static devices constructed with two or more windings used to transfer AC power from one circuit to another at the same frequency but with different values of voltage and current.

Magnetic fields are created due to the movement of electrical charge, and are present around permanent magnets and wires carrying current (electromagnet). In permanent magnets, spinning electrons produce a net external field. If a current carrying wire is wound in the form of a coil of many turns, the net magnetic field is stronger than that of a single wire. This electromagnetic field is further intensified if this coil is wound on an iron core. In many applications, the strength of magnetic fields must vary. Electromagnets are very commonly used in such applications. The magnetic field is represented by "lines of flux," helping us to visualize the magnetic field of any magnet even though they only represent an invisible phenomena. Magnetic field forms an essential link between transfer of energy from mechanical

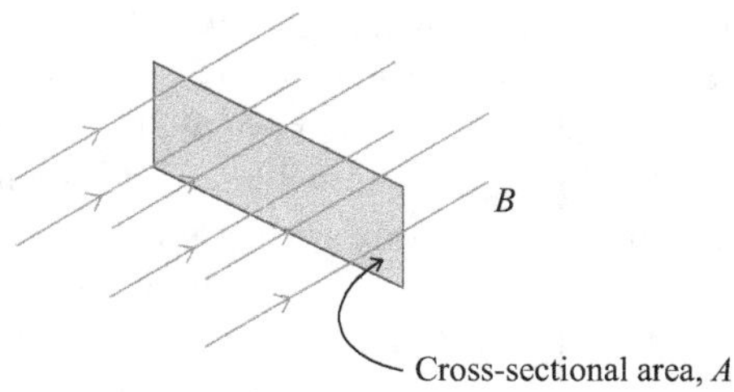

Figure 3.2 Magnetic flux density

to electrical form and vice-versa, forming the basis for the operation of transformers, generators, and motors. Examine the cross-sectional area of the magnet as shown in Figure 3.2 and assume that the flux is uniformly distributed over the area; the magnetic flux density is defined as the magnetic flux per cross-sectional area:

$$B = \frac{\phi}{A} \tag{3.1}$$

where A is the cross-sectional area and B is the magnetic flux density.

If a magnetic flux is passing through the surface bounded by a coil, we are saying that the magnetic flux is linking the coil. The flux passing through a coil is the product of number of turns, N, and flux passing a single turn, ϕ, and the product is called the magnetic flux leakage of the coil, λ, expressed as:

$$\lambda = N\phi \text{ weber-turns}$$

M. Faraday discovered, the electromagnetic induction (Faraday's law), which states that whenever a conductor is moved through a magnetic field, or whenever the magnetic field near is changed, current flows in the conductor. Voltage induced in a single loop, due to a time dependent magnetic flux ϕ is:

$$e(t) = \frac{d\phi(t)}{dt}$$

For a coil with N number of turns, the total induced voltage can be calculated by adding the voltage induced in all the turns:

$$e(t) = N\frac{d\phi(t)}{dt} = \frac{d\lambda}{dt} \text{ V} \tag{3.2}$$

Usually, power transformers are built and manufactured as three-phase units and may have multiple and complicated winding patterns. But three-phase transformers are at the core of things, understandable as three single-phase units with only a single-phase electric power, the discussion starts with single-phase transformers. Transformers are fairly complex electromagnetic systems, subject to involved analysis, but for the power system learning purpose, the model of ideal transformer is sufficient.

A transformer contains two or more windings linked by a mutual magnetic field flowing through its core. The simplest one consists of two windings on a core. Figure 3.3 show an ideal transformer and its symbol. The windings are commonly referred as the HV and LV windings, high- and low-voltage windings, respectively. The designations *primary* and *secondary* are also common. Often additional windings (*tertiaries*) are added. In short, the action of a transformer is a particular case of the principle of mutual induction; a transformer consists essentially of the primary and the secondary windings on a common magnetic core. A transformer is either core-type or shell-type construction, as shown in Figure 3.4. In a core-type construction, the primary and the secondary windings are wound as a pair of concentric coils on each limb, whereas in a shell-type construction, the primary and the secondary windings form interleaved layers on a single limb. In all cases, the core is a laminated construction in order to keep iron losses to a minimum.

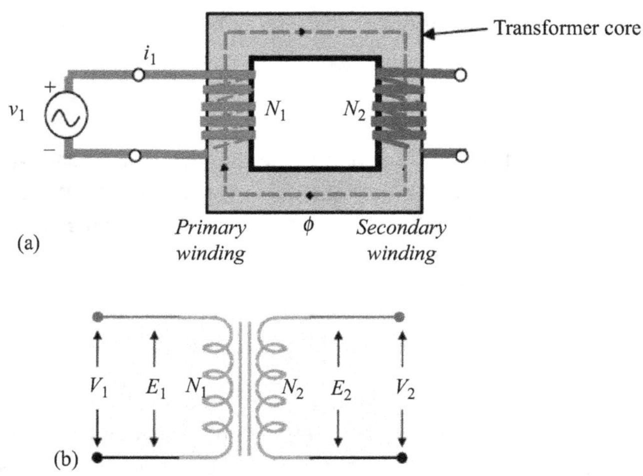

Figure 3.3 (a) Ideal transformer schematic and (b) the ideal transformer symbol

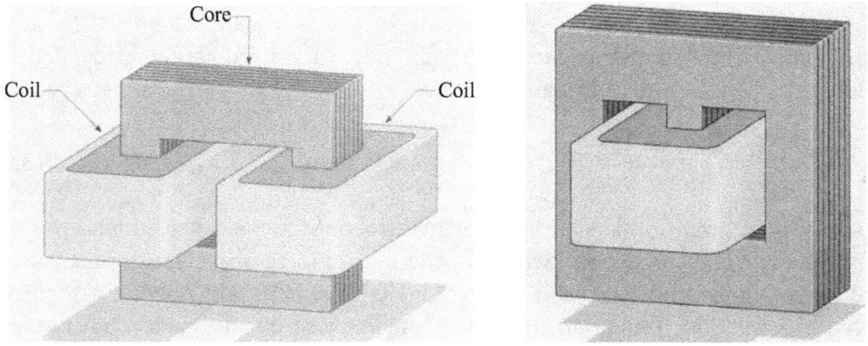

Figure 3.4 Core- and shell-type transformer constructions

An ideal transformer is the one with negligible winding resistances and reactances and no exciting losses, infinite magnetic permeability of the core and all magnetic flux remains into the transformer core. It consists of two conducting coils wound on a common core, made of high grade iron. There is no electrical connection between the coils, coupled to each other through magnetic flux. The coil on input side is called the primary winding (coil) and that on the output side the secondary. The essence of transformer action requires only the existence of time-varying mutual flux linking two windings. Such action can occur for two windings coupled through air. However, coupling between the windings can be made much more effective through the use of a core of iron or other ferromagnetic material because most of the flux will be confined to a definite, high-permeability path linking the windings. Such a transformer is commonly called an iron-core transformer. Most transformers are of this type. The following discussion is concerned almost wholly with iron-core transformers. When an AC voltage is applied to the primary winding, time-varying current flows in the primary winding and causes an AC magnetic flux to appear in the transformer core. The arrangement of primary and secondary windings on the transformer core is shown in Figure 3.3. The primary is connected to AC voltage sources, resulting in an alternating magnetic flux whose magnitude depends on the voltage and the number of turns of the primary winding. The magnetic flux links the secondary winding and induces a voltage in it with a value that depends on the number of turns of the secondary windings. The transformer operation is subjected to Faraday's and Ampere's Laws. If the primary voltage is $v_1(t)$, the core flux $\phi(t)$ is established such that the counter-emf $e(t)$ equals the impressed voltage (neglecting winding resistance), as:

$$v_1(t) = e_1(t) = N_1 \frac{d\phi(t)}{dt} \tag{3.3}$$

Here N_1 is the number of turns of the primary winding. The emf $e_2(t)$ is induced in the secondary by the alternating core magnetic flux $\phi(t)$, expressed as:

$$v_2(t) = e_2(t) = N_2 \frac{d\phi(t)}{dt} \tag{3.4}$$

Taking the ratio of (3.3) and (3.4), we obtain:

$$\frac{v_1}{v_2} = \frac{N_1}{N_2} = a \tag{3.5}$$

Here a is the transformer turns ratio. If $a > 1$, the transformer is step-down, if $a < 1$, the transformer is step-up, while if $a = 1$, the transformer is the so-called impedance transformer, used to separate two electric circuits. If a load is connected across the secondary terminals, as shown in Figure 3.5, it will result in current i_2. This current will cause the change in the mmf in the amount $N_2 i_2$. Ohm's law for magnetic circuits must be satisfied. The only way in which this can be achieved is for the primary current to arise, such as:

$$N_1 i_1 - N_2 i_2 = \Re\phi$$

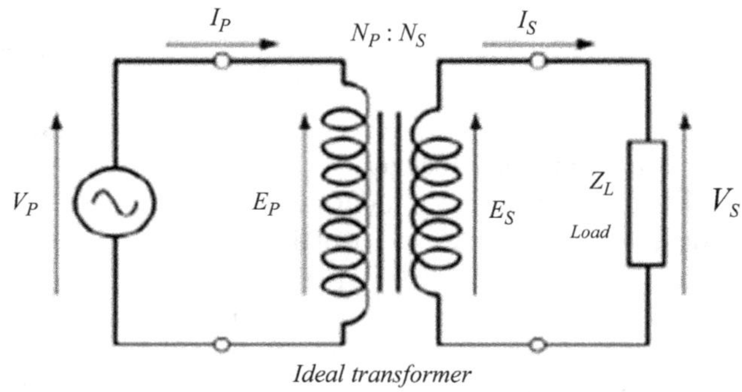

Figure 3.5 Loaded ideal transformer

Here $\Re$ is the core magnetic reluctance. The reluctance or magnetic resistance for a magnetic core is simply calculated as:

$$\Re = \frac{l_c}{\mu A} = \frac{l_c}{\mu_r \mu_0 A} \tag{3.6}$$

Here, l_c is the mean (average) core length (m), A is the core cross-sectional area (m^2), μ, and μ_r, are the magnetic permeability, relative magnetic permeability of the core, respectively, and $\mu_0 = 4\pi \times 10^{-7}$ H/m is the magnetic permeability of the free space. For an ideal transformer, we assumed infinite permeability or $\Re = 0$. Neglecting losses (ideal transformer), the instantaneous power is equal on the both transformer sides (power conservations):

$$v_1 i_1 = v_2 i_2 \quad \text{or} \quad V_P I_P = V_S I_S$$

Combing above relationship with (3.5), we get:

$$\begin{aligned} \frac{i_1}{i_2} &= \frac{N_2}{N_1} = \frac{1}{a} \\ \frac{I_P}{I_S} &= \frac{1}{a} \end{aligned} \tag{3.7}$$

If all variable are sinusoidal this equation applies also to the voltage and current phasors:

$$\frac{V_P}{V_S} = \frac{I_S}{I_P} = a \tag{3.8}$$

Example 3.1: A 220/20-V transformer has 50 turns on its low-voltage side. Calculate:

1. The number of turns on its high side.
2. The turns ratio a, when it is used as a step-down transformer.
3. The turns ratio a, when it is used as a step-up transformer.

Solution: Transformer, used as the step-down one, turns ratio is:

$$a_{SD} = \frac{220}{20} = 11$$

The number of turns in the high-voltage side is then:

$$N_P = aN_S = 11 \times 50 = 550\,\text{turns}$$

The turns ratio, when the transform is used as step-up is:

$$a_{SU} = \frac{1}{a_{SD}} = \frac{1}{11} = 0.091$$

Depending on the transformer turns ratio; the RMS secondary voltage can be greater or lesser than the RMS primary voltage. Equations (3.5) and (3.6) are showing that any desired voltage ratio, or ratio of the transformation, can be obtained by adjusting the number of turns of the transformer windings. Transformer action requires a magnetic flux to link two windings (coils), obtained more effectively if an iron (or iron based) core is used because it confines the magnetic flux to a definite path linking the windings. However, a magnetic material such as iron undergoes losses of energy due to the application of alternating voltage in the B-H loop. These losses are composed of two parts, the first one is the *eddy-current loss*, and the second one is the *hysteresis loss*. Eddy-current loss is basically an I^2R loss due to induced current in magnetic materials of the core due to the alternating magnetic flux linking the windings. To reduce these losses, the magnetic core is usually made by a stack of thin iron-alloy laminations. For analyzing an ideal transformer, we make the following assumptions. The resistances of the windings can be neglected, while the reluctance of the core is negligible. All the magnetic flux is linked by all the turns of the coil and there is no leakage of flux. We can write the equations for sinusoidal voltage in this ideal transformer as follows. The primary winding of turns N_1 is supplied by a sinusoidal voltage:

$$v_1(t) = V_{1m}\cos(\omega t) \tag{3.9}$$

Maximum value of voltage and the RMS value are related through:

$$V_1 = \frac{V_{1m}}{\sqrt{2}} = 0.707V_{1m} \tag{3.10}$$

In the case of sinusoidal excitation with supply frequency f Hz, the RMS value of the primary emf, from (3.3) is given by:

$$V_1 = 4.44fN_1\Phi_m \tag{3.11}$$

Here Φ_m is the peak value of the magnetic flux. In a similar manner, the RMS value of the secondary emf will be given by:

$$V_2 = 4.44fN_2\Phi_m \tag{3.12}$$

Example 3.2: Suppose a coil having 100 turns is wound on a core with a uniform cross-sectional area of 0.25 m^2. A 5 A, 60 Hz current is flowing into this coil. If the maximum magnetic flux density is 0.75 T, find the mmf and the voltage induced into the coil.

Solution:

$$mmf = NI = 100 \times 5 = 500 \text{ At}$$

The induced voltage is computed by using (3.11):

$$V = 4.44 \times 60 \times 100 \times 0.75 \times 0.25 \times 10^{-4} = 0.4995 \simeq 0.5 \text{ V}$$

By dividing (3.11) with (3.12), the transformer voltage relationship (3.7) is obtained. Consider now an arbitrary load (Z_2) connected to the secondary terminals of the ideal transformer as shown.

The input impedance seen looking into the primary winding is given by:

$$Z_1 = \frac{V_1}{I_1} = \frac{aV_2}{\frac{I_2}{a}} = a^2 Z_2 \tag{3.13}$$

The impedance seen by the primary voltage source of the ideal transformer is the secondary load impedance times the square of the transformer ratio. Using this property, the secondary impedance of an ideal transformer can be reflected to the primary side, expressed by (3.13). In a similar fashion, a load on the primary side of the ideal transformer can be reflected to the secondary.

$$Z_2 = \frac{V_2}{I_2} = \frac{\frac{V_1}{a}}{aI_1} = \frac{Z_1}{a^2} \tag{3.14}$$

Another important property of an ideal transformer, (3.7) is the power conservation:

$$V_1 I_1 = V_2 I_2 \tag{3.15}$$

The power conservation states that *the primary and secondary apparent powers (volt-amperes) are equal in an ideal transformer.*

Example 3.3: Determine the primary and secondary currents for the ideal transformer, supplied by a 120 V, if the source impedance is $Z_s = (18 - j4)$ Ω and the load impedance is $Z_2 = (2 + j1)$ Ω. The transformer ratio is $a = 4$.

Solution: The secondary voltage is:

$$V_2 = \frac{V_1}{a} = \frac{120}{4} = 30 \text{ V (RMS)}$$

The load impedance seen in the primary side is:

$$Z_2' = a^2 Z_2 = (4)^2(2 + j1) = 32 + j16\ \Omega$$

The primary and secondary currents are computed as:

$$I_1 = \frac{V_1}{Z_S + Z_2'} = \frac{120\angle 0^\circ}{18 - j4 + 32 + j16} = 2.33\angle -13.5^\circ \text{ A (RMS)}$$

$$I_2 = aI_1 = 9.32\angle -13.5^\circ \text{ A (RMS)}$$

While the primary and the secondary voltages are:

$$V_1 = Z_2' I_1 = (32 + j16)2.33\angle -13.5^\circ = 83.50\angle 13.07^\circ \text{ V (RMS)}$$

$$V_2 = \frac{V_1}{a} = \frac{83.50\angle 13.07^\circ}{4} = 20.88\angle 13.07^\circ \text{ V (RMS)}$$

3.2.1 *Polarity of transformer windings*

The operation of the transformer depends on the relative orientation of the magnetic fields of the primary and secondary coils. By marking one of the terminals on the primary and secondary coils with a dot in order to denote that currents entering these two terminals produce magnetic flux in the same direction within the transformer core (as shown in Figure 3.6). If coil orientation is reversed, the dot positions are reversed and the current and voltage equations must include a minus sign. With power transformers, the polarity is important only if the transformers are paralleled to gain additional capacity or to hook up three single-phase transformers to make a three-phase transformer bank. The way the connections are made affects angular displacement, phase rotation, and rotation direction of the connected motors. Polarity is also important when hooking up current transformers for relay protection and metering. Transformer polarity depends on which direction coils are wound around the core (clockwise or counterclockwise) and how the leads are brought out. Transformers are often marked at their terminals with polarity marks. Often, polarity marks are shown as white paint dots (for plus) or plus-minus marks

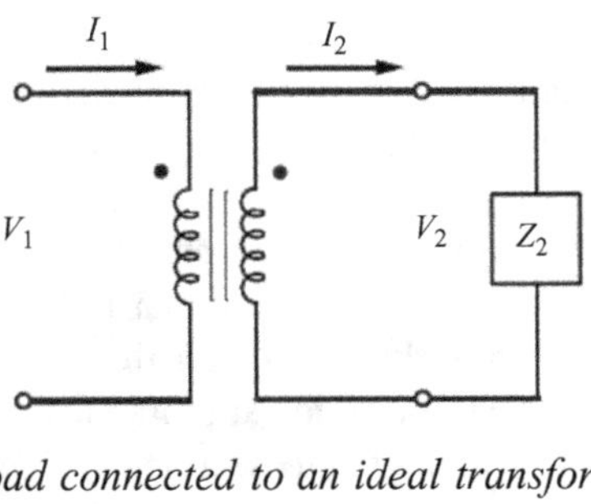

Figure 3.6 Load connected to an ideal transformer secondary

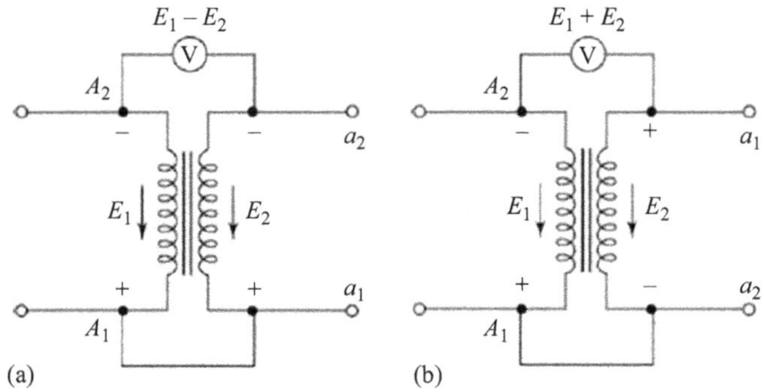

Figure 3.7 (a) Subtractive and (b) additive polarity

on the transformer and symbols on the nameplate. These marks show the connections where the input and output voltages (and currents) have the same instantaneous polarity.

Quite often, the transformer polarity is shown simply by the American National Standards Institute (ANSI) designations of the winding leads as H1, H2 and X1, X2. By ANSI standards, if the low-voltage side of a single-phase transformer (the side marked X1, X2), the H1 connection is always be on the far left. If the terminal marked X1 is also on the left, it is subtractive polarity. If the X1 terminal is on the right, it is additive polarity. Additive polarity is common for small distribution transformers. A transformer is said to have additive polarity if, when adjacent high and low-voltage terminals are connected and a voltmeter placed across the other high- and low-voltage terminals reads the sum of the high- and low-voltage windings. Figure 3.7 shows the high- and low-side voltage relationships for subtractive and additive polarity. It is subtractive polarity if the voltmeter reads the difference (subtractive) between the voltages of the two windings. If this test is conducted, use the lowest AC voltage available to reduce potential hazards. An adjustable AC voltage source is recommended to keep the test voltage low.

3.2.2 Practical (non-ideal) transformers

Certain assumptions made about ideal transformer are not valid for an actual transformer. In a practical transformer, the windings have resistances, not all windings link the same magnetic flux, the core permeability is not infinite, and the core losses occur when core material is subject to time-varying magnetic flux. In the actual transformers analysis, all these factors must be considered. In each of the transformer configuration, most of the flux is confined to the core and therefore links both windings. The windings are also producing additional magnetic flux, the leakage flux, which links one winding without linking the other. Although the leakage flux is a small fraction of the total magnetic flux, it plays an important role in determining the transformer behavior. In practical transformers, leakage is reduced by subdividing the windings into sections placed as close as possible. In the core-type construction, each winding consists of two sections, one on each of

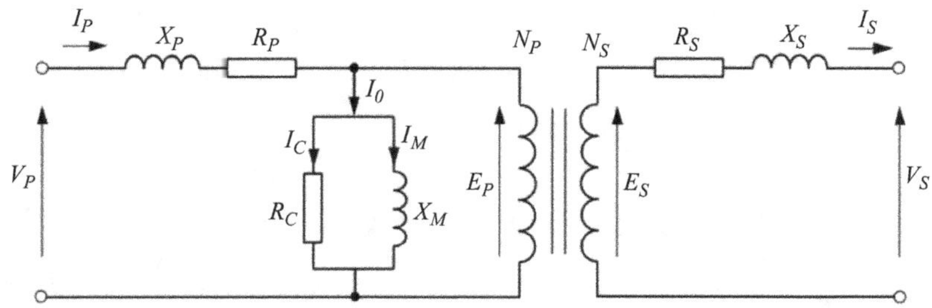

Figure 3.8 The equivalent circuit of a non-ideal transformer

the two core legs, the primary and secondary windings being concentric coils. In the shell-type construction, variations of the concentric-winding arrangements are used or the windings may consist of a number of thin "pancake" coils assembled in a stack with primary and secondary coils interleaved. A non-ideal transformer accounts for all of the losses that are neglected in the ideal transformers. The individual losses are (a) primary and secondary winding resistances (losses due to the resistance of the wires), (b) primary and secondary leakage reactances (losses due to magnetic flux leakage out of the transformer core), and (c) core resistance (core losses due to hysteresis loss and eddy current loss), and magnetizing reactance (magnetizing current needed to establish magnetic flux into the transformer core). The equivalent circuit diagram of a non-ideal transformer is shown in Figure 3.8. Given that the voltage drops across the primary winding resistance and the primary leakage reactance are typically small, the shunt branch of the core loss resistance and the magnetizing reactance (excitation branch) can be shifted to the primary input terminal. The primary voltage is then applied directly across the shunt impedance and allows for the winding resistances and leakage reactances to be combined.

The transformer equivalent circuit (Figure 3.8) is determined experimentally from three tests: (a) a DC test in which the primary and secondary windings can be measured, (b) an open-circuit test, where the secondary is left unconnected and the normal rated voltage is applied to the primary; and (c) a short-circuit test, where the secondary terminals are short-circuited and a low voltage is applied to the primary, sufficient to circulate the normal full-load current. Because of the nonlinear magnetic properties of iron, the waveform of the exciting current differs from the waveform of the flux; the exciting current for a sinusoidal flux waveform is not sinusoidal. This effect is especially pronounced in closed magnetic circuits, as ones found in the transformers. In magnetic circuits where the reluctance is dominated by an air gap with its linear magnetic characteristic, as is the case in many electric machines, the relationship between the net magnetic flux and the applied magnetomotive force (mmf) is almost linear and the exciting current is much closer to a sinusoidal one. A further approximation to the equivalent circuit is made by eliminating the excitation branch. This approximation removes the core losses and the magnetizing current from the transformer model, resulting equivalent circuit is shown in Figure 3.9, the so-called cantilever approximate circuit of a transformer.

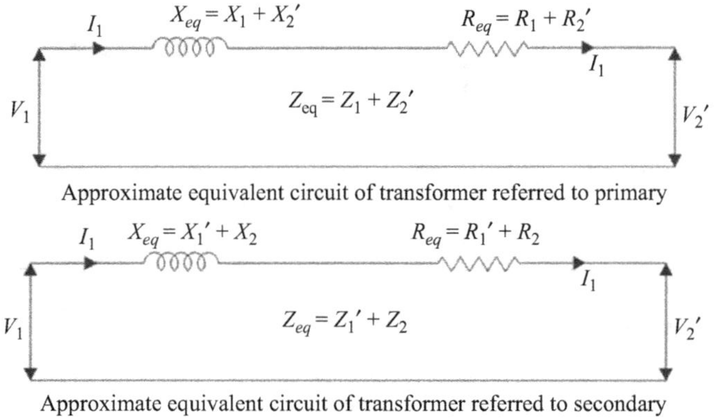

Figure 3.9 Approximate equivalent circuit of a transformer

The *equivalent resistances* of the transformer, referred to the primary side and the secondary side, respectively are:

$$R_{eq\text{-}1} = R_1 + R_2^{'} = R_1 + a^2 R_2 \tag{3.16}$$

$$R_{eq\text{-}2} = R_1^{'} + R_2 = \frac{R_1}{a^2} + R_2 \tag{3.17}$$

While the equivalent reactances are expressed in similar manner as:

$$X_{eq\text{-}1} = X_1 + X_2^{'} = X_1 + a^2 X_2 \tag{3.18}$$

$$X_{eq\text{-}2} = X_1^{'} + X_2 = \frac{X_1}{a^2} + X_2 \tag{3.19}$$

When no confusion is likely to arise, the adjective "equivalent" and its subscript are often omitted. In power engineering applications, it is convenient to specify the equivalent resistance and reactance in a way that is indicative to the voltage across them relative to the rated transformer voltage. *Percent resistance* of the transformer is defined as the voltage across the equivalent resistance referred to the primary (or secondary) when the primary or secondary windings are carrying the rated current, expressed as percent of the rated primary (or secondary) voltage. The *percent reactance* is defined in a similar manner by inserting the reactance in place of resistance in the earlier definition. The percent values are the per-unit values, which we learned in the previous chapter, they are normalized values of specific quantities.

Example 3.4: A single-phase step-up transformer, rated at 1,000 kVA, 60 Hz, and 11/110 kV. Transformer equivalent resistance and reactance are 1.452 Ω and 12.1 Ω, respectively. Compute the percent resistance and reactance, as seen from primary and secondary transformer sides. (Comment: The transformer rated voltage ratio is always its turn ratio and no-load voltage ratio, unless specified otherwise.)

Solution: The transformer turns ratio is:

$$a = \frac{V_P}{V_S} = \frac{11}{110} = 0.1$$

Rated primary current is:

$$I_P = \frac{1,000\,\text{kVA}}{11\,\text{kV}} = 90.9\,\text{A}$$

Now, the primary percent equivalent resistance and reactance are:

$$R_{eq}(\%) = \frac{R_{eq}(\text{in}\,\Omega) \times \text{Rated current}}{\text{Rated voltage}} = \frac{1.452 \times 90.9}{11,000} \times 100 = 1.2\%$$

$$X_{eq}(\%) = \frac{X_{eq}(\text{in}\,\Omega) \times \text{Rated current}}{\text{Rated voltage}} = \frac{12.1 \times 90.9}{11,000} \times 100 = 10.0\%$$

The secondary resistance and reactance are computed as:

$$R_{eq(\text{secondary})} = \frac{R_{eq(primary)}}{a^2} = \frac{1.452}{(0.1)^2} = 142.5\,\Omega$$

$$X_{eq(\text{secondary})} = \frac{X_{eq(primary)}}{a^2} = \frac{12.1}{(0.1)^2} = 1,210\,\Omega$$

In order to determine the secondary percent quantities, we need first to estimate the rated secondary current:

$$I_S = \frac{1,000\,\text{kVA}}{110\,\text{kV}} = 9.09\,\text{A}$$

And the secondary percent (per-unit) resistance and reactance are:

$$R_{eq}(\%) = \frac{R_{eq}(\text{in}\,\Omega) \times \text{Rated current}}{\text{Rated voltage}} = \frac{145.2 \times 9.09}{110,000} \times 100 = 1.2\%$$

$$X_{eq}(\%) = \frac{X_{eq}(\text{in}\,\Omega) \times \text{Rated current}}{\text{Rated voltage}} = \frac{1,210 \times 9.09}{110,000} \times 100 = 10.0\%$$

3.2.3 Voltage regulation

For a given input (primary) voltage, the output (secondary) voltage of an ideal transformer is independent of the load attached to the secondary. As seen in the transformer equivalent circuit, the output voltage of a realistic transformer depends on the load current. However, it is safely can be assumed that the current through the excitation branch of the transformer equivalent circuit is small compared to the current flowing through the winding loss and leakage reactance components. The percentage voltage regulation (VR) is defined as the percentage change in the magnitude of the secondary voltage as the load current changes from the no-load to the loaded condition.

$$VR(\%) = \frac{|V_S|_{NL} - |V_S|_{FL}}{|V_S|_{NL}} \times 100 \tag{3.20}$$

The transformer equivalent circuit, shown earlier, gives only the reflected secondary voltage. The actual loaded and no-load secondary voltages are equal to the loaded and no-loaded reflected secondary values divided by the transformer turns ratio. Thus, the percentage voltage regulation may be written in terms of the reflected secondary voltages, as:

$$VR(\%) = \frac{|V'_S|_{NL} - |V'_S|_{FL}}{|V'_S|_{NL}} \times 100 \tag{3.21}$$

Here, the prime quantities are the secondary voltages reflected in the primary of the transformer. According to the approximate transformer equivalent circuit, the reflected secondary voltage under no-load conditions is equal to the primary voltage, while the secondary voltage for the loaded condition is taken as the rated voltage, so:

$$|V'_S|_{NL} = V_P$$

$$|V'_S|_{FL} = |V'_S|_{rated}$$

Inserting the previous two equations into the percentage voltage regulation equation gives:

$$VR(\%) = \frac{|V_P| - |V'_S|_{rated}}{|V'_S|_{rated}} \times 100 \tag{3.22}$$

Example 3.5: Compute the voltage regulation of the transformer in Example 3.3 for: (a) unit power factor, (b) 0.8 lagging power factor, and (c) 0.8 leading power factor.

Solution: The impedance of the transformer (Figure 3.9) is:

$$Z_S = R_S + jX_S = 145.2 + j1210\,\Omega$$

For unit power factor:

$$V'_1 = V_2 + I_2Z_2 = 110{,}000 + 9.09\angle 0^\circ \times (145.2 + j1210) = 111{,}320 + j11{,}100 \text{ V}$$

And

$$|V'_1| = \sqrt{1{,}113{,}206^2 + 11{,}100^2} \simeq 111{,}870 \text{ V}$$

The voltage regulation, by using (3.19) is:

$$VR = \frac{111{,}870 - 110{,}000}{110{,}000} \times 100 = 1.7\%$$

For 0.8 lagging power factor

$$V_1' = V_2 + I_2 Z_2 = 110{,}000 + 9.09\angle{-36.9^\circ} \times (145.2 + j1210) = 117{,}805 + j8078 \text{ V}$$

And

$$|V_1'| = \sqrt{117{,}805^2 + 8{,}078^2} \simeq 118{,}070 \text{ V}$$

The voltage regulation, by using (3.19) is:

$$VR = \frac{118{,}070 - 110{,}000}{110{,}000} \times 100 = 7.3\%$$

For 0.8 leading power factor

$$V_1' = V_2 + I_2 Z_2 = 110{,}000 + 9.09\angle 36.9^\circ \times (145.2 + j1210) = 104{,}355 + j9662 \text{ V}$$

And

$$|V_1'| = \sqrt{104{,}355^2 + 9{,}662^2} \simeq 104{,}800 \text{ V}$$

The voltage regulation, by using (3.19) is:

$$VR = \frac{104{,}800 - 110{,}000}{110{,}000} \times 100 = -4.7\%$$

3.2.4 *Multi-winding transformer*

Transformers which have more than one secondary winding on the same core are commonly known as multiple winding or multi-winding transformers. A three-winding transformer has a primary winding and two secondary windings. The principal of operation of a multi-winding transformer is the same as the one of an ordinary transformer. Primary and secondary voltages, currents, and turns ratios are all calculated the same, the difference is to pay special attention to the voltage polarities of each coil winding, the dot convention marking the positive or negative winding polarities, when they are connected together. Multi-winding transformers can also be used to provide a step-up, a step-down, or any combination between the various windings. In fact, a multiple windings transformer can have several secondary windings on the same core with each one providing a different voltage or current level output. A four-winding transformer is showed in Figure 3.10.

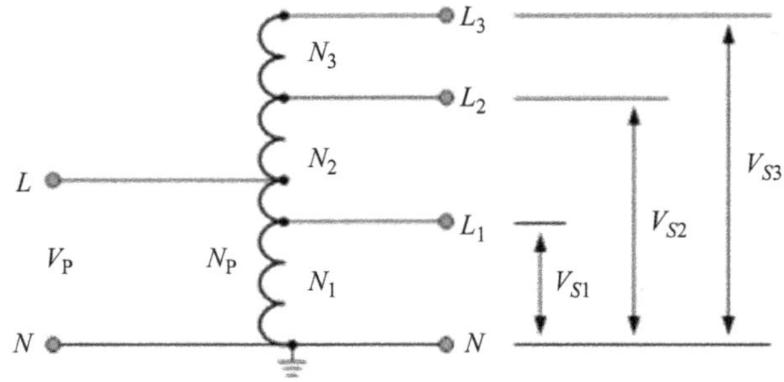

Figure 3.10 Multi-winding transformer

Since the magnetic flux linking the four windings with assumption of an ideal transformer is the same, we have the following voltage relationship:

$$\frac{V_P}{N_P} = \frac{V_{S1}}{N_1} = \frac{V_{S2}}{N_2} = \frac{V_{S3}}{N_3} \tag{3.23}$$

With the same ideal transformer assumption, the conservation of power in a multi-winding transformer is valid, so:

$$N_P I_P - N_1 I_1 - N_2 I_2 - N_3 I_3 = 0$$

Here I_P, I_1, I_2, and I_3 are the primary current and the currents in the secondary windings, respectively.

3.2.5 Transformer ratings, categories, types, and tap changers

Transformer ratings are usually given on the nameplate, which indicates the normal operating conditions. The nameplate includes the following parameters: primary-to-secondary voltage ratio, design frequency of operation, and apparent rated output power. Transformers ratings are related to its primary and secondary windings. The ratings refer to the power in kVA and primary and secondary voltage ratios. A rating of 10 kVA, 1,100/110 V means that the primary is rated for 1,100 V while the secondary is rated for 110 V ($a = 10$). The kVA ratings give the power information for the transformer. Power generated at a generating station (usually at a voltage in the range of 11–25 kV) is stepped up by a transformer to a higher voltage (220, 345, 400, or 765 kV) for transmission. It is one of the most important and critical components of the power system, usually it has a fairly uniform load. Generator transformers are designed with higher losses since the cost of supplying losses is cheapest at the generating station. Generator transformers are usually provided with off-circuit tap changer with a small variation in voltage (e.g., ±5%) because the voltage can always be controlled by the generator field circuit. Unit auxiliary transformers are step-down transformers with primary connected to generator output directly. Its secondary voltage is of the order of 6.9 kV for fitting to the power requirements of the auxiliary generating station equipment. Station transformers are required to supply auxiliary equipment during setting-up of the generating station and during each start-up operation. The ratings of these transformers are small, their primary being connected to high voltage transmission lines, resulting in smaller conductor sizes for HV winding, necessitating special measures for increasing the short-circuit strength. Interconnecting transformer are normally autotransformers used to interconnect two grids/power systems operating at two different system voltages (e.g., 400 and 220 kV or 345 and 138 kV). They are normally located in the transmission system between the generator transformer and receiving end transformer, in this case reducing the transmission voltage (400 or 345 kV) to the subtransmission level (220 or 138 kV). In autotransformers, there is no electrical isolation between primary and secondary windings, some volt-amperes are conductively transformed and remaining is inductively transformed. Autotransformer design becomes more economical as the secondary-to-primary voltage ratio approaches unity. They are characterized by a wide tapping range and an additional tertiary winding which may be loaded or unloaded. Unloaded tertiary acts as a stabilizing winding by providing a path for the third harmonic currents. Synchronous condensers or shunt reactors are connected to the tertiary winding, if required, for reactive power compensation. In the case of an unloaded tertiary, adequate conductor area and proper supporting arrangement are provided for withstanding short-circuits under asymmetrical fault conditions.

Example 3.6: Determine the turns ratio and the rated currents of a transformer from its nameplate data, 480 V/120 V, 48 kVA, and 60 Hz.

Solution: Assuming ideal transformer, the transformer ratio is:

$$a = \frac{480}{120} = 4$$

The primary and the secondary currents are:

$$I_P = \frac{|S|}{V_P} = \frac{48000}{480} = 100\ \text{A}$$

$$I_S = \frac{|S|}{V_S} = \frac{48000}{120} = 400\ \text{A}$$

Receiving station transformers are step-down transformers reducing the transmission or sub-transmission voltages to primary feeder levels (e.g., 33 kV). Some of them may be directly supplying an industrial facility. Loads on these transformers vary over wider limits, and their losses are expensive. The farther is the location of transformers from the generating end, the higher is the cost of the supplying losses. Automatic tap changing on load is usually needed, and tapping range is higher to account large voltage variations. Lower noise level is also desirable in residential areas. Distribution transformers are used to adjust the primary feeder voltage to the actual utilization voltages (415 V or 460 V) for domestic or industrial uses. A large variety of transformers fall into this category due to various arrangements and connections. Their loads vary widely, being often overloaded. Lower or no-load losses are desirable to improve all-day efficiency, because the no-load loss is usually capitalized with a high rate at the tendering stage. Since very little supervision is possible, so the users are expecting the least maintenance on these transformers. The cost of supplying losses and reactive power is the highest for these transformers. Such transformer classification is based on their location and broad function in the power system. Power transformers for utility applications, single-phase or three-phase service have fixed voltage ratings, and limited capabilities to adjust the transformer voltage ratio for applications in which the system voltage is slightly off the nominal values. When a transformer is required to give a constant load voltage despite changes in load current or supply voltage, the transformer turns ratio is altered. In such situations where the system voltage differs from the nominal values, voltage taps are used to accomplish such tasks, via a tap-setting mechanism. The taps are usually on the HV windings to adjust for the variations in the supply voltage, a function of tap changers. There are two basic types: (a) *off-load*, and (b) *on-load*. For tap operation understanding, let's consider how a standard tap changer works by considering this example.

Example 3.7: A 13,800 V/4,160 V transformer has five taps on the primary winding giving −5%, −2 1/2%, nominal, +2 1/2% and +5% turns. If, on-load, the secondary voltage reduces to 4,050 V then, which tap, should be used to maintain 4,160 V on-load (assuming the supply voltage remains constant)?

Solution: To keep the secondary voltage at (or as close as possible to) 4,160 V, either primary supply voltage or the HV winding tap position must be altered. Examining the transformer relationship, indicates that, in order to keep the equation in balance with primary voltage and secondary winding turns fixed, either V_2 or N_1 must be adjusted. Since the objective is to raise V_2 back to nominal, then N_1 must be reduced. To raise V_2 from 4,050 V to 4,160 V requires an increase in secondary volts of:

$$Tap = \frac{4{,}160}{4{,}050} = 1.027 \text{ or } 102.7\%$$

N_1 must be reduced to $1/1.027 = 0.974$. Therefore N_1 must be reduced by $(1 - 0.974) = 0.026$ or 2.6%.

Reducing N_1 by 2.6% is accomplished the increase in secondary voltage output. The nearest tap to select is −2 1/2%, as specified in the example statement.

Most transformers associated with medium level voltage distribution for stations have off-circuit tap *changers. A tap changer requires the transformer to be switched out of circuit before any tap* changing can be done. The contacts on the tap changer are not designed to break any current, even the no-load current. If an attempt is made to change the tap positions while on-line, severe arcing can result which may destroy the tap changer and/or the transformer. On-load tap changers permit tap changing and hence voltage regulation with the transformer on-load operation. Tap changing is usually done on the HV side for two reasons. First, because the currents are lower, the tap changer contacts can be smaller. Second, as the HV winding is wound outside the LV winding, the tap connections are easier to be set out to the tap changer. The tap changer has four essential features. *Selector switches*, selecting the physical tap position on the transformer winding and, because of their construction, cannot and must not make or break the load current. The load current must never be interrupted during a tap change. Therefore, during each tap change, there is an interval where two voltage taps are spanned. *Reactors (inductors)* are used in the circuit to increase the selector circuit impedance and limit the current level circulating due to this voltage difference. Under normal load conditions, equal load current flows in both halves of the reactor windings, the fluxes balance out giving no resultant core flux. With no magnetic flux, there is no inductance and, therefore, no voltage drop due to inductance. However, a very small voltage drop due to the resistance exists. During the tap change, the selector switches are selected to different taps and circulating current flows in the reactor circuit, creating a magnetic flux, which are resulting in inductive reactance that limits the circulating currents. The vacuum switch device performs the duty of a circuit breaker that breaks current during the tap changing sequence. The bypass switch operates during the tap changing sequence but, at no time, does it make or break load current, though it only makes *before break* each connection.

3.2.6 Transformer connections

Single-phase transformers can be connected in many configurations. Two single-phase can be connected in four different combinations provided that their polarities are observed. When transformer windings are connected in parallel, the transformer having the same voltage and polarities are paralleled. When connected in series, windings of opposite polarity are joined in one junction. Coils of unequal voltage may be series-connected with polarities either adding or opposing. In many sections of the power system, three windings transformers are used, with the three windings housed on the same core to achieve economic savings. Three-phase (3-phase or 3φ supplies) are used for electrical power generation, transmission, and distribution, as well as for all industrial uses. Three-phase supplies have many electrical advantages over single-phase power and when considering three-phase transformers, we have to deal with three alternating voltages and currents differing in phase-time by 120°. A transformer cannot act as a phase changing device and change single-phase into three-phase or three-phase into single phase. To make the transformer connections compatible with three-phase supplies, we need to connect them together in a particular way to form a three-phase transformer configuration.

A three-phase transformer can be constructed either by connecting together three single-phase transformers, thereby forming a so-called three-phase transformer bank, or by using one preassembled and balanced three-phase transformer which consists of three pairs of single phase windings mounted onto one single laminated core. The advantages of building a single three-phase transformer is that for the same kVA rating it will be smaller, cheaper, and lighter than three individual single-phase transformers connected together because the copper and iron core are used more effectively. The methods of connecting the primary and secondary windings are the same, whether using just one three-phase transformer or three separate single-phase transformers. The primary and secondary windings of a transformer can be connected in different configuration as shown to meet practically any requirement. In the case of three-phase transformer windings, three forms of connection are possible: "star" (wye), "delta" (mesh), and "interconnected-star" (zig-zag). The combinations of the three windings may be with the primary delta-connected and the secondary star-connected, or star-delta, star-star, or delta-delta, depending on the transformers use. When transformers are used to provide three or more phases, they are generally referred to as a poly-phase transformer. But what do we mean by "star" (also known as wye) and "delta" (also known as mesh), when dealing with three-phase transformer connections. A three-phase transformer has three sets of primary and secondary windings. Depending upon how these sets of windings are interconnected, determines whether the connection is a star or delta configuration. The three available voltages, which themselves are each displaced from the other by 120 electrical degrees, not only decided on the type of the electrical connections used on both the primary and secondary sides, but determine the flow of the transformers currents. With three single-phase transformers connected together, the magnetic flux in the three transformers differs in phase by 120 time-degrees. With a single three-phase transformer, there are three magnetic fluxes in the core differing in time-phase by 120°.

The standard method for marking three-phase transformer windings is to label the three primary windings with capital (upper case) letters A, B, and C, used to represent the three individual phases of RED, YELLOW, and BLUE (see Figure 3.11 for details). The secondary windings are labeled with small (lower case) letters a, b, and c. Each winding has two ends normally labeled 1 and 2 so that, for example, the second winding of the primary has ends which will be labeled B_1 and B_2, while the third winding of the secondary will be labeled c1 and c2, as shown. We now know that there are four ways in which three single-phase transformers may be connected together between primary and secondary three-phase circuits. The configurations are delta-delta, star-star, star-delta, and delta-star. Transformers for high voltage operation with the star connections has the advantage of reducing the voltage

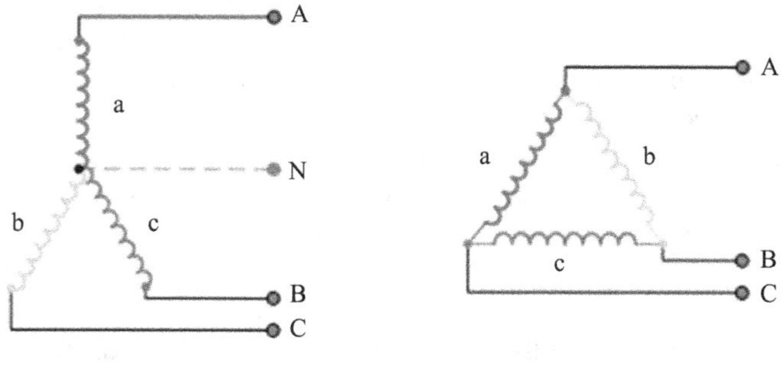

Figure 3.11 Star (Y), left panel and delta (Δ), right panel transformer connections

on an individual transformer, reducing the number of turns required and an increase in the size of the conductors, making the coil windings easier and cheaper to insulate than delta transformers. The delta-delta connection nevertheless has one big advantage over the star-delta configuration, in that if one transformer of a group of three should become faulty or disabled, the two remaining ones will continue to deliver three-phase power with a capacity equal to approximately two thirds of the original output from the transformer unit. One disadvantage of delta-connected three-phase transformers is that each transformer must be wound for the full-line voltage, (in our example earlier 100 V) and for 57.7%, line current. The greater is the number of turns in the windings and the required insulation between turns necessitate larger and more expensive coils than the star connection. Another disadvantage with delta-connected three-phase transformers is that there is no "neutral" or common connection. In the star-star arrangement (Y-Y, wye-wye), each transformer has one terminal connected to a common junction, or neutral point with the three remaining ends of the primary windings connected to the three-phase mains supply. The number of turns in a transformer winding for star connection is 57.7%, of that required for delta connection. The star connection requires the use of three transformers, and if any one transformer becomes fault or disabled, the whole group might become disabled. Nevertheless, the star-connected three-phase transformer is especially convenient and economical in electrical power distributing systems, in that a fourth wire may be connected as a neutral point, (n) of the three star-connected secondaries, as shown in Figure 3.11.

3.2.7 Transformer efficiency

The efficiency (η) of a transformer is defined as the ratio of the output power (P_{Out}) to the input power (P_{In}). The output power is equal to the input power minus the transformer losses. The transformer losses have two components: core loss (P_{Core}) and so-called copper loss (P_{Cu}) associated with the winding resistances. The transformer efficiency in percentage is given by:

$$\eta = \frac{P_{Out}}{P_{In}} \times 100 = \frac{P_{Out}}{P_{Out} + Losses} \times 100 = \left(1 - \frac{Losses}{P_{In}}\right) \times 100 \tag{3.24}$$

Or

$$\eta = \frac{P_{Out}}{P_{Out} + P_{Core} + P_{Cu}} \times 100 \tag{3.25}$$

Assuming a relatively constant voltage source on the primary of the transformer, the core loss can be assumed to be constant and equal to power dissipated in the core loss resistance of the equivalent circuit for the no-load test. The copper loss in a transformer may be written in terms of both the primary and secondary currents, or in terms of only one of these currents based on (3.7). The copper losses are function of the load current, while the core losses depend on the peak core magnetic flux density, which depends on the transformer applied voltage. The core losses are in fact constant because the supply voltage is constant. From the equivalent circuit, one can show that the transformer efficiency depends on the load current (I_2) and the load power factor (θ_2), expressed as:

$$\eta = \frac{V_2 I_2 \cos(\theta_2)}{V_2 I_2 \cos(\theta_2) + P_c + R_{2eq} I_2^2} \times 100 \tag{3.26}$$

Example 3.8: Compute the efficiency of a transformer that has core losses of 1 kW, load current 17.7 A, and load phase angle of 30°. The output voltage is 900 V, and the equivalent resistance from the secondary side is 1.2 Ω.

Solution: The copper losses are:

$$P_{cu} = R_{2eq} I_2^2 = 1.2 \cdot (17.7)^2 = 375.95 \simeq 376 \text{ W}$$

The output power is:

$$P_{out} = V_2 I_2 \cos(30°) = 900 \cdot 17.7 \cdot 0.866 = 13795.4 \text{ W}$$

Using (3.26) the efficiency is:

$$\eta = \frac{V_2 I_2 \cos(\theta_2)}{V_2 I_2 \cos(\theta_2) + P_c + R_{2eq} I_2^2} = \frac{13795.4}{13795.4 + 1{,}000 + 376} = 0.909 \text{ or } 90.9\%$$

By taking the derivative of the efficiency versus load current, the transformer maximum efficiency is determined. The condition for maximum efficiency is that the copper losses equal the core losses. Power distribution transformers usually operate near maximum capacity for 24 h and are taken out of the power network when are not required. In order to account for the distribution transformer efficiency performance, a merit figure, the "all-day" or "energy" efficiency is used. This is expressed as:

$$\eta_{All-day} = \frac{\text{24-h energy output}}{\text{24-h energy input}} \tag{3.27}$$

3.3 AC electrical motors

Electric motors are found in almost every production process or equipment. Getting the most out of your application is becoming more and more important in order to ensure cost-effective operations. An electrical motor is an electromechanical device which converts electrical energy into a mechanical energy, being essentially inverse generators: a current through coils of wire causes some mechanical device to rotate. The core principle underlying motors is the electromagnetic induction. By Ampere's law, the current induces a magnetic field, which can interact with another magnetic field to produce a force, and that force can cause mechanical motion. A motor is basically a generator run backwards (using current to produce motion rather than motion to produce current), and in fact the modern era of practical motors was initiated by accident when one DC generator was accidentally connected to another in 1873, producing motion and leading Zenobe Gramme to realize that his generators could also be used as motors. The first AC motors (synchronous and then induction) were invented by Tesla in the 1880s. All electrical motors exploit the force which is exerted on a current-carrying conductor placed in a magnetic field. The magnitude of the force depends directly on the current in the wire, and the strength of the magnetic field, and the angle between the conductor and the magnetic field. The force on a wire of length l, carrying a current I, and exposed to a uniform magnetic flux density B throughout its length is given by:

$$F = BIl \tag{3.28}$$

where F is in newton (N), B is in tesla (T), I in ampere (A), and l in meter (m). In electrical motors, we intend to use the high magnetic flux density to develop force on current-carrying conductors.

3.3.1 Electric motor fundamentals

Electric motors are estimated to now consume over 25% US electricity use (though some estimates are even higher, to up to 50%, and over 20% of US total primary energy). While large electric motors can be extremely efficient at converting electrical energy to kinetic energy (with efficiencies higher than 90%). However, these efficiencies are only achieved when the motors are well-matched to their loads. Actual efficiencies in normal usage practice are substantially suboptimal (motors are often oversized for the loads they drive). Small electric motors tend to be inherently less efficient. Motor design and, even more importantly, motor choice and use practices are an important area of potential energy conservation. Before the examination of the drive functions, a good understanding of the motor basic operation is needed. A driver is used to convert the electrical energy, supplied by the controller, to mechanical energy to a load. There are two main types of motors, AC and DC. The basic principles are alike for both. Magnetism is the basis for all electric motor operation. It produces the force required to run the motor. There are two types of magnets, the permanent magnet and the electromagnet used in motors. Electromagnets have the advantage over permanent magnets in that the magnetic field can be made stronger and adjustable, while the polarity of the electromagnet can easily be reversed. When a current passes through a conductor, lines of magnetic flux are generated around the conductor. The direction of the flux is dependent on the direction of the current flow. In terms of conventional current flow (positive to negative) then, using your right hand point your thumb in the direction of the current flow, the fingers wrap around the conductor in the same direction of the flux lines. There are basically two types of AC motors: *synchronous* and *induction motors*.

3.3.2 Synchronous motors

The synchronous motor has the special property of maintaining a constant running speed under all conditions of load up to full load. This constant running speed is maintained even under variable line voltage conditions. Therefore, they are useful in applications where the running speed must be accurately known and unvarying. It should be noted that, if a synchronous motor is severely overloaded, its operation speed is suddenly losing its synchronous properties and the motor is coming to a halt. The synchronous motor gets its name from the term synchronous speed, which is the natural speed of the motor rotating magnetic field. This natural speed of rotation is strictly controlled by the number of pole pairs and the frequency of the applied power. Like the induction motor, the synchronous motor makes use of the rotating magnetic field. In a synchronous machine, the rotor is magnetized and it runs at the same speed as the rotating magnetic field. Permanent-magnet rotors are common in small machines, their structure being similar to that of the brushless DC motors. Unlike the induction motor, however, the torque developed does not depend on the rotor induction currents. Briefly, the principle of operation of the synchronous motor is that a three-phase AC source is applied to the stator windings for generating a rotating magnetic field. A DC current is applied to the rotor windings and a fixed magnetic field is produced. The motor is constructed such that the two magnetic fields interact causing the rotor to rotate at the rotating magnetic field speed. If a load is applied to the rotor shaft, the rotor falls momentarily behind the rotating field but continues to rotate at the same synchronous speed. Once the rotor's north and south poles line up with the stator's south and north poles, the stator current is reversed changing the orientation in the stator and the rotor is pushed again. This process repeats until the current in the stator stops alternating or stops flowing. In a three-phase motor, the stator magnetic flux rotates around the motor and the rotator actually follows this rotating magnetic field. This type of motor is called a synchronous motor because it always runs at synchronous speed (rotor and magnetic field of stator are rotating at exactly the same speed). Maximum torque is achieved when the stator flux vector and the rotor flux vector are 90° apart. Synchronous motors operate at synchronism with the line frequency and maintain a constant speed regardless of load without sophisticated electronic control. The synchronous motor typically provides up to a maximum of 140% of rated torque. These designs start like an induction motor but quickly accelerate from approximately 90% of the rated synchronous speed. When operated from an AC drive, they require boost voltage to produce the required torque to synchronize quickly after power application. A synchronous machine can be represented by a simple equivalent circuit, as shown in Figure 3.12.

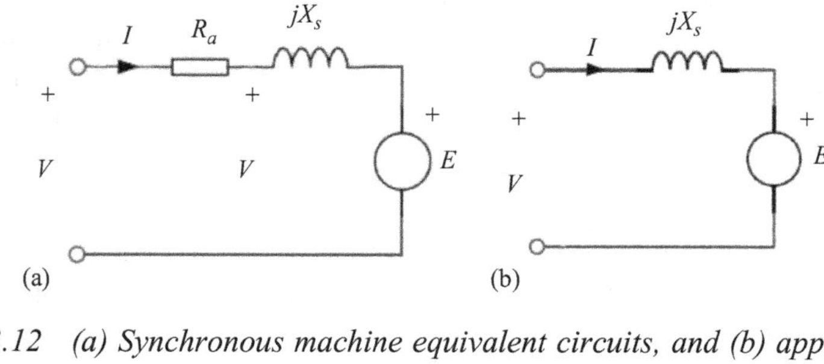

Figure 3.12 (a) Synchronous machine equivalent circuits, and (b) approximate equivalent circuit

There is no fundamental difference between a synchronous motor and a synchronous generator. In a motor, the magnetic axis of the rotating magnetic field is ahead of the magnetic axis of the rotor, resulting in a positive torque that depends on the displacement between the two axes. In a synchronous generator, the displacement is reversed: the magnetic axis of the rotor is ahead of the magnetic axis of the rotating field, so the torque is negative. Most of the AC generators in electric power systems are synchronous machines. High-speed turbine generators normally have two poles. The rotor is made from a cylindrical steel forging, with the field winding embedded in slots machined in the steel. Apart from the slots for conductors, the active surfaces of the stator and rotor are cylindrical, so these are uniform air-gap or nonsalient machines. Low-speed hydro generators have many poles. These are salient-pole machines, where the poles radiate like spokes from a central hub. The circuits in Figure 3.12 represent one phase of a three-phase synchronous machine. The voltage V is the phase voltage at the machine terminals, and the current I is the corresponding phase current. Other elements in the circuit have the following significance. The voltage E is termed as the excitation voltage. It represents the voltage induced in one phase by the rotation of the magnetized rotor, so it corresponds to the rotor magnetic field. The reactance X_S is termed as the synchronous reactance. It represents the magnetic field of the stator current in the following way: the voltage jX_SI is the voltage induced in one phase by the stator current. This voltage corresponds to the stator magnetic field. The resistance R_a is the resistance of one phase of the stator, or armature, winding. The resistance R_a is usually small in comparison with the reactance X_S, usually being neglected in most calculations from the equivalent circuit (Figure 3.12(b)). Voltage V represents the voltage induced in one phase by the total magnetic field and is expressed as:

$$V = E + (R_a + jX_S)I \tag{3.29}$$

The approximate equivalent circuit, shown in Figure 3.12(b), is described by the simplified equation:

$$V = E + jX_SI$$

For operation as a motor, V leads E by an angle δ (the torque or power angle), as shown in Figure 3.13. The phase angle is now less than 90°, indicating a flow of electrical power into the machine. The developed torque is thus given by:

$$T_d = \frac{3VE}{\omega_S X_S}\sin(\delta) \tag{3.30}$$

This has a maximum value when the torque angle $\delta = 90°$, given by:

$$T_{dMax} = \frac{3VE}{\omega_S X_S}$$

Then the three-phase synchronous motor power is given by:

$$P_d = \frac{3VE}{X_S}\sin(\delta) \tag{3.31}$$

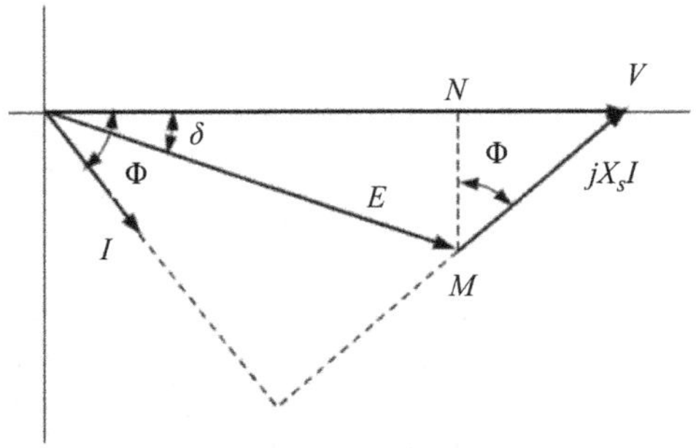

Figure 3.13 Synchronous motor phasor diagram

The magnitude of internal generated voltage induced in a given stator is:

$$E = K\Phi\omega \tag{3.32}$$

Since the magnetic flux, $\mathbf{\Phi}$ in the machine depends on the field current, the internal generated voltage is a function of the rotor field current. If the mechanical load on a synchronous motor exceeds T_{dMax}, the rotor is pulled out of synchronism with the rotating field, and it stalls. T_{dMax} is therefore known as the pullout torque. The synchronous motor torque characteristic has an important practical consequence. If the rotor loses synchronism with the rotating magnetic field, the load angle is changing continuously. The torque is alternately positive and negative, with a mean value of zero. Synchronous motors are therefore not self-starting. Induction machines do not have such limitation, and the induction principle can be used for synchronous motor starting. If the mechanical load is removed from an over-excited synchronous motor, then $\delta = 0$, and the phase angle is now 90°, the machine behaves as a three-phase capacitor, known as a synchronous compensator. The magnitude of the current, and hence the effective value of the capacitance, depends on the difference between E and V. The ability of a synchronous motor to operate at a leading power factor is extremely useful. It will be shown in next chapter subsection that induction motors always operate at a lagging power factor. Many industrial processes use large numbers of induction motors, with the result that the total load current is lagging. It is possible to compensate for this by installing an over-excited synchronous motor. This may be used to drive a large load, such as an air compressor, or it may be used without a load as a synchronous compensator purely for power factor correction.

Example 3.9: A synchronous generator stator reactance is 190-Ω and the internal voltage (open circuit) generated is 35 kV. The machine is connected to a three-phase bus whose line-to-line voltage is 35 kV. Find the maximum possible output power of this synchronous generator.

Solution: The line-to-neutral input voltage is:

$$V_{\text{ln}} = \frac{V_{LL}}{\sqrt{3}} = \frac{35\ \text{kV}}{\sqrt{3}} = 20.2\ \text{kV}$$

The maximum power is when the torque angle is 90°, so from (3.24):

$$P_{dMax} = \frac{3VE}{X_S} = \frac{3 \times 20.2 \times 20.2}{190} = 6.3\text{ MW}$$

Efficiency of a synchronous motor is computed by using a well-known relationship. It is defined in the normal way as the ratio of useful mechanical output power P_{OUT} to the total electrical input power P_{IN}:

$$\eta = \frac{P_{OUT}}{P_{IN}} = 1 - \frac{Losses}{P_{IN}} \tag{3.33}$$

Example 3.10: A 1,492 kW, unity power factor, three-phase, star-connected, 2,300 V, 50 Hz, synchronous motor has a synchronous reactance of 1.95 Ω/phase. Compute the maximum torque in N-m which this motor can deliver if it is supplied from a constant frequency source and if the field excitation is constant at the value which would result in unity power factor at rated load. Assume that the motor is of cylindrical rotor type. Neglect all losses.

Solution: Rated three-phase apparent power, at $PF = 1$ is $S_{3\text{-}\Phi} = 1{,}492$ kVA, while the rated per-phase apparent power is $S = 1{,}492/3 = 497.333$ kVA. The rate voltage per-phase is:

$$V_{ph} = \frac{V_{LL}}{\sqrt{3}} = \frac{2300}{\sqrt{3}} = 1327.906\text{ V}$$

The rated per-phase current is:

$$I_{ph} = \frac{S}{V_{ph}} = \frac{497{,}333}{1327.906} = 374.52\text{ A}$$

The induced per-phase voltage for unit power factor is:

$$E = \sqrt{V_{ph}^2 + \left(X_S I_{ph}\right)^2} = 1515.489\text{ V}$$

The maximum power is computed using (3.24) for a torque angle of 90°:

$$P_{Max} = \frac{EV_{ph}}{X_S} = \frac{1515.489 \times 1327.906}{1.95} = 1032.014\text{ kW/phase}$$

The maximum per-phase torque is computed is then:

$$\tau_{Max} = \frac{P_{Max}}{\omega_m} = \frac{1032{,}014}{2\pi 50} = 3285\text{ N}\cdot\text{m/phase}$$

The three-phase maximum torques of this synchronous motor is 9,855 Nm.

In general, larger synchronous machines have the higher efficiencies because some losses do not increase with machine size. Losses are due to: rotor resistance, iron parts moving in a magnetic field causing currents to be generated in the rotor body, resistance of connections to the rotor (slip rings), stator: resistance, magnetic losses (e.g., hysteresis and eddy current losses), mechanical losses (windage, friction at bearings, friction at slip rings), and stray load losses, due to nonuniform current distribution.

Synchronous motors are usually used in large sizes because in small sizes they are costlier as compared with induction machines. The principal advantages of using synchronous machine are as follows:

1. Power factor of synchronous machine can be controlled very easily by controlling the field current.
2. It has very high operating efficiency and constant speed.
3. For operating speed less than about 500 rpm and for high-power requirements (above 600 KW) synchronous motor is cheaper than induction motor.

In view of these advantages, synchronous motors are preferred for driving the loads requiring high power at low speed; for example, reciprocating pumps and compressor, crushers, rolling mills, pulp grinders etc. Synchronous motors are used for constant speed and steady loads. For high power factor operations, these motors are sometimes exclusively used for power factor improvement. These motors find application in driving low speed compressors, slow speed fans, pumps, ball mills, metal rolling mills, and process industries.

Example 3.11: A factory takes 600 kVA at a lagging power factor of 0.6. A synchronous motor is to be installed to raise the power factor to 0.9 lagging when the motor is taking 200 kW. Calculate the corresponding apparent power (in kVA) taken by the motor and the power factor at which it operates.

Solution: Load (factory) power factor angle is:

$$\theta = \cos^{-1}(0.6) = 53.2^\circ$$

Load active and reactive powers are:

$$P = S \cdot \cos(53.2) = 600 * 0.6 = 360\,\text{kW}$$
$$Q = S \cdot \sin(53.2) = 600 * 0.8 = 480\,\text{kVAR}$$

With the addition of the synchronous motor (200 kW), the overall factory power factor is 0.9 then:

$$\alpha = \tan^{-1}(0.9) = 42.0^\circ$$
$$\tan(\alpha) = \frac{Q_{old} - Q_{SM}}{P_{old} + P_{SM}}$$

Solving for the new reactive power, we are getting:

$$Q_{SM} = Q_{old} - (P_{old} + P_{SM}) \cdot \tan(\alpha) = 480 - (360 + 200) \cdot \tan(42°)$$
$$= 208.78\,\text{kVAR}$$

The synchronous motor apparent power is:

$$S_{SM} = \sqrt{P_{SM} + Q_{SM}} = 289.118\,\text{kVAR}$$

3.3.3 *Induction motors*

An **induction** or **asynchronous motor** is an AC electric motor in which the electric current in the rotor needed to produce torque is obtained by electromagnetic induction from the magnetic field of the stator windings. An induction motor can therefore be made without electrical connections to the rotor as are found in universal, DC and synchronous motors. An induction motor's rotor can be either wound type or squirrel-cage type. The electrical section of the three-phase induction motor as shown in Figure 3.14 consists of the fixed stator or frame, a three-phase winding supplied from the three-phase mains and a rotor. There is no electrical connection between the stator and the rotor. The currents in the rotor are induced via the air gap from the stator. Stator and rotor are made of highly magnetizable core sheet providing low eddy current and hysteresis losses. The *stator winding* consists of three individual windings which overlap one another and are offset by an electrical angle of 120°. When it is connected to the power supply, the incoming current will first magnetize the stator. This magnetizing current generates a rotary field which turns with synchronous speed N_S. The *rotor* in induction machines with squirrel-cage rotors consists of a slotted cylindrical rotor core sheet package with aluminum bars which are joined at the front by rings to form a closed

Figure 3.14 State-of-the-art closed squirrel-cage three-phase motor

cage. The rotor of three-phase induction motors sometimes is also referred to as an anchor. The reason for this name is the anchor shape of the rotors used in very early electrical devices. In electrical equipment, the anchor's winding would be induced by the magnetic field, whereas the rotor takes this role in three-phase induction motors. The stopped induction motor acts like a transformer shorted on the secondary side. The stator winding thus corresponds to the primary winding, the rotor winding (cage winding) to the secondary winding. Because it is shorted, its internal rotor current is dependent on the induced voltage and its resistance. The interaction between the magnetic flux and the current conductors in the rotor generates a torque that corresponds to the rotation of the rotary field. The cage bars are arranged in an offset pattern to the axis of rotation in order to prevent torque fluctuations.

Three-phase squirrel-cage induction motors are widely used in industrial drives because they are rugged, reliable, and economical. Single-phase induction motors are used extensively for smaller loads, such as household appliances like fans. Although traditionally used in fixed-speed service, induction motors are increasingly being used with variable-frequency drives (VFDs) in variable-speed service. VFDs offer especially important energy savings opportunities for existing and prospective induction motors in variable-torque centrifugal fan, pump, and compressor load applications. Squirrel cage induction motors are widely used in both fixed-speed and variable-frequency drive (VFD) applications. Variable voltage and variable frequency drives are also used in variable-speed service. The induction motor operates much the same way that the synchronous motor does. It uses the same magnetic principles to couple the stator and the rotor. However, one major difference is the synchronous motor uses a permanent magnet rotor and the induction motor uses iron bars arranged to resemble a squirrel cage. As the stator magnetic field rotates in the motor, the lines of flux produced will cut the iron bars and induce a voltage in the rotor. This induced voltage will cause a current to flow in the rotor and will generate a magnetic field. This magnetic field will interact with the stator magnetic field and will produce torque to rotate the motor shaft; which is connected to the rotor. The torque available at the motor shaft is determined by the magnetic force (flux) acting on the rotor and the distance from the center of rotation that force is. The flux is determined by the current flowing through the stator windings. Another factor determining torque and another difference between the induction motor and the synchronous motor is slip. Slip is the difference between the stator magnetic field speed and the rotor speed. As implied earlier, in order for a voltage to be induced into a conductor, there must be a relative motion between the conductor and the magnetic lines of flux. Slip is the relative motion needed in the induction motor to induce a voltage into the rotor. If the induction motor ran at synchronous speed, there would be no relative motion and no torque would be produced. This implies that the greater the slip, the greater the torque. This is true to a limit, as we can see in the figure later. The earlier curve shows the speed/torque characteristics that the typical induction motor would follow, excited by a given voltage and frequency. We can see by this curve that the motor produces zero torque at synchronous speed because there is no slip. As we apply a load, the rotor begins to slow down which creates slip. At about 10% slip (at the knee of the curve), we get maximum torque and power transfer from the motor. This is really the best place on the curve to operate the motor. Vector control (slip control) from a closed loop drive system can be used to keep the motor operating at this optimum point on the curve. Vector control is implemented using a microprocessor-based system that has a mathematical model of the motor in memory and a position transducer on the motor to indicate rotor. The mathematical model allows the microprocessor to determine what the speed versus torque curve the motor will follow with any applied voltage and frequency, will be. This will allow the system to control the slip in the motor to keep it operating at the knee of the speed/torque curve. This technology achieves extremely high performance. Now that we have a basic understanding of the operation of the motor, we can better understand the function and operation of the high performance drive. The RPM (rotation per minute) speed and the torque of an induction motor are given by the following equations:

$$N_S = \frac{120f}{P} \tag{3.34}$$

and

$$T = K_{IM} I_{RMS} \tag{3.35}$$

Here T is the motor torque, K_{IM} is the torque constant, I_{RMS} is the RMS motor current, N_S is the motor synchronous (rpm), f is the frequency of stator current (power supply frequency), and P is the motor number of poles. The magnetic field produced in the rotor because of the induced voltage is alternating in nature. To reduce the relative speed, with respect to the stator, the rotor starts running in the same direction as that of the stator magnetic flux, trying to catch up with the rotating flux. However, the rotor never succeeds in "catching up" to the stator field. An essential feature of induction motors is the speed difference between the rotor and the rotating magnetic field, which is known as slip. There must be some slip for currents to be induced in the rotor conductors, and the current magnitude increases with the slip. It follows that the developed torque varies with the slip, and therefore with the rotor speed. The rotor runs slower than the speed of the stator field. This speed is called the base or mechanical speed (N_m). The difference between N_S (or angular synchronous velocity, ω_{syn}) and N_m (or actual angular velocity of the motor, ω_m) is called the slip, s and is expressed as:

$$s = \frac{N_S - N_m}{N_S} = \frac{\omega_{sun} - \omega_m}{\omega_{syn}} \tag{3.36}$$

The slip varies with the load. An increase in load will cause the rotor to slow down or increase slip. A decrease in load will cause the rotor to speed up or decrease slip. The frequency of the rotor currents is proportional to the slip speed, and therefore proportional to s. When the rotor is stationary, $s = 1$, and the rotor frequency must be equal to the stator supply frequency, so we have the important result:

$$f_r = sf_s \tag{3.37}$$

where f_r is the frequency of rotor currents and f_s is the stator supply frequency.

Example 3.12: An induction motor has four poles, the supply frequency is 60 Hz, and the actual speed of rotation is 1,775 RPM. Determine the synchronous speed, the slip, and the rotor frequency.

Solution: The synchronous speed is:

$$N_S = \frac{120f}{P} = \frac{120 \times 60}{4} = 1{,}800\,\text{RPM}$$

Then the slip is:

$$s = \frac{1{,}800 - 1{,}775}{1{,}800} = 0.014 \text{ or } 1.4\%$$

By using (3.32), the rotor frequency is:

$$f_r = 0.014 \times 60 = 0.833\,\text{Hz}$$

The rotating magnetic field exerts a torque T_d on the rotor, and does work at the rate $\omega_s T_d$. This represents an input of power to the rotor. The rotor revolves at an angular speed ω_r, and therefore does work at the rate $\omega_r T_d$, which represents the mechanical output power of the rotor. The difference between these two powers represents power lost in the resistance of the rotor. As in any electrical motors, there are mechanical losses in the motor, so the shaft torque T is less than T_d. We have the following set of rotor power relationships. If the rotor electromagnetic input is:

$$P_{em} = \omega_{syn} T_d \tag{3.38}$$

Then the rotor power output is:

$$P_{rot} = \omega_r T_d = (1 - s)\omega_{syn} T_d \tag{3.39}$$

So the rotor copper loss is then given by the difference between relationships (3.38) and (3.39):

$$P_{Loss} = \left(\omega_{syn} - \omega_r\right) T_d = s\omega_{syn} T_d \tag{3.40}$$

Thus, a fraction (1−s) of the rotor electromagnetic input power is converted into mechanical power, and a fraction s is lost has heat in the rotor conductors. The quantity (1−s) is termed as the rotor efficiency. Since there are other losses in the motor, the overall efficiency must be less than the rotor efficiency. For high efficiency, the fractional slip s should be as small as possible. In large motors, with power ratings of 100 kW or more, the value of s at full load is about 2%. For small motors, with power ratings below about 10 kW, the corresponding value is about 5%. When the rotor is stationary, the induction motor behaves as a three-phase transformer with a short-circuited secondary. When the rotor moves, the voltage induced in the rotor will depend on the relative motion. It is shown that this can be represented by a simple change to the equivalent circuit: the secondary resistance is

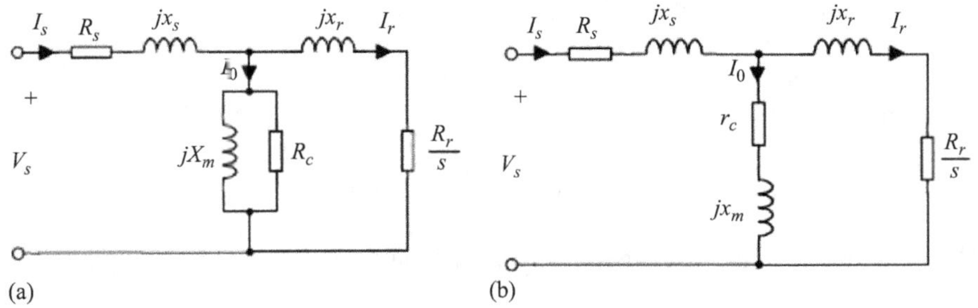

Figure 3.15 (a) Induction motor equivalent circuit and (b) modified equivalent circuit

not constant, but depends on the fractional slip s. The circuit for one phase takes the form shown in Figure 3.15(a). The parameters in Figure 3.15(a) have the same significance as in a transformer: R_S is the stator winding resistance, x_S is the stator leakage reactance, representing stator flux that fails to link with the rotor, R_r is the rotor resistance referred to the stator, x_r is the rotor leakage reactance referred to the stator, R_C represents core loss, mainly in the stator, and X_m is the magnetizing reactance. Similar to a transformer, the equivalent circuit is simplified by moving the shunt elements to the input terminals. However, this is a poor approximation for induction motors, because the magnetizing reactance X_m is much smaller in comparison with the leakage reactances x_S and x_r. The reason for this is the presence of an air-gap between the stator and the rotor, which increases the reluctance of the magnetic circuit. The developed torque is obtained by equating the power absorbed in the resistance Rr/s to the rotor input power from (3.39), giving the result:

$$T_d = \frac{3}{\omega_{syn}} \frac{R_r}{s} I_r^2 \text{ N} \cdot \text{m} \tag{3.41}$$

It is necessary to solve the equations of the equivalent circuit (Figure 3.13(a)) for the currents, I_S and I_r, and hence determine the torque from (3.41). This process is simplified by transforming the equivalent circuit to the form shown in Figure 3.15(b). Here, the parallel combination of X_m and R_C has been replaced by the series combination of x_m and r_C. The series elements in Figure 3.15(b) are related to the parallel elements in Figure 3.15(a) by the following equations:

$$\begin{aligned} r_C &= \frac{X_m^2}{R_C^2 + X_m^2} R_C \\ x_m &= \frac{R_C^2}{R_C^2 + X_m^2} X_m \end{aligned} \tag{3.42}$$

The value of r_C depends on X_m, and therefore on the frequency. When the speed of an induction motor is controlled by varying the frequency, the resistance R_C is approximately constant. Under these conditions, r_C is proportional to the square of the frequency, so the modified equivalent circuit is less useful. In this section, however, the frequency is assumed constant, so the modified circuit can be used. Values of the stator and rotor currents are easily determined by first defining impedances as follows:

$$Z_S = R_S + jx_S, \quad Z_r = \frac{R_r}{s} + jx_r, \quad Z_m = r_C + jx_m \tag{3.43}$$

The parallel combination of the magnetizing branch and the rotor branch is:

$$Z_P = \frac{Z_m Z_r}{Z_m + Z_r} \tag{3.44}$$

So the stator and the rotor currents are now estimated as:

$$I_S = \frac{V_S}{Z_S + Z_P} \tag{3.45}$$

And

$$I_r = \frac{Z_P I_S}{Z_r} \tag{3.46}$$

Example 3.13: A 4-pole 3.6 kW, wye-connected induction motor operates from a 50 Hz supply with a line voltage of 400 V. The equivalent-circuit parameters per phase are as follows:

$$R_S = 2.27\ \Omega, \quad R_r = 2.28\ \Omega, \quad x_S = x_r = 2.83\ \Omega$$
$$X_m = 74.8\ \Omega, \quad r_C = 3.95\ \Omega$$

If the full-load slip is 5%, determine: (1) the no-load current, (2) the full-load stator current, (3) the full-load rotor current, (4) the full-load speed in rev/min, and (5) the full-load developed torque.

Solution: The per-phase voltage is:

$$V_P = \frac{400}{\sqrt{3}} = 231\ \text{V}$$

The motor impedances are:

$$Z_S = R_S + jx_S = 2.27 + j2.83\ \Omega$$
$$Z_r = \frac{R_r}{s} + jx_r = \frac{2.28}{0.05} + j2.83\ \Omega = 45.6 + j2.83\ \Omega$$
$$Z_m = r_C + jx_m = 3.95 + j74.8\ \Omega$$
$$Z_P = \frac{Z_m Z_r}{Z_m + Z_r} = 31.1 + j20.3\ \Omega$$

The no-load current is:

$$I_0 = \frac{V_P}{|Z_S + Z_m|} = \frac{231}{|6.22 + j77.6|} = \frac{231}{77.9} = 2.97 \text{ A}$$

The full-load current and its magnitude are:

$$I_S = \frac{V_P}{Z_S + Z_P} = \frac{231}{33.4 + j23.1} = 4.68 - j3.23 \text{ A}$$

$$|I_S| = |4.68 - j3.23| = 5.69 \text{ A}$$

The full-load rotor current and its magnitude are:

$$I_r = \frac{Z_P I_S}{Z_r} = \frac{211 - j5.89}{45.6 + j2.83}$$

$$|I_r| = \frac{|211 - j5.89|}{\left|\dfrac{211 - j5.89}{45.6 + j2.83}\right|} = \frac{211.1}{45.7} = 4.62 \text{ A}$$

The synchronous speed and the supply angular speed are:

$$N_s = \frac{120f}{P} = \frac{120 \times 50}{4} = 1{,}500 \text{ RPM}$$

$$N_m = N_S(1 - s) = 1500(1 - 0.05) = 1{,}425 \text{ RPM}$$

$$\omega = 2\pi f = 2 \times 50 \times \pi \simeq 314 \text{ rad/s}$$

The developed torque is then:

$$T_d = \frac{3}{\omega_S}\frac{R_r}{s}I_r^2 = \frac{3 \times 2.28 \times (4.62)^2}{314 \times 0.05} = 37.2 \text{ N} \cdot \text{m}$$

Notice: The maximum torque is known as *the breakdown torque*. If a mechanical load torque greater than this is applied to the motor, it will stall. The torque is zero at the synchronous speed of 1,500 RPM. When the rotor is stationary, the fractional slip is $s = 1$. When the rotor is running at the synchronous speed, the fractional slip is $s = 0$. The rotor frequency is:

$$f_r = sf = 0.05 \times 50 = 2.5 \text{ Hz}$$

Most induction motors are started by connecting them straight to the AC mains supply. This is known as direct on line starting, however it may result in a large starting current. Direct on line starting may be unacceptable, either because the supply system cannot support such a large current, or because the transient torque could damage the mechanical system. In depth presentation of the induction motor starting methods are discussed later in this book. The efficiency of an induction motor, as well as of any power system equipment is of great importance to the user.

It is defined in the normal way as the ratio of useful mechanical output power P_{OUT} to the total electrical input power P_{IN}:

$$\eta = \frac{P_{OUT}}{P_{IN}} = 1 - \frac{Losses}{P_{IN}} \tag{3.47}$$

The induction motor losses are considered to have five components as follows:

1. Stator I^2R loss (stator copper loss), P_{SCL}
2. Rotor I^2R loss (rotor copper loss), P_{RCL}
3. Core losses
4. Friction and windage loss (rotational loss): P_{w+f}
5. Stray load loss: P_{Stray}

The total loss is the sum of items from 1 to 5. Core loss is the eddy-current and hysteresis loss in the magnetic core of the machine, mostly in the stator, which is represented by the resistance r_C. Friction and windage loss is the total mechanical power loss within the motor, from bearing friction and aerodynamic drag on the rotor. Stray load loss is an additional loss under load, which is not included in the other four categories. It may be attributed to departures from a purely sinusoidal winding distribution and to effects of the stator and rotor slot openings on the magnetic field distribution into the machine air gap. The full set of power relationships for poly-phase induction motors are:

$$\begin{aligned}
P_{IN} &= \sqrt{3}V_{LL}I_L \cos\theta = 3V_{ph}I_{ph}\cos\theta \\
P_{SCL} &= 3R_S I_S^2 \\
P_{RCL} &= 3R_r I_r^2 \\
P_{AG} &= P_{IN} - (P_{SCL} + \text{Core Losses}) \\
P_{Conv} &= P_{AG} - P_{RCL} \\
P_{OUT} &= P_{Conv} - \left(P_{w+f} + P_{Stray}\right)
\end{aligned}$$

The parameters of the induction motor equivalent circuit in Figure 3.15(b) are determined from three tests: (a) DC test of the stator phase resistance, (b) no-load test, essentially for efficiency determination, and (c) a locked-rotor (or blocked-rotor) test, where the rotor is prevented from revolving. These tests fully resemble the transformer open-circuit and short-circuit tests for determining its equivalent circuit.

3.4 DC machines

The armature of the DC motor is a loop of wire (current carrying conductor), free to rotate. The field magnets are permanent or electromagnets with their N and S poles facing each other to set-up the magnetic lines into the air gap (Figure 3.16 for details of the DC motors). The armature is connected to the commutator which rides along the brushes, connected to a DC power source. The current from the DC power source

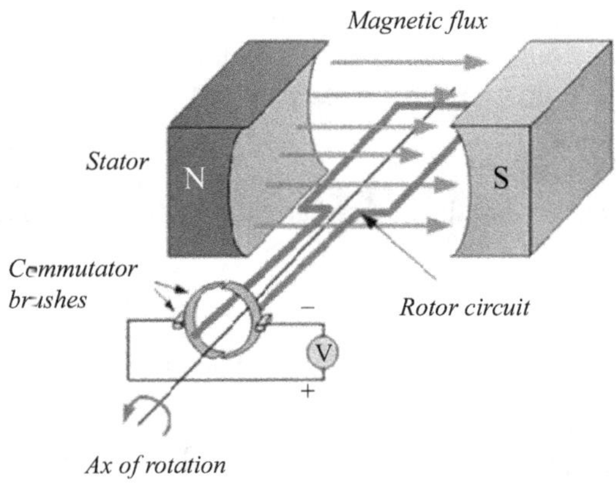

Figure 3.16 Schematic diagram of a DC motor

flows, through the brush through one commutator section, through the armature coil, through the other commutator section, back to the negative supply lead. This current generates flux lines around the armature and affecting the lines of flux in the air gap. On the side of the coil where the lines of flux oppose each other, the magnetic field will be made weaker. On the side of the coil where the lines of flux are riot opposing each other, the magnetic field is made stronger. Because of the strong field on one side of the coil and the weak field on the other side, the coil will be pushed into the weaker field and, because the armature coil is free to rotate, it will rotate. The torque available at the motor shaft (turning effort) is determined by the magnetic force (flux) acting on the armature coil and the distance from the renter of rotation that force is. The flux is determined by the current flowing through the armature coil and strength of the field magnets, so the DC motor torque can be expressed as:

$$T = K_{DC}\Phi_F I_c \tag{3.48}$$

Here, T represents the DC motor torque, K_{DC} is the torque constant, Φ_F is the magnetic field flux, and I_a is the motor armature current. The total emf induced in the motor by several such coils wound on the rotor can be expressed as:

$$E_b = K\Phi_F \cdot \omega_m \tag{3.49}$$

where K is an armature constant, and is related to the geometry and magnetic properties of the motor, and ω_m is the rotor angular velocity (rad/s). The rotational (RPM) speed (N) of the motor is determined by the voltage applied to the armature coil and is expressed as:

$$N = \frac{V_t - R_a I_a}{K_V \Phi_F} \tag{3.50}$$

where V_t is the terminal (supply) voltage, K_V is the motor voltage constant, I_a is the armature current, and R_a is the armature resistance. The DC motor converts electrical power from the adjustable DC voltage source to rotating mechanical force and power. Motor shaft rotation and direction are proportional to the magnitude and polarity of the DC voltage applied to the motor. The electrical power (motor developed mechanical power) generated by the machine is given by:

$$P_{developed} = E_b I_a = \omega_m T_{developed} = K\Phi_F \omega_m I_a \tag{3.51}$$

This is the power delivered to the induced armature voltage (counter-voltage) and is given by:

$$E_b I_a(\text{Electrical power}) = \omega_m T_{developed}(\text{Mechanical power}) \tag{3.52}$$

The following are the basic types of DC motors and their operating characteristics:

1. Shunt-wound motors have the field controlled separately from the armature winding. With constant armature voltage and constant field excitation, the shunt-wound motor offers relatively flat speed-torque characteristics.
2. The series-wound motor has the field connected in series with the armature. Although the series wound motor offers high starting torque, it has poor speed regulation. Series-wound motors are generally used on low speed, very heavy loads.
3. The compound-wound DC motor utilizes a field winding in series with the armature in addition to the shunt field, to obtain a compromise in performance between a series and a shunt wound type motor. The compound-wound motor offers a combination of good starting torque and speed stability.
4. In a separately excited DC motor, the field winding is independent of the armature winding. Such DC motors offer flexible speed-torque control. They are often used as actuators. This type of motors are used in trains and for automatic traction purposes.
5. The permanent magnet motor has a conventional wound armature with commutator and brushes. Permanent magnets replace the field windings. This type of motor has excellent starting torque, with speed regulation slightly less than that of the compound motor. Peak starting torque is commonly limited to 150% of rated torque to avoid demagnetizing the field poles. Typically, these are low horsepower.

DC Machines can be classified according to the electrical connections of the armature winding and the field windings. The different ways in which these windings are connected lead to machines operating with different characteristics. The field winding can be either self-excited or separately-excited, that is, the terminals of the winding can be connected across the input voltage terminals or fed from a separate voltage source (as in the previous paragraph). Further, in self-excited motors, the field winding can be connected either in series or in parallel with the armature winding. These different types of connections give rise to very different types of machines, as we will study in this section. In separately excited machines, the armature and field winding are electrically separate from each other, and the field winding is excited by a separate DC source. Applying KVL in the armature and field circuits of Figure 3.17:

$$V_T = E_b + R_a I_a \tag{3.53}$$

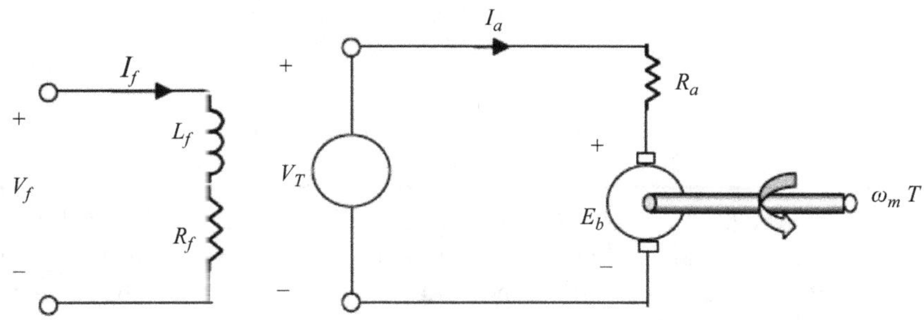

Figure 3.17 Separately excited DC motor

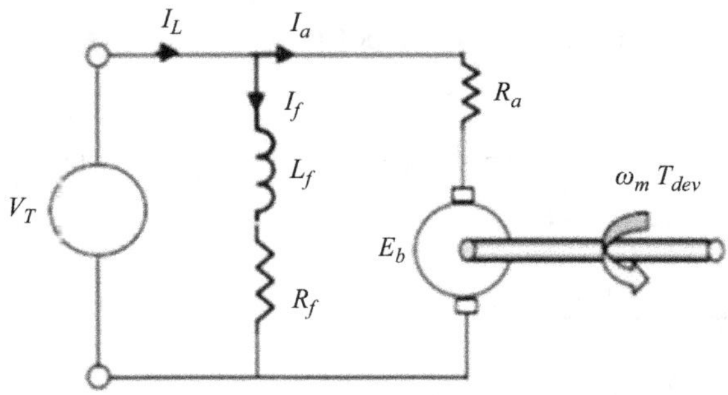

Figure 3.18 Shunt DC motor

And

$$V_f = R_f I_f \tag{3.54}$$

where V_T is voltage applied to the armature terminals of the motor and R_a is the resistance of the armature winding, V_f is voltage applied to the field winding (to produce the magnetic field), R_f is the resistance of the field winding, and I_f is the current through the field winding. Note that the total input power of a separately excited DC motor is:

$$P_{IN} = V_f I_f + V_T I_a \tag{3.55}$$

In the self-excited DC electrical machines, instead of a separate voltage source, the field winding is connected across the main voltage terminals. Schematic diagram of shunt DC machine is shown in Figure 3.18. The armature and field windings are connected in parallel, and the same terminal voltage is across each of them. The total current (I_L) drawn from the power supply is:

$$I_L = I_f + I_a \tag{3.56}$$

While the total input power of a DC shunt motor is given by:

$$P_{IN} = V_T I_L \tag{3.57}$$

Voltage, current, and power equations are given in (3.51), (3.54), and (3.55)

Example 3.14: A 230 V, 10 HP DC shunt motor delivers power to a load at 1,200 r/min. The armature current drawn by the motor is 200 A. The armature circuit resistance of the motor is 0.2 Ω and the field resistance is 115 Ω. If the rotational losses are 500 W, what is the value of the load torque?

Solution: The back EMF induced in the armature is:

$$E_b = V_T - R_a I_a = 230 - 0.2 \cdot 200 = 190 \text{ V}$$

Power developed (in the rotor), and the power delivered to the load

$$P_{developed} = E_b I_a = 190 \times 200 = 38{,}000 \text{ W}$$

$$P_{OUT} = P_{developed} - \text{Rotational losses} = 38{,}000 - 500 = 37{,}500 \text{ W}$$

The angular velocity is:

$$\omega_m = \frac{2\pi N}{60} = \frac{2\pi 1200}{60} = 40\pi \text{ rad/s}$$

The load torque is given by:

$$T_{Load} = \frac{P_{OUT}}{\omega_m} = \frac{37{,}500}{40\pi} = 298.4 \text{ Nm}$$

In a series DC machine, the field winding and armature winding are connected in series, carrying the same current. A series DC wound motor is also called a universal motor. It is universal in the sense that it will run equally well using either an AC or a DC voltage source. Reversing the polarity of both the stator and the rotor cancel out. Thus the motor will always rotate the same direction regardless of the voltage polarity. By applying KVL in Figure 3.19, the terminal voltage, armature

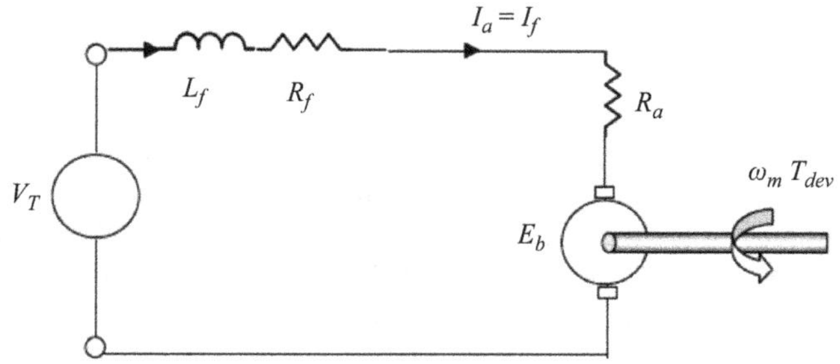

Figure 3.19 Series DC motor

current, back EMF, field, and armature resistance are related to the following equation:

$$V_T = E_b + (R_f + R_a)I_a \tag{3.58}$$

Compound DC machine—If both series and shunt field windings are used, the motor is said to be compounded. In a compound machine, the series field winding is connected in series with the armature, and the shunt field winding is connected in parallel. Two types of arrangements are possible in compound motors: (a) *cumulative compounding*, if the magnetic fluxes produced by both series and shunt field windings are in the same direction (i.e., additive), and (b) *differential compounding*, if the two fluxes are in opposition. In both these types, the connection can be either short shunt or long shunt.

In most applications, DC motors are used for driving mechanical loads. Some applications require that the speed remain constant as the load on the motor changes. In some applications the speed is required to be controlled over a wide range. It is therefore important to study the relationship between torque and speed of the motor. The performance measure of interest is the speed regulation, defined as the change in speed as full load is applied to the motor, expressed as:

$$VR(\%) = \frac{N_{NL} - N_{FL}}{N_{FL}} \times 100 \tag{3.59}$$

where N_{NL} is the speed at no load, and N_{FL} is the speed when full load is applied. In order to effectively use a DC motor for an application, it is necessary to understand its characteristic curves. For every motor, there is a specific Torque/Speed curve and Power curve. The relation between torque and speed is important in choosing a DC motor for a particular application. A separately excited DC motor equivalent circuit was shown in Figure 3.17. From voltage and torque equation, we can get a relationship for developed torque, voltage, and magnetic flux for separately excited and shunt DC motors, expressed as:

$$T_{developed} = \frac{K\Phi}{R_a}(V_T - K\Phi\omega_m) \tag{3.60}$$

This equation shows the relationship between the torque and speed of a separately excited DC motor. If the terminal voltage V_T and flux Φ are kept constant, the torque-speed relationship is a straight drooping line. Note that torque is inversely proportional to the speed of the output shaft. In other words, there is a tradeoff between how much torque a motor delivers, and how fast the output shaft spins. However, the motor load determines the final operating point on the torque curve. The DC series motor characteristics can be analyzed in much the same way as the shunt motor discussed earlier. In series motors, the series field winding is connected in series with the armature (refer to the Figure 3.19). Therefore, the torque developed in the rotor can be expressed as:

$$T_{developed} = \frac{K_m V_T^2}{(R_a + R_f + K_m\omega_m)^2} \tag{3.61}$$

Here K_m is the series DC motor constant. From this equation, if the terminal voltage V_T is kept constant, the speed is almost inversely proportional to the square root of the torque. A high torque is obtained at low speed and a low torque is obtained at high speed. As power flows from DC motor input terminals to the output (shaft), some losses take place. Figure 3.17 shows the flow of power in a separately excited DC motor. Efficiency of the motor can be calculated as the ratio of output power to the total input power. For example, for a separately excited DC motor, the input power is given by the relation in (3.49). The total copper losses are given by the power dissipated in the field and armature windings:

$$\text{Copper losses} = \frac{V_f^2}{R_f} + R_a I_a^2 = R_f I_f^2 + R_a I_a^2 \tag{3.62}$$

The developed power is the difference between input power, copper losses, and core losses (hysteresis and eddy-current losses). The output power and torque are less than the developed values because of rotational losses, which include friction and windage losses. Rotational power loss is approximately proportional to motor speed. The efficiency of the DC motor can be calculated as:

$$\eta = \frac{P_{OUT}}{P_{IN}} = \frac{P_{OUT}}{P_{OUT} + \text{Copper losses} + \text{Core losses} + \text{Rotational losses}} \tag{3.63}$$

It is important to mention that the total input and output power can be calculated in many different ways using the power flow diagram of a specific type of DC, depending on the information given. Also note that the torque developed inside the rotor is different from the final (output) torque supplied to the load due to rotational losses. DC motors are typically rated in terms of: (a) rated voltage (the operating voltage on the input side of the motor), (b) rated power (in horsepower (HP) or watts) that the motor is designed to deliver to the load (i.e., output power) for continuous operation, (c) rated speed (in revolutions per minute, denoted by r/min or rpm) for which the motor is designed to operate for continuous operation, and (d) rated load (the load which the motor is designed to carry for (theoretically) infinite period of time. "Full load" or "rated load" operating condition refers to the operation of motor when it is delivering rated power to the load).

Note: A motor may not always operate at its rated power and/or speed. Operation above these values is not advisable due to overloading. (Note that 1 HP = 746 W)

Example 3.15: A series-connected DC motor has an armature resistance of 0.5 Ω and field winding resistance of 1.5 Ω. In driving a certain load at 1,200 RPM, the current drawn by the motor is 20 A from a voltage source of $V_T = 220$ V. The rotational loss is 150 W. Find the output power and efficiency.

Solution: Total input power is given by:

$$P_{IN} = V_T I_a = 220 \times 20 = 4{,}400\ \text{W}$$

The induced voltage is:

$$E_b = V_T - (R_f + R_a)I_a = 220 - (0.5 + 1.5)20 = 180 \text{ V}$$

Power developed in the armature can be calculated as:

$$P_{dev} = E_b I_a = 180 \times 20 = 3{,}600 \text{ W}$$

Output power delivered to the load:

$$P_{OUT} = P_{dev} - \text{Rotational losses} = 3{,}600 - 150 = 3{,}450 \text{ W}$$

Therefore, efficiency can be calculated as:

$$\eta = \frac{P_{OUT}}{P_{IN}} = \frac{3{,}450}{4{,}400} = 0.784 \text{ or } 78.4\%$$

Example 3.16: A 440 V, 20 HP, 500 RPM, DC shunt motor has rotational losses of 780 W at rated speed. The armature resistance and the field resistance are 0.5 Ω and 220 Ω, respectively. The armature current is 38 A. Compute: (a) the electromagnetic torque, (b) the shaft (output) torque, and (c) the motor efficiency.

Solution: The output power and the developed (electromagnetic) power are:

$$P_{OUT} = 20 \times 746 = 14{,}920 \text{ W}$$
$$P_{em} = P_{OUT} + \text{Rotational losses} = 14{,}920 + 780 = 15{,}700 \text{ W}$$

(a) The electromagnetic torque is then: $\tau_{em} = \dfrac{P_{em}}{\omega} = \dfrac{15{,}700}{2\pi\frac{500}{60}} = 300 \text{ Nm}$

(b) The shaft(output) torque is: $\tau_{OUT} = \dfrac{P_{OUT}}{\omega} = \dfrac{14{,}920}{2\pi\frac{500}{60}} = 285 \text{ Nm}$

(c) Input current is:

$$I_L = I_a + I_f = I_a + \frac{V_T}{R_f} = 38 + \frac{440}{220} = 40 \text{ A}$$

The efficiency is then:

$$\eta = \frac{P_{OUT}}{P_{IN}} = \frac{P_{OUT}}{V_T \times I_L} = \frac{14{,}920}{440 \times 40} = 0.847 \text{ or } 84.7\%$$

3.5 Chapter summary

Motors and transformers are the key driving force for industrial, commercial and residential equipment, and appliances. In industry, all types of linear or rotational force, movement, torque, etc. are applied largely by electrical motors. Industries are getting automated day by day, hence the use of electrical motors are increasing with the same pace. The power supply to any medium or large scale industry comes through transformer as the utilities prefer to supply at higher grid voltage. The transformers are common and indispensable component of power and electronics systems where it is used to transform voltages, currents, and impedances to appropriate levels for optimal use. The transformers offer us opportunities to investigate the properties of magnetic circuits, including the concepts of mmf, magnetizing current, and magnetizing, mutual, and leakage fluxes and their associated inductances. In both transformers and rotating machines, a magnetic field is created by the combined action of the currents in the windings. In an iron-core transformer, most of this flux is confined to the core and links all the windings. This resultant mutual flux induces voltages in the windings proportional to their number of turns and is responsible for the voltage-changing property of a transformer. In rotating machines, the situation is similar, although there is an air gap which separates the rotating and stationary components of the machine. Directly analogous to the manner in which transformer core flux links the various windings on a transformer core, the mutual flux in rotating machines crosses the air gap, linking the windings on the rotor and stator. As in a transformer, the mutual flux induces voltages in these windings proportional to the number of turns and the time rate of change of the flux. A significant difference between transformers and rotating machines is that in rotating machines there is relative motion between the windings on the rotor and stator. An electromechanical energy conversion device is essentially a medium of transfer between an input side and an output side. Three electrical machines (DC, induction, and synchronous) are used extensively for electromechanical energy conversion. Electro-mechanical energy conversion occurs when there is a change in magnetic flux linking a coil, associated with mechanical motion. Maximum portion of power that is consumed in any industry is by electrical motors. So the efficiency is a great issue for an industry owner to think about. The efficiency of the major consumer, the motor must be of as high as possible. The efficiency of the transformer, through which all the power is consumed, must also be near 100%. So, every personnel related to the decision making in industry must have the knowledge regarding the energy efficiency issue for electrical motors and transformers. The basic principles of transformer and motors, their operation, applications, and uses are also discussed in this chapter.

Further readings

[1] P. Schavemaker and L. van der Sluis, Electrical Power System Essentials, John Wiley and Sons, Chichester, UK, 2008.

[2] A.R. Daniels, Introduction to Electrical Machines, Macmillan Press, London, UK, 1976.

[3] H. Majmudar, Introduction to Electrical machines, Allyn and Bacon, Boston, USA, 1969.
[4] O. Elgerd, Basic Electric Power Engineering, Addison-Wesley, Massachusetts, 1977.
[5] V. del Toro, Electric Machines and Power Systems. Prentice Hall, New Jersey, 1985.
[6] R.D. Schultz and R.A. Smith, Introduction to Electric Power Engineering, John Wiley & Sons, New York, 1988.
[7] R.A. Pearman, Electrical Machinery and Transformer Technology, Harcourt College, Publisher, California, 1994.
[8] J.R. Cogdell, Foundations of Electrical Power, Prentice Hall, New Jersey, 1999.
[9] B.S. Guru and H.R. Hizioglu, Electric Machinery and Transformers, Oxford University Press, New York, 2001.
[10] S.J. Chapman, Electrical Machinery and Power System Fundamentals (2nd ed.), McGraw Hill, New York, 2002.
[11] J.J. Winders, Power Transformers—Principles and Applications, Marcel Dekker, Inc. New York, 2002.
[12] C.I. Hubert, Electric Machines—Theory, Operation, Applications, Adjustment and Control (2nd ed.), Prentice Hall, New Jersey, 2002.
[13] M.A. Pai, Power Circuits and Electro-mechanics, Stipes Publishing L.L.C., Champaign, IL, 2003.
[14] A.E. Fitzgerald and C. Kingsley, Electric Machinery (6th ed.), McGraw Hill, New York, 2003.
[15] M.E. El-Hawary, Introduction to Power Systems, Wiley-IEEE Press, USA, 2008.
[16] M.A. El-Sharkawi, Electric Energy: An Introduction (2nd ed.), CRC Press, Boca Raton, 2009.
[17] J.L. Kirtley, Electric Power Principles: Sources, Conversion, Distribution and Use, Wiley, USA, 2010.
[18] F.M. Vanek, L.D. Albright, and L.T. Angenent, Energy Systems Engineering, McGraw Hill, New York, 2012.
[19] T. Gonen, Electric Machines with MATLAB (2nd ed.), CRC Press, Boca Raton, 2012.
[20] P. Sen, Principles of Electric Machines and Power Electronics (3rd ed.), Wiley, USA, 2014.
[21] J.E. Fleckenstein, Three-Phase Electrical Power, CRC Press, Boca Raton, 2016.
[22] S.L. Herman, Electrical Transformers and Rotating Machines (4th ed.), Cengage Learning, UK, 2017.

Questions and problems

1. Core-losses of a transformer operated from a constant-voltage supply are assumed to constant and independent of the load. Why?
2. How much force will be created on a wire that is parallel to the magnetic field?

3. What is meant by leakage flux? How is kept to a minimum?
4. What are the two types of core construction used in transformers?
5. What is the function of a transformer?
6. Why the capacity of a transformer is rated in kilo-volt-amperes?
7. What are the three-phase transformer connections?
8. What are the common efficiencies of a transformer?
9. State the purpose of a tap-changer in an electrical system.
10. What is the starting torque of a motor?
11. What determine the speed of a synchronous motor?
12. Explain how a synchronous motor can improve the power factor of a load with low lagging power factor.
13. What are some applications of synchronous motors?
14. What factors determine the torque developed by a DC motor?
15. What effect does the slip on the rotor reactance of an induction motor?
16. What are the differences between three-phase squirrel-cage induction motors and wound-rotor induction motors? List the main advantages and disadvantages of each type.
17. Upon what factors does the torque of an induction motor depend?
18. What happens if the direction of current at the terminals of a DC series motor is reversed?
19. What happens when the load from a series motor is suddenly taken off?
20. What happens when a DC motor is connected across an AC supply?
21. How would the field winding inductance affect the operation of a DC motor under steady-state?
22. What happens if the direction of current at the terminals of a DC series motor is reversed?
23. A transformer is rated at 500 kVA, 60 Hz, and 2,400/240 V. There are 200 turns on the 2,400-V winding. When the transformer supplies rated load, find (a) the ampere-turns of each winding, and (b) the current in each winding.
24. A transformer is made up of a 1,200-turn primary coil and an open-circuited 80-turn secondary coil wound around a closed core of cross-sectional area 45 cm^2. The core material can be considered to saturate when the RMS applied flux density reaches 1.50 T. What maximum 60-Hz RMS primary voltage is possible without reaching this saturation level? What is the corresponding secondary voltage? How are these values modified if the source frequency is lowered to 50 Hz?
25. A single-phase transformer has 1,200 turns on primary and 400 turns on secondary. The primary winding is connected to 240 V supply and the secondary winding is connected to a 6.40 kVA load. If the transformer is considered ideal determine: (a) the load voltage, (b) the load impedance; and (c) the load impedance referred to the primary side.
26. A single-phase core-type transformer is designed to have primary voltage 33 kV and a secondary voltage 6.6 kV. If the maximum flux density permissible is 1.24 Wb/m^2 and the number of primary turns is 1,350, calculate the number of secondary turns and the core cross-sectional area when operating at 50 Hz frequency.

27. A transformer is to be used to transform the impedance of a 75-Ω resistor to an impedance of 225-Ω. Calculate the required turns ratio, assuming the transformer to be an ideal one.
28. Prove that the maximum efficiency of a transformer is when the core-losses are equal to RI^2 copper losses (due to the winding resistance) if the secondary voltage and the power factor are assumed constant.
29. A single-phase step-up transformer, rated at 1,800 kVA, 60 Hz, and 13.5/135 kV. Transformer equivalent resistance and reactance are 1.560 Ω and 15.6 Ω, respectively. Compute the percent resistance and reactance, as seen from primary and secondary transformer sides.
30. The high-voltage side of a step-down transformer has 800 turns, and the low-voltage side has 100 turns. A voltage of 240 V is applied to the high side, and the load impedance is 3 Ω (low side). Find:
    1. The secondary voltage and current.
    2. The primary current.
    3. The primary input impedance from the ratio of primary voltage and current.
    4. The primary input impedance.
31. Compute the voltage regulation of the transformer in problem 7 for: (a) unit power factor, (b) 0.85 lagging power factor, and (c) 0.85 leading power factor.
32. A single-phase transformer with a nominal voltage ratio of 4,160–600 V has an off-load tap changer in the high voltage winding. The tap changer provides taps of 0, ±2½, and ±5%. If the low voltage is found to be 618 V, what tap would be selected to bring the voltage as close to 600 V as possible?
33. Determine the rotor speed (RPM) of the following three-phase synchronous machines: (a) $P = 4$ and $f = 60$ Hz, (b) $P = 12$ and $f = 50$ Hz, and (c) $P = 4$ and $f = 400$ Hz.
34. If the torque angle of Example 3.4 is limited to 45°, find the generator power output.
35. What is the speed (RPM) of a 30-pole, 60 Hz 440-V synchronous motor? Is this motor classed as a high- or low-speed motor?
36. A synchronous motor with an input of 480 kW is added to a system that has an existing load of 720 kW at 0.82 lagging power factor. What are the new system active power, apparent power, and power factor, if the new motor is operated at: (a) 0.85 lagging power factor, (b) unit power factor, and (c) 0.85 leading power factor?
37. A three-phase induction motor, 50 Hz has a synchronous speed of 1,500 RPM and runs at 1,450 RPM at full load. (1) How many poles have the motor? (2) What is the slip when it runs at full load? (3) What is the rotor frequency?
38. A 208-V, 10-HP four-pole, 60 Hz, Y-connected induction motor has a full-load slip of 5%. Find (a) the motor synchronous speed, (b) the actual motor speed at rated load, (c) the rotor frequency, and (d) the shaft torque at rated load.
39. What is the rotor frequency of an eight-pole 60 Hz squirrel cage motor operating at 850 RPM?

40. A 50-HP 230-V three-phase induction motor require a full-load current of 130 A per terminal at a power factor of 0.88. What is tis full-load efficiency?
41. A 480-V, 60 Hz, three-phase induction motor is drawing 60 A at 0.85 PF lagging. The stator copper losses are 2 kW, the rotor copper losses are 0.7 kW, the windage and friction losses are 0.6 kW, the core losses are 1.8 kW, while the stray losses are 0.2 kW. Find: the air gap power, the converted power, the output power, and the motor efficiency.
42. A two-pole, 50-Hz induction motor supplies 15 kW to a load at a speed of 2,950 rpm.
    1. What is the motor's slip?
    2. What is the induced torque in the motor in Nm under these conditions?
    3. What will be the operating speed of the motor if its torque is doubled?
    4. How much power will be supplied by the motor when the torque is doubled?
43. Find the counter EMF of a DC shunt motor when the terminal voltage is 240 V and the armature current is 60 A. The armature resistance is 0.1 Ω. What is the developed power by this motor?
44. A 240 V DC shunt motor on no load runs at 900 RPM and takes 5 A. Compute rotational losses if the field resistance and the armature resistance are 240 Ω and 0.2 Ω, respectively.
45. A separately excited DC motor has a variable armature voltage supply that can vary from 50 V to 250 V. The field circuit is supply from a 250 V constant voltage source, the field and the armature resistances are 250 Ω and 0.2 Ω, respectively. The motor runs at 1,000 RPM at no load at takes 4 A from the armature supply when it is 250 V. Compute the rotational losses at 1,000 RPM, and shaft power, if the armature takes 50 A when the armature supply is 250 V.

Chapter 4
Load characteristics, wiring, and power cables

Objectives and abstract

Electrical distribution networks, transmission lines, electrical service, wiring devices, protection, and equipment are essential building subsystems and components. Power engineers are concerned with every step and aspects in the process of electricity generation, transmission, distribution, and utilization. Adequate electricity amount and its efficient utilization are essential for the growth and development of any country. Past developments of the power distribution often resulted in higher system losses and poor power quality services. Consequently, an efficient and effective power distribution network, building, or industrial electric systems have become important issues. By optimizing the power distribution, reducing the capital cost, power losses, and improving the power quality are critical issues in power system operation and management, resulting in substantial savings of energy. However, the electric load varies with time and place, such as the load variation customer types and the power production and distribution system must respond to the customers' load demand at any time. Therefore, modern electricity distribution utilities need accurate load data for pricing and tariff planning, distribution network planning and operation, power generation planning, load management, customer service and billing, and finally to provide information to customers and public authorities. After completing this chapter, students and readers are able to learn and understand the power distribution network structure, configurations, operation and management, and the impacts on the building electrical, mechanical, thermal, energy and lighting systems, as well as a good understanding of the building electrical system operation, components and equipment, and related issues. They will also learn to estimate and compute the demand load, apply demand factors, determine demand load for motor, equipment and appliances, understand methods to calculate cable and conductor sizing and capacity, voltage drop calculations and service entrance, operation, parameters and characteristics of wiring devices and their applications, and to develop an understanding and appreciation of the importance of codes and standards.

4.1 Introduction, building energy analysis, and electrical design procedure

Power supply system exploitation for industry, mining, transportation, agricultural, commercial, or residential use, is often characterized with significant overestimate in the installed power capacity, leading to overestimation of power use and lower efficiency leading to increased supply of cable cross sections, wires, conductors, circuit breakers, fuses, or wiring devices, which leads to problems, ranging from power losses to power protection malfunctions. This is the result of objective factors, such as reduced generation and power consumption and proactively over-estimated parameter values and electrical loads. All electricity supply system elements, such as transformers, power lines, cables, wires, switches, protection devices, etc., are selected based on the electrical load estimates. Small errors in the load calculations may easily lead to the incorrect selection of all system components, which can lead to all kinds of design, operation, and management issues. It is not a coincidence that special attention for the load calculations is paid in any electric design. Electrical loads vary with time and place, so power generation and distribution must respond to any load demand at any time. Therefore to operate efficiently, the modern power distribution systems need accurate load estimates and determination, power distribution planning, design, operation and management, building electrical system design, providing information to the costumers, public, or authorities. However, the problem of electrical load calculation cannot be considered solved, since significant and nontolerable discrepancies between calculated and actual loads still exist. Reason for this lays both in imperfection of calculation methods and incorrect use of regulatory factors.

Building electrical system design is an important part of the overall building design processes. With very few exceptions, all modern mechanical, air condition, heating, ventilation, and appliances are electrical powered, and together with communication, monitoring, control, and security systems are making the building electrical system complex and are demanding proper electrical power to operate. The design of building electrical system and its component selection are strongly influenced by the mechanical, heating, security, control, monitoring, ventilation, air conditioning, and control systems. Even the standards and electrical codes which are in use for decades ago need to be updated and completed. However, modern building electrical systems require less space, compared, for example, with mechanical and heating counterparts. Most of the electrical operating devices are normally exposed into the building occupied spaces. Therefore their location, configuration, appearance, and design must be closely coordinated with the building architecture and interior design. Unfortunately, quite often these electrical operating or wiring devices (switches, outlets, receptacles, lighting devices, controls, and alarms) are installed without considering the location, size, shape, or color. Building electrical system design involve, analyzing building requirements, electrical load estimates, electrical system selection, component selection, preparation of electrical plans, diagrams, and specifications, as well as coordinate with other building design groups, and with overall building design processes. Electrical system design is an integral and important part of the overall building design and analysis. Most of the mechanical, air conditioning, thermal, heating, building equipment, and appliances are electricity based or are using electric power in order to operate. Design, planning, and selection of building electrical systems are strongly influenced by the selection and characteristics of these systems and equipment.

This chapter concentrates on the most common and practical methods of electric load estimates, specifications and provisions of electric codes, standards and utilities, mainly for building electric design and analysis purposes. It presents the overall issues, methods and procedures for load estimate and calculation, conductor and cable selection and sizing, building electrical system design, operation, upgrading, expansion, or restructuring. Special attention is given to the standards, codes, and utility regulations. Energy conservation, efficiency, and the new technical developments and trends are also presented. Wiring devices, power ratings, selection, and sizing procedures are also discussed. These devices are used to control the power flow and to allow the operation and control of other devices and building equipment. The concepts of unit load, and demand load, as well as practical examples are presented and discussed in this chapter. The overall chapter objectives of this paper are (1) to present information on the calculation of loads for the electrical system of a building so that one can read or prepare an electrical plan for the building; (2) as part of the construction plans, one is able to understand and take the electrical plan for building permit applications; and (3) with a full understanding of the electrical plan, the contractor can better estimate the construction cost of a project.

4.1.1 Electrical design procedure and building energy analysis

Design of the building electrical systems involve the following important steps; analyze and understand the building energy needs and requirements, determine and estimate electrical loads at highest degree of accuracy, select appropriate electrical systems and equipment, coordinate with other deign decisions, and prepare the detailed electrical plans and specifications. However, in practice, these steps may be bypassed or performed in order that is more appropriate with a specific project or application, which may depend on the local utility policy and procedures, etc. First step in any electrical system design consists of the accurate analysis of building or facility energy needs and requirements as part of the overall architectural program. Some of the major factors affecting building electrical systems are occupation factors involving building occupancy types, present and future number of occupants, installed and future equipment and appliances, types, etc. Architectural factors, such as building size, structure, number of floors, floor plans, and building footprint or elevation, all have strong effects on the electric system design and structure. Important design factors refer to illumination criteria; building environment, such as air conditioning and type of heating systems, either central or unitary; type of lighting systems; and other mechanical and thermals systems, such as hot water, sewage disposal, fire protection and monitoring systems, etc. Building equipment, such as, vertical transportation systems, food preparation and recreational facilities, computing and processing equipment and systems, production equipment requiring electric power are also important factors taken into account in electric system design process. Designers must also take into account the building auxiliary systems, fire and security protection systems, monitoring equipment, TV and communication systems. For hospitals, banks, hotels, and industrial facilities, auxiliary and special equipment, devices, and systems must be included into electric system design and analysis. However, for most of building types, the auxiliary equipment power requirements in general are minimal. Last but not the least, the cost and fund availability is also an important design factor.

Accurate estimates of demand load on each branch circuits, feeders, or services is critical and necessary to determine the ratings of these circuits, as well as, to set the number of branch circuits and feeders. Article 220 of NEC discusses the rules and procedures for minimum estimated demand load for branch circuits, feeders, and services. The basic rule is that each branch circuit, feeder, or service must have enough rating to supply the demand load need to be served. A circuit demand load must be equal to the minimum estimated load as determined by applying the NEC or the actual calculated load, whichever is greater. System design is based on the greater of these two estimates. Designer need to follow the information from NEC, IEEE, and IEC standards and technical guidelines to determine the minimum estimated loads but also must exercise the common sense, judgment, and technical experience when determining the load to be served. The number and location of the receptacles, wiring devices, and lighting fixtures depend on the nature of occupancy and use. For example, a general office area usually require only functional lighting, providing a certain illumination level, while a restaurant or a theater may have decorative fixtures necessary to create a specific ambiance. Future power and energy needs and requirements, more than any other factors and components, are the ones that are changing more frequently. The general trend and tendency was that building power and energy requirements keep increasing from year to year. The question is how much capacity and duration need to be included into the electric system design. However, with the advent of more efficient and less energy consuming building equipment and systems, smart and intelligent energy management, green design methods and the use of environment for lighting, heating or ventilation started to reverse the trend in energy use and conservation.

4.1.2 Branch circuits and feeders

Power, or more correctly the electrical energy, is transferred in any electrical system from the service equipment and network to the lighting systems, machines, equipment, and outlets. Regardless of the wiring methods and systems, power carrying conductors and cables are divided into feeders and branch circuit conductors. Service conductors are the one that are extending from the power company or utility terminals to main disconnects, while the feeders are the ones originated at the main disconnect device or main distribution and terminate at another distribution center, panel-board, or load center. Subfeeders are the conductors originating at the distribution center other than the main distribution and terminate to panel-board, load center, and disconnect switches and are supplying branch circuits. Panel-boards, in single or multiple configurations are equipment containing switches, fuses, and circuit breakers, intended for power switching and controlling and for circuit and equipment protection. Branch circuits are the section of wiring system extending beyond the final overcurrent device, usually originating from a panel and transferring power to load devices. They include circuits servicing single motors (individual circuits) and the ones servicing multiple lighting systems and receptacles (multiwire), and are operating usually at lower currents, 30 A or less but can also supply higher currents. A branch circuit consists of conductors originating at safety switches (disconnects) or more often at a panel-board. Most common types of branch circuits include *individual branch circuit* that are supplying a single load, *multioutlet branch circuit*, supplying multiple loads, *general purpose branch circuits*, multioutlet type that is supplying multiple outlets for lighting and appliances, *appliance branch circuit*, supplying a single appliance load, and *multiwire branch circuit* with two or more ungrounded conductors and one grounded conductor. Any branch circuit is sized in accordance to the supplied load, while sizing the circuit for additional future loads is good design practice. Its rating depends on the rating of the overcurrent device protecting the circuit. Branch circuit serving a single load can have any rating, while the ones that are serving multiple loads have limited ratings of 15, 20, 30, 40, or 50 A. Section 210.6 of the NEC specifies, in agreement to the supplied circuits, the branch circuit voltage limits. Regarding the feeders, the conductors between the service equipment and the branch-circuit overcurrent protection devices, codes are providing information regarding safe and adequate sizing and installations of these conductors. Feeder loading depends on the total system power requirements, assuming that all connected loads are operated as required and the feeder has sufficient ampacity to meet the load demands. Prior to installation, several factors must be taken into account to ensure the proper feeder size, type, and overcurrent protection is correct for that application. The most important factors are the conductor and cable types and characteristics. Feeder materials are copper or aluminum, can be single or multiconductor cable, feeder environment (damp, hot, corrosive) must be considered, as well as the voltage drop for determining the feeder length. Feeder can run in conduit, cable trays, or other systems, can be paralleled or individual, while the neutral may not be necessary. Conductor sizing includes several factors, such as conduit fill, ambient temperature, connected load demand, demand, and derating factors. Various overcurrent protection devices are used with feeders.

Circuit loads are classified into two main categories: *continuous* and *noncontinuous loads*. A continuous load is any load which is on for 3 h or more (NEC Article 100). Examples of continuous loads are lighting, computers, copy and printing machines, HVAC equipment, and other devices and equipment found in offices. Most of the industrial facility equipment and devices, and industrial processes are also considered to be continuous. Noncontinuous loads are general-purpose receptacle outlets or residential lighting outlets. If there is a doubt whether a load is continuous or noncontinuous, the best design practice is to consider it continuous. Overcurrent protection for continuous loads must be sized at no less than 125% of the load. The increase in the fuse or breaker size is not because the current from a load increases when a load operates for more than 3 h. Continuous loads must not exceed 80% of the circuit rating allotted to it. The problem with continuous loads is the heat that is generated in both conductors and in terminals where conductors are connected. Heat is produced when the current flows through conductors and terminals. If the conductors are sized correctly and the loads do not run continuously, then the heat produced is insignificant. However, the longer the current flows through a conductor or through terminals where it is connected the more they are heated. The NEC has found that loads operated for 3 h or more can cause significant heat to build-up, adversely affecting the conductors, terminations, and overcurrent devices. Increasing the rating of overcurrent devices and the ampacity of conductors for continuous loads by 125% compensates for the heat build-up at the overcurrent device terminals. Circuit breakers are thermal-magnetic devices, so increased heat at the breaker terminals can cause them to trip at a current below the breaker's ratings. A larger breaker compensates for the heat build-up of continuous loads by having a higher thermal trip point. A circuit breaker thermal setting is to protect the circuit from overloads, not for short circuits. Notice that if the overcurrent protective device is listed for continuous operation at 100% of this rating, then 80% of the correction factor is not used. Branch circuit loads are classified into the following categories: lighting loads, receptacle loads, equipment loads, heating and cooling loads, and motor loads. Branch circuit conductor current ratings must be greater than the maximum load supplied by the circuit. In the event of multiple loads, the conductor ampacity must correspond to the overcurrent protective device. However, the circuits supplying hard-wired devices (electric heaters, air-conditioning equipment, and water heaters), the fuse or circuit breaker is rated at the next higher rating.

4.2 Load estimate and calculations

Analysis of the building energy requirements, demands and usage is critical in the building electrical or thermal system design, restructuring, retrofitting, upgrading and/or expansion, or in the energy and electricity conservation and efficient use. An important phase of this process consists of the most accurate possible identification and estimation of the current and if possible future energy requirements and use. Major factors affecting building electrical or heating systems include present, anticipated and future building occupancy, electrical appliances and equipment, cost, economic and architectural factors, building environment, air condition, hot water, heating and ventilation types, illumination criteria, lighting system type, mechanical systems, equipment, building transportation, monitoring, control, security, and recreational systems, computing, production and processing equipment, as well as the electricity supply availability. The electric system analysis and assessment also include the most accurate evaluation of building auxiliary systems, building management, communication, specialty equipment (needed or used in hospitals, hotels, industrial facilities, schools, banks, etc.), and the most accurate possible future energy and electricity demand estimates and usage changes. As a rule of thumb, a minimum 25% spare electricity capacity should be included at building main or distribution centers. There was a significant increase in the building electricity usage in the past decades, especially in the office buildings, due to the increased use of computers, communication equipment, printers, copy machines, control, and monitoring systems. However, the new trends towards energy conservation, higher equipment efficiency, new lighting systems, high efficient electrical motors, sustainable design, and energy management have significantly decreased the uses and needs of building electricity.

Accurate estimates of the demand load on a particular branch circuit, feeder, or service are necessary to determine the required rate of these circuits. Actual number of branch circuits and feeders, conductors, wires, and cable size and selection are also based on the determination of the estimated demand load and electricity needs. The required rules for determining the minimum estimated demand load for branch circuits, feeders, and services are discussed in the Article 200 of the NEC (National Electric Code), and similar guideline in the IEC (International Electrotechnique Commission, in French Commission électrotechnique internationale—CEI) standards. *The basic rule is that the branch circuit, feeder, or service must have a rating sufficient to supply the demand load to be served.* The load demand on an electric circuit is equal to the minimum estimated load as determined by applying the *NEC minimum estimates, or the actual load, whichever is greater*. Electrical circuit load capacity represents the total amount of electrical power that a home, hospital, school, industrial, or commercial facility is actually using. In order to decide how large the electrical service is needed in a residential, commercial, or industrial facility, the most accurate load estimate possible is required. Older homes, for example, have a 60 A of electrical service, connected to a fused panel-board, while the newer ones have 100 A or 200 A electrical services connected to breaker panel-board. In order to calculate the amount of power required, we must have a clear idea about the existing equipment, electrical appliances, lightning, facility mechanical, heating, air conditioning, monitoring, processing equipment, etc. and amount of power usage during the operation periods. As technology continues to advance, it seems we add more and more electrical loads to our home. However, the general trends are the newer equipment, devices, and appliances that require less power and are more efficient in terms of energy use than the older ones. Although codes and standards, such as NEC or IEC provide necessary information to determine and estimate minimum loads, the designers and engineers must exercise common sense and judgment while estimating the load to be served, by taking into account not only the present power requirements but also future changes.

4.2.1 Convenience power, connected and demand loads

Electrical power and energy needs and demands in buildings, residential, industrial and commercial, especially in office buildings increased dramatically at the end of twentieth century, due to the extended use of computing, information technology

(IT), telecommunication and printing equipment, systems, and devices. On the other hand, recent trends and applications of sustainability principles, energy conservation and management, sustainable design, new and more efficient equipment, motors, and devices have reversed the energy use tendencies. Electrical system design is an integral and important part of the overall building design and analysis. Most of the mechanical, air conditioning, thermal, heating, building equipment, and appliances are electricity based or are using electric power for its operation. Design, planning, and selection of building electrical systems are strongly influenced by the selection and characteristics of these systems and equipment. Electrical power for mechanical equipment and systems varies widely with their types, and with building designation, location, and architectural design. Mechanical systems and equipment include HVAC, plumbing, and fire protection. Such equipment and systems require large amount of power in order to operate, being more efficient and economically designed at higher voltages, such as 208-, 240-, and 480-V and three-phase. However, such type of residential equipment is normally designed for 120 V, 240 V, and single-phase power.

Building equipment include elevators, escalators, and other vertical transportation systems, food service equipment, recreational, household, and miscellaneous equipment and systems. Power requirements and voltage levels for such equipment and systems vary widely in capacity and operating characteristics. Building auxiliary equipment and systems include the ones deigned for life safety, communication, monitoring, surveillance, and security, which do not require large amount of power to operate and are usually powered at single-phase 120 V or 240 V. Convenience loads refer to plug-in equipment and systems, such as personal computers, laptops, printers, laboratory instruments or equipment, portable lights, audio and video equipment, and some of the household appliances, etc. The recommendations are to reserve about 180 VA per duplex outlet for convenience power plus any, large and/or special equipment for sizing the service. Typical power ratings for some of the common household appliances are given in Table 4.1, while the recommended and suggested power ratings for various interior spaces and buildings are included in Table 4.2.

The electrical load requirements for commercial, industrial, and residential installations result in a great deal of diversity and complexity in usage, meaning that some types of equipment or electrical loads are in use for extended periods, others are only used occasionally or for short periods. In addition, there are often electrical loads on the same service or feeder that are not in service simultaneously by their nature, such as heating and air conditioning. For this reason, demand factors are applied when calculating service and feeder loads. Different sets of demand factors apply for different electrical loads, and even for different types of commercial, industrial, or residential buildings. Although most of the service and feeder commercial load calculation requirements are found in Art. 220 of NEC, other rules affecting these loads are scattered throughout the NEC. For example, NEC Chapter 3 is focusing with the wiring methods used, while other articles provide more in-depth requirements for particular equipment or applications, such as the specific requirements for motor circuits found in Art. 430. The load that an individual customer or a group of customers presents to the power distribution is constantly changing.

Table 4.1 Typical power ratings of some common household appliances

Appliance or equipment	Power range (VA)	Appliance or equipment	Power range (VA)
Window air conditioning (115 V, 0.5 ton)	500–700	Owen (115 V/230 V)	2,000–5,000
Window air conditioning (115 V, 1.0 ton)	1,200–1,400	Microwave Owen (115V)	300–1,000
Window air conditioning (230 V, 2.0 ton)	2,400–2,800	TV Set (115 V)	200–1,000
Refrigerators (115 V)	300–1,000	Desktop PCs	80–200
Freezers (115 V)	300–800	Printers	100–200
Washing machines (115 V)	800–1,200	Laptops	20–75
Dryers (115 V)	3,000–5,000	Scanners	10–20
Water heaters (115 V/230 V)	3,000–6,000	Coffee makers	800–1,400
Dishwasher (115 V)	1,200–1,500	Ceiling fans	20–100
Water filter and cooler (1,115 V)	75–100	Vacuum cleaner	200–700
		Toasters	800–1,800

Table 4.2 Typical power specifications for convenience power (portable and fixed loads)

Space type	Power allowance (W/ft^2)	Power allowance (W/m^2)
Offices	1–4	11–44
Classrooms (No PCs)	2–3	22–33
Classrooms (With PCs)	3–5	33–55
Meeting rooms	3–5	33–55

Every time a light bulb or an electrical appliance is switched on or off, the load seen by the distribution feeder changes. The building electrical load calculations consist of several components that constitute an electrical power plan, which is part of the building construction plans. In order to describe the changing load, the following terms and parameters used to characterize the load are defined in the next paragraph. Load is the power consumed in a closed circuit. The "average load" is the average power consumed in the circuit system during a specific time period. The "maximum load" is the maximum value of the consumed loads during the same time period under all conditions. The "demand load" is the required power to satisfy adequate device or circuit operation. Load at receiving terminal averaged over a specific period of time, and is representing the load demand. Load can be expressed in kW, kVA, or even in A, only.

Factors used for load calculations and analysis are referring to the following equations or ratios. *Demand factor* (DF) is the ratio of the maximum demand load to the total load connected. When the power input is connected to a circuit system, whether the switch is closed or open, the maximum demand load should always be equal or smaller than the connected load because of a system loss and other technical reasons. *Coincidence factor* is the maximum system demand divided by the sum of individual maximum demand, while diversity factor is its reciprocal, i.e., the sum of individual maximum demands divided by the maximum system demand. The power planning is more effective for a greater DF because the value of nominator is fixed but the value of denominator or the maximum system demand changes with different designs. Load factor (LF) is the ratio of the average load to the maximum load. The maximum load should occur under the worst case load event. In this context, the average load is always less than the maximum load. It would likely be more economical in developing a power plan if these two values are closer. This shall depend on the qualities of the devices and equipment in the building. *Maximum demand* represents the greatest of all demands that occur during a specific time, and must include the demand interval, period, and units in the definition. The average demand of a load curve in kW equals, and is given by:

$$P = \frac{W}{\Delta T} \tag{4.1}$$

Here W is the electrical energy consumption in period of ΔT is in hours, and is the time interval in hours (=24 h for the daily load curve, and 8,760 h for the annual load curve). *LF* is an indication how well the utility's facilities are being utilized. From utility stand point, optimal *LF* would be 1.0 (system consumption approaches the maximum). It is reflected on the electricity bill, the bigger *LF* value the better is for consumers or electricity users. *LF* is defined as:

$$LF = \frac{\text{Average demand}}{\text{Maximum demand}} = \frac{P_{Average}}{P_{Maximum}} \tag{4.2}$$

Demand factor is the ratio of the maximum demand of a system to the building or facility total connected load (maximum demand when all are used), being usually less than one. It gives the fractional amount of some quantity being used relative to the maximum amount that could be used by the same system. The lower the demand factor, the less system capacity required to serve the connected load. It is computed as:

$$DF = \frac{\text{Maxium load demand}}{\text{Maximum possible load demand}} \tag{4.3}$$

Example 4.1: A large industrial facility might have a connected load of 20 MW but if only 75% of its electrical equipment is operating, what is the demand factor?

Solution: The demand factor would be only 75% or 0.75 of the maximum load power.

Coincident factor is the ratio of the peak demand of a whole system to the sum of the individual peak demands within that system. The peak demand of the whole system is referred to as the peak diversified demand or as the peak coincident demand. The individual peak demands are the noncoincident demands. The coincident factor is less than or equal to one. Normally, the coincident factor is much less than one because each of the individual loads does not hit their peak at the same time (they are not coincident). Diversified demand represents the sum of demands imposed by a group of loads over a particular period. It must include load demand interval, period, and the units. *Diversity factor* represents the ratio of the sum of the individual peak demands in a system to the peak demand of the whole system. The diversity factor is greater than or equal to one and is the reciprocal of the coincident factor. DF [also known as Simultaneity Factor (ks)] is also expressed as the ratio of the maximum noncoincident load demand to the maximum diversified demand (the inverse of ks). For a quantitative measure of the inherent diversity of individual load peaks, a diversity factor is defined as:

$$Div\,F = \frac{\sum_{i=1}^{N} P_{\max,i}}{P_{\max,T}} \tag{4.4}$$

Greater the diversity factor, lesser is the cost of generation of power. Term "simultaneity" is used by some and means:

$$ks = \frac{1}{Div\,F}$$

This factor is thus higher than 1. The residential load has the highest diversity factor. Industrial loads have low diversity factors usually of 1.4, street light practically unity and other loads vary between these limits (Table 4.3). The *responsibility factor* is the ratio of load's demand at the time of the system peak to its peak demand. A load with a responsibility factor of one has a peak at the same time as the overall system. The responsibility factor can be applied to individual customers, customer classes, or circuit sections. The loads of certain customer classes tend to

Table 4.3 Diversity factors. Adapted from IEC

System component	Residential	Commercial	General power	Large industrial
Between individual users	2.00	1.46	1.45	—
Between transformers	1.30	1.30	1.35	1.05
Between feeders	1.15	1.15	1.15	1.05
Between substations	1.10	1.10	1.10	1.10
From users to transformers	2.00	1.46	1.44	—
From users to feeder	2.60	1.90	1.95	1.15
From users to substation	3.00	2.18	2.24	1.32
From users to generating station	3.29	2.40	2.40	1.45

vary in similar patterns. Commercial loads are usually highest from 8 a.m. to 6 p.m. Residential loads peak in the evening, with second peak (usually smaller than the evening one) earlier in the morning. Weather significantly changes loading levels. On hot summer days, for example, the air conditioning increases the demand and reduces the diversity among loads. Loads are separated into (a) *fixed loads*, which are permanently wired loads, and (b) *convenience loads*, the plug-in ones. In general, each load has a nameplate specifying its current rating (A), real (active) power (W), or apparent power (VA). These power ratings are used to estimate the total system and/or building loads and power requirements. *Connected load* of a system or building represents the sum of all connected electrical loads, regardless the way that they are used, expressed by a simple relationship between connected load (CL) of various load groups, lighting, equipment, and convenience loads and the total (gross) connected load of a power system or a building, as:

$$TCL = \sum_{k=1}^{N} CL_k \tag{4.5}$$

Demand load (DL) of an electrical power system or of a facility is the net load that is likely used at the same time by each load groups. The demand load is always smaller than the connected load, except when all connected loads are used at the same time, when the demand load is equal to the connected load. Straight and simple relationships between the connected load (CL), demand load, demand factor (DF), and the total (gross) demand load (TDL) are given by the following two relationships, respectively:

$$DL = CL \times DF \tag{4.6}$$

and

$$TDL = \sum_{k \geq 1} DL_k \tag{4.7}$$

Diversity factor (sometimes called diversity coefficient, to avoid confusion with demand factor) can be defined in terms of parameters introduced in (4.6) and (4.7). It accounts on the demand diversity of different load groups, being a time-consuming parameter to calculate. As a rule of thumb, recommended in the literature (see the end of chapter references), value equal to 1.0 is considered for systems that do not have load diversification, and value of 1.2 in the case of large systems and/or ones with diversified loads. In the case, when the diversification is greater than suggested by the diversification factor, a complete analysis of the behavior of the loads at each hour, or even at every minute is needed, under all operating conditions. If the diversity coefficient is neglected during the design stage, it may lead in the system oversizing and increased cost. Net demand load (NDL), total demand load, and diversity factor (coefficient) are related by the following relationship:

$$NDL = \frac{TDL}{Div\,\mathrm{F}} \tag{4.8}$$

Example 4.2: A building electrical system is calculated to have the following load groups: lighting 150 kW and diversity factor 0.9, receptacles 105 kW and 0.2, building equipment 200 kW and 0.6, mechanical equipment 240 kW and 0.85, and auxiliary equipment 50 kW and 0.2, respectively. Calculate the connected load, and if the diversity factor is 1.15 the net load demand of the system.

Solution: Total connected load and total demand load are calculated as:

$$TCL = 150 + 105 - 200 + 240 + 50 = 745 \text{ kW}$$

$$TDL = 0.9 \times 150 + 0.2 \times 105 + 0.6 \times 200 + 0.85 \times 240 + 0.2 \times 50 = 490 \text{ kW}$$

$$NDL = \frac{TDL}{Div\ F} = \frac{490}{1.15} = 426.1 \text{ kW}$$

The recommendation is to use 430 kW.

4.2.2 Lighting load estimate methods

Lighting accounts for a significant part of the overall electrical load for most of the buildings or facilities. In broad sense, lighting loads are categorized in general lighting, show-window, track, sign, outline, and other lighting. General lighting is the overhead lighting within the building and facility, with an adequate intensity for any type of work performed in that area. Lighting fixtures are generally designed for 120 V and single-phase power service, but may be designed for 208-V, 240-V, and 277-V single-phase power systems. Estimation of the general lighting load is based on either load per area method or the actual full-load of the fixtures used, whichever is greater. Within a building or facility there are usually several different types of area, such as storage, office, hallways, cafeterias, etc.—and these must be considered separately. Continuous improvements and advances into the lighting technology and systems have increased the electricity to light conversion efficiency, reducing the needs for electrical power for building lighting. Back in 1980s was common to require 3–5 VA/ft^2 for office areas but only 2 VA/ft^2 in 1990s, being now down to about 1 VA/ft^2. For planning electrical power for building lighting NEC has prescriptive requirements tabulated for various space types. In general, these are in excess to the actual needs and requirements of the ASHRAE/IEC Standard 90-2007. The general lightning load applied to a particular occupancy is based on specific unit load per unit of area (sq. foot or sq. meter) expressed in volt-ampere per unit area. The total minimum estimated load is to multiply it with the total occupancy area. The unit loads for various occupancies are given in Table 4.4 (Table 220.3(A) of the NEC). If the type occupancy is not listed in Table 220.3(A), the load is based on the actual connected equipment. In addition, only habitable areas of dwelling are used in the square-footage determination. Unfinished basements, garages, or porches are not considered habitable and are not usually included in the square-footage determination. If the load is continuous, the computed load is multiplied by 1.25 to determine the circuit requirements. However, the general lighting loads is not required, if the

Table 4.4 General lightning loads. Adapted from Table 220.3(A) of the NEC

Occupancy type	Unit load (VA/m^2)	Unit load (VA/ft^2)
Atrium	11	1
Auditorium	11	1
Banks	39	3.5
Barber shops	33	3
Churches	11	1
Clubs	22	2
Court rooms	22	2
Dwelling units	33	3
Garages (commercial)	6	0.5
Hospitals	22	2
Hotels and motels	22	2
Industrial buildings	22	2
Lodge rooms	17	1.5
Museums	22	2
Office buildings	39	3.5
Restaurants	22	2
Schools	33	3
Stores	33	3
Warehouses	6	0.25

load of each fixture is determined, separately with the same provision, if it is considered continuous load.

Example 4.3: Determine the lighting load for an office area of 6,300 ft^2 using actual lights and the information of Table 4.1 (Table 220.3 (A) of NEC) The office contains 30 fluorescent lights having a ballast of 280 V and 0.8 A.

Solution: By using actual office lightning, the estimate lighting load is:

$$\text{VA lighting load} = 30 \times 280 \times 0.8 = 6{,}720 \text{ VA}$$

By using the Table 220.3(A) of NEC the estimate load is:

$$\text{VA lighting load (NEC)} = 6{,}300 \times 3.5 = 22{,}050 \text{ VA}$$

Note: Either the actual load estimate is 6,720 VA, so the estimate load demand is 22,050 VA, based on the NEC specifications.

If the occupancy type is not listed in Table 220.3(A), the load is determined by using the power requirements of the actual connected loads. In addition, only the habitat area of buildings (dwelling units) are included in the area determination, other spaces, such as unfinished basement, garages, porches, etc. are not included in such estimates. For a dwelling unit consisting of a single unit that provides complete and independent living facilities, according to the NEC definition, Art. 100. Table 220.3 is applied for lighting load determination, however, the load from general purpose receptacle outlets in habitable areas (residential buildings, hotels, motels, and dwelling units) is permitted to be included with the general lighting load estimates, being taken into account in the unit lighting load. For office buildings, banks, designed school rooms, and data centers, an additional 1 VA/ft^2 needed to be added, if the number of actual receptacles is not known, due to the fact that in such areas, the final furniture layout and the plug-in equipment are not known, and extra provision for the load must be included.

Example 4.4: Compute the estimated lighting load of an office area of 250 m^2, containing 30 fluorescent lighting fixtures, each having a ballast of 277 V and 0.7 A.

Solution: From Table 4.4, the total lighting load is:

$$S_{lighting} = 250 \times 39 = 9{,}750 \text{ VA}$$

The lighting load of the connected equipment (florescent lighting fixtures) is calculated form actual power requirements:

$$S_{fixture} = 30 \times 277 \times 0.7 = 5{,}817 \text{ VA}$$

Actual connected load,\! based on the connected equipment is only 5,817 VA, the estimated demand load is 9,750 VA based on the NEC minimum requirements, and so the number of the branch circuits for this area. The greater of the load estimates applies.

Show-window lighting is any window designed to the good display and advertising materials whether it is entirely open at the rear, fully or partially enclosed, or has or not a platform raised higher than the street level. Show-window lighting load is not considered part of the general lighting load, Section 220.43(A) and 220.14 (G) of the NEC requires that a 200 VA per linear foot of the show window or the maximum volt-ampere rating of the equipment, whichever greater to be used to estimate this type of lighting load. NEC also requires, in addition to the show-window lighting load at least one receptacle of 180 VA for every 12′ of the show-window space measured horizontally. Chain-supported fixtures in a show window are permitted to be externally wired, with on other externally wired fixture allowed.

Example 4.5: Computed lighting load for two show windows of store, 36′ and 48′ long, respectively.

Solution: Total show-window length is: $36' + 48' = 84'$. The show-window lighting load is:

$$\text{Show-window load} = 84' \times 200 = 16{,}800 \text{ VA}$$

This lighting is a continuous load and the circuit load requirement is: $16{,}800 \times 1.25 = 21{,}000$ VA. In addition, receptacles are required at every 12′ of window and total of 7 (3 and 4, respectively) receptacles are required, and the receptacle load is: $7 \times 180 = 1{,}260$ VA. The total load is then:

$$\text{Total load} = 16{,}800 + 1{,}260 = 18{,}060 \text{ VA}$$

Sign and outline lighting, discussed in Article 600 of the NEC, which requires that a structure have at least one circuit, exclusively used to supply such lighting, and are considered continuous loads. This circuit must be designed for a minimum load of 1,200 VA. Track lighting, often used in commercial buildings is used for accent lighting, and is discussed in Article 410 (part XV) of the NEC. It is a manufactured assembly designed to support and energize luminaires (lighting fixtures) that are readily to be repositioned on the track. Notice that the track length can be altered by the addition of reduction of the track sections. The minimum track lighting load requirement is to assign 150 VA for every 2′ of the track length. Any other and additional lighting loads are computed separately from the general lighting, being added to the total lighting load. Such lighting loads include security, parking area, sidewalk, roadside, and stadium lightings and are calculated using the actual loads of equipment and devices.

Example 4.6: A furniture store has a total area of $120' \times 250'$, uses for storage $100' \times 60'$, a small office $10' \times 20'$ and the remainder the building is used as showroom. There are also a total of 48′ show windows, a 40′ track and two outdoor sign. What is the total lighting load?

Solution: The storage, office and the showroom areas, and the corresponding lighting loads are:

Storage area: $100 \times 60 = 6{,}000$ sq. ft
The storage lighting load: $6{,}000 \times 0.25 = 1{,}500$ VA

Office area: $10 \times 20 = 200$ sq. ft
The storage lighting load: $200 \times 3.5 = 700$ VA

Store total area: $250 \times 120 = 30{,}000$ sq. ft
Show-room area: $30{,}000 - 6{,}000 - 200 = 25{,}800$ sq. ft
The show-room lighting load: $25{,}800 \times 3 = 77{,}400$ VA

General lighting load: $1{,}500 + 700 + 77{,}400 = 79{,}600$ VA

Show-window lighting load: $200 \times 48 = 9{,}600$ VA

Sign lighting load: $1{,}200 \times 2 = 2{,}400$ VA

Track lighting load: $0.5 \times 40 \times 150 = 3{,}000$ VA

The sign lighting load is estimated on the minimum 1,200 VA, so a total of 2,400 VA.

Total lighting load: $79{,}600 + 9{,}600 + 2{,}400 + 3{,}000 = 94{,}600$ VA
Circuit requirement (continuous load): $94{,}600 \times 1.25 = 118{,}250$ VA

Table 4.5 Derated factors for lighting loads

Occupancy type	Load portion (VA)	Demand factor (%)
Dwelling unit	0–3,000	100
	3,001–120,000	35
	Over 120,000	25
Hotel and motel	0–20,000	50
	20,001–100,000	40
	Over 120,000	30
Hospital	0–50,000	40
	Over 50,000	20
Warehouses	0–12,500	100
	Over 12,500	50
All others	Total VA	100

The lighting loads have a high degree of diversity, especially on commercial and industrial facilities, or hospitals. For example, it is very unlikely that that every light in a hospital or a department store to be operated in the same time. The lighting derating factors are specified in Section 220.42 of the NEC, allowing the power reduction (derating) of the feed, panel-board or service. However, there areas where the derating factors are not recommended to be applied in a specific structure, while some special areas may require lighting all the time, for example, operating rooms in hospitals, emergency rooms, intensive care units, data centers, stairways, etc. In Table 4.5, an adapted version of Table 220.42, derating factors are given for some structures or facilities. Notice that these derating factors are not applied to the branch circuit conductor or branch circuit overcurrent protection device calculations.

4.2.3 Dedicated and general-purpose receptacle load estimates

Most of the receptacles installed in commercial facilities and apartment buildings are not supplying continuous loads, being quite difficult to predict the loads that they are supplying, unless the receptacle is dedicated (assigned to a specific purpose). NEC does not require a specific or minimum number of outlets for commercial or apartment buildings. Usually, many receptacles are required and installed. If a receptacle is the load of a branch circuit, its current rating must be equal or higher to that of the branch circuit, while there are multiple receptacles on a branch circuit, the receptacle ratings varies with the current rating, as shown in Table 4.6. A load of 180 VA is assigned to each receptacle either it is single type, duplex or triplex. If the receptacle is dedicated to a specific load, the actual load is used. If the dedicated load is continuous then a 125% correction factor is applied. To calculate the allowable number of receptacles on a branch circuit, the product of voltage and current is divided to 180 VA. When a receptacle is the load supplied by an individual branch circuit, the receptacle current (ampere) rating must be equal or greater than that of the branch circuit. In the case of multiple receptacles on a

Table 4.6 Receptacle rating as determined by the circuit rating

Circuit rating (A)	Receptacle rating (A)
15	Up to 15
20	15 or 20
30	30
40	40 or 50
50	50

branch circuit, the receptacle rating varies with the current rating. In such cases, NEC has specific requirements and recommendations, and the specific receptacle ratings are given Table 4.6 (adapted from Table 210.21 (B)(3) of the NEC). If a receptacle is dedicated for a specific device or equipment, the actual load is used, and in the case of continuous loads than a 125% overrate is appropriate. To calculate the allowable number of receptacles on a branch circuit, multiply the voltage and current (the circuit power) and divide the result to 180 VA, then round off the lower integer.

Example 4.7: How many receptacles can be placed on a 120 V, 15 A circuit? and how many on a 120 V, 20 A circuit?

Solution: The maximum circuit power in each of the two cases is:

$$P_{15\text{ A}} = 120 \times 15 = 1{,}800 \text{ VA}$$

$$P_{20\text{ A}} = 120 \times 20 = 2{,}400 \text{ VA}$$

The number of receptacles, in each case is:

$$Nr.\ receptacles = \frac{1{,}800 \text{ VA}}{180 \text{ VA}} = 10$$

$$Nr.\ receptacles = \frac{2{,}400 \text{ VA}}{180 \text{ VA}} = 13.3$$

A 120 V, 15 A branch circuit can supply 10 receptacles, while a 120 V, 20 A can supply 13 receptacles (we rounded off to the lower value).

The receptacle load can be included in the general lighting load by adding a value of 1 VA/ft^2 or 11 VA/m^2 to the general lighting unit loads of Table 4.4 (Table 220.12 of NEC). However, the recommendation is that this approach to be used only when the number of receptacles in unknown.

Example 4.8: Calculate the receptacle load for a 200′ × 150′ department store, and number of 15-A circuit needed to supply the load at 120-V power supply. The number of receptacles is unknown.

Solution: For unknown, a load of 1 VA/ft^2 is used, so:

$$\text{Area} = 200 \times 150 = 30{,}000 \text{ sq. ft.}$$

$$\text{Receptacle load} = 1 \times 30{,}000 = 30{,}000 \text{ VA}$$

Maximum single circuit load: $120 \times 15 = 1{,}800$ VA

$$\text{Number of circuits} = \frac{30{,}000}{1{,}800} = 16.67$$

Maximum number of circuit is 17.

Multioutlet assembles and systems are quite often installed in repair shops, lighting display areas, electronics departments, research and educational laboratories, data and computing centers, small industrial facilities, and other locations where multiple outlets are needed. These multioutlet assemblies require 180 VA for each 5′ of length. NEC allows, in laboratories, repair shops and stores to be derated in accordance with Table 220.42 (Table 4.5 in this chapter), if the load exceeds 10 kVA. The diversity and loading intermittency and inconstancy allows that the total receptacle load to be derated, as specified in Section 220.44 of the NEC. If the load is exceeding 10 kVA, the first 10 kVA are counted at 100%, while the additional load is derated (reduced) at 50%. However, this procedure may not be used if NEC requires that specific appliances or equipment cannot be derated (refers to Sections 22.12 and 220.444 of the NEC, and Table 220.44).

Example 4.9: Determine the receptacle load of a 120′ by 200′ hardware store that has one duplex receptacle at every 12′ of wall around the store area, a total of 24 floor receptacles and 24′ and 30′ show windows.

Solution: The number of receptacles placed on each walls is:

$$\text{Receptacles per } 120' \text{ wall} = \frac{120}{12} = 10$$

$$\text{Receptacles per } 200' \text{ wall} = \frac{200}{12} = 16.67$$

There are 10 receptacles or each of the 120′ walls and 17 receptacles for each of the 200′ walls, a total of:

$$\text{Wall receptacles} = 2 \times 10 + 2 \times 17 = 54$$

One receptacle is need for every 12′ of show window, so the 24′ show window requires 2 receptacles, while the 30′ show wind requires 3 receptacles, a total of 5. The total number of store receptacles is then:

$$\text{Total receptacles} = 54 + 5 + 24 = 83$$

The total receptacle load is:

Total receptacle load $= 83 \times 180 = 14{,}940$ VA

Or feeder sizing, the first 10 kVA must count at 100%, and the excess of 4,940 VA is 50% derated, so the feeder load is:

Feeder load $= 10{,}000 + 0.5 \times 4{,}940 = 12{,}470$ VA

NEC and IEC standards require that certain branch circuits, the dedicated branch circuits in dwelling units and apartment buildings be designated for receptacle outlets only, prohibiting the connection of lighting outlets to them. Included into the dedicated branch circuits are a minimum of two 20 A branch circuits to supply small appliance loads in the kitchen areas, as well as the laundry equipment of the dwelling units. Section 220.16 of the NEC limits the computed loads on each of this type of branch circuit to 1,500 VA, which is in addition to the unit VA load computed from the general lighting. In addition, NEC Section 220.3(B).9 requires that a minimum 180 VA be assigned to the general-use receptacles, not located in the habitable areas or connected to dedicated branch circuits. Notice that the 180 VA applies to a device located on a single yoke or mounting strap. However, the estimated load can be always greater than the minimum NEC load determination. Dwelling units have special requirements for load calculations. Most of the actual load calculation requirements are in Art. 220, others are scattered throughout the Code and come into play when making certain calculations. In addition to the branch circuits required for dedicated appliances and those needed to serve the general lighting and receptacle loads, a dwelling unit must have the following branch circuits: (a) a minimum of two 20 A, 120 V small-appliance branch circuits for receptacles in the kitchen, dining room, breakfast room, pantry, or similar dining areas, as specified in Section 220.11(C)(1). These circuits must not be used to serve other outlets, such as lighting outlets or receptacles from other areas, as described in the 210.52(B)(2). These circuits are included in the feeder/service calculation at 1,500 VA for each circuit [220.52(A)]. (b) One 20 A, 120 V branch circuit for the laundry receptacle(s). It can't serve any other outlet(s), such as lighting, and can serve only receptacle outlets in the laundry area, in agreement to the 210.52(F) and 210.11(C)(2). In the feeder or service load calculation, must include 1,500 VA for the 20 A laundry receptacle circuit, as described in 220.52(B).

4.2.4 *Equipment, auxiliary, industrial, and motor load calculations*

Equipment, such as appliances, water heaters, washers, dryers, cooking equipment, some laboratory, industrial and commercial equipment and devices, usually used short periods of time are considered as noncontinuous loads and are included into the branch circuit loads. Such equipment can be hard-wired but quite often be cord-and-plug type connected to a receptacle. Branch circuits for appliances and equipment must have ratings equal or higher than the appliance or equipment rating. The current rating is marked on the appliance by the manufacturer or is found on the equipment nameplate. If the appliance or equipment has an electrical motor, the current rating of the branch circuit conductors must be 125% of the motor current rating.

Motor and motor load calculations are some of the most challenging calculations, performed when dealing with the NEC, IEC, and other codes and standards. While they are challenging, they are easier to understand if we deal with each section separately rather than as a whole. For the purpose of this textbook, we will concentrate on more common calculations pertaining to motors rather than try and understand the entire article of philosophy and technical details. If you have one available, you should also follow along with your NEC book. If you turn your book to Article 430, you will see that there are 13 sections that deal with all of the code requirements for motors. These sections are made easier to understand with the addition of some easy to use tables and formulas for sizing overcurrent protection, conductors, and motor full load currents. It also contains a very useful chart that gives a graphical depiction of all of the relevant code sections. Most calculations dealing with a motor circuit are based on the full load amperage (FLA) of the motor or motors connected. Therefore, it is important that we start with the correct amperage. To determine this, NEC provides several tables that are to be used to find the motor amperage. The code reference that explains this is discussed in further sections.

Highest rated or smallest rated motor are determined in compliance with Section 430.24, 430.53(B) and 430.53(C) of NEC, the highest rated or smallest rated motor shall be on the rated full-load current as are selected from Table 430.247, Table 430.248, Table 430.249, and Table 430.250. There are six NEC tables, displaying motor full-load currents, however, the two used most often are Table 430.248 and Table 430.250. Electric motor loads are caculated based on the FLA, with 25% added to allow for overload operating conditions. For load estamating purposes, the FLA for various motors are found in these tables of the NEC. In addition, for the feeders or branch circuits supplying multiple electrc motor loads, 25% of the load othe largest motor in the group must be added to the total motor load estmate. Tables 4.7–4.9, adapted from Tables 430.248 and 430.250 of the NEC are providing the FLA values for most common single-phase and three-phase AC motors.

Example 4.10: Determine the estimated load for the following single-phase motors, supplied by a dedicated branch circuit: (a) 1/3 HP at 115 V, (b) 3 HP at 230 V, and (c) 7.5 HP at 208 V.

Solution: From Table 4.7 (Table 430.248 of NEC), we can find the full-load currents and the estimated load are calculated as:

(a) $FLA = 7.2$ A, and the estimated load is: $S = 7.2\ \text{A} \times 115\ \text{V} = 828$ VA
(b) $FLA = 17.0$ A, and the estimated load is: $S = 17.0\ \text{A} \times 230\ \text{V} = 3{,}910$ VA
(c) $FLA = 44.0$ A, and the estimated load is: $S = 44.0\ \text{A} \times 208\ \text{V} = 9{,}152$ VA

Table 4.7 Full-load current (A) for single-phase AC motors (Table 430.248 of the NEC); Listed volatges are the rated motor voltages, and the listed currents are permitted for system voltage range of 110–120 and 220–240 V. FLA values are for motor running at usual speeds and motors with normal torque charactersitics

Horsepower	115 V	200 V	208 V	230 V
1/6	4.4	2.5	2.4	2.2
1/4	5.8	3.3	3.2	2.9
1/3	7.2	4.1	4.0	3.6
1/2	9.8	5.6	5.4	4.9
3/4	13.8	7.9	7.6	6.9
1.0	16.0	9.2	8.8	8.0
1.5	20.0	11.5	11.0	10.0
2.0	24.0	13.8	13.2	12.0
3.0	34.0	19.6	18.7	17.0
5.0	56.0	32.2	30.8	28.0
7.5	80.0	46.0	44.0	40.0
10.0	100.0	57.5	55.0	50.0

Example 4.11: Determine the estimated load for the following three-phase motors, supplied by a dedicated branch circuit: (a) 30 HP at 460 V, induction motor, (b) 20 HP at 200 V, (c) 200 HP, 2800 V synchronous motor, and (d) 10 HP at 208 V, induction motor.

Solution: From Tables 4.6 and 4.7 (Table 430.50 of NEC), we can find the full-load currents and the estimated load are calculated as:

(a) $FLA = 40.0$ A, and the estimated load is: $S = 40.0\ \text{A} \times 460\ \text{V} = 18{,}400$ VA
(b) $FLA = 62.1$ A, and the estimated load is: $S = 62.1\ \text{A} \times 200\ \text{V} = 12{,}420$ VA
(c) $FLA = 400.0$ A, and the estimated load is: $S = 400.0\ \text{A} \times 230\ \text{V} = 92{,}000$ VA
(d) $FLA = 30.8$ A, and the estimated load is: $S = 30.8\ \text{A} \times 208\ \text{V} = 6{,}406.4$ VA

4.2.5 Heating, cooling, electric cooking, and laundry equipment

Section 220.56 of the NEC is discussing the commercial cooking equipment related load calculations. Total feeder or panel load is simply the nameplate ratings of such type of appliances, with derating factors (listed in Table 4.10) applied in the case of three or more cooking equipment units. However, the branch circuit loads are not derated using these values. Oven, fryers, grills, food heaters, large vat blending machines, large mixers, booster heaters, conveyors, and try assembles are considered kitchen equipment and may be derated in agreement with Table 220.56. Auxiliary equipment, such as exhaust fans, space heaters, and air conditioning units are not included in this category and cannot be derated. Notice that derating factors are applied to kitchen equipment that has either thermostatic control or intermittent use.

Table 4.8 Full-load current (A) for three-phase squirrel-cage induction motors (Table 430.250 of the NEC); Listed volatges are the rated motor voltages, and the listed currents are the ones permitted for system voltage range of 110–120, 220–240, 440–480, and 550–660 V. FLA values are for motor running at usual speeds for belted motors and motors with normal torque charactersitics

Horsepower	115 V	200 V	208 V	230 V	460 V	575 V	2,300 V
1/2	4.4	2.5	2.4	2.2	1.1	0.9	—
1/3	6.4	3.7	3.5	3.2	1.6	1.3	—
1.0	8.4	4.8	4.6	4.2	2.1	1.7	—
1.5	9.8	6.9	6.6	6.0	3.0	2.4	—
2.0	12.0	7.8	7.5	6.8	3.4	2.7	—
3.0	13.6	11.0	10.6	9.6	4.8	3.9	—
5.0	—	17.5	16.7	15.2	7.6	6.1	—
7.5	—	25.3	24.2	22.0	11.0	9.0	—
10	—	32.2	30.8	28.0	14.0	11.0	—
15	—	48.3	46.2	42.0	21.0	17.0	—
20	—	62.1	59.4	54.0	27.0	22.0	—
25	—	78.2	74.8	68.0	34.0	27.0	—
30	—	92.0	88.0	80.0	40.0	32.0	—
40	—	120.0	114.0	104.0	52.0	41.0	—
50	—	150.0	143.0	130.0	65.0	52.0	—
60	—	177.0	169.0	154.0	77.0	62.0	16.0
75	—	221.0	211.0	192.0	96.0	77.0	20.0
100	—	285.0	273.0	248.0	124.0	99.0	26.0
125	—	359.0	343.0	312.0	156.0	125.0	31.0
150	—	414.0	396.0	360.0	180.0	144.0	37.0
200	—	552.0	528.0	480.0	240.0	192.0	49.0
250	—	—	—	—	302.0	242.0	60.0
300	—	—	—	—	361.0	289.0	72.0
350	—	—	—	—	414.0	336.0	83.0
400	—	—	—	—	477.0	382.0	95.0
450	—	—	—	—	515.0	412.0	103.0
500	—	—	—	—	590.0	472.0	118.0

Note: For 80% and 90% power factor, the values must be multiplied with 1.25 and 1.1, respectively.

Electric cooking equipment usually consists of electric ranges, counter-mounted cooktops, or well-mounted electric ovens. The counter-mounted cooktops and well-mounted electric ovens are separated units and are permitted to be supplied from the branch circuit; however, it is preferable from separate circuits. Free-standing electric ranges have both a cooktop and an oven built in a single unit and therefore is serviced from a separate branch circuit, determined based on the estimated load. The ratings for branch circuits supplying counter-mounted cooktops or well-mounted electric ovens are based on the estimated load of the individual units. For purpose of load estimation, electric ranges are considered individually. A counter-mounted cooktop and up two wall-mounted electric ovens supplied from the same branch circuit may be considered as a single unit for load

Table 4.9 Full-load current (A) for three-phase synchronous motors, running at unity power factor (Table 430.250 of the NEC); Listed volatges are the rated motor voltages, and the listed currents are permitted for system voltage range of 110–120, 220–240, 440–480, and 550–660 V. FLA values are for motor running at usual speeds for belted motors and motors with normal torque charactersitics

Horsepower	230 V	460 V	575 V	2,300 V
25	53.0	26.0	21.0	—
30	63.0	32.0	26.0	—
40	83.0	41.0	33.0	—
50	104.0	52.0	42.0	—
60	123.0	61.0	49.0	12.0
75	155.0	78.0	62.0	15.0
100	202.0	101.0	81.0	20.0
125	253.0	126.0	101.0	25.0
150	302.0	151.0	121.0	30.0
200	400.0	201.0	161.0	40.0

Table 4.10 Feeder demand factor for multipieces of commercial kitchen equipment

Equipment units	Demand factor (%)
1–2	100
3	90
4	80
5	70
6+	65

estimation; the equivalent rating is the sum of the individual ratings. The demand load for the household electric cooking appliances is taken from Table 4.8, adapted from the Table 220.19 of the NEC. NEC recommendations for equally rated ranges over 12 kW but not more than 27 kW, an adjustment of 5% for each additional kW of rating (including major fractions) is added to the maximum demand in Column C, while for unequally rated ranges, an average value is computed by adding all range ratings to obtain the total connected load (by using 12 kW for any range rated less than 12 kW) and dividing by the range numbers, then the maximum demand (Column C) is increased by adding 5% for each kW or major fraction exceeding 12 kW. For power ratings between 1.75 and 8.75 kW, it is permissible to add nameplate ratings of all household cooking appliances in this rating range and multiply the sum by the demand factors specified in Columns A and B (Table 4.11). It is also permissible, according to the NEC to calculate the branch circuit load for one range using Table 220.19, using nameplate ratings, adding them for more than one and considering the total rating.

Table 4.11 Household cooking appliance over 1.75 kW, demand loads. Adapted from Table 220.18 of the NEC

Number of appliances	A. Demand load (%) less than 3.5 kW	B. Demand load (%) 3.5–8.75 kW	C. Maximum load (kW) Not over 12 kW
1	80	80	8
2	75	65	11
3	70	55	14
4	66	50	17
5	62	45	20
6	59	43	21
7	56	40	23
8	53	36	23
9	51	35	24
10	49	34	25
11	47	32	26
12	45	32	27
13	43	32	28
14	41	32	29
15	40	32	30
16	39	28	31
17	38	28	32
18	37	28	33
19	36	28	34
20	35	28	35
21	34	26	36
22	33	26	37
23	32	26	38
24	31	26	39
25	30	26	40
26–30	30	24	15 kW + 1 kW
31–40	30	22	For each range
41–50	30	20	25 kW + 0.75 kW
51–60	30	18	For each range
61 and over	30	16	

Example 4.12: Determine the estimated load for a branch circuit supplying: (a) both a 4.5 kW counter-mounted cooktop and a 6 kW wall-mounted oven; and (b) both a 7.5 kW counter-mounted cooktop and an 8 kW wall-mounted oven.

Solution: (a) It is permitted to consider both cooking appliances as a single unit with the combined rating of $4.5 + 6 = 10.5$ kW. From Column C of Table 4.8, treating as a single range of 10.5 kW, the estimated load demand is 8 kW. (b) Again it is permitted to consider both as a single range (unit) for the purpose of determining the load demand. The combined rating is $7.5 + 8 = 15.5$ kW, and the excess over 12 kW is 3.5 kW, meaning an increase of 17.5% added to the 8 kW. The estimated load demand is $8 \times 1.175 = 9.4$ kW for the units.

Example 4.13: Determine the estimated demand load for following household equipment: (a) 15 kW range, (b) 8 kW oven, (c) 20 kW range, (d) 3.5 kW electric dryer, and (e) 7.5 kW electric dryer.

Solution:

(a) The rating is 3 kW in excess of 12 kW, therefore the demand load is:

$$\textit{Demand load} = 8\text{ kW} + 8\text{ kW} \times 0.15(15\%) = 9.2\text{ kW}$$

(b) The branch circuit demand load in this case is the unit rated power, or 8 kW.
(c) For a 20 kW range, there is an excess of 8 kW of 12 kW, therefore the demand load is:

$$\textit{Demand load} = 8\text{ kW} + 8\text{ kW} \times 0.40(40\%) = 11.2\text{ kW}$$

(d) A minimum of 5 kW apply even the actual power rating is 3.5 kW.
(e) The actual dryer power rating applies, since is larger than 5 kW, the demand load is 7.5 kW.

According to the NEC Section 220.18, the required demand load for household clothes dryer is equal to 5,000 VA or nameplate, whichever is larger, and the branch circuit must have a rating sufficient to supply the dryer load. Demand factors for more than one dryer connected to a feeder or service is shown in Table 4.12, adaptation of the Table 220.54 of the NEC.

Table 4.12 Demand factors for household electric clothes dryers. Adapted from Table 220.54 of the NEC

Number of dryers	Demand factors (%)
1	100
2	100
3	100
4	100
5	80
6	70
7	65
8	60
9	50
10	47
11	47% minus 1% for each dryer exceeding 11
12–23	35% minus 0.5% for each dryer exceeding 23
24–42	43 and over 25%
>42	

Example 4.14: Determine the estimated load for two household dryers: (a) one 4.5 kW and (b) another rated at 8 kW.

Solution:

(a) The minimum load demand of 5,000 VA is applying, even the nameplate ratings is 4.5 kW.
(b) The nameplate rate for the second dryer is 8 kW and according to NEC this is also applying, being larger than 5,000 VA.

NEC requires that regardless of the building or facility type, commercial, industrial, or residential, the heating loads are computed at the rated value of the nameplate unit, while the supplying branch circuit requires specific considerations. Article 424 of the NEC covers fixed electric space heating equipment, including central heating, boilers, heaters cables, and unit heaters (baseboard, panel, and duct heaters). NEC also requires disconnects for heater and motor controllers and supplementary overcurrent protection for any fixed electric space heating units. The nameplate rating is used to determine the load and if the current operate more than 3 h (continuous load), its rating is increased by a factor of 1.25. Fixed electric space heating are considered continuous loads. Similar, air-conditioning system ratings are determined from the nameplate values. The nameplate rated current is used to size the branch circuit supplying the unit. However, the recommendation is to use the larger of the branch circuit current rating or the full-load current rating to size the branch circuit. If a unit has two or more motors, the circuit rating is computed at 125% of the largest motor plus the sum of other motors, while for a single motor unit disconnect, we must use a 115% of the full-load current.

4.2.6 Load and correction factors estimate applications

Receptacle outlets (if known) and fixed multioutlet assemblies (if any) are entered into the calculation and if the load is great enough, a demand factor is applied. As with the dwelling-unit load calculations, general lighting is computed using outside dimensions, as discussed in the subsection of the previous chapter. Other items, such as a sign outlet (where required) and show window(s) (if present) are part of the nondwelling load calculation. Continuous loads are very important analysis and design computation parameter. All continuous load ratings must be *increased by 25% for load estimate purpose*. While kitchen equipment is not included in every type of calculation, be aware that kitchen equipment is not limited to restaurants. For instance, kitchen equipment could be a portion of a load calculation for a school. All other loads are referring to any load not included otherwise in the calculation. The load calculation form has little room for listing individual items, such as motors, equipment, etc. Depending upon the size of the occupancy, the calculation could contain hundreds, if not thousands, of individual items. Utilization factor, another load computation and characterization parameter is the time that equipment is in use divided by the total time that it could be in use. In normal operating conditions, the power consumption of a load is sometimes less than that indicated as its nominal power rating, a fairly common occurrence that justifies the application of a utilization factor (ku) in the estimation of realistic values. For example, a motor may only be used for 8 h/day, 50 weeks/year. The hours of operation would then be 2,000 h, and the motor utilization factor for a base of 8,760 h/year would be 2,000/8,760 = 22.83%. With a base of 2,000 h/year, the motor utilization factor would be 100%. The bottom line is that the use factor is applied to get the correct number of hours that the motor is in use. In an industrial installation this factor is estimated on an average at 0.75 for motors. For incandescent-lighting loads, the factor always equals 1, while for socket-outlet circuits, the factors depend on the type of appliances being supplied from the sockets concerned. IEC recommendations for estimating the diversified peak demand of residential building consists of multidwelling units are:

1. Illumination: 50% of total connected load.
2. Small appliance circuits: 100% of the rated load for maximum outlet wattage in the circuit plus 40% of the total connected loads of other outlets in the circuit.
3. Fixed appliance circuits and fixed electric ranges: 100% of the rated load of largest equipment plus 50% of the rated load for the 1st equipment following the largest one plus 33% of the 2nd equipment following the largest load plus 20% of the total connected load of other equipment.
4. Electric water heaters: 100% of the rated load of largest equipment plus 100% of the rated load for the 1st equipment following the largest one plus 25% of the total connected load of other equipment.
5. Air-conditioning units: 100% of the total connected load in all cases.

This way of calculating the DF is per standard and does NOT depend on social category or environment.

Example 4.15: Compute the diversity factor for the connected loads of a high load diversity apartment of an area of 200 m^2. Loads are given in the table below.

Solution: The load values are given in the above table. Applying IEC recommendations for the connected loads of the high load diversity apartment building gives:

$$L_1 = 0.5 \times 6{,}000 = 3{,}000 \text{ VA}$$
$$L_2 = 1{,}000 + 0.4(500 + 900 + 2{,}000) = 2{,}360 \text{ VA}$$
$$L_3 = 2{,}500 + 0.5 \times 2{,}500 + 0.33 \times (2{,}000 + 2{,}000 + 1{,}000) = 5{,}400 \text{ VA}$$
$$L_4 = 1{,}500 \text{ VA}$$
$$L_5 = 9{,}000 \text{ VA}$$
$$P_{max} = 3{,}000 + 2{,}360 + 5{,}400 + 1{,}500 + 9{,}000 = 22{,}160 \text{ W}$$
$$\sum \text{Connected load demand} = 30{,}900 \text{ VA}$$
$$Div\,F = \frac{22{,}160}{30{,}900} = 0.72$$

This is a very common value for the diversity factor in residential buildings. Notice that the more the equipment is included, the lower the DF is.

Load type	Specifications	Unit load	Total load
L_1: General	Lighting and general use receptacles	30 VA/m^2 at 200 m^2	6,000 VA
L_2: Small appliance circuits	Vacuum cleaner	1000 W	4,400 VA
	Refrigerator	500 W	
	Small oven	900 W	
	Kitchen appliances	2,000 W	
L_3: Fixed appliance circuits	Washing machine	2,500 W	10,000 VA
	Dishwasher	2,500 W	
	Fixed appliance	2,000 W	
	Water heater	2,000 W	
	Ironing	1,000 W	
L_4: Electric cooker	Electric cooker	1,500 W	1,500 VA
L_5: HVAC	2 Units @ 4,500VA	4,500 VA	9,000 VA

To understand this concept, if we have a 15 A circuit, the safe operating amperage would be no greater than 12 A. The total wattage would be 1,800 W, means the safe wattage usage would be 1,440 W If you had a 1,100 W hair dryer plugged into this circuit, noticing that just one device uses almost the entire desired load capabilities. For a 20 A circuit, the safe operating amperage would be no greater than 16 A. If the total wattage is 2,400 W, the safe wattage usage is 1,920 W. In this instance, a hair dryer, radio, and electric razor running can be connected on the same circuit but not much else, and this way should be additional bathroom circuits to cover lighting, exhaust fans, and heat lamps for drying. On a 30 A circuit, the safe operating amperage would be no greater than 24 A. The total wattage is 3,600 W, means the safe wattage usage would be 2,880 W. Similar information comes with central air conditioners, electric dryers, electric ranges, and electric ovens. Table 220.3(A) applies to general lightning loads. The load from general-purpose receptacle outlets in habitable areas of dwelling units, motels is permitted to be included with general lighting load for the purpose of load estimate. For office building and banks, an additional load of 1 VA/ft^2 shall be included, if the actual number of general-purpose receptacles is unknown. The designer may not know at the time of design what the final layout is, so this extra provision for load must be included. Section 220.3(B) of the NEC requires a minimum of 180 VA be assigned to other general-use receptacle outlets that are not located in habitable areas or connected to a dedicated branch circuit. The estimated load is usually greater than the required NEC minimum of 180 VA. We must use common sense and judgment while estimating loads on circuits where large loads might be connected. In addition to lighting and receptacle loads, other loads may be present in the occupancy. These loads may consist of electric heating, electric hot water systems, electrical motors and other loads, and/or other receptacle loads. For nonelectrical motor operated appliances and equipment, the apparent power ratings are calculated by simply multiplying the actual voltage and current ratings of the device. Motor loads are calculated using the full-load current rating with the addition of 25% to allow slight overload conditions on the motor. NEC specifications can be used for load estimated purpose of the electric motors (the full-load current). In addition, for feeders or branch circuits supplying multiple motor loads, a 25% of the load of the largest motor in the group must be added to the total motor load.

4.3 Conductors and cables

Two important parameters that need to be determined to estimate the size of the electrical system that will power any building environment are the total critical loads and the total noncritical loads. In general, the electrical supply must be large enough to support the sum of these two numbers, plus the related building additional electrical loads and services. The steady-state power consumption of the loads within a building establishes the power consumption for purposes of determining electricity management, electrical costs, design or restructuring, and upgrading. However, the electrical service and the backup generator power sources, if any, that provide power to the building cannot be sized to the steady-state values, being sized to the peak power consumption of the loads, plus any derating or oversizing margins required by code or standard engineering practice. Once the total electrical capacity is estimated from the process described earlier, two critical determinations can be made: first an estimate of the electrical service needed to supply the building or facility; and the second is the size of any standby generator capacity needed to achieve the desired availability. Component parts of an electric circuit or service and their protection are determined such that all normal and abnormal operating conditions are satisfied and electricity is delivered at the highest level of quality and reliability. The cabling and its protection at each level must satisfy several conditions at the same time, in order to ensure a safe and reliable installation, such as carry the permanent full-load current and normal short-time overcurrents; not have voltage drops that are resulting in an inferior performance of certain loads; protect the cabling and busbars for all levels of overcurrent, up to and including short-circuit currents; and ensure personnel protection against indirect contact hazards. The methods for the calculation of conductor cross-sectional areas are described in sections of the next chapter. Apart from this method, some national standards may prescribe a minimum cross-sectional area to be observed for reasons of mechanical endurance. Particularly, some loads may require that the cable supplying them must be oversized, and that the protection of the circuit be likewise modified.

4.3.1 Conductor types and sizes

A wire is metal drawn or rolled to long lengths, normally understood to be a solid wire. Wires may or may not be insulated. A conductor is one or more wires suitable for carrying electric current. Often the term wire is used to mean conductor. Conductors used in electrical power distribution and building electrical systems are made of copper or aluminum (see Table 4.13 for material characteristics).

The conductors, usually from copper or aluminum, may be solid or stranded, depending on the size and the required flexibility. They are packaged in several ways to form electric power cables. The most common cable construction type for low-voltage

Table 4.13 Nominal or minimum properties of conductor wire materials. Adapted from IEC standard

Cable type/ property	Int. annealed copper stranded	Hard-drawn copper wire	Standard 1350-H19 aluminum wire	Galvanized steel core wire	Aluminum clad steel
Resistivity (20 °C Ω in^2/1,000 ft.)	0.008145	0.008397	0.01331	0.101819	0.04007
Thermal resistivity coefficient per °C	0.00393	0.00381	0.00404	0.00327	0.00360
Density at 20 °C lb./in^3	0.3212	0.3212	0.0977	0.2811	0.2381
Linear expansion coefficient 10^{-6} per °C	16.9	16.9	23.0	11.5	13.0

(600 V or less) power is the single-conductor cable covered with single layer insulation, or by an outer nylon jacket. They are typically installed in conduit or other suitable raceway systems. Insulation materials are usually extruded along the electrical conductors to provide insulation between the conductors and the cable. Several variations exist with respect to the number and size of the conductors in the cable assembles. The thickness of the insulating material is generally determined by the voltage rating of the cable. Common voltage classes for cables are 600 V, 2 kV, 5 kV, 15 kV, 25 kV, and 35 kV. Various insulation thicknesses are permitted with a given class.

A multiconductor cable is an assembly of two or more conductors (often four conductors), having an outer jacket and insulation for easy installation or replacement. Utilities use aluminum for almost all new overhead installations. However, aluminum alloys are used only in larger conductors, provided that it is an approved alloy. Copper has very low resistivity and is widely used as a power conductor, although its use as an overhead conductor has become rare because copper is heavier and more expensive than aluminum. It has significantly lower resistance than the aluminum ones, so a copper conductor has equivalent ampacity (resistance) of an aluminum conductor that is two American wire gage (AWG) system sizes larger. Copper has very good resistance to corrosion. It melts at 1,083 °C, starts to anneal at about 100 °C, and anneals most rapidly between 200 °C and 325 °C, depending on the presence of impurities and amount of hardening. Different sizes of conductors are specified with gage numbers or area in circular mils. Smaller wires are normally referred to using the AWG system. The gage is a numbering scheme that progresses geometrically. A number 36 solid wire has a defined diameter of 0.005 in. (0.0127 cm), and the largest size, a number 0000 (referred to as 4/0 and pronounced "four-ought") solid wire has a 0.46-in. (1.17-cm) diameter. The larger gage sizes in sequence of increasing conductor size are: 4, 3, 2, 1, and 0 (1/0), 00 (2/0), 000 (3/0), 0000 (4/0). Going to the next bigger size (smaller gage number) increases the diameter by 1.1229. Some other useful rules are (a) an increase of three gage sizes doubles the area and weight and halves the dc resistance; and (b) an increase of six gage sizes doubles the diameter.

The smallest wire size used in power distribution networks and building electrical systems is #14 AWG, consisting typically of a single conductor having outside diameter of 0.0641 in., or 64.1 mils (one mil is equal to 1/1000 of an inch). The largest AWG designation is #4/0 AWG, having a diameter of 0.072 in. or 72 mils. Each individual strand of a seven-strand #4/0 AWG conductor has a diameter of 173.9 mils. Larger conductors are specified in circular mils of cross-sectional area. One circular mil is the area of a circle with a diameter of one mil. Conductor sizes are often given in kcmil, thousands of circular mils. In the past, the abbreviation MCM was used, which means thousands of circular mils (M is thousands, not mega, in this case). By definition, a solid 1000-kcmil wire has a diameter of 1 in. The diameter of a solid wire in mils is related to the area in circular mils by $d = \sqrt{A}$.

Outside America, most conductors are specified in square millimeter. Some useful conversion relationships are:

$$1\ \text{kcmil} = 1000\ \text{cmil} = 785.4 \times 10^{-6}\ \text{in}^2 = 0.5067\ \text{mm}^2 \tag{4.9}$$

Stranded conductors have better flexibility than the regular ones. Typically a two-layer arrangement has 7 wires, while a three-layer arrangement has 19 wires. A four-layer arrangement has 37 wires. The cross-sectional area of a stranded conductor is the metal cross-sectional area, having a larger diameter than a solid conductor of the same area. The ACSR conductor area is defined by the conductor aluminum area. Utilities with heavy tree cover often use covered conductors, with a thin insulation covering, which is not rated for full conductor line-to-ground voltage, being thick enough to reduce the chance of flashover, if a tree branch falls between conductors. Covered conductors are also called tree wire or weatherproof wire. Tree wire also helps with animal faults and allows utilities to use armless or candlestick designs or other tight configurations. Tree wire is available with a variety of covering types. The insulation materials polyethylene, XLPE, and EPR are most common. Insulation thicknesses typically range from 30 to 150 mils (1 mil = 0.001 in. = 0.00254 cm). From a design and operating viewpoint, covered conductors must be treated as bare conductors according to the National Electrical Safety Code (NESC) or IEEE C2-2000, with the only difference that tighter conductor spacing are allowed. Some of the conductor characteristics of aluminum and copper conductors are given in Table 4.13.

Multicore and multiconductor cables are the ones comprising more than one conductor, which may eventually include bare conductors. The term three-core cable is used to designate the cable making up the phases of a three-phase system. A single-core cable comprises a single insulated conductor. The term single-core cable is used to designate a cable making up one of the phases of a three-phase system. A wiring system is an assembly made up of one or more electric conductors and the devices ensuring their fixation and, if necessary, their mechanical protection. Conductor sizes are specified in terms of American Gauge Wire (AWG) or thousand circular mills (kcmil). The AGW designation for conductors ranges from #14 up through #4/10, while the kcmil designation is used from sizes from 250 up to 2,000 kcmil. For power distribution circuits including feeders and branch circuits, wire sizes in the range of #14 AWG to 500 kcmil are the most commonly specified. Wire sizes larger than 500 kcmil are frequently avoided due to difficulty with installation into the raceways and conduits. The electrical resistance of an electrical conductor is a measure of the difficulty to pass an electric current through that conductor. The inverse quantity is electrical conductance and is the ease with which an electric current passes. Electrical resistance shares some conceptual parallels with the notion of mechanical friction. The SI unit of electrical resistance is the Ohm (Ω), while electrical conductance is measured in Siemens (S). The resistivity and conductivity are proportionality constants, and therefore depend only on the material the wire is made of, not the geometry of the wire. Resistivity is a measure of the material's ability to oppose electric current. The conductor resistance has an effect on the current-carrying capability is given by:

$$R = \frac{\rho l}{A} = \frac{l}{\sigma A} \tag{4.10}$$

Here, $\sigma = 1/\rho$ is the material conductivity, while ρ is the conductor material resistivity, in Ω·m or Ω·cmil/ft. The resistivity values at 25 °C are 2.65×10^{-8} Ω·m, or 17.291 Ω·cmil/ft. for hard-drawn aluminum, and 1.724×10^{-8} Ω·m, or 10.571 Ω·cmil/ft. for hard-drawn copper, respectively. Conductor resistance changes with temperature, expressed by the following relationship:

$$R_{t2} = R_{t1} \frac{M - t_2}{M - t_1} \tag{4.11}$$

where R_{t2} is the resistance at temperature t_2 given in °C, R_{t1} = resistance at temperature $t1$ given in °C, and M is the temperature coefficient for the given material, 228.1 for aluminum, and 241.5 for annealed hard-drawn copper. For a wide range of temperatures, resistance rises almost linearly with temperature for both aluminum and copper. The resistivity, so the resistance of a material being temperature dependent, is expressed as:

$$R = R_{25}[1 + \alpha(T - 25°)] \tag{4.12}$$

where R is the resistance at the new temperature, R_{25} is the resistance of the conductor at 25 °C, T is the new temperature in Celsius degrees, and α is the temperature coefficient of the resistivity, 0.00385 for copper, and 0.00395 for aluminum, respectively.

Example 4.16: Determine the resistance of conductor of 100-m-long, having the diameter 0.5 cm at 25 °C, (a) made from hard-drawn copper and (b) hard-drawn aluminum. Determine the resistances of these conductors at 45 °C.

Solution: The area of both conductors is:

$$A = \frac{\pi D^2}{4} = \frac{3.14 \times \left(0.5 \times 10^{-2}\right)^2}{4} = 0.0019625\ \text{m}^2$$

From (4.11), the resistances are:

$$R_{Cu} = \frac{1.724 \times 10^{-8} \cdot 100}{0.0019625} = 0.000879\ \Omega$$

$$R_{Al} = \frac{2.65 \times 10^{-8} \cdot 100}{0.0019625} = 0.001351\ \Omega$$

By using (4.12), we can estimate the resistances at 45 °C, as:

$$R_{Cu(45)} = 0.000879 \cdot [1 + 0.00385 \cdot (45 - 25)] = 0.000947\ \Omega$$

$$R_{Al(45)} = 0.0013510 \cdot [1 + 0.00395 \cdot (45 - 25)] = 0.001458\ \Omega$$

We can also linearly interpolate using resistances provided at two different temperatures as:

$$R(t_c) = R(t_1)\frac{R(t_2) - R(t_1)}{t_2 - t_1}(t_c - t_1) \tag{4.13}$$

where $R(t_c)$ is the conductor resistance at temperature t_c, $R(t_2)$ is the resistance at the higher temperature, t_2, and $R(t_1)$ is the resistance at the lower temperature, t_1. With alternating current, skin effects raise the resistance of a conductor relative to its dc resistance. A conductor offers a greater resistance to a flow of alternating current than it does to the direct current. This increased resistance is generally expressed as the AC/DC resistance ratio. At 60 Hz, the resistance of a conductor is very close to its DC resistance except for very large conductors, and such effects are ignored in this textbook. Skin effects are much more important for high-frequency analysis, such as switching surges and power-line carrier problems, playing a larger role in larger conductors. For most distribution power-frequency applications, we can ignore skin effects (and they are included in AC resistance tables). There is a large variety of sizes and types of conductors used in power distribution and building electrical systems. Several electrical, mechanical, and economic characteristics affect conductor selection, such as:

- Ampacity, the peak current-carrying conductor capability limits the current, and the power capability.
- Economics, often a conductor operates well below its ampacity rating, the cost of the extra aluminum or copper pays for itself with lower I^2R losses. The conductor runs cooler, leaving also room for expansion.
- Mechanical strength, especially on lines with long span lengths, plays an important role in size and type of conductor. Stronger conductors like ACSR are used more often, while ice and wind loadings must be also take into account.
- Corrosion, while not usually a problem, sometimes can limit certain types of conductors in certain applications.

4.3.2 Cable impedance calculations

This chapter section provide details on the calculation of cable impedances—DC resistance, AC resistance, and inductive reactance. The DC and AC resistance of cable conductors can be calculated based on IEC 60287-1 Clause 2.1. The DC cable conductor resistance is calculated by using:

$$R_{DC} = \frac{1.02 \times 10^6 \rho_{20}}{A}[1 + \alpha_{20}(\theta - 20°)] \tag{4.14}$$

Here, A is the conductor cross-sectional area (mm^2), R_{DC} is the DC resistance at the conductor operating temperature θ (Ω/m), ρ_{20} is the conductor resistivity, at 20 °C (Ω.m). For copper conductors, $\rho_{20} = 1.7241 \times 10^{-8}$, while for the aluminum conductors, $\rho_{20} = 2.8264 \times 10^{-8}$, α is the temperature coefficient of the conductor material per K at 20 °C (for copper conductors, $\alpha = 3.93 \times 10^{-3}$, while for aluminum conductors, $\alpha = 4.03 \times 10^{-3}$), and θ is the conductor operating temperature (°C).

The AC resistance of cable conductors is the DC resistance corrected for skin and proximity effects. *Skin effect* is the AC current tendency to become distributed within a conductor such that the current density is largest near the surface of the conductor, and decreases with conductor depths. The electric current flows mainly at the "skin" of the conductor, between the outer surface and a level called the *skin depth*. The skin effect causes the effective resistance of the conductor to increase at higher frequencies where the skin depth is smaller, thus reducing the effective cross-section of the conductor. The skin effect is due to opposing eddy currents induced by the changing magnetic field resulting from the alternating current. At 60 Hz in copper, the skin depth is about 8.5 mm. In a conductor carrying alternating current, if currents are flowing through one or more other nearby conductors, such as within a closely wound coil of wire, the distribution of current within the first conductor will be constrained to smaller regions. The resulting current crowding is termed as the proximity effect. This crowding gives an increase in the effective resistance of the circuit, which increases with frequency. The AC conductor resistance is given by:

$$R_{AC} = R_{DC}(1 + y_{skf} + y_{prf}) \tag{4.15}$$

Here R_{AC} is the AC resistance at the conductor operating temperature θ (Ω/m), R_{DC} is the dc resistance at the conductor operating temperature θ (Ω/m), y_{skf} is the skin effect factor, and y_{prf} is the proximity effect factor. Skin effect and proximity effect factors, for a current frequency f (Hz), diameter D_{con} (mm), and the distance between conductor axes, x (mm) are calculated with the following relationships:

$$y_{skf} = \frac{x_s^4}{192 + 0.8x_s^4} \tag{4.16}$$

And for 2C and 2 × 1C cables:

$$y_{prf} = \frac{x_p^4}{192 + 0.8x_p^4}\left(\frac{D_{con}}{s}\right)^2 \times 2.9 \tag{4.17a}$$

Table 4.14 Skin effect and proximity effect factors

Type of conductor	Dried and impregnated?	k_s	k_p
Copper			
Round, stranded	Yes	1	0.8
Round, stranded	No	1	1
Round, segmental	—	0.435	0.37
Sector-shaped	Yes	1	0.8
Sector-shaped	No	1	1
Aluminum			
Round, stranded	Either	1	1
Round, 4 segment	Either	0.28	0.37
Round, 5 segment	Either	0.19	0.37
Round, 6 segment	Either	0.12	0.37

While for 3C and 3 × 1C cables:

$$y_{prf} = \frac{x_p^4}{192 + 0.8x_p^4}\left(\frac{D_{con}}{s}\right)^2 \left[0.312\left(\frac{D_{con}}{s}\right)^2 + \frac{1.18}{\dfrac{x_p^4}{192 + 0.8x_p^4} + 0.27}\right] \tag{4.17b}$$

Here:

$$x_s^4 = \left(\frac{8\pi f}{R_{DC}} k_s \times 10^{-7}\right)^2 \text{ and } x_p^4 = \left(\frac{8\pi f}{R_{DC}} k_p \times 10^{-7}\right)^2 \tag{4.18}$$

The factors k_s, and k_p used in (4.17) are given in Table 4.14. The above relationships for skin and proximity effects are accurate provided that k_s and k_p are lower or equal to 2.8. For shaped conductors, the proximity effect factor is two-thirds the values calculated above, and with D_{con} equal to the diameter of an equivalent circular conductor of equal cross-sectional area and degree of compaction (mm), and:

$$s = D_{con} + t$$

where t is the thickness of the insulation between conductors (mm).

The cable series inductive reactance X_C (Ω/km) can be approximated by the following equation:

$$X_C = 2\pi f \left[K + 0.2 \ln\left(\frac{2s}{D_{con}}\right)\right] \times 10^{-3} \tag{4.19}$$

Here s is the axial spacing between conductors (mm), D_{con} is the diameter of the conductor. For shaped conductors, the diameter of an equivalent circular conductor of equal cross-sectional area and degree of compaction (mm), K is a constant factor pertaining to conductor formation, with typical values given in Table 4.15. For 3C and 3 × 1C cables, the axial spacing parameter depends on the geometry of the conductors, as shown in Figure 4.1.

Table 4.15 Typical K values

Number of wire strands in conductor	*K*
3	0.0778
7	0.0642
19	0.0554
37	0.0528
>60	0.0514
1 (solid)	0.0500

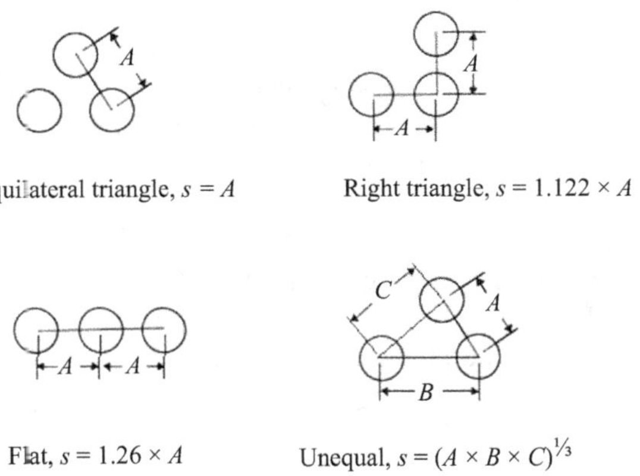

Figure 4.1 Typical three-wire cable configurations

4.3.3 Conductor ampacity

The ampacity is the maximum designed current of a conductor, and is given in amperes. A conductor has several ampacities, depending on its application and the assumptions used. It is also determined by the weather conditions for overhead cables. When higher current flows, it results in higher conductor and insulation temperatures as the conductor power losses increase. One limiting factor of the conductor ampacity is the current need to bring the conductor to a certain temperature. Internal generated heat, due to power losses must be dissipated for the temperature to reach the equilibrium condition. Moreover, different types of insulation designations reflect their ability to withstand various temperatures, such as 60 °C, 75 °C, or 90 °C. The conductor and insulation ability to dissipate heat is affected by the ambient temperature and the proximity of other current-carrying conductors. Higher ambient temperatures reduce the conductor heat dissipation capability, while the current-carrying conductors in close proximity to one another generate heat and rise the surface temperature of the conductors.

The most used tables of the NEC used to determine the conductor ampacity are 310.16 and 310. 17 are included in Appendix B of this book. These tables are listing the conductor ampacity for copper and aluminum conductors, for an ambient temperature of 30 °C. First table applies where there are no more than three current-carrying conductors in a raceway or conduit, while Table 310.17 refers to a single conductor in free air. The temperature rating and insulation type are shown at the table head, and the conductor size designation in the left-side column. The conductor ampacity for not more than three current-carrying conductors, in a conduit or raceway or for a single current-carrying conductor in free air are read directly from Table 310.16 and Table 310.17, respectively. However, selection of proper conductor size requires the value of load current supplied by the conductor. For applications involving not more than three current-carrying conductors in a raceway and 30 °C ambient temperature, the conductor ampacity must be equal or greater than the load current.

Example 4.17: Determine the ampacity of the following conductors: (a) #8 TW copper, (b) 600 kcmil THW copper, and (c) 300 kcmil RHH aluminum. There are not more than three conductors in the raceway.

Solution:

(a) For # 8 TW copper the ampacity is 40 A.
(b) For 600 kcmil THW copper the ampacity is 460 A.
(c) For 300 kcmil RHH aluminum the ampacity is 255 A.

4.3.4 Cable corrections factors

The power transmission capacity of an insulated cable system is the product of the operating voltage and the maximum current that can be transmitted. Power transmission systems operate at fixed voltage levels so that the delivery capability of a cable system at a given voltage is dictated by the current-carrying capacity of the conductors. The delivery capability is defined as the "ampacity" of the cable system. The operating voltage determines the dielectric insulation requirements of a cable, while the conductor size is dictated by the ampacity rating. These two independent parameters (insulation and conductor size) of the cable system are inter-related by thermal considerations; a bigger conductor size (less I^2R losses) results in higher ampacity, while increases in insulation material (lower heat dissipation) results in lower ampacity. The parameters of great influence in determining ampacity are the cable size, insulation characteristics, thermal resistivity of the soil, depth of burial, for underground cables, and the horizontal spacing between the circuits. However, a given conductor can have several ampacities, depending on its application and the assumptions used. Sun, wind, and ambient temperature change a conductor's ampacity. A conductor's temperature depends on the thermal balance of heat inputs and losses. Current driven through a conductor's resistance creates heat (I^2R). The sun is another source of heat into the conductor. Heat escapes from the conductor through radiation and from convection. Considering the balance of inputs and outputs, the ampacity of a conductor is expressed as:

$$I = f\left(q_{c,r,c}, R_{ac}\right) \tag{4.20}$$

where q_c is the convected heat loss, W/ft., q_r is the radiated heat loss, W/ft., q_s is the solar heat gain, W/ft, and R_{ac} is the nominal AC resistance (at low frequencies is the same as the conductor DC resistance) at operating temperature t, W/ft. The cables are designed to operate within the NEC prescribed operating characteristics. Ampacity is the current that a conductor can carry continuously under the conditions of use without exceeding its temperature rating. If the conductors get too hot, they can burn up and short out, as the conductor heats up the current-carrying capacity goes down. If the capacity of the conductors is overloaded, they are heating up and short out. Methodology for calculation of the cable ampacities is described in the NEC articles. Conductor ampacity is presented in the tables along with factors that are applicable for different laying configurations. An alternative approach to the NEC is the use of equations for determining the cable current rating. Some of the main factors impacting ampacity are: (1) the allowable conductor temperature, the ampacity increases significantly with higher allowed temperatures; (2) ambient temperature, the ampacity increases about 1% for each 1 °C decrease in ambient temperature; and (3) for the aerial conductors, the wind speed, even a small wind helps cool conductors significantly. With no wind, ampacities are significantly lower than with a 2-ft./s crosswind. Electrical wiring systems, especially for the ones built into walls having heating elements, it is necessary to reduce current-carrying capacities by applying the reduction factors specified in codes and standards. This supposes that the temperature distribution inside the heated walls in contact with the electrical wiring system is known.

A conductor's temperature depends on the thermal balance of heat inputs and losses. Current driven through a conductor's resistance creates heat (I^2R losses). Heat escapes from the conductor through radiation and from convection. The ability of an underground cable conductor to conduct current depends on a number of factors. The most important factor of the utmost concern to the designers of electrical transmission and distribution systems or building electrical installations are the following: weather conditions, thermal details of the surrounding medium, ambient temperature, heat generated by adjacent conductors, and heat generated by the conductor due to its own losses. Once the load has been sized and confirmed, the cable system must be designed in a way to transfer the required power from the generation to the end user. The total number of overhead or underground cable circuits, their size, the layout methods, crossing with other utilities, such as roads, telecommunication, gas, or water network are of crucial importance when determining design of the cable systems. In addition, overhead, aerial, or underground cable circuits must be sized adequately to carry the required load without overheating. Heat is released from the conductor as it transmits electrical current. Cable type, its construction details and installation method determine how many elements of heat generation exist. These elements can be Joule losses (I^2R losses), sheath losses, etc. Heat generated in these elements is transmitted through a series of thermal resistances to the surrounding environment. The foregoing adjustment factors apply where all current-carrying conductors carry current continuously. Load diversity is involved in cases, such as in numerous industrial applications, for more than nine conductors in a raceway or cable, Table B310-11 (NEC) provides factors with less severe reduction in ampacities than the values shown above. Conductor sizes and types have an influence on the amount of current carried by the conductor and where the conductor is installed in close proximity to other current-carrying conductors. For practical reasons, the numbers given for the adjustment factors are not exact. However, they serve well to ensure minimum levels of safety that can be achieved by design, installation, and verification.

Example 4.18: Suppose that a cable has an ambient temperature derating factor of $k_{amb} = 0.94$, and a grouping derating factor of $k_g = 0.85$. If the base current rating is $I_b = 45$ A, what is the installed current rating?

Solution: The overall derating factor is:

$$k_{d(total)} = k_{amb} \times k_g = 0.94 \times 0.85 = 0.799$$

And

$$I_c = k_{d(total)} \times I_b = 0.799 \times 45 = 36.0 \text{ A}$$

As discussed in Section 4.3.3, and in earlier paragraphs, conductor ampacities are determined based on the ambient temperature of 30 °C. If the ambient temperature is higher than 30 °C, temperature correction factor must be applied to the table-listed ampacity to determine the derated conductor ampacity. The correction factors are found at the bottom of Table 310.16 and Table 310.17. By adapting (4.19), the derated conductor ampacity is calculated as:

$$\text{Derated ampacity} = \text{Table-listed ampacity} \times k_{Amb} \tag{4.21}$$

Here k_{Amb} is the ambient temperature correction factor, as found in the earlier-mentioned tables.

Current-carrying conductors in raceways and conduits are dissipating heat, and the more such conductors are in the proximity of each other more heat is disputed. Table 310.16 was developed on the assumption of not more than three current-carrying conductors in the raceway, which is reasonable for three-phase balanced circuits, with the load current flowing into the ungrounded conductors. If there are more than three current-carrying conductors in the raceway, a fill or grouping adjustment factor, listed in Table 4.15 must apply to reduce the table-listed conductor ampacity, as:

$$\text{Derated ampacity} = \text{Table-listed ampacity} \times k_{grp} \tag{4.22}$$

Here k_{grp} is the grouping adjustment factor, as found in Table 4.16. The number of current-carrying conductors in a raceway or conduit consists of all ungrounded

Table 4.16 Bundle correction factor. Adapted from Table 310-15(b)(2)(a): Adjustment factors for more than three current-carrying conductors in a raceway or cable [Applies also to single conductors or multiconductor cables in free air, stacked, or bundled more than 24 in. (0.61 m)]

Number of percent of current-carrying	**Values in tables conductors 310-16 or 316-17 (%)**
1–3	100
4–6	80
7–9	60
10–20	50
21–30	45
31–40	40
Above 40	35

(phase) conductors, plus any neutral conductor that is considered to carry current under normal operating conditions. For example, the neutral conductor of a three-phase, four-wire, wye-connected circuit supplying linear loads, is carrying only the unbalanced load current in the circuit and is not included in the counting. However, if this system is supplying nonlinear loads, such as motor drivers, fluorescent lamps, rectifiers, etc. that may generate harmonics that add up in the neutral, the neutral current may have enough magnitude to be counted as current-carrying conductor. Also, any equipment grounding conductor occupying the same raceway as ungrounded conductors is not counted as current-carrying conductor. When derating is required for both ambient temperature and raceway fill, a total derating factor, the product of the temperature correction factor and adjustment factor is applied. The derated conductor ampacity is now given by:

$$\begin{aligned} &\text{Derated ampacity} = \text{Table-listed ampacity} \times CF_{Total} \\ &CF_{Total} = k_{Amb} \times k_{grp} \end{aligned} \tag{4.23}$$

Example 4.19: Determine the derated ampacity of a 300 kcmil THW copper conductor used in a three-phase, four-wire circuit supplying an induction motor drive, operating at 42 °C.

Solution: Because of nonlinear load, the grounded conductor is counted as current-carrying conductor, and the raceway fill adjustment factor is 80%. For Table 310.16, the ambient temperature correction factor for this type of conductor is 0.82, while the table-listed ampacity is 285 A. The derated ampacity is given by (4.22).

$$\text{Derated ampacity} = 285 \times 0.8 \times 0.82 = 197 \text{ A}$$

4.3.5 Voltage drop calculation

Voltage drop is an important subject that unfortunately gets little attention in the NEC since it is perceived to be more important to equipment performance than to safety. Some manufacturers specify the minimum voltage to be supplied to their equipment. This becomes a requirement under NEC 110.3(B). In addition, special precautions must be taken to provide adequate voltage during startup of certain equipment utilizing a large motor. Completely ignored is energy efficiency, which is a major consideration today. NEC states in two Informational Notes, a maximum voltage drop of 3% for branch circuit or feeder conductors and 5% for branch circuit and feeder conductors together will provide reasonable efficiency of operation for general use circuits. For sensitive electronic equipment operating within its scope, NEC 647.4(D) requires that the voltage drop on any branch circuit shall not exceed 1.5%, and that the combined voltage drops of branch circuit and feeder conductors shall not exceed 2.5%. Correct voltage is critical for optimal load operation. Low operating voltage causes some equipment to draw higher than normal load current. For constant wattage loads, load current increases to make up the difference from low voltage to maintain the power output. Consequently, low voltage can cause motors and certain equipment and components to run hotter than normal, and can cause components to fail prematurely. In addition, significant voltage drop across phase and neutral conductors can result in incorrect operation of computers and other sensitive electronic equipment. Manufacturers of air conditioning and refrigeration equipment, fire pumps, submersible pumps, and many other types of motor-driven equipment specify acceptable ranges of operating voltages. During startup, motors typically draw several times their rated operating currents. Sizing conductors based only on operating currents may not provide sufficient voltage to allow motors to start. Sensitive electronic equipment is also particularly susceptible to maloperation or failure, if voltage drop is excessive. In electrical wiring circuits, voltage drops also occur from the distribution board to the different subcircuit and final subcircuits but for subcircuits and final subcircuits, the value of voltage drop should be half of that allowable voltage drops (i.e., 2.75 V of 5.5 V in the above case). Normally, voltage drop in tables is described in Ampere per meter (A/m), e.g., what would be the voltage drop in a 1 m cable carrying 1 A current? There are two methods to define the voltage drop in a cable which we will follow. In SI (International System), voltage drop is described by volt per meter. In foot pound system (FPS), voltage drop is described in V/100 ft.

Example 4.20: If the supply voltage determine the voltage drop for a combined branch circuit and feeder.

Solution: The combined voltage drop, as required by NEC 647.4 (D) is:

$$\textit{Allowable voltage drop} = 220 \times \frac{2.5}{100} = 5.5\ \mathrm{V}$$

Current flowing through a conductor with a finite resistance causes a voltage drop over the length of the conductor. Voltage drop can be calculated using a variation of the familiar Ohm's Law ($E = IR$), as follows:

$$VD = kRi(t) \tag{4.24}$$

where *VD* is the voltage drop across the length of the conductor in volts, "*k*" is a constant depending upon whether the system is single-phase or three-phase, *i*(*t*) is the load current flowing through the conductor in amperes, and *R* is the direct-current resistance of uncoated copper conductors at 75 °C (167 °F), *R* is calculated by multiplying ohms per 1,000 ft. (the value given in Table B3-II later, which is adapted from NEC Chapter 9, Table 8) by the one-way circuit length in thousands of feet. For single-phase loads, the constant "*k*" is 2. For three-phase loads, the constant "*k*" is the square root of 3, or 1.732. For simplicity, the examples in this chapter use direct current resistance *R* values, as shown in Table 9, Chapter 9 of NEC. However, if the geometrical dimensions and material of the conductor are known, (4.10) can be used. For alternating current systems which are found in practice, the values are slightly different, and for loads that are not purely resistive, impedance values *Z* should be used, taking power factor into consideration. This is also covered in NEC Chapter 9.

4.3.6 Cable construction

The basic characteristics of the cable's physical construction, which includes conductor materials, shapes, types, surface coating, insulation types, number of conductors, installation conditions and arrangements (underground or above the ground, cable bunching, and spacing), ambient temperature, soil characteristics, depth of laying, etc. Conductor material normally used are copper or aluminum, while the conductor shape, such as circular, rectangular, or square shaped. Most common conductor types are of stranded or solid types. Conductor surface coating can be plain (no coating), tinned, silver, or nickel, while the most used insulation types are PVC, XLPE, or EPR. Number of conductor cores includes single core or multicore (e.g., 2, 3, or 4 conductors) types. The installation conditions refer to how the cable will be installed, which includes above the ground or underground cables. Installation arrangements refer for example for underground cables, if it is directly buried or buried in the conduit, while for the above ground cables, if it is installed on cable tray, conduit or ladder, against a wall, by hanging on the supports, into the air, etc. Ambient or soil temperature of the installation site is important in the cable sizing, ampacity calculations, and sizing the protective devices. Cable bunching is the number of cables that are bunched together, without affecting the cable performances and characteristics. Cable spacing refers to the fact that whether cables are installed are touching or are spaced. Soil thermal resistivity is an important factor for the calculation of the characteristics of the underground cables.

Cable selection is based on the current rating estimates. Current flowing through a cable generates heat through the resistive losses in the conductors, dielectric losses through the insulation and resistive losses from the current flowing through any cable screens/shields and armoring. The component parts that make up the cable (e.g., conductors, insulation, bedding, sheath, armor, etc.) must be capable of withstanding the temperature rise and heat emanating from the cable. The current-carrying capacity of a cable is the maximum current that can flow continuously through a cable without damaging the cable's insulation and other components (e.g., bedding, sheath, etc.). It is sometimes also referred to as the continuous current rating or ampacity of a cable. Cables with larger conductor cross-sectional areas (i.e., more copper or aluminum) have lower resistive losses and are able to dissipate the heat better than smaller cables. Therefore, a 16 mm^2 cable will have a higher current-carrying capacity than a 4 mm^2 cable. International standards and manufacturers of cables will quote base current ratings of different types of cables in tables, such as the one shown on the right. Each of these tables pertains to a specific type of cable construction (e.g., copper conductor, PVC insulated, voltage grade, etc.) and a base set of installation conditions (e.g., ambient temperature, installation method, etc.). It is important to note that the current ratings are only valid for the quoted types of cables and base installation conditions. In the absence of any guidance, the following reference-based current ratings may be used. In general, there are several ways to construct a cable and not one standard to which all vendors are adhere to, most cables tend to have common characteristics, and from electrical system design point of view, the cables are divided into following main categories: (a) low voltage power and control cables pertain to electrical cables, (b) low voltage instrumentation cables pertain to cables for use in instrument applications, and (c) medium/high voltage cables pertain to cables used for electric power transmission at medium and high voltage.

Conductor consists usually of stranded copper (Cu) or aluminum (Al). Copper is denser and heavier, and better conductor than aluminum. Electrically equivalent aluminum conductors have a cross-sectional area approximately 1.6 times larger than copper but are half the weight (which may save on material cost) and lower support cost, also. *Annealing* is the process of gradual heating and cooling the conductor material to make it more malleable and less brittle. *Coating*, or the so-called surface coating (e.g., tin, nickel, silver, lead alloy) of copper conductors is common to prevent the insulation from attacking or adhering to the copper conductor and prevents deterioration of copper at high temperatures. Tin coatings were used in the past to protect against corrosion from rubber insulation, which contained traces of the sulfur used in the vulcanizing process. *Conductor screen* consists of a semi-conducting tape to maintain a uniform electric field and minimize electrostatic stresses (for MV/HV power cables). *Insulation* is critical and important characteristics of cables used in electrical systems. Commonly used materials are thermoplastic (e.g., PVC) or thermosetting (e.g., EPR, XLPE) type materials. Mineral insulation is sometimes used but the construction of such type of cables is entirely different to normal plastic/rubber insulated cables. Typically used are a thermosetting (e.g., EPR, XLPE) or paper/lead insulation for cables under 22 kV. Paper-based insulation in combination with oil or gas-filled cables, are generally used for higher voltages. *Plastics* are one of the more commonly used types of insulating materials for electrical conductors. It has good insulating, flexibility, and moisture-resistant qualities. Although there are many types of plastic insulating materials, thermoplastic is one of the most common. With the use of thermoplastic, the conductor temperature can be higher than with some other types of insulating materials without damage to the insulating quality of the material. Plastic insulation is normally used for low- or medium-range voltage. The designators used with thermoplastics are much like those used with rubber insulators. The following letters are used when dealing with NEC type designators for thermoplastics:

T—Thermoplastic
H—Heat-resistant
W—Moisture-resistant
A—Asbestos
N—Outer nylon jacket
M—Oil-resistant

Paper has little insulation value alone. However, when impregnated with a high grade of mineral oil, it serves as a satisfactory insulation for extremely high-voltage cables. The oil has a high dielectric strength, and tends to prevent breakdown of the paper insulation. The paper must be thoroughly saturated with the oil. The thin paper tape is wrapped in many layers around the conductors, and then soaked with oil. *Enamel* is the wire used on the coils of meters, relays, small transformers, motor windings, and so forth, is called magnet wire. This wire is insulated with an enamel coating. The enamel is a synthetic compound of cellulose acetate (wood pulp and magnesium). In the manufacturing process, the bare wire is passed through a solution of hot enamel and then cooled. This process is repeated until the wire acquires from 6 to 10 coatings. For thickness, enamel has higher dielectric strength than rubber. It is not practical for large wires because of the expense and because the insulation is readily fractured when large wires are bent. *Mineral-insulated* (MI) *cable* was developed to meet the needs of a noncombustible, high heat-resistant and water-resistant cable. MI cable has from one to seven electrical conductors. These conductors are insulated in a highly compressed mineral, normally magnesium oxide, and sealed in a liquid-tight, gastight metallic tube, normally made of seamless copper. *Silk and cotton* are used in certain types of circuits (for example, communications circuits), a large number of conductors are needed, perhaps as many as several hundreds. Because the insulation in this type of cable is not subjected to high voltage, the use of thin layers of silk and cotton is satisfactory. Silk and cotton insulation keeps the size of the cable small enough to be handled easily. The silk and cotton threads are wrapped around the individual conductors in reverse directions. The covering is then impregnated with a special wax compound for extra protection and strength.

Insulation screen represents a semi-conducting material that has a similar function as the conductor screen (i.e., control of the electric field for MV/HV power cables). *Conductor sheath* represents a conductive sheath/shield, typically of copper tape or sometimes lead alloy, is used as a shield to keep electromagnetic radiation in, and also provide a path for fault and leakage currents (sheaths are earthed at one cable end). Lead sheaths are heavier and potentially more difficult to terminate than copper tape but generally provide better earth fault capacity. *Filler* is the interstices of the insulated conductor bundle are sometimes filled, usually with a soft polymer material. *Bedding* or *inner sheath*, typically a thermoplastic (e.g., PVC) or thermosetting (e.g., CSP) compound, is included to keep the bundle together and to provide a bedding for the cable armor. *Individual screen*, typically for instrument cables is an individual screen which is occasionally applied over each insulated conductor bundle for shielding against noise/radiation and interference from other conductor bundles. Screens are usually a metallic (copper, aluminum) or semi-metallic (PETP/Al) tape or braid. Typically used in instrument cables but not in power cables.

Drain wire (instrument cables) are the ones that have an individual screen that has an associated drain wire, designed to assist in the termination of the screen, and are typically used in instrument cables, not in power cables. *Overall screen* (instrument cables) is represented by an overall screen applied over all the insulated conductor bundles for shielding against noise/radiation, interference from other cables and surge/lightning protection. Screens are usually a metallic (copper, aluminum) or semi-metallic tape or braid. Typically used in instrument cables but not in power cables. *Armor* is cable part designed for mechanical protection of the conductor bundle. Steel wire armor or braid is typically used. Tinning or galvanizing is used for rust prevention. Phosphor bronze or tinned copper braid is also used when steel armor is not allowed. There are two categories of cable armor: (a) steel wire armor, typically used in multicore cables (magnetic) and (b) aluminum wire armor, typically used in single-core cables (nonmagnetic). When an electric current passes through a cable, it produces a magnetic field (the higher the voltage the stronger the field), which induces an electric current in steel armor (eddy currents), which can overheat the AC systems. The nonmagnetic aluminum armor prevents this from happening. *Outer sheath* is applied over the armor for overall mechanical, weather, chemical, and electrical protection. Materials typically used are a thermoplastic (e.g., PVC) or a thermosetting (e.g., CSP) compound, and often is used the same material as the bedding. Outer sheath is normally color coded to differentiate between LV, HV, and instrumentation cables. Manufacturer's markings and length markings are also printed on the outer sheath. *Termite protection* is used for underground cables, consisting of a nylon jacket can be applied for termite protection, although sometimes a phosphor bronze tape can be used.

As discussed earlier, a multiconductor cable is an assembly of two or more conductors having an outer jacket, for holding the conductors and easy installation. Common examples are the service entrance cable and nonmetallic sheathed cable. Type AC armored cable consists of insulated conductors contained in a flexible metal raceway. There are several variations with respect of the conductor size and numbers, for example, an armored cable assembly is available with two, three, or four insulated conductors, solid or stranded, and in sizes from #14 AWG to #2 AWG. Article 320 of the NEC permits that this type of cable to be installed in dry areas, embedded in masonry or plaster, or running to the masonry hollows. Its outer armor may be used for equipment grounding. A metal-clad (MC) cable consists of three or four individual insulated conductors with one or more grounding conductors grouped together and enclosed in an outer sheath. Variations include cables containing power and control conductors in the same sheath and cables used for #8AWG or larger for easy installation. Fillers are used between conductors to keep the conductors in place within the cable. Article 320 permits MC cables to be installed indoors or outdoors, in wet (not submerged) or dry locations, as cable open runs (not exceeding 6 ft.), in cable tray, in raceway or conduit, or directly buried where listed for direct burial.

The nonmetallic (NC), sheathed cable is the most used for residential branch circuits, being available with two or three insulated conductors, and with or without a bare copper ground, jacketed with an outer polyvinylchloride sheath. It is usually rated at 90 °C with nylon jacket, having common sizes from #14 AWG to #2 AWG. Article 334 of the NEC restricted NM cables to be used in any structure more than three floors above grade, for service entrance cable, in commercial garages, or embedded in poured concrete. Underground feeded (UF) cable has a similar construction to the NM cable, being used in residential applications to supply lamp posts, pumps, or other similar outdoor equipment, as discussed in Article 340. Variations of NM cables are permitted to be used in dry, damp, moist, or corrosive locations, or consist of assembly, containing power, signaling, and other communication conductors in the same sheath. Service entrance cables, usually identified as SE, USE, SEU, or SER generally consist of two or three insulated conductors wrapped by a concentrically neutral bare, jacketed with a PVC outer covering. Article 338 describes such cables, while Article 230 of the NEC discusses service entrance requirements. Referring to cable insulation, there are two major types of insulation used for buildings, thermoplastic, and thermoset. First one is begin to melt at temperatures higher than rated, and the common examples are polyethylene and polyvinylchloride. Thermoset insulations, such as rubber are not starting to melt at temperature reasonable higher than the rated; however, they are deteriorated faster at these higher temperatures. As we mentioned, heat produced by the conductor current is transmitted to the insulation, so the conductor and the insulation temperatures are basically the same. There are typically three temperature ratings assigned to the conductor insulations: 60 °C, 75 °C, and 90 °C, and operation above these rated temperature causes premature insulation degradation. Table 4.17 summarizes the most common cable designations used in building electrical systems.

Table 4.17 Cable insulation types and designations

Cable identifier	Designation
THW-2	Thermoplastic insulation (PVC), Heat resistant (90 °C), Wet locations
THHN	Thermoplastic insulation (PVC), High heat resistant (90 °C), Dry locations, Nylon jacket
XHHW-2	Cross-linked polyethylene insulation (X), High heat resistant (90 °C), wet, and dry locations
RHH	Rubber insulation, High heat resistant (90 °C), dry locations
RHW-2	Rubber insulation, Heat resistant (90 °C), wet locations
NM-B	Nonmetallic sheathed cable, ampacity limited at 60 °C, thermoplastic, overall PVC insulation, nylon jacketed
SEU	Service entrance cable, unarmored, usually XHHW type, rated at 90 °C for dry locations, and 75 °C for wet locations

4.4 Wiring devices

This section is focusing on the description of various types of wiring devices used in power distribution to control the flow of electrical current, such as switches, receptacles, and disconnects. These devices allow the operation of other devices and equipment or are used to control and isolate the specific equipment. There are various types of wiring devices that are used in power distribution networks, such as switches, conductors, cables, receptacles, and disconnects. They are used to control the flow of electrical current and electricity to allow for desired operation of other devices and equipment. Types and ratings of wiring devices are discussed in this section. The use of disconnect switches to control and isolate electric equipment. Wiring (connection) diagram represents a diagram that shows the connection of an installation or its component devices or parts. It shows, as closely as possible, the actual location of each component in a circuit, including the control circuit and the power circuit. A line (ladder) diagram is a diagram that shows the logic of an electrical circuit or system using standard symbols, helping in operation, maintenance or restructuring of the actual electric circuits in a building or industrial facility. A line diagram is used to show the relationship between circuits and their components but not the actual location of the components. Line diagrams provide a fast, easy understanding of the connections and use of components. Selecting the correct cable for the application is imperative to ensure a satisfactory life of conductors and insulation subjected to the thermal effects of carrying current for prolonged periods of time in normal service. Choosing the minimum size cross sectional area of conductors is essential to meet the requirements for the protection against electric shock; protection against thermal effects; overcurrent protection; voltage drop; and limiting temperatures for terminals of equipment to which the conductors are connected. A properly engineered and installed cable tray wiring system provides some highly desirable safety features that are not obtained with a conduit wiring system.

4.4.1 Switches

An electrical switch is a device for making or breaking an electrical circuit. The definition suggests the ultimate in simplicity that a switch needs to be no more than the bare ends of two wires that can be touched to make circuit or separated to break circuit. There are many different types of switches: toggle, rotary, pushbutton, "rocker," "pull-chain," slide, magnetic, mercury, timer, voice-activated, "touch-sensitive," and many others. Each one with specific characteristics and performances are suitable for specific applications. The switches used in most residential and commercial applications, for controlling the light fixtures, and occasionally to control small motor loads are toggle switches. Most of these switches have two positions, and many of them are marked with the ON and OFF positions. Many toggle switches are self-contained with no externally visible switch contacts, with the exception of disconnect switches, while the switch arrangement is specified by the number of poles (the number of inputs to the switch) and by the number of throws

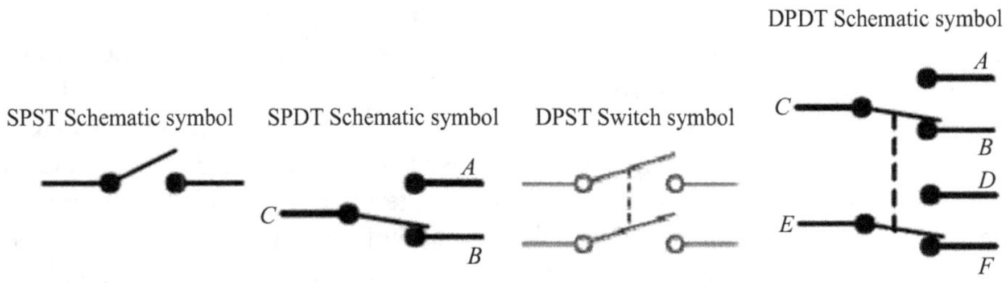

Figure 4.2 Toggle switches, contact, and arrangements

(the number of outputs affected by the switch operation). SPST (single-pole single-throw) is the most common switch arrangement for toggle switches, with a single conductor entering and leaving the switch. The switch has two states: ON or OFF. SPST switches are commonly used to control the switching operation of light fixtures from a single location. The double-pole single-throw (DPST) has two separate inputs that are switched to two separate outputs by its operation, so two separate outputs can be switched ON or OFF from two separate inputs, usually switching two ungrounded (hot) conductors in a circuit. The single-pole double-throw (SPDT) switching arrangements is designed to switch a single input to two outputs, is also known as a *three-way switch*. In the DPST, two distinct inputs are connected to two distinct outputs. Figure 4.2 is showing the diagrams of these toggle switches. The most common applications of the toggle switches are to connect, so they are used to energize or to deenergize a load from a source. The switches must be connected to the ungrounded conductor in order to compliance with the NEC requirements.

Three major terms designate a switch's functions: POLE, THROW, and BREAK. The term POLE refers to the number of circuits that can be controlled by the switch. In the example below, the single-pole switch is capable of interrupting the current in a single circuit. A double-pole switch, on the other hand, is capable of simultaneously interrupting the current in two separate circuits. The term THROW indicates the number of conductors or paths the switch can control. In the example below, the movable contact member of the single-throw switch completes a circuit to only one conductor. However, a circuit to a double-throw switch permits its movable contact element to alternately complete two different paths. The term POSITION refers to the number of stops the switch actuator will make when moved from one extreme position to the opposite position. For example, an ON-NONE-OFF is a two POSITION switch and an ON-OFF-ON is a three POSITION switch. The term BREAK is self-explanatory. It refers to the breaking or opening of a circuit. For example, single-break means that the contacts are separated at only one place. A double-break switch has two pairs of contacts that open the circuit at two places. A double-break switch provides greater volume of contact material, permitting greater heat dissipation and thereby longer switch life. The double-break switch also has twice the voltage breaking capacity, a desirable feature for DC circuit applications. Despite of the great number of switches, these devices have a common denominator in basic components, i.e., the operator which initiates switch operation; the contacts, low-resistance metal that make or break the electrical circuit; and the switch mechanism which is linked to the operator and opens and closes the contacts. The next type of switch (no diagram) is the DPST. These switches are used when there are two "live" lines to switch but can only turn on or off (single throw). These switches are not used much and are usually found in 240 V applications. SPDT switch with a common terminal is the middle terminal in the SPDT Knife Switch or if you are using a household switch, it would be the brass colored terminal (the other two would be silver colored). This circuit clearly demonstrates what happens when the SPDT switch is moved back and forth. Light A goes on and B goes off, B goes on and A goes off and so forth. You can see that this popular switch would have many practical applications: the transmit/receive button on a "2-way" radio, the "high/low beam" switch for your car headlights, the pulse/tone dialing switch on your telephone, and so on. If you are using the SPDT knife switch, you have a "center off" position, which an ordinary wall switch would NOT have in which case you will need to add an SPST switch for shutting this circuit off. (In electronics work, many SPDT switches have a middle position in which the electricity is turned off to BOTH circuits. It is an SPDT center off switch. Also, some electronic SPDT switches have a "center on" position. The best example of this type of switch is the "pickup" selector on an electric guitar which can choose the rhythm, treble or both pickups for three varieties of sounds).

4.4.2 Receptacles

Receptacles are wiring devices designed to allow portable appliances and other systems, the so called *cord-and-plug connected equipment* to be connected to power supply. The majority of receptacles in commercial and residential structures are not supplying continuous loads, so it is quite difficult to predict the load size supplied by a receptacle unless is dedicated (assigned to a specific purpose). Receptacle requirements are covered in the 406 Section of the NEC. Notice that the code does not require a minimum number of outlets for commercial buildings. However, many receptacles are usually required. Load connections to the power sources are done through a variety of receptacle configurations. The load ungrounded (hot) conductor is connected to the shorter receptacle slot. The cord plug that is connected must have the same configuration as the receptacle. The connections to the receptacles are usually made by screw terminal connections on the receptacle sides, in order to allow for feed-through receptacle wiring. Typical receptacle current ratings are 15 A and 20 A. These types of receptacles are grounded, and the requirement is not to use them if the circuit is not actually grounded. The equipment-grounding conductor terminal, green in color is located near the receptacle top or bottom, while ungrounded conductor terminals are bronze in color and the grounded conductor terminals are silver in color. NEMA has established standard slot and prong configurations for various ampere- and voltage-rated receptacles. To prevent accidental plug disconnection from the receptacle, in a locking-type receptacles and plugs, after the plug is inserted in the receptacle, is then slightly twisted to lock into place. The common receptacles are nonlocking-type.

Ground fault circuit interrupted (GFCI) receptacles are designed to minimize the electrical shock hazards. Typical rating for GFCI receptacle is 20 A feed-through (allowing standard receptacles to be used downstream) and 15 A through its own terminals. The main component of a GFCI receptacle is the current sensor that is monitoring the current through the ungrounded and grounded conductors. Under normal operation conditions, the currents through the ungrounded and grounded conductors are equal in magnitude, and the net sensor current is zero. When a fault occurs between ungrounded conductor and the ground, the load current returns to the source through the equipment ground, bypassing the ground fault sensor, the current through the sensor is unbalanced, resulting in opening of the GFCI contacts, deenergizing the load. The GFCI receptacles respond only to the load faults and short-circuit faults involving ground, and not to short-circuit between the ungrounded and grounded circuit conductors. A typical GFCI is tripping at a current of 5 mA in approximately 25 ms time range. Once the GFCI is tripped due to a fault or testing, the device must be reset. Notice that a GFCI receptacle can be used as a replacement for an ungrounded receptacle outlet.

Other special types are the isolated ground receptacles, identified by an orange triangle on the face plate designed to power the sensitive electronic equipment, computers, or data processing systems by eliminating the ground loops, formed by the interconnections of equipment grounding conductors. In this way, the electromagnetic interference, generated in these current loops is minimized. The receptacle mounting strap, metal boxes, and metal faceplate (if used) are grounded through raceway system, by using a separate bare or insulated ground. Hospital-grade receptacles are heavy-duty receptacles used in areas where they may be subject to impact and sudden cord removal or insertion. This receptacle is identified by a green dot on the faceplate. Its internal contact action usually provides for multiple levels of wiping action between the plug prongs and contacts, improving the receptacle-plug conductivity and extending its service lifetime. Temper resistant receptacles, designed to prevent the insertion of small objects into receptacle have usually some form of sliding barrier incorporated into their design. They are usually installed in pediatric care areas, kindergartens, or schools. To protect sensitive electronic equipment against transient overvoltages, transient voltage surge suppressor receptacles are used. Transient overvoltages result from lightning strikes to power lines or switching surges. These receptacles come in a variety of forms and structures, often they incorporate metal oxide varistors (MOV) to absorb the surge transient energy and divert it to the ground. Some of them can have light indicators to show when the suppression feature is functional, other have sound alarms that sounds when the surge suppression is no longer functioning. They are available in standard duplex, isolated ground, hospital grade, and isolated ground hospital grade receptacle types. In residential facilities, receptacles that are dedicated for stationary appliances where the receptacle is inaccessible or for those located 2 m above the floor or finished grade are not required to be tamper-resistant.

Receptacles are wired to permit the parallel load connections, so the rated voltage is applied across each load connected. The ungrounded (hot) conductor is connected to the terminals on receptacle side, while the grounded (neutral) conductor is connected to the other set of terminals. The grounding conductor is connected to the green grounding conductor. Duplex receptacles are often split-wired in order to allow for a load division between two circuits supplying the receptacles. The receptacles, in this case may be supplied by two circuits from two pole-breaker circuits to ensure that both ungrounded conductors are deenergized in the event of short circuit on either of the tow supplying circuits. The voltage between the two hot terminals of a split-wire duplex receptacle is 240 V for single-phase three-wire system. The grounding conductor is pigtailed to each receptacle grounding terminals, as well as the neutral conductor is pigtailed to each receptacle terminal. If one receptacle is removed, the pigtail is disconnected from the receptacle, with continuity of the grounded conductor maintained to other receptacle in the duplex. As previously stated, a load of 180 VA is assigned to each general purpose receptacle, single, duplex, or triplex. The multioutlet receptacle assemblies, such as plug strips that are common in commercial applications, require that for each 5 ft. (or fraction thereof) of multioutlet receptacle assembly, use 180 VA in your feeder/service calculations. This is assuming that it's unlikely for the appliances plugged into this assembly to operate simultaneously, as stipulated in Article 220.14(H). If we are expecting several appliances to operate simultaneously from the same multioutlet receptacle assembly, consider each foot (or fraction of a foot) as 180 VA for feeder/service calculations. A multioutlet receptacle assembly is not generally considered a continuous load.

4.4.3 Disconnect switches

Disconnect switches are used to connect or disconnect a load from power supply, and compared with conventional switches as they are heavy-duty devices, having higher ratings than the toggle switches. The most common applications are in motor circuits. They are usually enclosed in a metal housing, with operating handle located on the right side of the enclosure. The disconnect switch contacts are usually of a knife-blade type, with both movable and stationary contact assembly. Switch contact mechanism is designed for quick-make and break operation in order to minimize the arcing. Disconnect switches can be unfused or fused. The fuse-type disconnects have a fuse-holder in addition to the switch contacts, allowing the switch to accommodate appropriate rated fuse for overcurrent and short-circuit protection of controlled equipment. A rejection kit is also available from some of the manufacturers to permit installation of current-limiting fuse only, to limit let-through fault current in the short-circuit events. Physical construction of the last one differs from that of the standard noncurrent-limiting fuses, making impossible to install a noncurrent-limiting fuse in fuse-holder with rejection feature. Disconnect switches are typically rated at 240 V for general duty or 600 V for heavy-duty applications, with provision that voltage ratings must be equal to or exceed the line-to-line system voltage. Common disconnect switches current ratings are 30 A, 60 A, 100 A, 200 A, 400 A, 600 A, 800 A, 1,200 A, 1,600 A, 2,000 A, 400 A, and 6,000 A. Disconnect installed fuse rating must be equal to or less than its rating. Notice that the physical fuse dimensions make it impossible to insert it into a fuse-holder having a rating lower than that of the fuse. In addition to the current ratings, disconnects have also horsepower ratings. The disconnect horsepower power rating must exceed the horsepower rating of any motor connected to the load side of the switch.

The disconnect enclosure types, as specified in NEMA standards are: NEMA 1, NEMA 3R, NEMA 4, NEMA 4X, NEMA 12, and NEMA 13, with the first two the most common ones. NEMA 1 is designated for indoor use, while the NEMA 3R for general outdoor use, both providing protection against contact with internal energized parts, and against the falling dirt inside, in addition NEMA 3R is waterproof under normal rainfall conditions but not watertight so is unsuitable in applications where the water spray are resulting. NEM 4 and 4X protect not only against the physical contact with energized parts but also falling dirt, dust, and wash-down with noncorrosive elements, while the last also protect against corrosive having gasketed doors. Both may be useful for indoor and outdoor applications. MEMA 12 and 13 enclosures are protecting against the contact with energized parts, falling dirt, dust, and oil seepage, while the last one has also oil spray proof capabilities. Both are designed for indoor applications.

The basic disconnect contact arrangements include: three-pole-single-throw (3PST), four-pole-single-throw (4SPT), three-pole-double-throw (3PDT), and four-pole-double-throw (4PDT). First two configurations can be used to connect three or four distinct outputs to three or four distinct inputs, while the last two can switch either three or four distinct inputs to three or four distinct outputs. For example, a 4PST switch can be used to disconnect three ungrounded circuit conductors and the neutral (grounded) conductor simultaneously. A 3PDT switch can be used to switch three hot conductors of a three-phase system simultaneously, while a 4PDT could be used to switch all three ungrounded conductors and neutral of a three-phase, four-wire system simultaneously. The most common applications are three-phase loads, while for 3PDT and 4PDT switches are the switching of hot and neutral conductors of a three-phase system. 3PDT and 4PDT switches are often used in transfer switches applied to system with ground fault protection. This type of switching arrangement is often used in conjunction with emergency generator to switch from utility supply to the emergency supply. An application not usually found on the 4PST disconnect is in two-phase, four-wire systems.

4.5 Summary of the load computation procedure and cable sizing

The proper sizing of an electrical cable, the cable load bearing is very important to ensure that the cable can operate continuously under full load without being damaged, withstand the worst short circuits currents flowing through the cable, provide the load with a suitable voltage (and avoid excessive voltage drops), and optional, can ensure operation of protective devices during an earth fault. This calculation can be done individually for each power cable that needs to be sized, or alternatively, it can be used to produce cable sizing waterfall charts for groups of cables with similar characteristics (e.g., cables installed on ladder feeding induction motors). Cable sizing methods more or less follow the same basic five step process, involving (1) gathering data about the cable, its installation conditions, the load that it will carry, etc., (2) determine the minimum cable size based on continuous current-carrying capacity and the minimum cable size based on the voltage drop considerations, (3) determine the minimum cable size based on short circuit temperature rise, (4) determine the minimum cable size based on earth fault loop impedance; and (5) select the cable based on the highest of the sizes calculated in step 2, 3, 4, and 5. The first step is to collate the relevant information that is required to perform the sizing calculation. Typically, you will need to obtain the load details. The characteristics of the load that the cable will supply, which includes load type (motor or feeder), three-phase, single-phase or DC system/source voltage, the full load current (A), actual or calculate this if the load is defined in terms of power (kW), full load power factor, locked rotor or load starting current (A), starting power factor, and distance/length of cable run from source to load—this length should be as close as possible to the actual route of the cable and include enough contingency for vertical drops/rises and termination of the cable tails.

The component parts that make up the cable (e.g., conductors, insulation, bedding, sheath, armor, etc.) must be capable of withstanding the temperature rise and heat emanating from the cable. The current-carrying capacity of a cable is the maximum current that can flow continuously through a cable without damaging the cable's insulation and other components (e.g., bedding, sheath, etc.). It is sometimes also referred to as the continuous current rating or ampacity of a cable. Cables with larger conductor cross-sectional areas (i.e., more copper or aluminum) have lower resistive losses and are able to dissipate the heat better than smaller cables. Therefore, a 16 mm^2 cable has a higher current-carrying capacity than a 4 mm^2 cable. International standards and cable manufacturers are quoting the base current ratings of different cable types, in tabular format. Each of these tables pertains to a specific type of cable construction (e.g., copper conductor, PVC insulated, voltage grade, etc.) and a base set of installation conditions (e.g., ambient temperature, installation method, etc.). It is important to note that the current ratings are only valid for the quoted cable types and base installation conditions. In the absence of any guidance, the following reference based current ratings, from standards or codes may be used. When the proposed installation conditions differ from the base conditions, derating (or correction) factors can be applied to the base current ratings to obtain the actual installed current ratings. International standards and cable manufacturers will provide derating factors for a range of installation conditions, for example, ambient/soil temperature, grouping or bunching of cables, soil thermal resistivity, etc. The installed current rating is calculated by multiplying the base current rating with each of the derating factors.

The main steps needed in the load service computation procedure are the following ones. First determine the general lighting load base on the total square footage and space designation (Table 220.12), or on the actual lighting fixtures if are known, and select whichever is greater, as the NEC specify. If the lighting is continuous, as is the case for most of the commercial structures, adjust the estimated value with 125%. For feeder and service load, use Table 220.42 and apply the lighting demand factor for that specific building type. Compute the heating and air-conditioning and discard the smaller value, as specified in Section 220.60 of the NEC. Compute the load of the receptacle outlets, as specified in Section 220.14 of the NEC, as for general purpose ones, and 200 VA per show window foot, 600 VA per heavy-duty lamp holders, and for others (Article 220.14). Apply demand factors as specified, add sign lighting load (1,200 VA minimum), motor loads (if any), and finally size the service and conductors.

Example 4.21: A three-story office structure ($100' \times 60'$) is supplied with 120/208 V service. Loads consist of 85 kVA air conditioning, 105 kVA heating, 270 duplex receptacles, 30 linear feet of show window, 32 exterior lighting fixture (at 175 VA each), 4 blower motors (at 3.5 A), and kitchen equipment of a total of 6.3 kW. Copper THW conductors are used. What is the conductor size?

Solution: The total load is computed by types. For the motors the current is 3.5 A, and the line-to-line voltage is $1.73 \times 208 = 360$ V, three-phase service.

1. General lighting load: $3 \times 100 \times 60 \times 3.5 = 63{,}000$ VA
2. Show window load: $30 \times 200 = 6{,}000$ VA
3. Exterior lighting load: $32 \times 175 = 5{,}600$ VA
4. Heating/air conditioning load: 105,000 VA (keeping the largest of non-coincidental loads)
5. Receptacle load: $270 \times 180 = 48{,}600$ VA
6. Motor load: $4 \times 3.5 \times 360 = 5{,}040$ VA
7. Overall kitchen load: 6,300 VA

Total load (sum of the individual ones): 239,540 VA. The total current is then:

$$I = \frac{P}{V_{LL}} = \frac{239{,}540}{360} = 665.3 = 665 \text{ A}$$

From Table 310.16, a 1,750 kcmil THW cable is suitable for this application.

4.6 Chapter summary

Accurate demand load estimates on a particular branch circuit, feeder, or service are critical and necessary to determine the required ratings of these circuits. The actual and required number of branch circuits or feeders is also based on the estimated demand load determination. This chapter provides a comprehensive overview of load calculation and estimation methods, steps required for accurate load estimates for commercial, industrial and residential facilities and buildings, cable sizing, types and construction. The calculation of electrical loads to be served in a specific building, concepts of unit loads, load demand, load correction, and derating factors are also presented in details. Several examples for load calculation to determine service entrance requirements for residential, commercial, and industrial facilities and occupancies are also included. The specification of the type, proper size, and insulation type for conductors supplying electrical loads is an important part of electrical system design. The most important factor in the conductor selection is that the conductor must be sized to safely carry the load currents, which are accurate, calculated if the demand load is accurate estimated. Likewise, the overcurrent protection devices must also be rated to carry the load currents. An important part of this chapter is designed to the presentation of wiring devices, their characteristics and applications.

References and further readings

[1] D. Maxwell and R. Van der Vorst, 2003. Developing sustainable products and services, J. Clean. Prod. Vol. 11, pp. 883–895.
[2] D.T. Allen and D.R. Shonnard, Green Engineering, Prentice Hall, Upper Saddle River, NJ, 397416, 2003.
[3] A.R. Daniels, Introduction to Electrical Machines, The Macmillan Press, London, 1976.
[4] V. Del Toro, Basic Electric Machines, Prentice Hall, Englewood Cliffs, NJ, 1985.
[5] R.D. Schultz and R. A. Smith, Introduction to Electric Power Engineering, John Wiley & Sons, New York, 1988.
[6] A.C. Williamson, Introduction to Electrical Energy Systems, Longman Scientific & Technical, London, 1988.
[7] K. Chen, Industrial Power Distribution and Illuminating Systems, M. Dekker, New York, 1990.
[8] J.R. Cogdell, Foundations of Electric Power, Prentice-Hall, Upper-Saddle River, NJ, 1999.
[9] T.R. Bosela, Electrical Systems Design, Prentice Hall, Upper-Saddle River, NJ, 2003.
[10] A.S. Pabla, Electric Power Distribution, McGraw-Hill, New York, 2005.
[11] P. Schavemaker and L. van der Sluis, Electrical Power System Essentials, John Wiley and Sons, Chichester, UK, 2008.
[12] M.E. El-Hawary, Introduction to Power Systems, Wiley—IEEE Press, NJ, 2008.
[13] Y. Goswami and F. Keith (eds.), Energy Conversion, CRC Press, Boca Raton, FL, 2008.
[14] M.A. El-Sharkawi, Electric Energy: An Introduction (2nd ed.), CRC Press, Boca Raton, FL, 2009.
[15] W.T. Geondzik et al., Mechanical and Electrical Equipment for Buildings (7th ed.), Wiley, USA, 2010.
[16] J.L. Kirtley, Electric Power Principles: Sources, Conversion, Distribution and Use, Wiley, Chichester, UK, 2010.
[17] N. Mohan, Electric Power Systems—A First Course, Wiley, USA, 2012.
[18] F.M. Vanek. L.D. Albright, and L.T. Angenent, Energy Systems Engineering, McGraw Hill, New York, 2012.
[19] R. Fehr, Industrial Power Distribution (2nd ed.), Wiley, NJ, 2015.
[20] S.L. Herman, Electrical Transformers and Rotating Machines, Cengage learning, UK, 2017.
[21] M.W. Earley (ed.), National Electrical Code (NEC) Handbook, NFPA 70, Quincy, MA, 2014.
[22] M.W. Earley (ed.), National Electrical Code (NEC) Handbook, NFPA 70, Quincy, MA, 2017.
[23] R.R. Janis, and W.K.Y. Tao, Mechanical and Electrical Systems in Buildings (5th ed.), Pearson, Upper Saddle River, NJ, 2013.

[24] W.T. Grondzik, and A.G. Kwok, Mechanical and Electrical Equipment for Buildings (12th ed.), Wiley, Hoboken, NJ, 2015.
[25] G.F. Moore, Electric Cables Handbook (3rd ed.), Blackwell Science, Cornwall, 1997.
[26] H.W. Beaty, Handbook of Electrical Power Calculations, McGraw Hill, New York, 2001.

Questions and problems

1. Define the diversity factor and the load demand factor.
2. Define the net load demand and the total connected load.
3. Explain the importance of the derating factors in the load determination.
4. What color insulation is used on equipment grounding conductor?
5. Define a single-conductor and a multiconductor cable.
6. Define the continuous and noncontinuous loads.
7. How the rating of a branch circuit is estimated?
8. What are load types of the branch circuit?
9. Describe the most common types of receptacles.
10. What are the general types of branch circuit loads?
11. Describe the construction of (a) type AC cable, (b) type MC cable, and (c) type NM cable. For each type of cable describe also typical applications.
12. What types of receptacles are used to connect sensitive electronic equipment?
13. What are the major disconnect applications?
14. What type of conductor extends beyond the final overcurrent protection device?
15. List the special receptacle types and their applications.
16. Explain the operation of GFCI receptacle.
17. Calculate the minimum estimate lightning load for: (a) a school building with a total floor area of 36,000 ft^2, being illuminated with 150 fluorescent light fixtures, each having a ballast input of 277 V, 0.8 A; and (b) an office area that is 125 ft. by 180 ft., illuminated with 120 fluorescent light fixtures, each having a ballast input of 277 V, 0.8 A.
18. An electric water heater is rated 120 V, 3.6 kW. Determine the estimated demand load and the heater rated current.
19. Determine the branch circuit estimated load supplying both: (a) a 5 kW counter-mounted cooktop unit and a 6 kW wall-mounted oven; and (b) a 6.5 kW counter-mounted cooktop unit and a 9 kW wall-mounted oven.
20. Determine the demand load (VA) for the following motors: (a) single-phase 10 HP, 230 V induction motor, (b) single-phase 1.5 HP, 115 V induction motor, and (c) three-phase 100 HP, 460 V induction motor.
21. A hardware store has a total area of 100′ × 200′, used for storage 100′ × 50′, a small office 15′ × 20′, and the remainder the building is used as showroom. There are also a total of 36′ show windows, a 50′ track and three outdoor sign. What is the total lighting load?
22. Determine the load on a 120 V, 20 A branch circuit that has eight general use duplex circuits connected.

23. Determine the minimum estimated lighting load for a school that has a total floor area of 28,000 ft^2, illuminated by 135 fluorescent light fixtures, each having a ballast of 277 V, 0.85 A.
24. Compute the receptacle load for a grocery store having 210 duplex and 90 single receptacles.
25. Compute the general lighting load of large 3,000 ft^2 hallway of a hotel, assuming continuous load.
26. What is the general lighting load for a 25-room motel, each room is $15' \times 18'$.
27. Repeat problem 20 for the following three-phase motors: (a) 30 HP, 208V, induction motor, (b) 75 HP, 575 V, induction motor, and (c) 40 HP, 230 V synchronous motor.
28. Determine the resistance of a conductor of 200 ft. length at 25 °C made up of: (a) cooper; and (b) aluminum.
29. Determine the resistance of the cable of Problem 10 at 60 °C.
30. A high school is supplied with 208/120 V, single-phase power. Calculate the total load, assuming continuous load. The school has the following dimensions and loads:
    (a) 36,000 ft^2 classroom space
    (b) 7,500 ft^2 auditorium
    (c) 5,500 ft^2 cafeteria
    (d) 12.5 kW outside lighting
    (e) 250 duplex receptacles (120 V)
    (f) Kitchen equipment: 15 kW ovens, 12 kW ranges, 4.5 kW fryers, 10 kW water heater, and 3.6 kW dishwasher
31. A residential building has a living space of 3,200 ft^2. There are six small-appliance branch circuits supplying kitchen and dining areas and two laundry circuits. There are total of 24 duplex receptacles for miscellaneous use in the residence unfinished areas. Determine the estimated load demand.
32. A 24-h restaurant has 105 duplex receptacles and 45 single receptacles. Calculate the total load for feeder calculations.
33. Sketch the schematic wire diagrams, showing the connection of the ungrounded, grounded, and equipment grounding conductors and wiring devices for: (a) single lighting outlet controlled by two switches, (b) two lighting outlets controlled by two switches, and (c) two duplex receptacles with of receptacle of each duplex hot and the other receptacle outlet on each duplex switched simultaneously from one switching location.
34. Specify all situations in which a grounded conductor must be included among the current-carrying conductor for the ampacity adjustment purpose.
35. A small retail store ($90' \times 50'$) is supplied with a single-phase 120/240 V service. Loads consist of: 45 kVA air conditioning, 55 kVA heating equipment, 1 HP, 240 V ventilating unit, 90 duplex receptacles, 30 linear feet of show window, 1,200 VA exterior lighting sign, and auxiliary equipment of a total of 4.2 kW. Copper THW conductors are used. What is the conductor size?
36. What is the ampacity of #1 AWG THW copper conductor in a raceway containing four current-carrying conductors, 41 °C, dry location?

Chapter 5

Power distribution, load, and motor centers

Outline and abstract

Electric power distribution is the portion of the power delivery infrastructure, taking the electricity from the highly meshed, high-voltage transmission systems and delivers it to customers. Primary distribution lines are medium-voltage circuits, usually in the range of 600 V to 35 kV. Close to the end-users, transformers step-down the primary distribution voltages to the low-voltage secondary distribution levels, commonly 120/240 V or other utilization voltages used mainly in industrial facilities. From the power distribution transformer, the secondary distribution circuits connect to the end-users, where the connection is made at the service entrance. Distribution infrastructure is extensive and complex, the electricity has to be delivered to customers concentrated in cities, suburbs or rural areas, industrial and commercial facilities, schools, hospitals, military bases, or communication infrastructure. In industrial and large commercial facilities, electricity is provided to the loads from specialized power distribution units and the load centers containing equipment necessary to protect, operate, and control the loads. The terms *switchgear* and *load centers* are used to describe combinations of enclosures, busbars, circuit breakers, power contactors, fuses, protective relays, controls, and indicating devices. There are several load center types, with their selection based primarily on the electrical requirements and installation environment. The essential parts of electrical distribution systems are discussed in this chapter, as well as in other sections of the book. After completing this chapter, the readers will have a good understanding and knowledge of several aspects of power distribution networks, load, and motor functionalities and requirements, such as: understanding the role, configurations and topologies of power distribution, ratings and characteristics, the purpose, basic construction, and configurations of load centers, switchgear and motor control centers, purpose, specifications of switchgear and motor control centers, their ratings, structure, and major applications. Readers will also understand the basics of circuit breaker and motor starter operation and applications, as well as protection requirements for switchgear and motor control centers, and learn about the most important provisions of standards and codes used in power distribution networks and load centers.

5.1 Introduction, power distribution, and electrical services

Power plants and generation stations may be located near the energy sources, cities, or large industrial facilities where larger amounts of electrical energy are consumed. Electric energy is delivered to the sites and users by means of the utility electric networks at high-voltage (HV) or medium-voltage (MV) levels, being distributed to the end-users through the medium- and low-voltage (LV) electrical networks. Thus, the efficient production and distribution of electrical energy represents a critical infrastructure. Depending on the load power requirements and electric grid layout, transformer substations (stepping-down the voltage downstream of the utility-delivering node) are located at a site boundary or distributed around it. The choice of different power distribution systems is based on technical and economic criteria, while the energy-saving targets are not the most critical parameter, due to the fact that quite small energy losses are involved. There are many different types of residential, commercial, and industrial customers of electricity. To meet this vast demand for electricity, power companies need to produce tremendous quantities of electrical power. This vast quantity of electricity is supplied by power-generating plants. Individual generating units of hundreds of MW or even larger of electrical power are in operation in many of the power plants. The vast majority of electrical energy is produced by power plants located throughout the country, often remote from major cities, converting into electricity the energy produced by burning coal, oil, or natural gas, the falling of water, or from nuclear reactions. Electrical energy must be supplied at the same time that it is demanded by consumers. There is no simple storage system to store electricity, which later may be used to supply additional electrical energy at peak demand times. This situation is quite unique and necessitates the production of sufficient quantities of electrical energy to meet the demand of the consumers at any time.

The electrical energy that must be produced by the power systems varies greatly according to several factors, such as year, season, week, day, and time of the day. Electrical power supply and demand is much more difficult to predict than most commodities that are bought and sold. Electricity must be readily available in sufficient quantity whenever it is required. Its supply and demand problem is something that is taken for granted until the electrical power is interrupted. The US electrical power systems are interconnected on a regional basis so the power stations can support one another in meeting the high variable energy demands. Industrial use of electrical power accounts for over 40% of the total kilowatt-hour (kWh) consumption and the industrial use of electricity is projected to increase at a rate similar to its present rate. Major increases in the residential power demand have been due to an increased use by customers. A smaller increase was accounted for by an increase in the customers. Commercial use of electricity accounts for less than 25% of the total usage. Commercial power consumption includes office buildings, apartment complexes, schools, shopping establishments, and motel or hotel buildings. The prediction of the electrical power demand by these facilities is somewhat similar to the residential demand. Commercial use of electrical power is also expected to increase at a declining rate in the future. There is no difference between a transmission line and a distribution line except for the voltage level and power handling capability. Transmission lines are usually capable of transmitting large quantities of electric energy over long distances, and are operating at HV levels. Power distribution lines, on the other hand, are carrying limited quantities of power over significantly shorter distances.

Electrical distribution systems are an essential part of the electricity systems, is the grid sections which distribute the electricity to the consumers. It is the portion of the power delivery infrastructure that takes the electricity from the highly meshed, HV transmission networks and delivers it to the end-users. In order to transfer the electrical power from an AC or a DC power supply to the place where it is used (loads), specific type of power distribution network must be utilized. The complex power distribution systems are used to transfer electrical power from power plants to industries, homes, and commercial buildings. Power distribution systems are usually broken down into three components, which are *distribution substation*, *primary*, and *secondary power distribution networks*. At a substation, the voltage is lowered as needed and power is distributed to the customers, one substation supplying power to thousands customers. Thus, the number of transmission lines in the power distribution systems is several times that of the transmission systems. Furthermore, most customers are connected to only one of the three phases in the power distribution system. Therefore, the power flow on each of the lines is different and the system is typically unbalanced. Primary distribution lines are MV networks, ranging from 600 V to 35 kV. At a distribution substation, transformers take the incoming transmission-level voltage (35–230 kV) and steps it down to several distribution primary circuits, fanning-out from the substation. Close to each end-user, power distribution transformers step-down the primary voltages to LV secondary circuits at 120/240 V or other utilization voltages. From the power distribution transformer, the secondary distribution circuits connect to the end-users at the service entrance. Power distribution employs equipment, such as transformers, circuit breakers, monitoring equipment, metering, and protective devices in order to deliver a safe and reliable power. The power distribution in the US is usually in the form of three-phase 60-Hz AC current. The electric power is changed in many ways through the employment of electrical special circuitry and devices. The distribution of electricity involves a system of interconnected transmission lines, originating at the electrical power-generating stations located throughout the US or any other country. The ultimate purpose of these power transmission and distribution systems is to supply the electricity needed for industrial, residential, and commercial uses. A typical electrical power distribution system is shown in Figure 5.1.

This chapter discusses the major components of the power distribution system, the distribution network, substations, and associated electrical equipment and controls, feeders, transformers, monitoring, control, and protection. A part of this chapter is reserved to how incorporating protection devices and controls into the system can create the so-called *smart grid* capable of handling the integration of

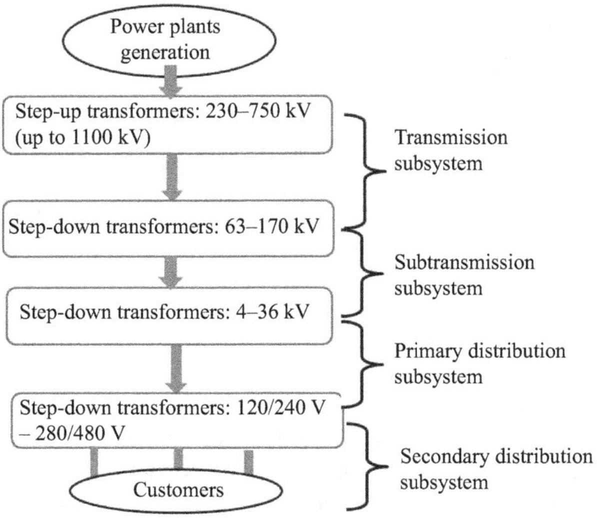

Figure 5.1 Electricity infrastructure from generation to power distribution

large amounts of distributed generation of sustainable, renewable energy sources, smart loads, intelligent control, management, and monitoring of the power systems. The primary purpose of an electricity distribution is to meet the customer's demands and requirements for energy after receiving the bulk electrical energy from power transmission and power subtransmission substations. The primary power distribution substations serve as load centers and the customer substation interfaces to the LV networks. Customer power distribution substation are referred to a power distribution center normally provided by the customers, accommodating a number of switchgear panels and the power distribution transformers to enable LV connections to the customer incoming switchboard or panel boards.

5.1.1 Electrical services and industrial power distribution

Electric energy, delivered to sites by means of the electric grids at HV and/or MV power lines, is distributed to the end-users through MV and LV power distribution networks, that are delivering electricity from the distribution substations to the service-entrance equipment located at residential, commercial, and industrial facilities. Depending on the power demand of loads and on the network layout, transformer substations, whose main task is to step-down the supply voltage downstream of the utility-delivering node, are located at a site, area boundary, or distributed around that area. The choice among different power distribution systems, such as radial or loop-feeder systems is based on technical and economic evaluations that are not generally considering energy-saving targets, because very low energy losses are involved. Most of the US distribution systems operate at usually at primary voltages from 12.5 to 24.9 kV, while some of them operate at 34.5 kV or LV distributions, such as 4 kV. However, most of the LV distribution systems are being phased out. Distribution transformers convert the primary voltage to secondary consumer voltages to a LV secondary circuit, usually 120/240 V, the secondary distribution circuits connect to the end-users at the service entrance. National Electrical Code defines service as, "The conductors and equipment for delivering electric energy from the serving utility to the wiring system of the premises served," indicating that the service is where a utility company provides electricity to a building or a structure, most buildings being served directly by the utility company, and the service includes a revenue meter. The requirements of NEC Section 130.5(A) refer to this service and to this meter. However not all buildings or structures are connected directly to the utilities and not all services have revenue-measuring meters. For example, a college campus can purchase bulk power from an electric utility company, in order to save energy costs. Usually, the revenue meters are located where the electric utility company is connecting the customers to the power distribution system. From this point of view, the customer owns the electrical system and becomes the serving utility for the NEC, IEEE, or IEC code purposes. Codes are also recommending that the customer's service conductors shall be continuous from the point of electricity delivery to the metering equipment. Conductors and cable installed in raceways or conduits comply with NEC Article 300.3 for safety and supply security reasons. The rating of the service entrance equipment must satisfy the general requirements stated by the NEC and local building codes. We have to keep in mind that considerable efforts are required in order to maintain the power supply within the requirements of various consumers, such as proper voltage, availability of the power on demand or supply, reliability, and security.

The electric power grid operates as a three-phase network to the level of the service point to residential, small, medium, and large commercial and industrial loads. Feeders are usually three-phase overhead pole lines or underground cables, and closer to the loads (many of which are single phase), three-phase, or single-phase laterals provide connections to the customers. Load centers, switchgears, and motor control centers receive the electrical energy through complex power distribution networks (distribution lines, protection, control, or metering equipment), delivering it to large loads in a safe and reliable way. Power transformers, used with the three-phase power are connected in either wye or delta configurations. Transformer type and the actual voltage levels depend on the power company requirements and capabilities, as well as the customer needs. Three-phase voltage is used throughout large commercial and industrial facilities to run AC motors and other large power equipment and devices. Installation, operation, branch and feeder calculation, and seizing are conducted based on power engineering principles, and the code and standard specifications. There are several organizations that are involved in establishing standards for the design, construction, and setting of load centers, switchgears, branch circuits and feeders, while the major standards and codes are established by UL (Underwriters Laboratories), NEMA, IEC, IEEE, and NFPA. It is not the scope of this book to discuss all the standards or codes in full details; however, references are made throughout the book to the important standards with which power distribution, electrical services, load and motor centers, switchgears, panel boards, protection, cables, or monitoring equipment and devices comply. The National Electrical Manufacturers Association (NEMA) is an organization that, among other things, develops standards for electrical equipment. The National Fire Protection Association (NFPA) is a nonprofit organization which publishes the National Electrical Code (NEC). The intent of the NEC is to describe safe electrical practices. The International Electrotechnical Commission (IEC/CEI) is an organization based in Geneva, Switzerland with over 50 member nations. IEC writes standards for electrical and electronic equipment practices.

5.2 Power distribution networks

Electricity distribution involves a very complex system of interconnected transmission lines, power transformers, monitoring, control, and protective equipment and devices. The ultimate purpose of these power transmission and distribution systems is to supply the electrical power necessary for industrial, residential, and commercial uses. Often electrical power systems are interconnected with one another in parallel circuit arrangements. These interconnections of power generation systems are monitored, controlled, and operated by the computerized control centers. The control centers provide means for data collection, recording, analysis, system monitoring, frequency and voltage control, and signaling. The electrical energy transmission requires long transmission lines from where it is produced to where it is used, requiring optimum planning to assure best use of land and minimal environmental impacts, while minimizing the overall cost and investment. The location of transmission lines is limited by zoning laws and by the populated areas, highways, railroads, waterways, topographical, and environmental factors. Overhead power transmission lines usually operate at voltage levels from 12 to 750 kV, with the most common range from 50 to 350 kV. Underground transmission methods are used primarily in urban areas and in suburban areas, where the right-to-way for overhead power lines is limited. Major advantages of overhead transmission lines are their ability to dissipated heat, lower costs, and easier installation. Underground cables are confined to short distances, being much more expensive than the overhead ones. To improve underground cables power-handling capacity, the research is focusing on forced-cooling techniques, such as circulating-oil and with compressed-gas insulation or on the cryogenic cables or superconductors operating at extremely low temperatures and having large power-handling capabilities. Notice that facilities using large quantities of power are direct connected to transmission power systems.

Power distribution infrastructure, taking the electricity from the HV transmission networks and delivers it to the customers, is an extensive and complex electric network. Primary distribution lines are MV circuits, in the range of 600 V to 35 kV. At a distribution substation, transformers take the incoming transmission-level voltage (often in the 35–230 kV range), stepping it down to the voltage levels of primary circuits, which are fanning out from the substation. Close to end-users, power distribution transformers step-down the primary-distribution voltages to lower secondary circuit voltages, commonly 120/240 V, other utilization levels.

From the power distribution transformer, the secondary distribution circuits connect to the end-users at the service entrance. However, the choice of voltage to be used on any particular distribution section is influenced, among other factors, by decisions associated with voltage drops resulting from large current loads, capital cost of transformers used to change voltage levels and construction of distribution lines and associated switchgear to operate at the chosen voltage, and environmental aspects of the system installation. Urban distribution networks are often underground, while the rural constructions are mainly overhead type. Suburban structures are a mix, with a good deal of new construction going underground. The power distribution systems are capital-intensive businesses, with about 10% of the all utility capital investments. Cost lowering, simplification, and standardization are all important power distribution design characteristics. However, few components or installations are individually engineered on power distribution circuits. Standardized equipment and designs are used wherever possible and recommended by engineering practices, codes and standards. Power distribution planning is the study of future power delivery needs. Its goals are to provide service at lowest cost and highest reliability and power quality possible. Planning requires a mix of geographic, engineering, and economic analysis skills. New circuits must be integrated into the existing distribution system within a variety of economic, political, environmental, electrical, and geographic constraints.

5.2.1 Power distribution configurations

Power distribution infrastructure is very extensive and complex, the electricity has to be delivered to the customers concentrated in cities, suburbs, rural areas and even in very remote places, while few places in the industrialized world do not have electricity from a power distribution system. Industrial facilities are using almost 50% of all the electrical energy. The three-phase power is distributed directly to most large industrial and commercial facilities. Electrical substations use massive three-phase power transformers and associated equipment (circuit breakers, HV conductors, and insulators) to distribute the electricity. From these substations, electricity is distributed to industrial sites, residential and commercial users. Facilities and buildings must provide maximum supply safety, consume few resources during construction and operation and be flexibly to adapt to any future power requirements. The intelligent integration of all building service installations offers an optimum to be attained for safety, energy efficiency, flexibility, and environmental compatibility, offering maximum comfort. Buildings are responsible for about 40% of the total energy consumption within the US and developed countries. In large buildings, the type of distribution depends on the building type, dimension, the length of supply cables, and the loads. The distribution system can be divided into (a) the vertical supply system (rising mains) and (b) the horizontal supply (distribution at each floor level). Power distribution circuits and networks are found along most of the secondary roads and streets. Urban construction is mainly underground power distribution, while the rural power distribution is mainly as overhead transmission type. Suburban structures are a mixture, with a large part of the new construction going underground.

Power distribution feeder circuits consist of overhead and underground transmission line networks (circuits) in a mix of branching circuits (laterals) from the station to the various customers. Each circuit is designed around various requirements, such as the required peak load, voltage levels, distance to customers, and other local conditions, such as terrain configurations, street layout, visual and environmental regulations, or user requirements. The secondary voltage in North America consists of a split single-phase service that provides the customer with 240 V and 120 V, the customer is then connected to devices depending on their ratings. This is served from a three-phase distribution feeder normally connected in a Y-configuration consisting of a neutral center conductor and a conductor for each phase. In most other parts of the world, the single-phase voltage of 220 V or 230 V is provided directly from a larger neighborhood power distribution transformer, also providing a secondary voltage circuit often serving hundreds of customers. The various branching laterals are operated in radial or in looped configurations, where two or more feeder parts are connected together usually through an open distribution switch. Power distribution networks are overhead or underground, highly redundant and reliable complex networks. Overhead lines are mounted on concrete, wooden, or steel poles arranged to carry power distribution transformers or other needed equipment, besides the conductors. Underground distribution networks use conduits, cables, manholes, and necessary equipment installed under the street surface. The choice between the two systems depends of factors, such as safety, initial, operation and maintenance costs, flexibility, accessibility, appearance, life, fault probability, location and repairs, and interference with communication systems. For example, the underground distribution systems comparing to the overhead systems are more expensive, require high investment, maintenance and operation costs, lower fault probabilities but more difficult to locate and repair a fault, or lower cable capacities and voltage drops. Each system has advantages and disadvantages, while the most important is the economic factors. However, non-economic factors are sometimes more important as the economic ones.

5.2.2 Feeder voltage drops, electric distribution losses, and power factor control

Electrical energy is generated at power plants or generation stations and transferred over HV transmission line to the utilization points. The employment of HV lines is motived by the increased line capacity and loss reduction, an important aspect of energy conservation. An electrical transmission line is modeled using series resistance, series inductance, shunt (parallel to ground) capacitance and conductance, while the series resistance and inductance are the most important and parallel elements are often neglected in calculations. In power engineering, the term *wire* refers to one or more insulted conductors of small size (solid or stranded for flexibility), while the term *cable* refers to one or more insulated conductors of large size grouped in common insulated jacket, quite often with a ground shield. A wire or combination of wires not insulated from one another is called a *conductor*. A stranded conductor consists of a group of wires, often twisted or

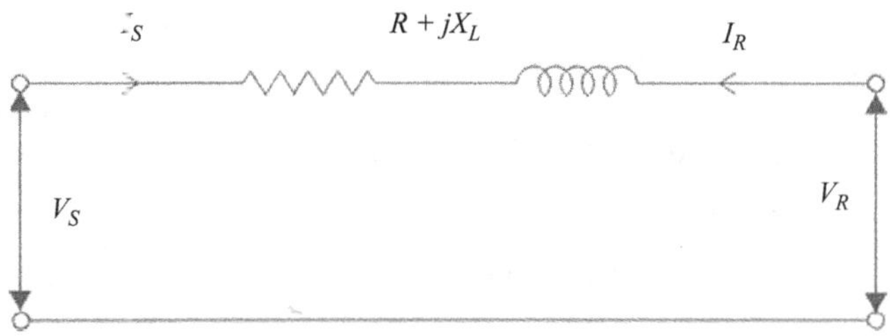

Figure 5.2 Cable impedance between source and load ends

braided together to minimize magnetic coupling. In the most common model, used in power engineering each of the cable phase conductor is represented by a resistance, R in series with leakage reactance, X_L ($= \omega L$), where L is the leakage inductance between the current carrying conductors, as shown in Figure 5.2. A cable design or calculation requires selecting the conductor size with required cable capacity (ampacity) at the operating temperature that is meeting the voltage limitations over the feeder length under the steady-state and inrush motor or equipment current. In such calculations the per-phase current, I, power factor, resistance, and inductive reactance are required. The voltage drop of the line in Figure 5.2 is then computed as:

$$V_{drop} = I \times (R \cdot \cos\theta + X_L \cdot \sin\theta) = I \times Z_{cable} \tag{5.1}$$

Here θ is the phase angle ($PF = \cos\theta$), and all calculations can be performed in standard units or in percent or p.u. values, taking θ positive for lagging power factor and negative for leading power factor. Equation (5.1) is convenient for the definition of the effective cable impedance because its product with the cable current is simply giving the voltage drop magnitude or the cable voltage drop in amperes:

$$V_{drop/A} = Z_{eff\_cable)}\,\mathrm{V/A}$$

Example 5.1: A 4.8 kV three-phase, line-to-line wye-connected feeder has a per-phase resistance of 50 mΩ and 0.33 mΩ inductive reactance per 1,000-ft. Calculate the feeder voltage drops for 0.85 lagging power factor, unit power factor, and 0.85 leading power factor at an apparent power of 1.2 MVA.

Solution: The feeder phase voltage and the current at three-phase apparent power of 1.2 MVA are:

$$V_\phi = \frac{4{,}800}{\sqrt{3}} = 2{,}775.6 < 0^\circ\ \mathrm{V}$$

$$I = \frac{1.2 \times 10^6}{\sqrt{3} \times 4{,}800} = 144.5\ \mathrm{A}$$

By using (5.1) the voltage drops for 0.85 lagging power factor, unit power factor and 0.85 leading power factor are:

$$V_{drop} = 114.5 \times \left(50 \times 10^{-3} \cdot 0.85 + 0.33 \times 10^{-3} \cdot 0.527\right) = 6.86\ \text{V}$$

$$V_{drop} = 114.5 \times \left(50 \times 10^{-3} \cdot 1.0 + 0.33 \times 10^{-3} \cdot 0.0\right) = 7.73\ \text{V}$$

$$V_{drop} = 114.5 \times \left(50 \times 10^{-3} \cdot 0.85 - 0.33 \times 10^{-3} \cdot 0.527\right) = 2.88\ \text{V}$$

The cable must be selected and manufactured to operate in the application specific requirements and conditions. Notice that the power factor has quite a significant effect on the cable voltage drop.

Power distribution systems have automatic voltage regulation schemes and devices at one or usually at several control power stations, such as area substation or facility switchgears to maintain the constant voltage regardless of the load currents. Voltage drops in line are in relation to the resistance and reactance of line, length, and the current drawn. For the same quantity of power handled by a transmission, lower the voltage, the higher is the current drawn and higher is the voltage drop. The current drawn is inversely proportional to the voltage for the same power handled, while the power loss in the line is proportional to the resistance and square of current (i.e., $P_{Loss} = I^2R$). Higher voltage transmission and distribution helps to minimize line voltage drop in the ratio of voltages, and the line power loss in the ratio of square of voltages. The feeder cables deliver power from the controlled *sending end* (power source) to the *receiving end* (the load); with the voltage gradually dropping form the initial maximum value to the minimum value at the load. For this reason, the power distribution system is seized not only for the required ampacity (capacity) but also to maintain the required voltage levels during the steady-state operation and during the transients, such as inrush current during motor starting. If the load current of a power factor, *PF* (usually lagging the voltage) flows from the switchboard is controlled to maintain the constant bus voltage, the magnitude of the voltage drop between the ends, (5.1) can be written for the cable impedance as:

$$V_{drop} = I \times \left(R \cdot PF + X_L \cdot \sqrt{1 - PF^2}\right) \tag{5.2}$$

Often the cable manufacturers are listing the cable effective impedance at typical 0.85 lagging power factor, therefore the catalog cable impedance values are:

$$Z_{cable} = R \times 0.85 + X_L \times \sqrt{1 - 0.85^2} = 0.85 \cdot R + 0.527 \cdot X_L\ \Omega/\text{phase} \tag{5.3}$$

However, if the power factor value is quite different than 0.85 lagging, (5.2) must be adjusted consequently. The cable size is usually selected to meet the capacity (ampacity) with up to 30% margin, limiting the solid-state voltage drop at

3%–5% from the switchboard to the load. We have noticed that besides the cable impedance, the feeder voltage drop also depends on the lad power factor. Capacitors can improve the load power factor correcting also the feeder voltage drops, be reducing or eliminating the reactive terms in (5.1) and (5.2). The voltage boost estimation produced by the capacitors (bank) placed at the end of the feeder (load), starts with the estimate the capacitor value, so the reactive power correction and values, Q, needed to improve power factor at the desired level as:

$$Q_{Cap} = Q_{Loaa} - Q_{Desired} \tag{5.4}$$

Expressing the reactive power, through the apparent power and the usually relations between power factor, the apparent power, reactive power, and active (real) power for a feeder supplying to a load, the load and the desired reactive power are expressed as:

$$\begin{aligned} Q_{Load} &= \sqrt{\left(\frac{P_{Load}}{PF_{Load}}\right)^2 - P_{Load}^2} \\ Q_{\text{Desired}} &= \sqrt{\left(\frac{P_{Load}}{PF_{Desired}}\right)^2 - P_{Load}^2} \end{aligned} \tag{5.5}$$

From the above relationships, the needed capacitor (bank) reactive power to improve the power factor at the desired level is given by:

$$Q_{Cap} = P_{Load} \times \left[\sqrt{\frac{1}{PF_{Load}^2} - 1} - \sqrt{\frac{1}{PF_{Desired}^2} - 1}\right] \tag{5.6}$$

The prephase capacitor (bank) value is then computed by using the well-known relationship:

$$C = \frac{Q_{Cap}}{2\pi f \cdot V_{LL}^2} \tag{5.7}$$

Here f is the electrical supply frequency and V_{LL} is the feeder line-to-line voltage. In addition to imprcve the network power factor, the capacitor (bank) is reduction the feeder voltage drop and the feeder cable losses by producing a voltage rise, due to the leading capacitor current. To understand the voltage rise calculation, we are considering the circuit of Figure 5.3. The rated capacitor (bank) current is given by:

$$I_C = \frac{Q_{Cap}}{\sqrt{3} \times V_{LL}} \tag{5.8}$$

The voltage rise due to the capacitor (bank) is equal to:

$$V_{Rise} = I_C \cdot X_L = X_L \frac{Q_{Cap}}{\sqrt{3} \times V_{LL}} \tag{5.9}$$

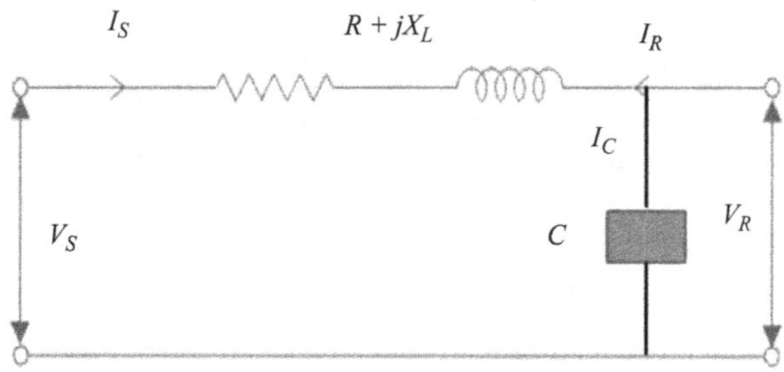

Figure 5.3 Voltage rise calculation circuit diagram

Example 5.2: An industrial facility has a power demand of 1,000 kW at a power factor of 0.75 lagging. Determine the capacitor bank reactive power ratings required to improve the power factor to 0.85 and 0.95, respectively. What is the voltage rise in each case, if the feeder inductive reactance is 0.015 Ω/phase and the line-to-line voltage is 460 V?

Solution: Direct application of (5.6) leads to:

$$Q_{Cap-0.85} = 1{,}000 \times \left[\sqrt{\frac{1}{0.75^2} - 1} - \sqrt{\frac{1}{0.85^2} - 1}\right] = 262.2\,\text{kVAR}$$

$$Q_{Cap-0.95} = 1{,}000 \times \left[\sqrt{\frac{1}{0.75^2} - 1} - \sqrt{\frac{1}{0.95^2} - 1}\right] = 553.2\,\text{kVAR}$$

The voltage boost in the case for the feeder characteristics (Equation (5.9)) are:

$$V_{Rise-0.85} = 0.015 \times \frac{262.2 \times 10^3}{\sqrt{3} \times 460} = 4.94\ \text{V}$$

$$V_{Rise-0.95} = 0.015 \times \frac{553.2 \times 10^3}{\sqrt{3} \times 460} = 10.4\ \text{V}$$

The voltage rise (boost) due to the capacitor can be expressed in per-phase values as:

$$V_{Rise-phase} = X_L \frac{Q_{Cap-phase}}{V_\phi}\ \text{V/phase and}\ \%V = \frac{V_{Rise-phase}}{V_\phi} \times 100 \qquad (5.10)$$

Notice that in (5.1) if the voltage drop is zero, then the $R/X_L = tan(\theta)$, therefore the reviving end voltage is constant. Typical R/X_L ratios for power distribution cables are in the range 0.2–0.3, with an average value of 0.25 and a power factor 0.97 leading, giving a flat receiving-end voltage value, regardless the load current. Even only a small part of the total electric power transferred through the transmission lines in a power distribution network (3% or less) is lost through Joule effect and additional losses (both related to the square of the line current) is still important to be able to estimate such losses. Cable manufacturers are usually providing the AC resistance and the series leakage reactance, including the skin and the proximity effects, as well as the conduit/raceway type. Notice that the difference between the DC and high frequency inductance is often neglected in power engineering studies. Electric distribution losses at any site are mostly due to the Joule effect, which depends on the square of the current and on the line resistance. However, most electric losses occur in end-users who can be grouped basically as electrical machinery and drives, electrically heated users, such as furnaces, ovens, boilers, induction heating equipment, resistors, and microwave equipment lighting equipment, and other load types, such as electrochemical equipment, monitoring, control, and communication systems. The basic relationship is expressed as follows:

$$P = n \times R \times I_{line}^2 \text{ (W)} \tag{5.11}$$

Here n is the number of phase conductors, R (Ω) is the phase conductor resistance and I_{line} the RMS (effective) line current (A). If the current is flowing through a series combination of N_S conductors the total resistance is given by:

$$R_{total} = \sum_{k=1}^{N_S} \frac{\rho_k \times l_k}{A_k} \ (\Omega) \tag{5.12}$$

In the case of N_P conductors connected in parallel sharing the total current, the lower the resistance the higher is the current flowing in each conductor. The total resistance of the parallel configuration of N_P conductors is expressed as:

$$R_{total} = \frac{1}{\sum_{k=1}^{N_P} \frac{A_k}{\rho_k \times l_k}} \ (\Omega) \tag{5.13}$$

In the above equations, ρ_k is the conductor material resistivity (Ωm), A_k is the cross-sectional conductor area (m^2), and l_k is the length of k conductor (m). The DC resistance, as we discussed in the previous book is also dependent on temperature and must be adjusted for AC system operation. These changes need also to be reflected in voltage drop or cable losses calculations. The AC resistance of the cable conductors is the DC resistance corrected for skin and proximity effects:

$$R_{AC} = R_{DC} \cdot \left(1 + F_{skin} + F_{proximity}\right) \tag{5.14}$$

Here R_{DC} is the cable DC resistance, while F_{skin} and $F_{proximity}$ are the skin effect and proximity effect correction factors. The skin effect depends on the electrical frequency, the size of the conductor, the amount of current flowing and the diameter of the conductor, through a rather complex relationship, having net results in an increase in the cable/conductor effective resistance. The proximity effect factor varies depending on the conductor or cable geometry. The proximity effect also increases the effective resistance and is associated with the magnetic fields of the cable conductors which are close together. The proximity effect decreases with increase in spacing between cables. There are a few empirical relationships used to calculate both these two factors and they are provided by the cable manufacturers. Notice that skin and proximity effects may be ignored with small conductors carrying low currents, being increasingly significant with larger conductors and it is often desirable for technical and economic reasons to design the conductors to minimize both these effects. AC resistances are important for calculation of current-carrying capacity. Values for standard designs of distribution and transmission cables are included in the tables in any cable design handbooks. The series inductive reactance of a cable, an important parameter in voltage drop calculations can be approximated by the following relationship:

$$X_L = 2\pi f \cdot \left[K + 0.2\ln\left(\frac{2\cdot s}{D_{cond}}\right)\right] \times 10^{-3}(\Omega/\text{km}) \tag{5.15}$$

where X_L is the conductor inductive reactance (Ω/km), f is the electrical supply frequency (Hz), s is the axial spacing between conductors (mm), D_{cond} is the diameter of the conductor, or for shaped conductors, the diameter of an equivalent circular conductor of equal cross-sectional area and degree of compaction (mm), and K is a constant factor pertaining to conductor formation and geometry. The most common values of the constant K are 0.0500 for solid conductors, 0.0778 for a 3-wire conductor and 0.0664 for a 7-wire conductor.

Example 5.3: Calculate the power losses per km of length, in a six-wire poly-phase system having the resistance per unit of length and phase of 0.018 Ω/km, carrying 2,000 A.

Solution: From (5.11), the power loss per length is:

$$P = 6 \times 0.018 \times (2{,}000)^2 = 0.432\ (\text{MW/km})$$

5.3 Power distribution system characteristics and components

Power distribution networks are coming in different configurations, topologies, circuit lengths, and characteristics, sharing common characteristics, as described in Table 5.1. A feeder represents one of the circuits originating out of a power substation, while the main feeder is the three-phase backbone of the circuit, often called the mains or mainline. Branching from the mains are one or more laterals, which are also called taps, lateral taps, branches, or branch lines. These laterals may be single-phase, two-phase, or three-phase circuits. Power distribution systems provide the infrastructure to deliver the electricity from the substations to the loads or users. This complex system provides the safe and reliable transfer of electric

Table 5.1 Typical distribution circuit parameters

Substation and feeder characteristics	Most common value	Other common values
Voltage	12.47 V	4.16, 4.8, 13.2, 13.8, 24.94, and 34.5 kV
Number of station transformers	2	1–6
Substation transformer size	21 MVA	5–60 MVA
Number of feeders per bus	4	1–8
Peak current	400 A	100–600 A
Peak load apparent power	7 MVA	1–15 MVA
Power factor		0.8 lagging to 0.95 leading
Number of customers	0.98 lagging	50–5000
Length of feeder mains	400	2–15 mi
Length including laterals		4–25 mi
Area covered		0.5–500 mi^2
		4–25 mi
Mains wire size	4 mi	4/0–795 kcmil
Lateral tap wire size		#4–2/0
Lateral tap peak current	8 mi	5–50 A
Distribution transformer	25 mi^2	10–150 kVA
size (1-ph)	500 kcmil	
	1/0	
	25 A	
	25 kVA	

energy to various customers throughout the service territory. The most common distribution primaries are four-wire, three-phase lines, and a multigrounded neutral. Single-phase loads are served by power distribution transformers connected between one phase and the neutral. The neutral conductor acts as a return conductor and equipment safety ground. A single-phase line has one-phase conductor and the neutral, and a two-phase line has two phases and the neutral. Some distribution primaries are three-wire systems (with no neutral). On these, single-phase loads are connected phase to phase, and single-phase lines have two of the three phases. Figure 5.4 shows the most common designs and configurations for power distribution feeders. Typically radial in nature, power distribution systems include feeders and laterals. The most common voltages are 34.5 kV, 14.4 kV, 13.8 kV, 13.2 kV, 12.5 kV, 12 kV, and sometimes even much lower voltages. The power distribution voltages in a specific service territory are likely similar because it is easier and more cost effective for maintenance and to stock spare parts when the system voltages are consistent and equipment is the same. The main-lines are usually conductors made of 500-kcmil or 750-kcmil aluminum conductors. Utilities often design the main feeder for 400 A and often allowing an emergency rating of about 600 A, while one or more laterals (taps, lateral taps, branches, or branch lines) are branching out from the main. These laterals may be single-phase, two-phase, or three-phase electrical circuits. The laterals normally have fuses or circuit breakers to separate them from the mainline in the event of faults or for other

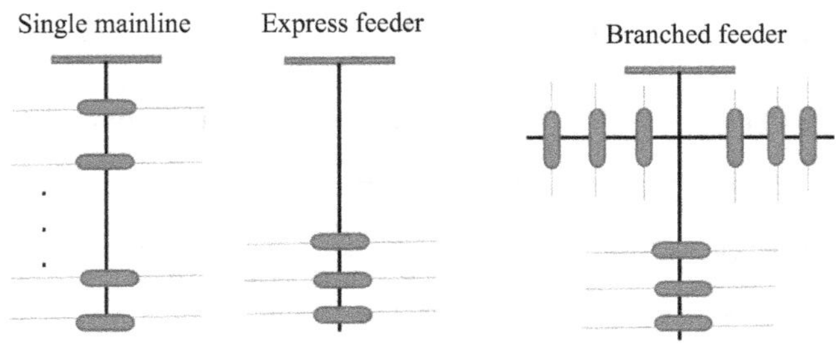

Figure 5.4 Common power distribution primary arrangements

reasons. There are also several distribution devices used to improve the safety, reliability, and power quality of the system, which a few of these types of devices are reviewed here.

The most common distribution primaries, are four-wire, multigrounded systems are often drawn as one-line drawings of the most common types of circuits to simplify the calculation and design. A single-phase line has one-phase conductor and the neutral, and a two-phase line has two phases and the neutral. Some distribution primaries are three-wire systems, with no neutral, and the single-phase loads are connected phase to phase (two of the three phases). There are several configurations of both primary and secondary power distribution. However, most of the power distribution networks are radial type configurations (both primary and secondary). Radial networks have many advantages over networked circuits including, among others: easier fault current protection, lower fault currents over most of the network, easier voltage control, easier prediction and control of power flows, and usually significantly lower costs. Power distribution primary systems are in a variety of shapes and sizes, as shown in Figure 5.4, depending on the street layouts and the area shape covered by the circuit, topography, and the large load locations. A common suburban layout has the main feeder along a street with laterals tapped downside streets or into developments. Radial distribution feeders may also have extensive branching, whatever it takes to get to the loads. Their major disadvantage is when a fault occurs all loads down the fault are deenergized. An *express feeder* serves load concentrations at some distance from the substation. A three-phase mainline runs a distance before tapping loads off to customers. A primary-loop scheme is a more reliable service that is sometimes offered for critical loads, such as hospitals or military facilities, in which the circuit is "routed through" each critical customer transformer. If any part of the primary circuit is faulted, all critical customers can still be fed by reconfiguring the transformer switches. Primary-loop systems are sometimes used on power distribution systems for areas needing higher reliability or limited long-duration interruptions. In the open-loop design, the loop being usually left open at some point, primary-loop systems have almost no benefits for momentary interruption or voltage sags, and are rarely operated in a closed loop. Radial distribution feeders may also have

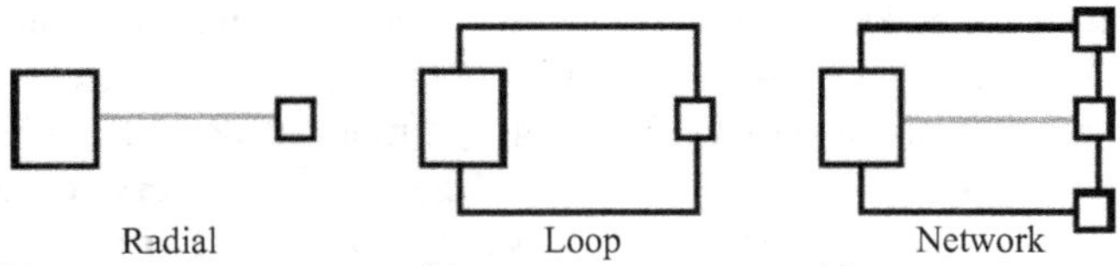

(a) (b) (c)

Figure 5.5 Typical distribution feeder configurations, while the radial design is the most common used

extensive branching, whatever it takes to get to the loads. An express feeder serves load concentrations some distance from the substation. A three-phase mainline runs a distance before tapping loads off to customers. With many circuits coming from one substation, a number of the circuits may have express feeders, some feeders cover areas close to the substation, and express feeders serve usually areas farther from the distribution substation.

The most common configurations of electrical power distribution systems are radial, ring, and network systems (as shown in Figure 5.5(a)–(c)). *Radial power distribution systems* (Figure 5.5(a)) are the cheapest topology and simplest distribution type since the electrical energy is coming from a single power source, usually located into the center of the load area. A generating system is supplying electricity from the substation to radial power lines which are extended to various areas. Most power distribution circuits are radial, both primary and secondary. Radial circuits have many advantages over networked circuits including (a) Easier voltage control, (b) easier prediction and control of power flows, and (c) lower cost. However, either the simplest power distribution system, the radial power distribution is the least reliable in terms of continuous service since there is no power backup. If any power line opens, one or more loads are interrupted, leading to a high probability of power outages. This type of power distribution is usually used in remote areas where other distribution systems are not economically feasible. *Ring (loop) power distribution systems* are used in heavily populated areas, and the power distribution lines encircle the service area (Figure 5.5(b)). Electricity is delivered from one or more power sources into substations near the service area. Then the power is delivered from the substations near the service area through radial lines. In the event that a power line is opened, through alternative lines the electricity is delivered to the loads. The ring power system provide a more continuous service that the radial systems. However, the rig systems are more expensive due to additional power lines and greater circuit complexity. *Network power distribution systems* are a combination of radial and ring systems (Figure 5.5(c)), usually result when one of the other two systems is expanded. Most of the US distribution systems are network power distribution type. This type of power distribution is more complex but it provides very reliable service to consumers. With a network power distribution each load is fed by two or more power circuits. Power flow on a line is a function of voltage and current, the current is inherently bidirectional, so the power can flow in either direction. However, other operational constraints, such as circuit breakers and control devices may not be able to accommodate bidirectional power flow without modification. The electrical grid operates as a three-phase network down to the level of the service point to residential and small commercial loads. Feeders are usually three-phase overhead pole lines or underground cables, while closer to the loads are usually single-phase. Three-phase lines provide advantageous characteristics for rotating machines by inducing a smooth rotating magnetic field with which the rotating magnetic field can be coupled, offering a significant advantage for power transmission and distribution. This makes it possible to string three conductors carrying voltage and rely on the mathematical cancellation of this power to provide a virtual neutral. Without a metallic return wire, significant cost savings can be achieved. Thus, nearly all transmission and distribution power lines have only three conductors. To ensure efficient operation, it is important to balance the phases so that they are approximately equal.

System voltage (level) is a term used to identify whether the reference is being made to the secondary or to the primary power distribution systems. Residential, commercial, and small industrial loads are usually served with 600 V or lower voltages. Manufacturers have standardized the provision of insulated wire to have a maximum 600 V AC voltage rating for "secondary" services. For example, household wire such as extension cords has a 600 V insulation rating. Other than changing the plugs and sockets on either end, such wire or conductor can be used for higher AC voltages, such as 240 V. The 34 kV system's voltage is used differently among electric power companies. Some companies use 34.5 kV power distribution system voltages to connect service transformers in order to provide secondary voltages to consumers, whereas other companies use 34.5 kV power lines between power distribution substations and not for connecting consumers. There is several common power distribution system voltages between secondary power distribution and 34.5 kV used in the industry. For example, many power companies have standardized power distribution at 12.5 kV while others use 25 kV. Some companies use 13.2 kV, 13.8 kV, 14.4 kV, 20 kV, and so on. There are several areas, where the utilities are still using the old 4.16 kV systems. However, these LV power distribution systems are being phased out due to their high losses and short-distance transfer capabilities. We have to keep in mind, that for any given electrical load (kVA), the higher the load voltage the lower is the resultant current required. As it is the current flowing in the supply cabling which creates the voltage drop and the heating (I^2R) loss, to minimize losses, the line current values must be kept as low as possible. In this way, the necessary distribution line voltage level can be determined, along with the resultant cost of constructing the line. In the last decades of the twentieth century, the primary distribution systems are moving toward HV levels. HV distribution systems have advantages and disadvantages, such as lower voltage drops and losses, higher line capacity, fewer substations and longer reach, on the expenses of lower reliability, higher equipment costs, and maintenance crew safety and acceptance. The voltage drop reduction, power and area coverage expansion due the use HV voltage distribution system are calculated as:

$$\Delta V_{HL-PD} = \left(\frac{V_{HV-PD}}{V_{LV-PD}}\right)^2 \cdot \Delta V_{LV-PD} \tag{5.16a}$$

$$P_{HL-PD} = \left(\frac{V_{HV-PD}}{V_{LV-PD}}\right) \cdot P_{LV-PD} \tag{5.16b}$$

And

$$A_{HL-PD} = \left(\frac{V_{HV-PD}}{V_{LV-PD}}\right) \cdot A_{LV-PD} \tag{5.16c}$$

Here, $V_{HV\text{-}PD}$ and $V_{LV\text{-}PD}$ are the voltages on HV and LV circuits, $P_{HV\text{-}PD}$ and $P_{LV\text{-}PD}$, the power on each circuits, $\Delta V_{HV\text{-}PD}$ and $\Delta V_{LV\text{-}PD}$, the voltage drops per unit length in percent on HV circuit and on the LV circuit, $A_{HV\text{-}PD}$ and $A_{LV\text{-}PD}$, the areas covered by the HV and LV circuit, respectively.

Example 5.4: A utility decided to increase the primary power distribution voltage level from 12.47 to 24.94 kV, estimate the voltage drop reduction, and the increases into the power and covered area.

Solution: The required voltage drop reduction, power and covered area increases are:

$$\Delta V(\%) = \left(\frac{24.94}{12.47}\right)^2 \times 100 = 400\%$$

$$P(\%) = \left(\frac{24.94}{12.47}\right) \times 100 = 200\%$$

$$A(\%) = \left(\frac{24.94}{12.47}\right) \times 100 = 200\%$$

The squaring effect on voltage drop is significant, meaning the doubling the system voltage quadruples the load that can be supplied over the same distance, or twice the load can be supplied over twice the distance, or the same load can be supplied over 4 times the distance. Resistive line losses are also lower on HV systems, especially in the voltage-limited circuits.

5.3.1 Power distribution equipment and components

The function of the electric power distribution system in a building or an installation site is to receive power at one or more supply points and to deliver the electricity to the individual loads (lighting devices, electrical motors, heaters, computers, and all other electrically operated devices and equipment). The critical importance of the power distribution system to the function of a building or an industrial site makes it almost imperative that the best system be designed, structured and installed. In order to design and structure the best and the most suitable power distribution system, the system design engineer must have the most complete information concerning the current and future loads, as well as the best knowledge of various types of power distribution systems that are applicable. Various categories of buildings have many specific design challenges but certain basic principles are common to all. Such principles, if followed, are providing a good executed power system design. In order to distribute and deliver electrical energy, several specialized equipment and devices are required and needed, including the overhead transmission lines, underground cables, power transformers, protection devices, circuit breakers, fuses, lightning arresters, power-factor correction capacitors, or power metering systems. The most common distribution primaries are four-wire, multigrounded systems: three-phase conductors plus a multigrounded neutral. Single-phase loads are served by transformers connected between one phase and the neutral. The neutral conductor acts as a return conductor and as an equipment safety ground. A single-phase line has one-phase conductor and the neutral, and a two-phase line has two phases and the neutral. Some distribution primaries are three-wire systems (with no neutral). On these, single-phase loads are connected phase to phase, and single-phase lines have two of the three phases. Modern electric power technologies may be useful to the designer, power engineer, and building owner in addressing these new challenges. The advent of microprocessor devices (smart devices) into power distribution equipment has expanded facility options and capabilities, allowing for automated communication of critical power system information (energy data and system operation information) and electrical equipment control. These building technologies may be grouped as: power monitoring and control, building management systems interfaces, lighting, heat, and air-conditioning control systems, automated energy management system, and predictive diagnostics. The design of an electrical distribution system for a given customer and facility, are requiring alternate design approaches that are best fitting the following overall goals, such as safety, minimum initial investment, maximum service continuity, maximum flexibility and expendability, maximum electrical efficiency, minimum operating and maintenance costs, maximum power quality, maximum supply security, consideration to the future building expansion and/or increased load requirements due to added utilization equipment when designing the electrical distribution system.

5.3.2 Three-phase power distribution, grounded, and ungrounded systems

Power distribution lines (feeders), as shown in Figure 5.6 are usually connected radially out of the power substation, meaning that only one end of the distribution power line is connected to a source. Therefore, if the source end becomes deenergized (open), the consumers connected to this feeder are out of service.

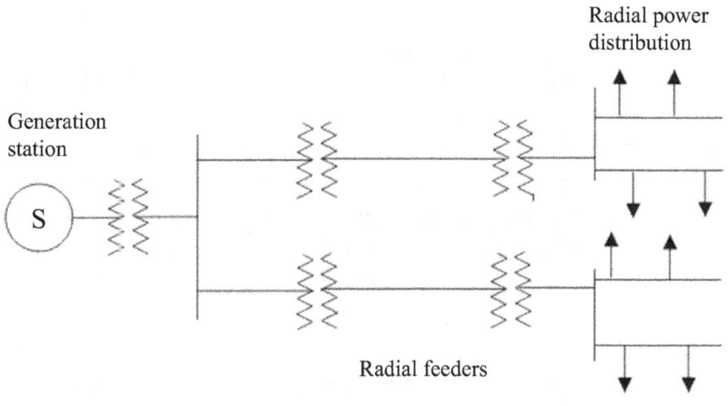

Figure 5.6 Schematic diagram of a power distribution system

The substation transmission side has usually multiple transmission lines feeding the substation. Distribution feeders might have several disconnect switches located throughout the line, allowing the load transfer capability among the feeders, the isolation of line sections for maintenance, and the visual openings for safety purposes while working on the lines or other HV equipment operations. Even though there might be several lines and open/closed disconnect switches connected throughout a power distribution system, the distribution lines are still fed radially. Most of the three-phase distribution feeders and transformer connections are using the wye (star) systems, having more advantages than disadvantages. Although delta-distribution systems exist, the trends are to convert them to star (Y) configurations. The wye connection has one wire from each coil connected together to form the neutral. Most of the time, this neutral is grounded. Grounded implies that the three common wires are connected together and then connected to a ground rod, primary neutral, or ground grid. The grounding system provides a low-resistance connection to earth. Grounding gives earth an electrical reference; in the case of the wye connection, this neutral reference is zero volts. Later chapters will cover grounding and protection issues in more details and include also some of the safety issues associated with proper grounding. Power transmission and distribution systems are used to interconnect electrical power production systems and to provide a means of delivering electrical power from the generating station to its point of utilization. Most electrical power systems cast of the Rocky Mountains are interconnected with one another in a parallel circuit arrangement. These interconnections of power production systems are monitored and controlled, in most cases, by a computerized control center.

However, both wye and delta configurations have distinguishable advantages and disadvantages when it comes to transmission or distribution systems. Transmission and subtransmission lines are built as three-phase, three-wire lines. The ends of the transmission lines are connected to either delta or source-grounded wye transformer connections. "Source-grounded wye" connection means that the transmission transformer in the substation is a four-wire wye transformer that has the three phases connected to the line conductors and the neutral connected to the substation ground grid. Note that the neutral is not provided on transmission lines, being not necessary to provide the neutral on the transmission line because all three phases are assumed to have balanced currents, so no current is flowing into the neutral conductor when the currents are balanced. With respect to power distribution lines, most systems use grounded-wye connections and current is usually present in the neutral because the three phase currents are normally not balanced. Three-wire delta distribution lines exist, primarily in rural areas, where a neutral is not present. Those lines are more vulnerable to stray currents and voltages through the earth as the earth tries to balance the current flow. The preferred standard for distribution is the grounded-wye configuration. From the perspective of distribution systems, the following predominant advantages and disadvantages apply. Advantages of grounded-wye configuration include:

1. The power company primary distribution neutral, the service transformer, and the customer's service entrance equipment are all grounded, all to the same reference voltage point, the common reference ground.
2. The common ground improves the voltage stability because the reference point is consistent, improving also the power quality.
3. Equipment can be connected to the line-to-neutral potential instead of the higher voltage line-to-line potential.
4. Since the equipment is connected at a lower voltage (line-to-neutral instead of line-to-line), bushings, spacing, and insulation requirements are all be smaller.
5. Since one side of the transformer winding is connected to the grounded neutral that connection does not need a bushing. Instead, single-bushing transformers have an internal connection to the neutral.
6. Easier to detect line-to-ground faults, if a phase conductor fall to the ground, a tree make contact with a phase, the short-circuit overcurrent condition significantly increases the current in the neutral back at the substation. Therefore, merely measuring the neutral overcurrent condition at the substation with a current transformer connected to the transformer neutral determines whether a line-to-ground fault condition exists out on the distribution feeder. This overcurrent condition initiates a trip signal to the feeder circuit breaker, and in the case of delta configurations, there is no true grounded neutral, making harder to detect line to ground faults.
7. Single-phase protection is better with fuses on transformers and distribution feeder lateral extensions are clearing the faults more reliably than fuses connected in delta configurations. Since deltas have equipment connected line to line, a line-to-ground fault could blow one or more fuses. Fuses in delta circuits are weakened from faults on other phases. Therefore, it is a common practice in delta systems to replace all three fuses in case one or more were weakened from a line-to-ground fault.

Major disadvantages of grounded-wye configuration includes, it requires four conductors, while delta systems require only three conductors for three-phase power. That is an advantage of the delta configuration that resulted in the majority of distribution lines being built with delta configurations early in the process of electrifying America. Today, most of these lines have been converted to four-wire wye systems due to the advantages that wye has over delta. Advantages of delta configuration are:

1. Three conductors versus four conductors (i.e., less expensive).
2. The third-order harmonics are eliminated due to a natural cancellation. In other words, the 60 Hz power sine wave is cleaner by nature. The 120° phase shift between phases acts to cancel out some unwanted interference voltages.
3. Lightning performance, mining that sometimes the isolated conductors in a delta from ground configuration minimize the effect that lightning has on a system. However, lightning arresters in delta systems are still connected line to neutral.

The major disadvantages of delta configuration are:

1. There is no ground reference, means that the service voltage may be less stable, fuse protection may be less effective, and there might be more overall power quality issues.
2. The distribution transformers can cause stray currents to flow in the earth when their LV secondary side is grounded. Although the primary side of the distribution transformer is not grounded, the secondary side is grounded. Therefore, a small but measurable voltage is inadvertently connected to ground, causing stray currents to exist.
3. Three-phase transformer banks can regulate the primary voltage or try to equalize the primary voltage. The delta connections along with the transformers having the same turns' ratios can cause the primary voltage to equalize, resulting in additional stray currents or unbalanced currents in the feeder.

Comparing all the advantages and disadvantages, the multigrounded neutral, four-wire wye-distribution feeder is the preferred method. Most of the delta distribution lines have been replaced with grounded wye systems, but some deltas still exist. The preference is for the power companies to use grounded wye systems on all distribution systems. However, converting delta to wye does require the cost of adding a conductor Conversion can be a slow process. Three-phase wye-connected power distribution systems are in three-wire without neutral or in four-wire with neutral, either in grounded or ungrounded configurations. Land-based systems are generally grounded, while ship-based or other mobile power systems are ungrounded. For neutral conductor can be in one of the following configurations, solidly grounded neutral (via strap of negligible resistance), high-impedance ground (via a series resistance and/or reactance with the ground strap to control the fault current to match with the protection specifications), and ungrounded systems with nongrounded neutral. It is worth to keep in mind that over 80% of the faults in three-phase systems are between the phase conductor and the ground. It is also important to understand the grounding the neutral and grounding the equipment chassis are not the same, the later one is important for personnel safety, while the former is for the equipment safety and protection.

5.3.3 Power distribution transformers and devices

A power distribution transformer is designed to reduce the primary voltage of the electric distribution system to the utilization voltage serving and required by the customer. It is a static device consisting of two or more windings used to transfer AC electric power by electromagnetic induction from one circuit to another at the same frequency but with different values of voltage and current. While power transformers are designed and optimized to function at high load continuously, power distribution transformers are operating at quite low loading for most of the day. This daily variation makes design parameters wider, optimizing for 10% to about 60% of peak loading, and results in lower efficiencies compared to power transformers. However, the new standards, released by DOE are increasing the minimum efficiency of these transformers. Advanced materials can be leveraged to achieve efficiencies in a cost-effective matter. It is also important to notice that as more distributed energy resources, with the smart grid advent, are deployed into the power distribution networks, it is expected that the loading variability on these power distribution transformers to increase, along with more dynamic voltage fluctuations and current flows. It is important to understand how these changes are impacting the efficiency, lifetime, performance, design, and protection of these critical grid components. Usually, the power distribution transformers are self-protected, being equipped with a lightning arrester, a weak-link or protective-link expulsion-type fuse (oil installed into the transformer tank), a secondary circuit breaker, and a warning light. The transformer primary bushing conductor is connected to one phase of the three-phase primary circuit through a partial-range current-limiting fuse. The transformer tank is grounded and connected to primary and secondary common-neutral ground wire. The self-protected transformer contains a primary fuse mounted on the primary bushing, a secondary terminal block, and an LV circuit breaker. Conventional overhead distribution transformers are also available. The key component distinguishing the conventional distribution transformer from the self-protected transformer is the lack of internal fusing in the conventional model. An external fuse/cut-out combination is mounted between the distribution primary and the conventional transformer. Pad-mounted transformers are usually used with underground systems. Three-phase pad-mounted transformers are used for commercial installations, and single-phase pad-mounted transformers are used for underground residential installations. Vault-type distribution transformers are installed for commercial customers where adequate space is not available for pad-mounted transformers. The vault-type transformer may be installed in a vault under a sidewalk or in a building. They are often used in the underground electric network areas. Submersible single-phase distribution transformers are used in some of the underground systems installed in residential areas. Corrosion problems and the high cost of installation are preventing the extended use of such transformers.

Specialized hardware systems are necessary to interconnect the elements and components of a power distribution system. Utility control and switching systems operated under heavy demanding conditions, including high voltage and current levels, lightning discharge exposure, higher safety and reliability requirements, and 24-h-a-day use. For reliability performance, large safety margins must be built in each element of the power distribution system. The primary control and switching elements are HV switches and protection devices. To protect distribution system equipment from damaging overloads, that equipment contains *protective devices*, which are in a variety of forms or types, usually through *fuses* or *circuit breakers*. These devices sometimes employed to provide the logic to decide when to trip a circuit breaker are referred to as *protective relays*, or simply *relays*. Relays are programmed for a variety of functions, including (but certainly not limited to) overcurrent, over- and under-voltage, over- and under-frequency, differential current, and reverse current. The primary objective of protective devices is to deenergize equipment when specific conditions occur. To protect equipment or maintain safety, these devices typically respond to faults (e.g., a short circuit) by isolating appropriate equipment. A goal of good protective coordination is to limit the isolation to the smallest portion of the power system, limiting the disruption and prevents adjacent portions of the system from being affected. A *fault* in an electric power system is an unintentional and undesirable creation of a conducting path (*short-circuit*) or a current blockage (*open-circuit*). The most common fault is the short-circuit fault and is usually implied when the people are using the term fault. The protection schemes must also account for the failure in isolating devices (e.g., fuses, circuit breakers, etc.), providing redundant protection capability, usually accomplished through protective zones whereby protective relays operate quickly for faults within their primary zones and more slowly for faults that are farther away in secondary or tertiary zones.

Switches are used for isolation, load interruption, and transferring service between different sources of supply. Distribution switches are devices used to disconnect various parts of the system from the feeder. These switches are manually, remotely, or automatically operated, being designed to break load current but not fault current and are used in underground circuits or tie switches. Isolating switches are used to provide visible disconnect to enable safe access to the isolated equipment. These switches usually have no interrupting current rating means that the circuit must be opened by other means (such as breakers). Load interrupting or a load-break switch combines the functions of a disconnecting switch and a load interrupter for interrupting at rated voltage and currents not exceeding the continuous-current rating of the switch. Load-break switches are of the air- or fluid-immersed type. The interrupter switch is usually manually operated and has a *quick-make, quick-break mechanism* which functions independently of the speed-of-handle operation, being typically used on voltages above 600 V. For services up to 600 V safety circuit breakers and switches are commonly used. Safety switches are enclosed and may be fused or un-fused. Distribution breakers, like switches are used to disconnect portions of the feeder. Distribution breakers have also the ability to interrupt fault current. Typically, these are tied to a protective relay, which detects the fault conditions and issues the open command to the breaker. Reclosers are a special type of breakers, typically deployed only on overhead feeders and are designed to reduce the outage times caused by momentary, caused by vegetation or temporary short circuits. During the reclose operation, the relay detects the fault, opens the switch, waits a few seconds, and issues a close. Many overhead distribution faults are successfully cleared and service is restored with this technique, significantly reducing outage times. Capacitors, three-phase types are designed to inject reactive power (VARs) into the power distribution circuit, to improve power factor or support system voltage (reducing the feeder voltage drops as discussed before). They are operated in parallel with the feeder circuit and are controlled by capacitor controllers, often connected to remote communications allowing for automatic or coordinated operation. Fuses are standard devices used to protect portions of the circuit when a breaker is too expensive or too large. Fuses are used as an over-current-protective device with a circuit-opening fusible link that is heated and severed as over-current passes through it. Fuses are available in a wide range of voltage, current, and interrupting ratings, current-limiting types, and for indoor and outdoor applications. Fuses perform the same function as circuit breakers, and there is no general rule for using one versus the other. Fuses can be used to protect single-phase laterals off the feeder or to protect three-phase underground circuits. The decision to use a fuse or circuit breaker is usually based on the particular application, and factors, such as the current interrupting requirement, coordination with adjacent protection devices, space requirements, capital and maintenance costs, automatic switching, etc. Lightning arresters are devices designed to reduce the surge on the line when lightning strikes the circuit, protecting equipment and people. Automated fault detection, localization, isolation, and load restoration (FDIR), part of modern distribution systems are designed to manage and solve the power system faults. They are detecting a fault, localize it to a segment of feeder, open the switches around the fault, and restore unfaulted sources via the substation and alternative sources as available. Modern systems use to communicate by using the Internet, secure broadband and wireless communication, which provide significant improvement over other systems, which include peer-to-peer communications, multiple access to tie switches, and remote access by communications and automation maintenance personnel.

5.4 Industrial power distribution and building power supply systems

Depending on the needs of the customers, the voltage supplied can be as low as 120 V single-phase or 120/240 V single-phase, where the 240 V secondary of the distribution transformer has a center tap, providing, if needed also two 120 V single-phase circuits. However, larger commercial customers usually utilize three-phase power, with the most common 208/120 V or 480/277 V services. However, very large commercial and industrial facilities can include higher three-phase services, such as 2.4 kV, 4.16 kV, or often higher voltage levels. Usually such industrial or large commercial customers are responsible for providing power transformers and other infrastructure to serve all of the lower voltage requirements needed by their facilities. In a residential area, a typical service transformer (ground mounted or pole mounted) converts the three-phase power, for example, at a distribution voltage of 13.8 kV (primary power distribution transformer) to power at a voltage level of 120/240 V, usually single-phase. Each of the 240 V, single-phase transformer secondaries has a center tap that provides two 120 V single-phase circuits. Therefore, an individual residence can be supplied with both 120 and 240 V single-phase service. The higher voltage is necessary for appliances, such as clothes dryers, water pumps, etc. Electricity for buildings can be supplied from the local utility distribution systems or can be generated partially or entirely on the facility site, using conventional energy generators or alternative energy conversion systems (wind turbines, photovoltaic systems, fuel cells, etc.). Of great interests among the building energy systems are the micro-combined cooling, heat and power generation (micro-CCHP) systems, designed for heating, cooling (air conditioning) and as by-product electricity generation, etc. There are several types of building power distribution systems that are used, and the most appropriate for a specific building depends on the building size, destination, and load characteristics (power ratings, voltages, frequency, etc.). The requirements for electrical power distribution systems apply to all nonresidential or residential buildings. The intention is to save energy and to allow future systems for power use, expansion, monitoring and control to be added when expected changes in the marketplace are occurring.

5.4.1 Switchgears, load, and motor centers

Power distribution systems used in large commercial and industrial applications are complex specific to the application requirements and provisions. Power may be distributed through switchgear, switchboards, transformers, and panel-boards. Power distributed throughout a commercial or industrial application is used for a variety of applications, such as heating, cooling, lighting, and motor-driven machinery. Motor control centers receive this power through complex distribution systems which include power distribution lines and related equipment. Transformers used with three-phase power require three interconnected coils in both the primary and the secondary. These transformers can be connected in either a wye or a delta configuration. The type of transformer and actual voltage depend on the requirements and capability of the power company and the customer needs. Motor Control Centers (MCCs) are centralized hubs containing motor control units sharing a common power bus. Used in LV (230–600 V) and MV (2.3–15 kV) three-phase applications, MCCs traditionally house startup and drive units. Modern MCC units go beyond these basic functions, being not uncommon for smart load centers to include programmable controllers, smart metering equipment for complex control schemes and safety features, in addition to the overload relays and contactors. Switchgear incorporates switches, circuit breakers, disconnects, and fuses used to route power and in the case of a fault, isolate parts of an electric circuit. In general, switchgear has three basic functions (1) protection and safety for equipment and workers, (2) electrical isolation to permit work and testing, and (3) local or remote circuit switching. Developments in switchgear design have led to the introduction of network support for monitoring and control as well as advanced diagnostic capabilities for the purposes of monitoring usage, loading, and a host of other operational parameters. The advantages of modern switchgear are (a) control and monitoring over a network, (b) remote control of the switchgear over a network, by removing the risk of arc flash to the operator, (c) individual analysis and power monitoring, involving the power monitoring and analysis of the main power and individual circuit breakers, and (d) advanced technology, the latest technology in circuit breakers and switching. Unlike other types of power distribution equipment, which are used with a variety of load types, MCCs primarily control the distribution of power to electric motors. Usually in industrial and large commercial facilities and structures, the electricity is delivered to loads from designated system, known as load centers, containing the necessary equipment to monitor, protect, and control the loads. There are different types of load centers, their selection based primarily on the electrical requirements of the loads and the installation environment. Several types of load centers are housed in metal enclosures to protect enclosed equipment, nearby objects, and personnel in the event of equipment malfunction. In terms of terminology, load centers supplying large motors and/or other large industrial loads are referred to as switchgear, while smaller load centers specialized to supply small to midsize motors are called motor control centers. Auxiliary equipment, such as control switches, indicator lights, monitoring, and metering equipment are installed in smaller metallic enclosures whose capabilities to withstand the external environment are determined by the specifications of the NEMA 250 Standard.

Switchgear is used to connect and disconnect electric power supplies and systems, being a general term which covers the switching device and its combination with associated control, measuring, protective and regulating equipment, together with accessories, enclosures, and supporting structures. Switchgear is applied in electrical circuits and systems from LV, such as domestic 220/240 V applications, right up to transmission networks up to 1,100 kV. Switchgears are specialized load center designed to supply large loads, such as large electric motors or smaller load centers. Secondary distribution switchgears are that which are connected directly to the electrical utility transformers which provide LV supplies to customers, distributing power from the primary substations can be conveniently divided into overhead circuits and underground cable networks. Switchgears come in two varieties: metal-enclosed and metal-clad. The later one is the more robust design, requiring shutters between the bus and the front of the equipment, compartments for live parts, and insulation of the bus and primary components. All switching and interrupting devices must be drawing out mounted, allowing the removal from the switchgear without unwiring. For metal-enclosed switchgears, these criteria are not required and often have lower interrupting ratings, lower breaker duty cycle, and may use fused or nonfused switches instead of circuit breakers. Metal-enclosed switchgear is governed by IEEE standard C37.20.3, while metal-clad switchgears are covered by multiple standards. Metal-clad switchgears meet or exceed the metal-enclosed switchgear requirements, but not vice versa. A switchgear lineup is made up of multiple sections or cubicles joined side by side. The front of the switchgear is hinged and opening the doors exposes the circuit breakers. When a breaker is removed from metal-clad switchgear, an insulated barrier separating the cubicle from the energized bus-work that runs the length of the switchgear is visible, an important safety feature, since metal-clad switchgears are usually maintained while the main bus is still energized. In industrial applications, where several electric motors are required, and it is often desirable to control some or all of the motors from a central location, through a motor control center.

Motor control centers are physical groupings of combination starters in one assembly. A combination starter is a single enclosure containing the motor starter, fuses or circuit breaker, and a disconnecting power device. Other devices associated with the motor, such as pushbuttons and indicator lights, may also be included. An MCC is a load center customized to serve several small to midsize motors. Since the combination starters are much smaller than switchgear breakers, stacking more than two units per horizontal section and reducing the width of the horizontal sections are saving considerable space. Motor control centers receive this power through complex distribution systems which include power distribution lines and related equipment. Transformers used with three-phase power are connected in either a wye or a delta configuration. MCCs are centralized hubs containing motor control units sharing a common power bus. Used in LV (230–600 V) and MV (2.3–kV) three-phase applications, MCCs traditionally house startup and drive units, are also including programmable controllers and metering equipment for complex control schemes and safety features, in addition to the overload relays and contactors. Switchgear incorporates switches, circuit breakers, disconnects, and fuses used to route power and in the case of a fault, isolate parts of an electric circuit. In general, switchgear has three basic functions: protection and safety for equipment and workers, electrical isolation to permit work and testing, and local or remote circuit switching. Developments in switchgear design have led to the introduction of network support for monitoring and control as well as advanced diagnostic capabilities for the purposes of monitoring usage, loading and a host of other operational parameters.

5.4.2 Switchgear and motor control center ratings

The standards controlling the design and testing of metal-clad switchgear were developed by the American National Standards Institute (ANSI) in conjunction with the IEEE. IEEE standard C37.20.2 stipulates metal-clad switchgear-rating criteria. LV switchgears (600-V class or lower) are governed by IEEE standard C37.20.1, while IEEE standards C37.04, C37.06, and C37.09 specify MV circuit breaker rating and testing criteria. There are several important ratings of the main bus-work in metal-clad switchgear. A continuous current rating is assigned to limit the temperature rise of the busbars to a value that does not compromise the bus insulation. The busbars typically are insulated with an epoxy-type material that can withstand fairly high-operating temperatures. Insulators, either porcelain or polymeric composite, support the bus bars. The insulators also have a maximum operating temperature. A main (horizontal) bus similar to that used in switchgear is implemented in MCCs. Since the loads supplied by an MCC are smaller than the loads fed from switchgear, the available bus ratings tend to be lower in an MCC. The MCC is made up of units which are stacked to form sections, while one or more sections are making a MCCA system, the busbar incorporated into the design to provide power to each unit in each section. Commonly available bus ratings range from 600 to 2,500 A for the horizontal bus, and 300 A to 1,200 A for the vertical bus although some manufacturers may offer other ratings. Interrupting ratings of 65 kA and 100 kA are common. Although MCCs are usually implemented at LVs (240 or 480 V), MV MCCs are also available. It is important to determine the continuous current rating of switchgear bus by temperature rise. While some equipment specifications cite a maximum allowable current density to determine the bus rating, this practice is not allowable. Short-circuit currents can exert tremendous mechanical forces on the busbar, when very large currents flow through the busbar. These currents are producing intense magnetic fields tend to deflect the busbar, forcing the parallel bus sections either toward or away from each other depending on the current direction. These mechanical forces give rise to *short-circuit current ratings*. In the past, the fault withstanding capability of switchgear and circuit breakers was rated in MVA. This meant that if the equipment was applied at a voltage lower than its rated voltage, an increase in fault current capability was also implied. However, because of the changes in circuit breaker technology, switchgear, and load centers are no longer manufactured with increased fault current capability voltages lower than rated voltage. Today, switchgears are rated in kA instead of MVA, which is some sort of confusion as a result of this change in rating methodology. The MVA-class nomenclature has remained in the switchgear industry. The electrical properties of the insulation system define a basic impulse level rating, which is indicating the maximum voltage that can be tolerated without insulation failure. The standards also specify overvoltages and impulse voltages which the switchgear must withstand, such as normal overvoltages which occur during switching or which may be transferred due to lightning striking exposed circuits. The switchgear also needs to provide sufficient isolating distances to allow personnel to work safely on a part of the system that has been disconnected. The lightning impulse withstand voltage and the normal power frequency withstand voltage are specified relative to the normal system voltage in the standards.

NEC, Section 110.26(A) is also specifying the working space around switchgears, panel-boards, and load centers for different current ratings. For proper safety, appropriate clearances must be maintained around switchgears, panel-boards and any loaf centers. The clearances are needed to provide for adequate work-space around electrical equipment to ensure safety in the event of the equipment failures. Rooms dedicated to the transformer and switchgear must be so arranged that they are easily accessible. The room dimensions are determined, taking noise, fire hazards, solid-borne sound, ability to replace equipment, operation and maintenance, and temperature rise into account. When installing transformers and/or switchgears, the requirements given below need attention: safety distances, access for transport, cooling/ventilation, ease of transformer replacement, easy to operate and maintain, locations of auxiliary equipment, installation of fire protection system; and galleys, pits, and sumps must be provided under transformers containing insulating liquids to protect against fire and water pollution. Also notice that the environment in which the switchgear is installed can dictate changes to the required ratings. Installing switchgear at higher altitudes, 1,000 m above sea level or higher for MV switchgear and 2,000 m above sea level or higher for LV switchgear requires that both the voltage and the current ratings must be adjusted, since the thinner air at higher altitudes has a lower dielectric strength and poorer cooling properties than the denser air at sea level. Installing switchgear in the areas with high ambient temperatures require an increase in continuous current rating.

5.5 Chapter summary

The basic layout, structure, and operations of the power distribution infrastructure remain quite the same over the last half of the twentieth century. However, in the past three decades, the equipment and power distribution structure have undergone steady improvements, transformers are more efficient, cables are much less expensive and easier to use, and metering infrastructure, control, monitoring and protection equipment and devices are better and computerized. Utilities operate more distribution circuits at higher voltages and use more underground circuits. But the concepts are much the same: alternating current, three-phase systems, radial circuits, fused laterals, overcurrent relays, and so on. Advances in computer technology have opened up possibilities for more automation and more effective protection, monitoring infrastructure, operation and control. In industrial environments, electricity is provided to loads from load centers, switchgear and motor control centers. These load centers have hinged doors for easy internal access for maintenance and operation purposes. Switchgear can be metal enclosed or metal clad. A bus system runs inside the load center in order to provide power to each unit, section, or cubicle. Switchgear with two power sources (double-headed switchgear) requires specific source transfer method to switch between sources, such as fast transfer, slow transfer, and parallel transfer. Power switchgears include equipment and devices, such as circuit breakers, disconnect switches, main bus conductors, interconnecting wiring, support structures with insulators, enclosures, secondary devices for monitoring and control. Power switchgear is used throughout the entire power system, from generation to industrial plants to connect incoming power supply and distribute power to users. Switchgear can be of outdoor or indoor types, or a combination of both. Reducing line losses in the modern electrical transmission and distribution system is a must and readily available option to enhance electrical efficiency and reduce generation-related emissions. Advances in technology and understanding have made possible significant efficiency gains through investments in improved grid components and, on the demand side, in load management at peak levels. Moreover, the increasing complexity of the power distribution grid is forcing needs to integrate the various systems, advanced metering infrastructure, intelligent control and monitoring, advanced communication, distributed generation and renewable energy systems, and smart loads on that grid. These are having the effect of reducing the costs and increasing the overall benefits of these technologies, while maintaining an improved quality and reliability of the electric energy provided to the customers on the distribution system. This stronger economic justification will drive the rate of advance of these new technologies causing a significant impact on the issues and elements of the design of the distribution system and associated automation systems. There are vast developments happening in the power industry changing whole transmission and distribution world including substations. Smart grid technologies make their way into transmission and distribution world to improve power supply, make it more efficient and reliable, and decrease greenhouse emissions. This became possible due to rapid developments and technological advances in computing, data sciences, information technologies, renewable energy, distributed generation, power electronics, Internet, and communications.

References and further readings

[1] T.R. Bosela, Electrical Systems Design, Prentice Hall, Upper-Saddle River, NJ, 2002.
[2] R.C. Dorf (Ed.), The Electrical Engineering Handbook, CRC Press, Boca Raton, Florida, 1993.
[3] G.F. Moore, Electric Cables Handbook (3rd ed.), Blackwell Science, London, UK, 1997.
[4] P. Schavemaker and L. van der Sluis, Electrical Power System Essentials, John Wiley and Sons, Chichester, UK, 2008.
[5] M.E. El-Hawary, Introduction to Electrical Power Systems, Willey-IEEE Press, NJ, 2008.
[6] A.R. Daniels, Introduction to Electrical Machines, The Macmillan Press, London, 1976.
[7] R.D. Schultz and R.A. Smith, Introduction to Electric Power Engineering, John Wiley & Sons, New York, 1988.
[8] A.C. Williamson, Introduction to Electrical Energy Systems, Longman Scientific & Technical, London, 1988.
[9] J.R. Cogdell, Foundations of Electric Power, Prentice-Hall, Upper-Saddle River, NJ, 1999.
[10] S.J. Chapman, Electrical Machinery and Power System Fundamentals (2nd ed.), McGraw Hill, New York, 2002.
[11] M.A. El-Sharkawi, Electric Energy: An Introduction (2nd ed.), CRC Press, Boca Raton, FL, 2009.
[12] J.L. Kirtley, Electric Power Principles: Sources, Conversion, Distribution and Use, Wiley, Chichester, UK, 2010.
[13] F.M. Vanek, L.D. Albright, and L.T. Angenent, Energy Systems Engineering, McGraw Hill, New York, 2012.
[14] J.E. Fleckenstein, Three-Phase Electrical Power, CRC Press, Boca Raton, FL, 2016.
[15] S.L. Herman, Electrical Transformers and Rotating Machines, Cengage Learning, UK, 2017.
[16] R. Fehr, Industrial Power Distribution (2nd ed.), Wiley, NJ, 2015.
[17] L.L. Grisby (ed.), Electrical Power Engineering, CRC Press, Boca Raton, 2001.
[18] W.A. Thue, Electrical Power Cable Engineering, Marcel Dekker, New York, 1999.
[19] EPRI 1000419, Engineering Guide for Integration of Distributed Generation and Storage into Power Distribution Systems, Electric Power Research Institute, Palo Alto, CA, 2000.

[20] EPRI 1016097, Distribution System Losses Evaluation, Electric Power Research Institute, Palo Alto, CA, 2008.
[21] EPRI 1023518, Green Circuits: Distribution Efficiency Case Studies, Electric Power Research Institute, Palo Alto, CA, 2011.
[22] EPRI 1024101, Understanding the Grid Impacts of Plug-In Electric Vehicles (PEV): Phase 1 Study—Distribution Impact Case Studies, Electric Power Research Institute, Palo Alto, CA, 2012.
[23] EPRI TR-109178, Distribution Cost Structure—Methodology and Generic Data, Electric Power Research Institute, Palo Alto, CA, 1998.
[24] EPRI TR-111683, Distribution Systems Redesign, Electric Power Research Institute, Palo Alto, CA, 1998.
[25] M.W. Earley (ed.), National Electrical Code (NEC) Handbook, NFPA 70, Quincy, MA, 2014.
[26] W.T. Grondzik, and A.G. Kwok, Mechanical and Electrical Equipment for Buildings (12th ed.), Wiley, Hoboken, NJ, 2015.
[27] IEEE Standard 141, IEEE Recommended Practice for Electric Power Distribution for Industrial Plants, 1993.
[28] IEEE C37.20.3-2001, IEEE Standard for Metal-Enclosed Interrupter Switchgear, 2001.
[29] IEEE C37.20 1-2015 - IEEE Standard for Metal-Enclosed Low-Voltage (1000 Vac and below, 3200 Vdc and below) Power Circuit Breaker Switchgear, 2015.
[30] J. Casazza and F. Delea, Understanding Electric Power Systems: An Overview of Technology and the Marketplace, Wiley-IEEE Press, Hoboken, NJ, 2003.

Questions and problems

1. What is purpose of power distribution systems?
2. Discuss the three classifications (configurations of power distribution systems).
3. List the advantages and disadvantages of overhead and underground distribution systems.
4. What is the purpose of power distribution substations?
5. Discuss the advantages and disadvantages of the overhead and underground power systems.
6. List the advantages and disadvantages of the three-phase delta (Δ) transmission lines.
7. List the advantages and disadvantages of the three-phase star(Y) transmission lines.
8. List the voltage levels used in the various stages of transmission and distribution in electric power systems.
9. What are the effects of high primary voltages on distribution systems?
10. What devices can be used to protect electric motors from overloads? What are the ones to protect from short circuits?
11. Briefly describe where is the single-phase power distribution system used? Where is the three-phase power distribution system used?

12. Why is the voltage drop important in electrical power distribution systems?
13. How the capacitor improve, not only the power factor but also the voltage regulation (i.e., by reducing the voltage drop)?
14. Three 25 kW loads are connected in a power distribution system, through a three-phase 600-V. The load power factors are 0.9 lagging. Compute the load currents?
15. List the NEMA MCC's enclosure types.
16. What the voltage levels of an MCC?
17. How many voltages can be set from a three-phase, wye-connected feeder transformer?
18. List the man devices used in power distribution protection, control, and operation.
19. Although a smart MCC can reduce operating costs, increase system efficiency, and provide more system information, it is typically more expensive to install and commission. It is this, true or false?
20. A three-phase 13.8 kV cable to power 24,000 HP induction motor has 95% efficiency and 0.93 PF lagging. Determine the voltage drop is the switchboard is 200 ft., if the cable has a resistance of 0.1 Ω and 0.085 Ω per phase and per 1,000 ft.
21. What are the difference and the purpose of the neutral grounding and equipment (chassis) grounding?
22. A utility decided to increases the primary power distribution voltage level from 12.47 to 34.50 kV, estimate the voltage drop reduction, and the increases into the power and covered area.
23. A three-phase 250 HP, 60 Hz, 480-V induction motor has an efficiency of 93% and is operating at a power factor of 0.86 lagging. If the motor cable reactance is 0.0165 Ω/phase, compute the capacitance needed to improve the motor power factor to 0.97. What is the voltage rise (boost) at the motor terminals?
24. A 450 kVAR, 480-V, three-phase capacitor bank is installed at the main switchgear of 1.25 kA service. The service transformer is rated at 1,500 kVA, 4.16 kV-480/277-V and has impedance referred to its LV side of $0.003 + j0.0135$ Ω/phase. Compute the percentage voltage boost due the capacitor bank installation.
25. A three-phase 180 HP. 460 V, 60 Hz induction motor is started direct from 480 V supply. The combined source and cable impedance is $0.01 + j0.028$ Ω/phase. If the motor starting power factor is 0.28 lagging, determined the voltage drop of the line voltage.
26. A 480/277 V, wye-connected, four-wire service is supplied by a 500 kVA service transformer. The transformer impedance is $0.0065 + j0.0285$ Ω/phase. If a 250 kVAR, 480 V capacitor bank is installed at the distribution switchgear, what is the voltage rise?
27. An industrial facility has a power demand of 1,500 kW at a power factor of 0.70 lagging. Determine the capacitor bank reactive power ratings required to improve the power factor to 0.85, 0.90, 0.95, and unit power factor, respectively. What is the voltage rise in each case, if the feeder inductive reactance is 0.015 Ω/phase and the line-to-line voltage is 460 V?

Chapter 6

Building electrical systems and industrial power distribution

Outline and abstract

The utilization of energy resources is considered one of the most challenging tasks, while finding the most optimal, proper, and efficient ways to effectively use these important resources is an essential ingredient of sustainable development. In any electrical system, power must be transferred from the service equipment to the lights, machines, electrical motors, equipment, appliances, and electrical outlets. Regardless of the wiring methods used, the electricity carrying conductors and cables fall into one of two categories: feeders or branch-circuit conductors. Important aspects of the electrical system design involve building electrical service, service entrance, branch circuits, feeders, panel-boards, switchboards, switchgears, and load centers, and the calculations and sizing of their associate equipment and devices, as well as the protection devices and conductors. Panel-boards, switchboards, feeders and branch circuits, and associated fittings and devices are important components of the power distribution inside the buildings, industrial, and commercial facilities. Cables are usually contained in raceways, conduits, ducts, or cable trays, protecting them from mechanical damage and influences of other cables. In addition to structural requirements, when designing cable tray systems, the electrical requirements must also be carefully considered, as well as to be complaint to the specifications and requirements of the codes and standards. Often the design information is presented in the form of cabling diagrams, an important communication tool between designer, engineers, and technicians. In order to properly develop cabling diagrams requires in-depth understanding of the NEC, codes and standards regarding branch circuits, feeders, loading receptacles and outlets, switching requirements, and specifications, etc. This chapter is exploring the characteristics of electrical service, feeders, and branch circuits. It introduces the design elements, code and standard requirements, and specifications for service entrance, and inside the utility metering practices. An important aspect of the electrical and industrial power system design involve the calculation and design of branch circuits and feeders to supply various loads in a given occupancy and facility. The general purpose of a conduit, duct, or a raceway is to provide a clear and protected pathway for a cable, or for smaller conduits (inner ducts). Advances in cable technologies, costs of repairing sensitive cable materials or to replace the cables as needed have driven preferences for protective conduits over direct cable burial into the ground or walls. In industrial facilities, the electricity is supplied to the loads from the load centers, containing the equipment necessary to protect and control the power flow and the loads. There exist different load center types, with their selection based primarily on the electrical requirements and installation environment. Load centers are housed in metal enclosures to protect enclosed equipment, nearby objects and personnel in the event of equipment malfunctions. Load centers supplying large motors and smaller load centers are referred to as switchgears, while smaller load centers specialized to supply small to midsize electrical motors are called motor control centers (MCCs). Load centers incorporate switches, circuit breakers, fuses, and disconnect devices to route power and in the event of faults, to isolate electric circuit sections. Switchgears have three basic functions: (a) protection and safety, (b) electrical isolation to permit work and testing, and (c) local or remote circuit switching. Developments in switchgear design have led to the introduction of network support for monitoring, control, and advanced diagnostic capabilities, loading and a host of other operational parameters. This chapter will introduce the most common and important aspects associated with panel-boards, switchboards, service, feeders, branch circuits, raceways, and cable trays. After completing this chapter, the readers are able to identify the feeder and branch circuit sections of power distribution, describe the branch circuits, feeders, and their characteristics and functions, conduits, and raceways are able to calculate them using requirements and appropriate specifications of codes and standards, size branch circuits and feeders in accordance with such specifications and requirements, and finally use the codes to size feeder conductors and cables.

6.1 Introduction, facility power supply calculations, and design

The function of the electric power distribution system in a building, industrial facility, or an installation site is to transfer power from one or more supply, in the most efficient way and at higher standards to the individual loads and to all other electrically operated devices. Today's industrial applications require the most technologically advanced support systems to ensure state-of-the-art power quality needs. Industrial facilities are becoming more and more dependent on computer control of their processes, and as a consequence, require an increase in cleanliness and reliability of the electrical power supply system. Electromechanical subsystems are being replaced by electronic logic. Harmonic interference, welding, variable speed drives, and other in-plant noise have reliable mitigation procedures. Use overhead distribution because it is generally less costly than underground. Where underground distribution is more cost effective, it should be used. When exceptions are considered, follow local requirements and practices. The building and facility electrical service consists of the conductors and cables connecting the building and facility electrical system to the local utility power source, the raceway system containing the electrical service entrance conductors and cables, metering equipment, the main service disconnect, and the main overcurrent protection devices. The main purpose of the power supply design is to provide the optimum power supply, equipment, feeder, branch circuits, protection, control and in the most economical way with the right selection of electrical equipment. Since the service is a connection point to the utility, it is providing the electric service location, metering equipment, pertinent information regarding the electrical service, and often special requirements for designer. The service and the utility-facility connections must comply with all NEC or IEC requirements and specifications.

There are in practice many cases where the electrical designer relies on the information and assistance from specialists in related but separate fields. This applies in particular to controls for heating and air conditioning, which are designed by specialists in that field and not by the consultant or contractor employed for the general electrical system. A description of them would, therefore, be out of place here. Many other services within a building include electrical equipment, electrical motors, heaters, air conditioning, and controls. From the user's point of view, the electricity service in a building consists of light switches, sockets, clock connectors, cooker control units, and similar outlets. Such fittings are collectively known as accessories; this name came about because they are accessory to the wiring, which is the main substance of the installation from the designer's and installer's point of view. To them, the way the outlets are served is the major interest, but it is quite secondary to the user who is concerned only with the appearance and function of the outlet. In the complete electrical installation of a building, the wiring and accessories are interdependent and neither can be fully understood without the other; a start has to be made somewhere however, and in this book it is proposed to consider accessories first.

To have a better understanding on how an electrical power supply system in a building, plant, or facility is designed, the following design elements needed to be analyzed and considered. Load analysis, which includes the estimation of the power absorbed by the loads and their relevant positions, the definition of the position of power centers (switchboards, load centers, panel-boards), path and calculation of the length of connection elements, and the definition of the total absorbed power, taking into accounts the utilization factors and demand factors, as defined in previous book chapters. After load analysis, the next phase consists of the dimensioning of transformers, feeders, load centers, and if the facility include the site power generation, the size of the local generators. As recommended by codes and standards, a 15%–30% margin should be considered for future expansion and changes. Dimensioning of the conductors, cables, and branch circuits is the next design phase. This phase consists of the estimate of the currents flowing through conductors, definition and selection of the conductor, cable type and insulation material, calculations of the cross section and the current carrying capacity, the voltage drops at the load current under normal and transient (motor starting, inrush current, overload, etc.) operations. Another important step is the verification of the voltage drop limits at the final load, and if the voltage drop is not in the limit, phase 3 must be modified. After the verification is completed, the short circuit current calculations are conducted, which consists of the estimations of the maximum value at the busbar and minimum values at the end of the line, followed by the phase of the selection of protective circuit breakers (CBs), fuses, and other protective devices, having breaking capacity higher than the maximum prospective short circuit current, rated current no lower than load current, and characteristics compatible with the type of protected loads (motors, capacitors, heaters, etc.). An important step of the verification of the protection of the conductors, cables, and branch circuits, consists of the verification of the protection against overload currents, the rated current or the set currents of the CB, fuses, or protection device that is higher than the load current but lower than the current capacity of the conductor or cable. The protection verification against short-circuit, consisting of the specific load through current by the CB under short-circuit condition must be lower than the specific current which can be withstood by the cable or conductor, expressed with standard notations in terms of power and energy ($I^2t \leq k^2S^2$, k is a design coefficient). In case of obtaining negative outcomes, all the earlier-mentioned stages must be repeated from third phase, sizing the conductors, cables, and branch circuits. The last step consists of the verification of the coordination with other equipment. Again in the case of obtaining negative outcomes, all the earlier-mentioned stages must be repeated from stage of the protective device selection.

After preparing load list(s) and understanding the demand, we need to design a stable network to supply our required power. During the design, feeding type mentioned in the load list and voltage level should be taken into account so that we would get general idea how to feed the loads. Normal loads are fed from the normal busbar and essential loads, such as emergency lightings are fed from the emergency busbar. While single line diagrams are prepared, the load balance studies shall simultaneously be done to calculate the required power supply, transformer sizing, and busbars sizing. In addition, active power, reactive power, and power factor for each bus and entire system need to be calculated. The base of the calculation is to obtain a sum of the active and reactive powers considering load factor from downstream (loads) to upstream (generator end or feeder). Load flow studies are carried out in order to calculate all bus voltages, branch power factors, currents, and power flows throughout the plant electrical system.

6.2 Building electrical system characteristics

The electrical service in a building consists of conductors and accessories connecting the building or facility to the utility power supply, the raceways containing electric service entrance conductors and cables, metering and monitoring infrastructure, main service disconnect, and main overcurrent protection device, usually a main CB and not often a fuse. Electrical service is the connection point with the utility or to grid in the case of large industrial facilities. Building electrical service must meet all requirements and provisions of the NEC code, similar national codes or IEC standards, as well as those of local utility. The service entrance includes all the wires, devices, and fittings that are transferring the electricity from the utility transformer to consumers. The type of equipment used for an electrical service entrance of a building may include high-current conductors and insulators, disconnect switches, protective equipment for each load circuit which are connected to the main power system, and the meters needed to measure power, voltage, current, or frequency. The service components protect, meter, monitor, and distribute the electrical energy to the feeders and branch circuits. Conductors brought to the building are run overhead, coming from a utility pole are called the service drop or routed underground, from either a pole or a transformer pad, called service laterals. In either case, these service conductors are connected to the service entrance conductors at the building. A single building is in general supplied by a single electrical service, as specified in the NEC Section 230.2, which means that only one set of service conductors between the utility and the building exists. A structure can have only one service, one service drop, or one service lateral. While this is the basic rule, there are practical exceptions, which are specified by the codes. However, if the building size and conditions require additional services to supply fire pumps, emergency and back-up power systems, or a parallel power supply are included, as specified in Section 230.2(A). If the building size requires, in multi-occupancy building or when the capacity requirements (over 2000 A and 600 V or less) dictate multiple or additional services are permitted as stated in the NEC Section 230.2. Other exceptions requiring additional service, which include generators, driven by an engine or wind turbines, may provide additional service, electric power is supplied by solar photovoltaic sources, or when different voltages and phases may be required within the same structure. As general guidelines, the service conductors must be kept as short as possible in order to minimize the voltage drops, service conductors must enter the building as close as possible to the service panel, and the service disconnect must be at or very near the building point of entry. NEC Section 230.2(D) permits additional services when different service characteristics are required, for example, a building supplied by a 120/240 V, single-phase, three-wire service, and a 240 V, three-phase four-wire service. It is quite unlikely that the utility is providing both services, and usually a transformer is installed to provide the needed services.

Service entrance conductors may be type RHH, THW, THWW, XHHW, or RHW, and can be run through conduit or enclosed in a cable assembly called service entrance cable. In the overhead service, the service entrance cable run from the service drop to the panel-board, via metering, while in underground service, usually the cable provides connection between meter and panel and is installed in proper raceway for the structure. The basic elements of a low-voltage service include the overhead service drop or lateral conductors, often a multiplexed cable, service mast or rack assembly, a grounded conductor for three-phase service, the service entrance raceway, panel-board, disconnect, and metering equipment. Regardless of the number of service drops, laterals, or sets of service entrance conductors, the service is usually planned so the service entrance conductors are not routed through the building interior unless encased in concrete. The service entrance's conductors terminate at disconnect outside or close to where they enter the structure. It is worth to mention that the underground service entrance is usually ending to the meter socket. The grounded service entrance conductors are sized based on the computed loading on the ground (neutral) conductor. The maximum loading is the result of unbalanced single-phase loading on the system. Special consideration must be given to the three-phase four-wire systems supplying non-linear loads, which are prone of harmonic generation. In this case, the conductor sizing is based on the estimate current level due to the harmonics. The outlets of the building electrical system are lightings, socket outlets, and fixed equipment. The wiring to each of them comes from an excess current protection device (fuse or CB) in a power distribution panel-board, but a single fuse or CB can serve several outlets. If the circuit supplies to any equipment, wiring from one fuse or CB is the final circuit, and all the outlets fed from the same fuse or CB are also the final circuit. The fuse or CB must be large enough to carry the largest steady current at any instant by the whole of the equipment on that final circuit. Since the fuse or CB protects the cables, no cable forming part of the circuit may have a current carrying capacity less than that of the fuse, unless the characteristics of the load or supply are such that an overcurrent cannot occur. The size of both the fuse or CB and cable is, therefore, governed by the number and type of outlets on the circuit. Temporary wiring designed to provide electrical power only during construction, remodeling, demolition, repair, or large-scale maintenance of a building, that must be removed when the work is completed, is specified by the Article 590 of the NEC. In addition, temporary power can be brought in for emergency situations, testing, and experimentation. One of the key factors for the installation of all temporary wiring is: all temporary wiring, devices, and equipment should be located in a safe place and should be as neat as possible, while the temporary wiring should be kept overhead as much as possible. Section 230.71(A) of the NEC requires that disconnect means or service entrance conductors cannot exceed six switches either in a single enclosure or a group of separate enclosures.

The MV electric service ranging from 4.16 to 35 kV, with the current ratings from 600 A to 2 kA is usually supplied from an underground power distribution network but outdoor services are also available. The MV cables are terminated in a set of medium-voltage switchgears, usually containing the MV service disconnect, instrument transformers and metering, control and monitoring equipment, and devices. For proper operation and safety reasons, there must be equipment that is disconnecting all wiring from the power source, if needed. This can be arranged in a single main disconnect switch or a main CB that is part of the service panel. Section 230.24 requires that service drops to not be easily accessible, with certain clearness to meet the inaccessibility specifications. Section 230.71(A) of the NEC requires that disconnect means or service entrance conductors cannot exceed six switches either in a single enclosure or a group of separate enclosures. Regardless of the method used, a disconnect must be located in an accessible place, as close as possible to the point where the service conductors enter the structure. The service cable capacity is determined by the load demand and in agreement with specifications and requirements of the NEC, codes and standards. Cable ampacity is determined before applying of ambient temperature correction or raceway fill correction factors, while separate calculations are required for phase and neutral conductors. Service conductor sizes are determined form the load estimations, and by following the guidelines from the codes and standards. NEC Section 79 requires that disconnect current rating to be higher than the computed occupancy or facility load supplied. Example for a single family occupancy minimum rating is 100 A. On the other hand, Section 230.80 requires that the sum of all disconnects to be equal or larger than a single disconnect ratings. For example, two 400 A can be used in place of an 800 A single disconnect. In general, the equipment is not permitted to be connected on the supply side of disconnect. However, the NEC Section 230.82 list the exceptions and types of equipment allowed to be connected on the disconnect supply side. The ground fault protection is required on each disconnect rated 1000 A or higher, on solidly grounded, wye-connected three-phase service of over 150 V to ground, and not exceeding 600 V, line-to-line. Such requirement is common for 480/277 V, Y-connected, three-phase, four-wire service consisting of a single 2 kA disconnect.

6.3 Branch circuits and feeders

Several definitions are essential to understanding branch circuits and feeders, their characteristics, purpose, and designations, from the power company terminals to the main service disconnect. Feeders are conductors and cables that originate at the main depower distribution or main disconnect device and terminates at another distribution center, panel-board, or load center, while the subfeeder are the conductors or cables originate at the power distribution centers other than the main power distribution center and extend to panel-boards, load centers, and disconnect switches that supply branch circuits. A panel-board can be a single panel or multiple panels containing switches, fuses, and CBs for switching, controlling, and protecting circuits. Branch circuits represent the section of the wiring system extending past the final overcurrent device. These circuits usually originate at a panel and transfer power to load devices. Any circuit that extends beyond the final overcurrent protective device is called a branch circuit, including the circuits servicing single motors (individual) and circuits serving several lights or lighting systems and receptacles (multiwire). As specified by codes and standards, ranch circuits are usually low current (30 A or less), but there are instances where they are also supplying higher currents. A basic branch circuit is made up of conductors or cables extending from the final overcurrent protective device to the load. Some branch circuits originate at safety switches (disconnects), but most originate at a panel-board. Based on their designation, the branch circuits are classified as: *individual branch circuit* is a branch circuit that supplies a single load, *multioutlet branch circuit*, a branch circuit with multiple loads, *general purpose branch circuit*, which is a *multioutlet branch circuit* that supplies multiple outlets for appliances and lighting, appliance branch circuit, a branch circuit that supplies a single appliance load, and a multiwire (conductor) branch circuit, a branch circuit with two or more ungrounded conductors and one grounded conductor, designed to supply specific loads.

A branch circuit is sized for the supplied load. Sizing the circuit for additional future loads is good engineering practice. The rating of a branch circuit depends on the rating of the overcurrent device protecting the circuit. Branch circuits serving only one device can have any rating, while a circuit supplying several loads is limited to ratings of 15 A, 20 A, 30 A, 40 A, or 50 A. Branch-circuit voltage limits are contained in Section 210.6 of the NEC Code, with similar provisions in the IEC standards. These limits are based on the equipment or loads that are supplied by the circuit. In residences and hotel rooms, circuits supplying lighting fixtures and small receptacle loads cannot exceed 120 V. Circuits that are 120 V and less may be used to supply lamp-holders, auxiliary equipment of electric-discharge lamps, receptacles, and permanently wired equipment. Branch circuits exceeding 120 V but less than 277 V may supply mogul-base screw-shell lamp-holders, ballasts for fluorescent lighting, ballasts for electric-discharge lighting, plug-connected appliances, and hard-wired appliances. Incandescent lighting operating over 150 V is permitted in commercial construction. Circuits exceeding 277 V and less than 600 V can supply mercury-vapor and fluorescent lighting. NEC Section 210.21 (B) specifies that receptacle ratings are permitted on multioutlet branch circuits, having a rating equal to or greater to the overcurrent device rating. NEC sections 210.52 through 210.63 specify the requirements for receptacle location and ratings. Section 210.70 lists the requirements and specifications for lighting outlets. NEC Section 210.70 is specifying the lighting outlet location for houses and other facilities. Ground fault circuit interruption (GFCI), as specified by Section 210.8 is required on 125 V, 15 A or 20 A receptacle outlets in certain locations. This NEC section specifies the GFCI locations, as well as the exceptions that apply. In the event that two energized conductors are coming in contact with each other or a different electrical potential surface an electric arc is initiated at the contact point, causing a fault current. This fault current may not be of sufficient magnitude to trip the CB or to blow the fuse, requiring a special electronic CB, arc fault circuit interrupter to be used. NEC Section 210.12(B) requires arc fault circuit interruption on all 15 A and 20 A, 125 V supplying outlets in dwelling unit bedrooms. NEC code also places load limitations on any branch circuit with continuous loads (loads with duration longer than 3 h, such as lighting loads). The continuous loads must not exceed 80% of the circuit rating allotted for it. If the overcurrent protective device is listed for continuous operation at 100% of its rating, the 80% factor is not used. Branch circuit loads are classified into five categories: lighting loads, receptacle loads, equipment loads, heating and cooling loads, and electrical motor loads.

The amperage rating of branch-circuit conductors must be greater than the maximum load the circuit is supplying. For multiple-load branch circuits, the conductor ampacity must correspond to the rating of the overcurrent protective device. However, for the branch circuits that is supplying hardwired devices, such as: electric heaters, air-conditioning units, and water heaters, the fuse or CB is usually rated at the next higher rating. The conductor is acceptable if its rating is at least that of the load current, even if the overcurrent protective device rating is higher. After the required receptacles, outlets, lightning outlets, branch circuit for appliances, equipment, and motor supply circuits are calculated and located, it is necessary to determine the minimum number of the feeders and the branch circuits needed to supply them. The actual number of branch circuit is usually exceeding the minimum specified by the codes and standards. Once the receptacle and lighting outlets are located, the number of the branch circuits is estimated and the branch circuits must be designed. The total number of the required branch circuits includes the ones supplying receptacle and lighting outlets for general lighting, small appliances, the laundry and bathroom branch circuits, dedicated equipment loads, outdoor, and unfinished areas receptacle outlets branch circuits. Usually the actual number of branch circuit exceeds the code or standard minimum number. NEC Section 210.11(A) requires that minimum number of branch circuits to supply the general lighting load, before the application of any demand factor is determined by:

$$N_{Min\text{-}BrCirts} = \frac{\text{General lighting load (VA)}}{\text{Maximum load per branch circuit (VA)}} \tag{6.1}$$

Here $N_{Min\text{-}BrCirts}$ is the minimum number of branch circuits as required by the NEC, while the maximum branch circuit load is based on the branch circuit rating. Usually, for residential applications, 15 A and 20 A branch circuits are used to supply general lighting, while 20 A circuits are used in commercial and industrial facilities. A good design practice is to limit the maximum load per branch circuit at 80% of the circuit rating. An alternative approach is to determine the permitted maximum area (m^2 or ft^2) to serve a branch circuit, and then by dividing the total to this value, the minimum number of the branch circuits supplying the general lightning as required by the code is found.

Example 6.1: What is the minimum number of branch circuits supplying the general lightning load for the following occupancies, a 9,600 ft^2 office (at 20 A rating) and a 200 m^2 residence (at 15 A)?

Solution: From Table 4.4 (Chapter 4), the unit load for square foot of office space is 3.5 VA plus 1 VA for general receptacles, while for a residence the load per unit area (m^2) is 33 VA. For the office space, the branch circuit rating is 120 V and 20 A, while for the house (residential space), the branch circuit rating is 120 V and 15 A. Therefore the maximum load permitted for first circuit is 1,920 VA (120·0.8·20), while for the second circuit is 1,440 VA (120·15·0.8). By applying (6.1), the minimum number of the branch circuits for the general lighting in each case is:

$$N_{office} = \frac{4.5\ (\text{VA/ft}^2) \times 9{,}600}{1{,}920\ \text{VA/circuit}} = \frac{43{,}200}{1920} = 22.5$$

$$N_{residence} = \frac{33\ (\text{VA/m}^2) \times 200}{1{,}440\ \text{VA/circuit}} = \frac{6{,}600}{1{,}440} = 4.58$$

Therefore, a minimum 23 branch circuits to supply the general lighting load are required for the offices space and 5 branch circuits for the residence space.

NEC Section 210.11 (C) states the requirements, provisions, and specifications for the branch circuits installed in residential specific areas, such as kitchen, laundry, bathroom, outdoor, and for dedicated loads and equipment. As general recommendation, these branch circuits need to supply only receptacles in their designated area, while the lighting outlets may not be connected to the space branch circuits. NEC Section 210.11(C) requires that in the laundry or bathroom to use 125 V and 20 A branch circuit ratings, and only receptacles in that area to be connected. NEC or IEC do not limit the number of receptacles or lighting outlets permitted on 15 A, 20 A, and 125 V, or IEC rating branch circuits, the designers must determine base on their experience and engineering practice. However, where the specific equipment ratings are known, these values can be used to properly design the branch circuits, where such values are not available, a load of 180 VA can be assigned to each general-purpose receptacle, while good design practice limit the maximum number of receptacles. Certain loads in industrial and small commercial facilities are connected to multiwire branch circuits. A multiwire circuit is one consisting of two or more ungrounded conductors and one grounded conductor, having the same voltage between each ungrounded and grounded conductors. The loads in a multiwire branch circuits can be connected between two ungrounded conductors (line-to-line) or between an ungrounded and the grounded conductors (line-to-neutral). Notice that for a fully balanced system, the neutral current is zero.

The conductors between the service equipment and the branch-circuit overcurrent devices are called feeders. The Article 215-Feeders of the NEC provide information regarding the safe, adequate sizing, and installation of feeders. This article also applies to subfeeders, which provide power to branch-circuit panels but originate at power distribution centers rather than the service equipment. Feeder loading is dependent on the total system power requirements. When all connected loads operate simultaneously, the feeder must be of sufficient ampacity to meet that load demand. If only 75% of the connected loads are operating at the same time, then the feeder is sized larger than the service conductors. Prior to installation, there are factors that must be considered to ensure the feeder size, type, and overcurrent protection is correct for that application. The feeder can be copper or aluminum, the environment around the feeder (damp, hot, corrosive) must be taken into consideration, the feeder can be run in conduits, cable trays, or other systems, and single or multiconductor cable can be used. Feeders can be paralleled or individual, while the voltage drop becomes a consideration in long run feeders. Conductor sizing includes several factors: conduit fill, ambient temperature, and connected load demand, and a neutral wire may not be necessary with the feeder, while the continuous loads are affecting the feeder size. Various overcurrent protective devices are also used with each feeder. Besides the previous adjusting factors, there are two important design factors for calculating the loads in electrical systems, demand factor, and diversity factors, as discussed in previous book chapters. The demand factor (always less than one) is the ratio of the maximum demand of an electrical system, or part of a system, to the total connected load on the system, or part of the system under consideration. The diversity factor is the ratio of the sum of the individual maximum demands of the various subdivisions of an electrical system, or part of a system, to the maximum demand of the whole system, or part of the system, under consideration. Diversity factor is usually higher than one. For example, these terms, when used in an electrical design, should be applied as follows. The total connected loads supplied by a feeder-circuit can be multiplied by the demand factor to determine the load used to size the system components. The sum of the maximum demand loads for two or more feeders is divided by the diversity factor for the feeders to derive the maximum demand load.

Example 6.2: Consider three individual feeder-circuits with connected loads of 200 kVA, 300 kVA, and 400 kVA and demand factors of 85%, 75%, and 80% respectively. If the diversity factor of 1.5 for all feeders, calculate the feeder load demands.

Solution: The load demand for each feeder circuits is:

$$200 \times 0.85 = 170\,\text{kVA}$$
$$300 \times 0.75 = 225\,\text{kVA}$$
$$400 \times 0.80 = 320\,\text{kVA}$$

The sum of the individual load demands is equal to 715 kVA. If the main feeder circuit is sized at unity diversity: kVA = 715 kVA $\div$ 1.00 = 715 kVA, so the main feeder circuit would have to be supplied by a 720 kVA transformer. However, using the diversity factor of 1.5, the kVA = 715 kVA $\div$ 1.5 = 476.7 kVA for the main feeder. For a diversity factor of 1.5, a 500 kVA transformer could be used. Notice that although feeder-circuit conductors should have an ampacity sufficient to carry the load, the ampacity of the feeder need not always be equal to the total of all loads on all branch-circuits connected to it. The demand factor permits a feeder circuit ampacity to be less than 100% of the sum of all branch-circuit loads connected to the feeder. Section 220.3(A) of the NEC states that the load on a service or feeder is the sum of all of the branch loads subject to their demand factors as permitted by the code rules.

6.4 Cable installations, raceways, and conduits

Cable installations make up a large portion of the initial distribution system investment, contribute to a lesser extent to the annual maintenance and operating costs, and affect system reliability. Power, control, and instrumentation cables and conductors are usually installed in raceways, such as conduit or cable tray to protect the cables from physical damage and for easy maintenance and replacement. Conduits are metallic or nonmetallic tubing, having circular shape, specifically designed to protect electrical cables. The metallic conduits can be of galvanized steel (magnetic) or nonmagnetic (aluminum or stainless steel). Electrical conduits are used to enclose, support and protect electrical conductors for power, control, and communication. Rigid metallic conduits (RMC) have a circular cross-section, trade size in the range 0.5–6″ and the heaviest wall thickness of all metallic conduits and a standard length of 10 ft. NEC Article 344 specifies the requirements for RMC installation and use. Such conduits have the outer finish galvanized, hot-dipped for corrosion protection and can be used in all environments and applications. Intermediate metallic conduit (IMC), a rigid steel type has a circular cross-section, trade size of 0.5″ to 4″ and lighter in weight than RMC. NEC Article 358 specifies the requirements for the electrical metallic tubing (EMT) is a circular conduit, made of steel or aluminum, trade size of 0.5–4″. EMT conduits are used in indoor applications and in noncorrosive environments. Nonmetallic conduit, made of such materials as fiber-based composites or polyvinyl chloride (PVC), is often encased in concrete to form a duct bank. Cable tray is a prefabricated structure often resembling a ladder. Aluminum is the most common material used to fabricate cable trays but steel (raw, galvanized, or stainless), fiber-reinforced plastic, and vinyl-coated steel trays are also be used.

The electric conduits are usually in a circular cross-section and may be rigid or flexible, and often are covered with outer PVC or other plastic materials for corrosion resistance and/or weatherproofing. Therefore, underground cables and their accompanying protective and operating devices should be selected in accordance with criteria set forth in the following paragraphs. The joint specifications of the Insulated Cable Engineers Association and National Electrical Manufacturers Association (ICEA-NEMA) and the specifications of the Association of Edison Illuminating Companies (AEIC) are used as covered by NFGS-16301. ICEA-NEMA specifications cover medium-voltage cables which are manufactured as stock items. Requiring medium-voltage cable to meet AEIC specifications are limited to medium-voltage cables which are not stock items (35 kV rating) or where the footage installed is large enough to make a special run. Single-conductor cables are usually used in power distribution having the installed cost is less than that of multiconductor cables. The use of multiconductor cables is recommended where justified by considerations, such as installation in cable trays, twisted to provide lower inductance for 400-Hz distribution systems, and for high-altitude electromagnetic pulse (HEMP) hardened systems. In harsh industrial environments, lids are often installed on cable trays to keep contaminants away from the cables. If lids are used, the natural cooling effect of the cables is reduced greatly, so the cables in a covered tray must be derated. A full range of accessories to join tray sections and support them from walls or ceilings is available, such as couplings to form three- and four-way intersections, supports or boxes. The NEC code states which kinds of cables can be installed in the same raceway, and which cables must be separated. In a conduit system, the cables are drawn into tubing called conduit. The conduit can be steel or plastic. The different sizes of conduit are identified by their inner diameter and in the case of electrical conduit, the nominal inner diameter is always the same as the outside diameter of the tube. Heavy gauge conduit is normally joined together by screwed fittings; there is a standard electrical thread which is different from other threads of the same nominal diameter. Conduit is thick enough for the cross-sectional area of the metal to provide a good ground (earth) low-impedance path. The conduit can be used as the earth continuity conductor and no separate cable or wire need be used for this purpose. Since metallic raceways can carry ground currents, grounding and bonding of the raceways are critical. It is essential that the conduit, with all its fittings and screwed joints, should form a continuous conducting path of low impedance and the safety of the installation depends on good electrical contact at all the joints. Even though it may be decided not to use the steel conduit as the circuit protective conductor, in preference for a separate protective conductor, usually copper, the conduit must be erected properly with tight joints. The final connection to machines and mechanical equipment, such as pumps, boilers, fans, fan heaters, workshop equipment and so on is usually made in flexible conduit. The fixed wiring terminates in a box either in the wall near the equipment to be connected or on the surface of the wall, and from this box a short length of flexible conduit is taken to the equipment. Solid conduit from this to the machine could involve a large number of bends in a short distance which would be difficult to make and impossible to pull cable through. Flexible conduit can take up a gentle curve and also serves to isolate the fixed wiring from any mechanical vibrations on the connected machine, and allows for belt tension adjustment of the motor. There are several types of flexible conduit. However, the flexible conduit cannot be used as a protective conductor. A conduit system must be completely installed before any cables are pulled into it.

Conduit systems are intended to be easy rewired, which means that at 20 or 30 years after the building has been erected, it should still be possible to pull all the cables out of the conduit and pull new needed cables into each conduit. If this is possible, then quite regardless of what happens when the building is first constructed, the layout of the conduit must be such that the cables can be drawn into it when it is complete and finished. The original reason for wanting to have electrical systems which could be recabled during the life of the building was that old cable types deteriorate in about 20 years or so to the stage at which it should be removed. However, the PVC cable appears to last indefinitely so that all modern installations which use this cable should not need rewiring. New cables then have to be run where there were no cables previously and the original conduit has at best to be added to and at worst abandoned altogether. Rewireability is then no help and in fact the need for a rewireable system is not as great as is often supposed. Notice, that there is always the possibility that a cable may become damaged during the construction of a building, and it is obviously an advantage if it can be replaced without difficulty after the building has been finished. If the conduit is installed so that the system is rewireable, repairs are always possible. To achieve rewireability, draw-in boxes must be accessible from the surface, or in other words their covers must be flush with the finished surface. When conduit and duct systems are designed, proper conduit size is selected. Conduits are sized properly so that the cables can be pulled without damaging the cable jacket or insulation and adequate cooling exists when the cables are energized. The NEC limits conduit fill, the sum of the cross-sectional cable areas in a conduit cannot exceed jam ratio of the inner conduit cross-sectional area, as shown in Table 6.1, as specified in the NEC.

Table 6.1 Conduit or tubing percentages are for conductors

Number of conductors	Maximum percentage fill (%)
1	53
2	31
3 or more	40

In addition to the percentage fill, a quantity called the jam ratio must also be determined. The jam ratio predicts the likelihood that the cables are jamming by lining up along the conduit inner diameter. Jamming occurs as the conductors that are installed or pulled around the bends, are twisted inside the conduit or tubing. Jam ratio (JR) is defined as the ratio of the conduit inner diameter (ID_{Cond}) over the cable outer diameter (OD_{Cable}). If the jamming ration has a value in the range 2.4–3.2, there is a good probability of cable jamming during the installation, and the large conduits are recommended.

$$JR = \frac{ID_{Cond}}{OD_{Cable}} \tag{6.2}$$

Example 6.3: Determine the jamming ratio for a 3 in. conduit and 500 kcmil XHHW-2 copper conductors.

Solution: The cross-section area of the conductors is 2.7936 sq. in., with a conductor outer diameter of 0.943 in. and inner conduit diameter is 3.09 in. (Appendix C table). The jam ratio is:

$$JR = \frac{ID_{Cond}}{OD_{Cable}} = \frac{3.09}{0.943} = 3.28$$

The jam ratio lies outside the jamming range in this case. A minimum bending radius must be determined according to the cables being pulled. Shielded cables contain a metallic layer just beneath the jacket to distribute evenly the electric field gradient throughout the insulation. This shield is sometimes provided in the form of a thin copper tape, or tape shield, which is grounded at the cable terminations. Typically, in three-phase circuits, each cable has enough copper strands in its concentric neutral to make up one-third of the system neutral. Conduits subjected to the large temperature variations, usually in outdoor and some of the industrial applications can expand or contract depending on the temperature fluctuations. The amount of the conduit expansion or contraction is computed by using thermal expansion relationship:

$$\Delta L_{cond} = L_{run} \times TEC_{Cond} \times \Delta T \tag{6.3}$$

Here, L_{run} is the run length (m or ft.), TEC_{Cond} is the coefficient of thermal expansion for the conduit material (m/°C or ft./°F), and ΔT is the change in temperature, in Celsius or Fahrenheit degrees, respectively.

Pull and junction boxes are used to provide conduit termination endings at specific locations and to enable cables and conductors to be pulled into a conduit system. Cable pulling calculations are needed to ensure that maximum permitted tension on the cable is not exceeded or to prevent the damage of the cable insulation or the cable. Pulling tension is the tensile force that must be applied to the cable to overcome friction as the cable is pulled through the raceway. Sidewall pressure is the crushing force applied to the cable by the conduit in the radial direction as the cable is pulled around a bend. Exceeding the maximum allowable pulling tension is not usually a problem for large power cables since that maximum value is usually quite large. Control and instrument cables, however, tend to have lower maximum allowable pulling tensions. When more than one cable is pulled in a conduit, a weight correction factor (w_c) is needed to account for the additional friction forces that exist between the cables or conductors. The most common case is to install three cables of the same size in a conduit for a three-phase power circuit. The weight correction factor depends on whether the cables are arranged in a triangular or cradled configuration. The tension required to pull a cable or group of cables through a straight section horizontal of conduit and an inclined section, up or down, respectively, is expressed as:

$$T_{OUT} = T_{IN} + w_c \mu_f LW \tag{6.4a}$$

And

$$T_{OUT} = T_{IN} \pm LW(\sin\theta \pm w_c \mu_f \cos\theta) \tag{6.4b}$$

where T_{OUT} is the pulling tension in pounds at the duct or conduit output end, T_{IN} is the conduit or duct input tensions, w_c is the weight correction factor (dimensionless), μ_f is the coefficient of dynamic friction (dimensionless), L is the length of straight section of conduit in feet, and W is the weight of cable in pounds per foot, θ is the incline angle of the conduit. Table 6.2 gives the representative dynamic friction coefficients for the most common type of insulation and conduit.

Table 6.2 Representative dynamic friction coefficients

Cable insulation	Conduit type			
	Metallic (steel or aluminum cement)	PVC	Fiber conduit	Asbestos
Polyvinyl chloride (PVC)	0.40	0.35	0.50	0.50
High-molecular weight (HMW) polyethylene (PE)	0.35	0.35	0.50	0.50
Cross-linked polyethylene (XLPE)	0.35	0.35	0.50	0.50
Hypalon (CSPE)	0.50	0.50	0.70	0.60
CPE (Chlorinated PE)	0.50	0.50	0.70	0.60
Nylon	0.40	0.35	0.50	0.50

The coefficient of friction is ranging from more than 0.5 for dry cable to less than 0.2 for well-lubricated cable. Suitable lubricants include wire soaps, waxes, and synthetic polymer compounds. Notice that if the coefficient of friction cannot be reduced sufficiently, reducing the length or changing the pull direction is strongly recommended. Installing pull boxes in the raceway can reduce the length of the pull. Equation (6.4b), the plus sign is in the case for pulling up a straight section of conduit, while the minus sign is for pulling down, a straight conduit section. The tension required to pull a cable through a horizontal bend is given by:

$$T_{OUT} = T_{IN} \exp(w_c \mu_f \theta) \tag{6.5}$$

where T_{OUT} is the tension out of the bend in pounds, T_{IN} is the tension coming into the bend in pounds, and θ is the bend angle in radians. A minimum bending radius must be determined according to the cables being pulled. As IEEE 576-2000 specify, when pulling an eye attached to copper conductors, the maximum pulling tension should not exceed 0.008 times circular mil area (C_m), q maximum pulling tension should not exceed 0.006 times C_m, when pulling an eye attached to aluminum conductors, expressed as:

$$T_m = k \times n \times C_m \tag{6.6}$$

Here, k is equal to 0.008 for copper and 0.006 for aluminum, when the maximum tension, T_m is expressed in lbf, k is 0.036 for copper and 0.027 for aluminum, n is the number of conductors, C_m is the circular mil area of each conductor. Maximum limitation for this calculation is 22,240 N (2,268 kgf or 5000 lbf) for a single conductor (1/C) cables and 44,480 N (4,536 kgf) (10,000 lbf) for three multiconductor cables. This limitation is due to unequal distribution of tension forces when pulling multiple conductors. When pulling cable through a vertical bend, the tension is calculated as for a horizontal bend, then the cable weight in the vertical section is either added (if the cable is pulled uphill) or subtracted (if the cable is pulled downhill) from the required tension. When pulling cable downhill, a negative tension can be calculated. The coefficient of dynamic friction depends on the cable insulation type and conduit. Note the significant effect on tension that small changes in μ_f (friction coefficient) can cause, especially in conduit bends where this friction coefficient is in the exponent. Inaccurate friction coefficients lead to poor correlation of tension calculations with actual tensions. Unfortunately, it is in multibend pulls, where the tension and sidewall pressure are of most concern, that the use of an inaccurate coefficient of friction produces the greatest error. For wiring applications in which three cables are pulled, the weight correction factor is typically in the range 1.15–1.35, while a value of 1.40 is recommended for situations involving four or more cables. Weight correction factors for cradled and triangular configurations are calculated using (6.7a) and (6.7b), respectively.

$$w_c = 1 + \frac{4}{3}\left(\frac{OD_{Cable}}{ID_{Cond} - OD_{Cable}}\right) \text{ (Cradled)} \tag{6.7a}$$

And

$$w_c = \frac{1}{\sqrt{1 - \left(\frac{OD_{Cable}}{ID_{Cond} - OD_{Cable}}\right)^2}} \text{ (Triangular)} \tag{6.7b}$$

Example 6.4: Determine the maximum pulling tension for a 250 kcmil, three-conductor copper cable.

Solution: The maximum pulling tension, (6.6) is given by:

$$T_m = k \times n \times C_m = 0.008 \times 3 \times 250{,}000 = 6{,}000 \text{ lbf}$$

When cable grip is used over non lead-jacketed cable, the pulling tension should not exceed 1,000 lbs or 1,000 lbs per grip (when used with multiconductor cables) and the tension calculated in (6.6). If the coefficient of friction is unknown a 0.5 friction coefficient is recommended. Sometimes physical constraints require that the cable be pulled in one specific direction. If no such constraints exist, it is a good engineering practice to attempt to pull the cable in the direction requiring the minimum pulling tension. The maximum permissible pulling length, L_m, for one straight cable section is given by:

$$L_{max} = \frac{T_{max}}{w_c \mu_f W} \tag{6.8}$$

Here T_{max} is the maximum tension in kgf (lbf), W is the linear weight of the cable(s) kg/m (lb/ft.), and μ_f is the coefficient of dynamic friction. Sidewall pressure, in a raceway or conduit is caused by the tension in the cable acting horizontally and the weight of the cable acting vertically. Sidewall pressure is the crushing force applied to the cable by the conduit in the radial direction as the cable is pulled around a bend, being a function of pulling tension and bending radius of the conduit, and differs according to the arrangement of the cables in the conduit. Sidewall pressure is typically the controlling factor in raceway design for large power cable. Sufficiently large conduit bending radii must be used, and pulling tension may need to be limited to values well below the maximum allowable pulling tension for the cable to keep the sidewall pressure below the maximum allowed by the cable. Generally, the tension of a cable immediately as it leaves a bend must not be greater than 300 times the bend radius (in feet), and the maximum sidewall pressure must not exceed 300 lbs/ft. Shown below are formulas to calculate the maximum allowable tension at a bend and the actual sidewall pressure. The maximum allowable pulling tension at bend, T_{bm} is the limit that the calculated pulling tension, T_b should be compared. If T_b, as computed by using (6.5) is greater than T_{bm}, the possibility of redesign or rerouting should be considered. The formulas to calculate sidewall pressure for various cable arrangements when the conduit bend radius is r, are given in (6.9a)–(6.9c).

$$P_{SW} = \frac{T_{bm}}{r} \text{ (One single-conductor cable)} \tag{6.9a}$$

$$P_{SW} = \frac{3w_c - 2}{3}\left(\frac{T_{bm}}{r}\right) \text{ (Three single-conductor cable)} \tag{6.9b}$$

And

$$P_{SW} = \frac{w_c}{2}\left(\frac{T_{bm}}{r}\right)\left(\begin{array}{l}\text{Three single-conductor cable triangular}\\ \text{or one three-conductor cable}\end{array}\right) \tag{6.9c}$$

As was stated earlier, a weight correction factor of 1.4 is usually used if four or more cables are installed in a conduit. Under these conditions, the sidewall pressure, given by (6.9a) may be estimated as:

$$P_{SW} = 0.75\left(\frac{T_{bm(out)}}{r}\right) \tag{6.10}$$

where P_{SW} is the sidewall pressure, in N/m (lbf/ft.) of radius, T_{bm} is the maximum tension (leaving the bend), N (lbf), r is the inside radius of conduit in m (ft.). One of the critical limitations to be considered in the installation of electrical cables is sidewall pressure. The sidewall pressure is the force exerted on the insulation and sheath of the cable at a bend point when the cable is under tension, and is normally the limiting factor in an installation where cable bends are involved. When installing single-conductor cables, or multiconductor cables in duct or conduit, the sidewall pressure acting on the cable at a bend is the ratio of the pulling tension out of the bend to the radius of the bend, (6.9a)–(6.9c). The normal maximum sidewall pressure per meter (foot) of radius is as given in Table 6.3. However, in order to minimize cable damage because of excessive sidewall pressure, the installer

Table 6.3 Recommended maximum sidewall pressure

Cable type	**Maximum sidewall pressure (lbf/ft.)**	**Maximum sidewall pressure (N/m)**
600 V Nonshielded multiconductor control	500	7,300
600 V and 1 kV single conductor (size 8 and smaller)	300	4,400
600 V and 1 kV single conductor (size 6 and larger)	500	7,300
5–15 kV power cable	500	7,300
25 and 35 kV power cable	300	4,400
Interlocked armored cable (all voltage classes)	300	4,400
Instrumentation cable (single pair)	300	4,400
Instrumentation cable (multipair)	500	7,300

Table 6.4 Minimum bending radius of single and multiconductor nonmetallic portable cable as a multiple of cable diameter

Cable type	Minimum bending radius (MBR)
0–5 kV	6
Over 5 kV	8
Control cable (seven conductors and over)	20
Single-conductor cable (metallic tape shielded cable	
Concentric neutral wire shielded cable	12
Lead sheath cable, metallic fine wire shield)	
Multiconductor cable (metallic tape shielded cable	
Concentric neutral wire shielded cable	7
Lead sheath cable, metallic fine wire shield)	

(technician) should check the cable manufacturer's recommendation for each type of cable to be installed. Although the normal maximum allowable sidewall pressure is as stated in Table 6.3, specific installation procedures and specific cable type or construction may cause this maximum to be increased or decreased. The weight correction factor for one cable (single or one multiconductor cable) is taken as 1.0.

During installation of the cable, it is recommended that the radius of bends be 1.5 times that of the minimum bending radius (MBR) for the final training. The minimum bend radius for the most common wire and cable under tension, as well as final training are given in Table 6.4. Notice that the minimum bending radius factor is applied to the overall cable diameter unless otherwise stated. These limits do not apply to conduit bends, sheaves, or other curved surfaces around which the cable may be pulled under tension while being installed. Larger radii bends may be required for such conditions. In all cases the minimum radius specified refers to the inner surface of the cable and not to the axis of the cable. The minimum values for the radii to which such cables may be bent while being pulled into an installation, being under tension can be determined by the formula given below. This value will greatly depend on the tension the cable is experiencing as it exits the bend in question. For instance, the greater the exiting tension, the greater the minimum bending radius is for the cable. In the case of dynamic condition, the minimum bending ratio, function of sidewall pressure, and maximum out tension is expressed as:

$$MBR = \frac{T_{bm}}{P_{SW}} \times 12 \tag{6.11}$$

Example 6.5: Determine the tension in the following cables, and the sidewall pressure where applicable for four 600 kcmil THWN-2 aluminum conductors, having a weight of 0.698 lb/ft., in a 6 in. (154.8 mm inner diameter) in a steel (metallic) conduit, friction coefficient 0.35. Assume a tension of 150 lb. at the input to each of these conduits, a weight correction factor of 1.35 for all calculations. The three conduits used in this application are (a) a 200 ft. straight horizontal conduit, (b) a 50 ft. pull up inclined at 45°, and (c) a bend of 90° having radius of 36 in. Calculate also the jamming ratio and fill factor for the first conduit.

Solution:

(a) Tension for this conduit is given by (6.4a), and being a horizontally straight, there is no sidewall pressure.

$$T_{OUT} = T_{IN} + w_c \mu_f LW = 150 + 1.35 \times 0.35 \times 200 \times 0.689$$

$$= 215.96 \text{ lbf} = 216 \text{ lbf}$$

The diameter of a 600 kcmil, THWN-2 conductor is 1.051 in. and 0.8676 sq. in. area, while the inner diameter of RMC of 6 in. is exactly 6.093 in., as found in the Table B2 of Appendix B. The jamming ratio is computed by (6.2) as:

$$JR = \frac{ID_{Cond}}{OD_{Cable}} = \frac{6.093}{1.051} = 5.80$$

This value is well out the jamming range, a 5 in. conduit (5.073 in.) may be used, and the jamming ratio in this case is 4.82 still well outside the jamming range. To estimate the percent fill the combine four-conductor area is divided by the conduit area, 29.158 sq. in. (as found in the same Table B2):

$$\%Fill = \frac{4 \times 0.8676}{29.158} = 0.12 \text{ or } 12\%$$

This value is less than the required 40% fill by the codes.

(b) For the second conduit, used in this application, the pull tension is computed by using (6.4b) as:

$$T_{OUT} = 150 + 50 \cdot 0.698(\sin 45^\circ + 1.35 \cdot 0.35 \cos 45^\circ) = 185.8647$$

$$= 185.9 \text{ lbf}$$

(c) The swept angle converter in radians is 1.5707 rad and the tension is given by (6.5):

$$T_{OUT} = T_{IN} \exp(w_c \mu_f \theta) = 150 \times \exp(1.35 \cdot 0.35 \cdot 1.5707) = 315.0688$$

$$= 315.1 \text{ lbf}$$

The sidewall pressure in this case, by (6.9b) is:

$$P_{SW} = \frac{3 \times 1.35 - 2}{3} \left(\frac{315.1}{3} \right) = 0.683 \times 105.033 = 71.77 \text{ lb/ft.}$$

This value for sidewall pressure is acceptable for this application.

6.5 Panel-boards and industrial power distribution

The enclosure of any equipment serves to keep out dirt, dust, moisture, intruders, and to protect equipment and personnel. This is a separate matter from protection against the explosions, while a piece of electrical equipment needed to be mounted outdoors and be protected against the weather, where there is no risk of explosion, or it may be indoors in a particularly dusty but nonflammable atmosphere. An internationally agreed system has been developed to designate the degree of protection afforded by any enclosure. It consists of the letters IP, which stands for International Protection followed by two digits, indicating the degree of protection. The first digit, 0–6, describes the protection against ingress of solids, while the second digit, 0–8, describes the protection against ingress of liquids. In both the cases, the higher the numbers the greater the degree of protection is. Panel-boards are metal cabinets, enclosing the main disconnect switch and the branch-circuit protective equipment. Distribution panel-boards are located between the power feed lines within a building and the branch circuits, connected to it. MV switchgears are used any time when a facility is directed connected to the MV section of power transmission and distribution. MV switchgears, the metal-clad switchgears have grounded metal barriers separating compartments and structures within the assembly. LV panel-boards and switchboards typically do not have such type of protection. Metal-enclosed LV switchgear is used in industrial and commercial buildings, as a power distribution control center to house the CBs, bus-bars, and terminal connections which are part of the power distribution system. NEMA standard 250-2003 defines criteria for electrical enclosures based on their ability to withstand external environments. Metal-enclosed switchgear is governed by IEEE standard C37.20.3, while metal-clad switchgears are covered by multiple standards and codes. Metal-clad switchgear meets or exceeds the requirements for metal-enclosed switchgear but not vice versa. Usually, a combination of switchgear and distribution transformers is placed in adjacent metal enclosures. This combination is referred to as a load-center unit substation since it is the central control for several loads. The rating of these load centers is usually 15 KV or lower for the high-voltage section and 600 V or less for the low-voltage section but higher values such 35 kV are available. Load centers provide flexibility in the electrical power distribution design of industrial plants and commercial buildings.

6.5.1 Panel-board and switchboards calculations and ratings

The panel-board and switchboard functions are to supply and distribute power to the branch circuits and feeders, through overcurrent protection devices. The panel-board structure must allow the settings of proper voltages to the supplied branch circuits and feeders. In addition to circuit protection, power distribution systems must have equipment which can be used to connect or disconnect the entire system or parts of the system. Safety switches are used only to turn a circuit off or on; however, fuses are often mounted in the same enclosure with the safety switch. Single-pole, double-pole, and three-pole CBs are available to be installed into the most common panel-board types. In order to understand how a panel-board distributed the power and voltage, its internal structure and the electrical diagram must be available. Panel-boards are classified as power panel-boards or as lighting and appliance branch circuit panel-boards. A panel-board that has less than 10% of its protection devices supplying lighting and appliance branch circuits is considered a power panel-board, as defined by the NEC Section 408.14(A). Otherwise is a panel-board has 10% or more of its protection devices supplying lighting and appliance branch circuits is considered a power panel-board is defined as lighting and appliance branch circuit panel-boards. NEC is defining as lighting and appliance branch circuit as one rated at 30 A or less, for example, a 20 A and 277 V branch circuit supplying lighting in commercial buildings.

Many of the panel-boards are provided with necessary overcurrent protection by using a main CB, or by the feeder overcurrent protection device. Power panel-boards which do not have system neutral to the panel and are supplying branch-circuit loads with only ungrounded conductors are not requiring overcurrent protection. However, good design practice recommended providing protection. Usually, two separate panels are installed for systems supplied by 240/120 V, three-phase, four-wire systems, one panel supplying 120/240 V, single-phase, three-wire loads, and the other supplying 240 V, three-phase, three-wire loads. Panel-boards supplied by three-phase, three-wire systems, with voltage ratings of 240 V and 480 V are supplying only three-phase loads, such as motors, insulation and step-down transformers, and electric heaters. In order to keep the record of the branch circuits, each panel-board has a panel schedule, designed to allow easy tabulation of the loads on the individual branch circuits. The load data are used to determine load balance among ungrounded conductors. Loads on the branch circuits maybe expressed in amperes (A) or volt-amperes (VA). The panel schedule also contains information about neutral and ground bus, very useful in application where the neutral must be separated from the ground bus. The neutral bus is usually rated to carry 100% of the panel-board rated current; however, there are panels with neutrals rated to carry 200% of the panel rated current.

Example 6.6: Determine the loading, expressed in A on the each of the phase (ungrounded) conductors for the specified loads (a) 120 V, single-phase, and 1,800 VA, (b) 277 V, single-phase, and 2,800 VA, and (c) 480 V, three-phase, and 36,000 VA.

Solution: The current (loading) in each case is:

$$I = \frac{1{,}800}{120} = 15.0\,\text{A}$$

$$I = \frac{2{,}800}{277} = 10.1\,\text{A}$$

$$I = \frac{36{,}000}{\sqrt{3} \times 480} = 43.35\,\text{A}$$

In order to ensure the safety of the electrical installation, proper, and required clearances must be ensured and maintained around panel-boards, switchboards, and switchgears. They are needed to provide adequate working space around electrical equipment and devices and safety protection space in the event of equipment failure. NEC Section 110.26(A) specifies a working space of at least 30 in., unless the equipment is wider than 30 in., in which case the required working space in the front of the equipment must be at least the width of the equipment. It is also required that hinged doors or panels to open at least 90° without any obstruction, while the height of working space must be at least 6.5 in. or equipment height, whichever larger. The required depth is a function of the nominal voltage to the ground and the installation equipment condition. In addition to the electrical equipment required working space, Section 110.26(C) specifies the clearance requirements and specification for access to the working spaces.

6.5.2 Load and motor centers

Power distribution systems used in large commercial and industrial and large commercial applications are complex. Power may be distributed through switchgear, switchboards, transformers, and panel-boards. Power distributed throughout a commercial or industrial application is used for a variety of applications, such as heating, cooling, lighting, and motor-driven machinery. Motor control centers (MCCs) receive this power through complex distribution systems which include power distribution lines and related equipment. Transformers used with three-phase power require three interconnected coils in both the primary and the secondary. These transformers can be connected in either a wye- or a delta-configuration. The type of transformer and the actual voltage depend on the requirements and capability of the power company and the customer needs. Unlike other types of power distribution equipment, which are used with a variety of load types, MCCs primarily control the distribution of power to electric motors. Usually in industrial and large commercial facilities and structures, the electricity is delivered to loads from designated system, known as load centers, containing the necessary equipment to monitor, protect, and control the loads. MCCs are centralized hubs containing motor control units sharing a common power bus. Used in low-voltage (230–600 V) and medium voltage (2.3–15 kV) three-phase applications, MCCs traditionally house startup and drive units. Auxiliary equipment is often found in almost all load centers, such as control switches, indicator lights, and metering equipment are sometimes installed in smaller metallic enclosures whose capabilities to withstand the external environment are determined by NEMA Standard 250.Today's units go far beyond these basic functions. It is not uncommon for modern "smart" centers to include programmable controllers and metering equipment for complex control schemes and safety features, in addition to the overload relays and contactors. Switchgear incorporates switches, CBs, disconnects, and fuses used to route power and in the case of a fault, isolate parts of an electric circuit. In general, switchgear has three basic functions (1) protection and safety for equipment and workers, (2) electrical isolation to permit work and testing, and (3) local or remote circuit switching. Developments in switchgear design have led to the introduction of network support for monitoring and control as well as advanced diagnostic capabilities for the purposes of monitoring usage, loading, and a host of other operational parameters. Among of the advantages of modern switchgear are (a) control and monitoring over a network, (b) remote control of the switchgear over a network, by removing the risk of arc flash to the operator, (c) individual analysis and power monitoring, involving the power monitoring and analysis of the main power and individual CBs, and (d) advanced technology, the latest technology in CBs and switching. There are different types of load centers, their selection based primarily on the electrical requirements of the loads and the installation environment. Several types of load centers are housed in metal enclosures to protect the enclosed equipment, nearby objects, and personnel in the event of equipment malfunction. Load centers supplying large motors and/or other industrial loads are referred to as switchgear, while smaller load centers specialized to supply small to midsize motors will be called MCCs. Auxiliary equipment, such as control switches, indicator lights, monitoring, and metering equipment are sometimes installed in smaller metallic enclosures whose capabilities to withstand the external environment are determined by the specifications of the NEMA 250 Standard. Switchgears are specialized load center designed to supply large loads, such as large electric motors or smaller load centers. Switchgears come in two varieties: metal enclosed and metal clad. Metal-clad switchgear is the more robust design, requiring shutters between the bus and the equipment front, compartmentalization of live parts, and insulation of the bus and primary components.

All switching and interrupting devices must be drawing-out mounted, allowing the removal from the switchgear without unwiring. Metal-enclosed switchgear does not need to meet these criteria and often has lower interrupting ratings, lower breaker duty cycle, and may use fused or nonfused switches instead of CBs. Metal-enclosed switchgear is governed by IEEE standard C37.20.3, while metal-clad switchgear is covered by multiple standards. Metal-clad switchgear meets or exceeds the requirements for metal-enclosed switchgear but not vice versa. A switchgear lineup is made-up of multiple sections or cubicles joined side by side. The front of the switchgear is hinged, and opening the doors exposes the CBs. When a breaker is removed from metal-clad switchgear, an insulated barrier separating the cubicle from the energized bus-work running the length of the switchgear is visible. This is an important safety feature, since metal-clad switchgear commonly is maintained while the main bus is energized. An MCC is a load center customized to serve small to midsize motors. CBs are replaced with combination motor starters. Since the starters are much smaller than switchgear breakers, stacking more than two units per horizontal section and reducing the width of the horizontal sections from 36″ to 20″ can save considerable space. MCCs are centralized hubs containing motor control units sharing a common power bus. Used in low voltage (230–600 V) and medium voltage (2.3–15 kV) three-phase applications, MCCs traditionally house startup and drive units. However, modern MCC units go far beyond these basic functions. It is not uncommon for modern *smart* centers to include programmable controllers and metering equipment for complex control schemes and safety features, in addition to the overload relays and contactors. MCCs are simply physical groupings of combination starters in one assembly. A combination starter is a single enclosure containing the motor starter, fuses or CB, and a disconnecting power device. Other devices associated with the motor, such as pushbuttons and indicator lights, may also be included. Switchgear incorporates switches, CBs, disconnects and fuses used to route power, and in the case of a fault, to isolate electric circuit sections. Switchgear has three basic functions (1) protection and safety for equipment and workers, (2) electrical isolation to permit work and testing, and (3) local or remote circuit switching. Developments in switchgear design have led to the introduction of network support for monitoring and control as well as advanced diagnostic capabilities for the purposes of monitoring usage, loading and a host of other operational parameters.

6.5.3 Load center, switchgear and motor control center ratings

The standards controlling the design and testing of metal-clad switchgear were developed by the American National Standards Institute (ANSI) in conjunction with the IEEE. IEEE standard C37.20.2 stipulates metal-clad switchgear-rating criteria. Low-voltage switchgear (600-V class and below) is governed by IEEE standard C37.20.1, while IEEE standards C37.04, C37.06, and C37.09 specify medium-voltage CB rating and testing criteria. Several important ratings are given to the main bus-work in metal-clad switchgear. Both copper and aluminum bus carry these ratings. A continuous current rating is assigned to limit the temperature rise of the busbars to a value that will not compromise the bus insulation. The busbars typically are insulated with an epoxy-type material that can withstand fairly high-operating temperatures. Insulators, either porcelain or polymeric composite, support the busbars. The insulators also have a maximum operating temperature. A main (horizontal) bus similar to that used in switchgear is implemented in MCCs. Since the loads supplied by an MCC are smaller than the loads fed from switchgear, the available bus ratings tend to be lower in an MCC. A system of vertical bus-work must be incorporated into the design to provide power to each unit in each horizontal section. Commonly available bus ratings range from 600 to 2,500 A for the horizontal bus, and 300–1,200 A for the vertical bus although some manufacturers may offer other ratings. Interrupting ratings of 65 kA and 100 kA are common. Although MCCs are usually implemented at low voltages (240 or 480 V), medium-voltage MCCs are also available. It is important to determine the continuous current rating of switchgear bus by temperature rise. While some equipment specifications cite a maximum allowable current density to determine the bus rating, this practice is not allowable. Facilities and structures requiring several panel-boards located through the building or facility are using main power distribution panel-boards (MDPs), serving as service entrance equipment for larger services and supply feeders for other panel-boards. The common MDP ratings are in the range 600 A to 5 kA. MDP schedules are providing pictorial information of the major MDP components, ratings, distributions, and subsections.

6.6 Chapter summary

The power distribution inside the buildings, structures, and industrial facilities is accomplished through the electric services, panel-boards, and switchboards. Electric services, feeders, branch circuits, and panel-boards are critical and essential components of the power distribution inside the buildings, structures, and industrial facilities. These elements are important aspects of the designing the electrical systems of buildings and facilities. The information required for feeder and branch circuit design is often presented as cabling diagrams. Feeders are conductors and cables that originate at the main depower distribution or main disconnect device and terminates at another distribution center, panel-board, or load center, while the subfeeder are the conductors or cables originate at the power distribution centers other than the main power distribution center and extend to panel-boards, load centers, and disconnect switches that supply branch circuits. A cable tray system is a unit or assembly of units or sections and fittings forming a rigid structural system used to securely fasten and/or support cables, conductors, and raceways, being part of facility structural system. The electrical considerations must be included in any design of cable tray systems. Several standards and guidelines exist for the design of cable tray systems, as the NEC Sections, NEMA and IEC specifications designed to address various aspects of cable tray systems. Most of the manufacturers provide cable trays in a variety of materials, designs, and shapes. In industrial facilities, the electricity is provided to the loads from the so-called load centers, switchgear supplying power to large electrical motors, equipment and smaller MCCs, and MCCs supplying electricity to mid-size motors and loads. Both of these load centers have metallic enclosures and hinged doors for easy internal access. Switchgears can be metal enclosed or metal clad. A system of busbar runs inside the load center to supply power to each of its unit, section, or cubicle. Switchgear with two power sources (double-headed switchgear) requires a method of source transfer to switch between sources. Several important ratings apply to both switchgear and MCCs. A continuous current rating limits the temperature rise of the load center components, particularly the insulation. A short-circuit current rating determines both the interrupting capability of the CBs and fuses and the mechanical strength of the bracing and support systems which must resist the severe mechanical forces exerted by the intense magnetic fields present during a fault. The basic impulse level (BIL) rating, determines the electrical strength of the insulation. This chapter gives a comprehensive presentation of the wiring and electrical protection systems in commercial and industrial building, fuses, CBs, instrument transformers and protective relays, grounding and ground-fault protection, feeder design and branch circuits for lighting, equipment, and electrical motors. A special notice for the engineers and contractors, the above demand factors are the most widely used on a regular basis because of their uniqueness to electrical design. With the application of demand factors, smaller components can be utilized in the electrical system and greater savings can be passed on to the consumer. Due to the high cost of the wiring, the designers need to utilize these techniques more than ever before to reduce the cost. Cables are contained in some type of raceway, typically conduit, duct, or cable tray.

Further readings

[1] T.R. Bosela, Electrical Systems Design, Prentice Hall, Upper-Saddle River, NJ, 2002.
[2] R.D. Schultz and R. A. Smith, Introduction to Electric Power Engineering, John Wiley & Sons, New York, 1988.
[3] R.C. Dorf (Ed.), The Electrical Engineering Handbook, CRC Press, Boca Raton, FL, 1993.
[4] G.F. Moore, Electric Cables Handbook (3rd Ed.), Blackwell Science, London, UK, 1997.
[5] P. Schavemaker and L. van der Sluis, Electrical Power System Essentials, John Wiley and Sons, Chichester, UK, 2008.
[6] M.E. El-Hawary, Introduction to Electrical Power Systems, Willey-IEEE Press, NJ, 2008.
[7] R.D. Schultz and R.A. Smith, Introduction to Electric Power Engineering, John Wiley & Sons, New York, 1988.
[8] A.C. Williamson, Introduction to Electrical Energy Systems, Longman Scientific & Technical, London, UK, 1988.
[9] J.R. Cogdell, Foundations of Electric Power, Prentice-Hall, Upper-Saddle River, NJ, 1999.
[10] R. Fehr, Industrial Power Distribution (2nd ed.), Wiley, NJ, 2015.
[11] L.L. Grisby (Ed.), Electrical Power Engineering, CRC Press, Boca Raton, FL, 2001.
[12] W.A. Thue, Electrical Power Cable Engineering, Marcel Dekker, New York, 1999.
[13] EPRI 1000419, Engineering Guide for Integration of Distributed Generation and Storage into Power Distribution Systems, Electric Power Research Institute, Palo Alto, CA, 2000.
[14] EPRI 1016097, Distribution System Losses Evaluation, Electric Power Research Institute, Palo Alto, CA, 2008.
[15] F.P. Hartwell, J.F. McPartland, and B.J. McPartland, McGraw-Hill's National Electrical Code 2017 Handbook (29th Ed), McGraw-Hill, New York, 2017.
[16] T.A. Short (Ed.), Handbook of Power Distribution, CRC Press, Boca Raton, FL, 2014.
[17] D.R. Patrick and S.W. Fardo, Electrical Distribution Systems (2nd ed.), CRC Press, Boca Raton, FL, 1999.
[18] J.E. Fleckenstein, Three-Phase Electrical Power, CRC Press, Boca Raton, FL, 2016.
[19] IEEE Standard 141, IEEE Recommended Practice for Electric Power Distribution for Industrial Plants, 1993.
[20] M.W. Earley (ed.), National Electrical Code (NEC) Handbook, NFPA 70, Quincy, MA, 2014.
[21] M.W. Earley (ed.), National Electrical Code (NEC) Handbook, NFPA 70, Quincy, MA, 2017.

[22] R.R. Janis, and W.K.Y. Tao, Mechanical and Electrical Systems in Buildings (5th ed.), Pearson, Upper Saddle River, NJ, 2014.
[23] W. T. Grondzik, and A. G. Kwok, Mechanical and Electrical Equipment for Buildings (12th ed.), Wiley, Hoboken, NJ, 2015.

Questions and problems

1. List the types of the branch circuits.
2. What are the purpose and what information is contained in a cabling diagram?
3. Describe the typical elements of a low-voltage service entrance and the ones of a low-voltage underground service entrance.
4. Describe the typical elements of a medium-voltage service entrance and the ones of a medium-voltage underground service entrance.
5. When are multiple services for a building permitted?
6. List the major components of a service entrance.
7. Explain what is meant by service drop and by service lateral.
8. Describe the purpose and location of the main service entrance components.
9. Determine the number of the 20 A branch circuits required to supply general lightning load in residence of 2,400 sq. ft.
10. Explain the operation and the purpose of an arc fault interruption device.
11. Why is important the short-circuit capabilities of a panel-board?
12. What is the minimum number of branch circuits supplying the general lightning load for the following occupancies, a 3,600 sq. ft. house (at 15 A rating) and a 1,500 m^2 office space (at 20 A)?
13. What is minimum bending permitted on a 2 in. rigid conduit?
14. The minimum number of 20 A and 120 V branch circuits required to serve the general lighting load for a 3,000 ft^2 one family dwelling unit is ______ circuits.
15. NEC Section 220-3(b) (9) requires that receptacles in other than dwelling units be computed at 180 VA each. The maximum number of receptacles that can be installed on a 20 V and 120 V branch circuit in a store is _______. The maximum number of receptacles that can be installed on a 15 A and 120 V branch circuit in a school is _____.
16. List the most common types of conduits.
17. What are the main reasons why the cables are installed into conduits?
18. What the purpose of calculating the jam ratio when pulling cables into conduits?
19. What is the size of rigid steel conduit that can be used to contain three cables, each with an outside diameter of 1.65 in.?
20. Compute the percent unbalance of three-phase, four-wire 480/277 V that has 80 A on phase A, 70 A on phase B, and 125 A on phase C. If unbalance is not acceptable, what is the suggested correction?
21. Determine the length change of a 300 ft. PVC conduit, if the temperature is changing from 10 °F to 90 °F (operating temperature range), and the exapsion coefficient is, 4.05×10^4 in./ft./°F.

22. How many cables, each with an outside diameter of 0.78 in., can be pulled in a 3-in. rigid steel conduit?
23. Determine which of the following service disconnects requires ground fault protection (a) 700 A, 480Y/277 V, (b) 2,000 A, 480Y/277 V, and (c) 2,000 A, 208Y/120 V.
24. What is common voltage and current range for MV switchgear?
25. Determine the loading, expressed in A on the each of the phase (ungrounded) conductors for the specified loads (a) 120 V, single-phase, 2,400 VA, (b) 277 V, single-phase, 4,800 VA, (c) 240, three-phase, 7,500 VA, and (d) 480 V, three-phase, 27,000 VA.
26. Determine the tension in the following cables, and the sidewall pressure where applicable for three 750 kcmil THWN-2 copper conductors in a 5 in. (128.2 mm inner diameter) PVC conduit, friction coefficient 0.35. Assume a tension of 200 lb. at the input to the conduit, a weight correction factor of 1.4 for all calculations. The three conduits used in this application are (a) a 300 ft. straight horizontal conduit, (b) a 60 ft. pull up inclined at 45°, and (c) a bend of 90° having radius of 36 in.
27. Consider a raceway layout, in horizontal plane consisting of 300 ft. linear run a 90° bend with 1.5 ft. radius, followed by a 150 ft. linear run. A four-conductor 500 kcmil copper power cable is used. The cable has an outer diameter 1.2 in. and weighs 1.83 pound per foot. Assuming a friction coefficient of 0.25, design the raceway.
28. Consider the conduit raceway layout shown in the figure below. All bends lie in the horizontal plane. Three single-conductor 500 kcmil tape-shielded 5 kV power cables need to be pulled through the conduit. The cable has a 1.093 in. outside diameter and weighs 1.83 lb/ft. A coefficient of friction of 0.25 anticipated and weight correction factor of 1.4. Determine the optimum conduit for this application.

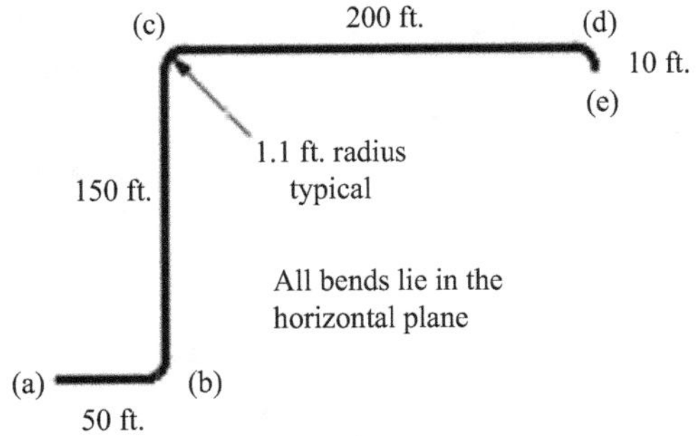

Chapter 7
Lighting systems

Objectives and abstract

Illuminating the work or living spaces, both interior and exterior, is a fundamental engineering function and requirement. In order to perform it properly, light knowledge and lighting equipment characteristics and behavior, and a complete understanding of the space use are necessary. Various activities that are taking place within the space have different illumination requirements and specifications. Illuminating Engineering Society of North America (IESNA) and other regulatory bodies have established criteria for lighting applications, and some of them are discussed in this chapter. Light is not only needed in visual task areas but also for the perception in the designated space. Rooms and designated spaces should be illuminated properly and in agreement with codes and standards. An extensive range of light sources and luminaires are available to provide adequate lighting. Significant lighting technical progresses have expanded its scopes, which in turn led to the development of specialized and efficient lighting systems and equipment. Lightning principles, concepts, parameters, lighting design process and methods, lighting equipment and systems, characteristics, and performances are discussed in this chapter. The factors involved in determining illumination requirements are analyzed in relation to the lighting levels for various tasks and the possible use of daylight. Lighting system design considerations related to luminaires are also addressed here. The purpose of a luminaire is 2-fold, i.e., to hold, protect, and connect a lamp(s) to the electrical system, and to control the light output. The needs for trade-offs and qualitative decisions when selecting luminaires for a particular space are emphasized. Standards, codes, and specifications for lighting systems are also included and discussed. Examples of lightning calculations, applications are also included. These examples demonstrate how the modern and advanced lighting technologies can be integrated, eventually with the daylight to ensure very efficient and high quality lightning applications and environment. After completing this chapter, the readers must have an understanding of the fundamentals of lighting system design, particularly for indoor areas, the lighting technologies available for commercial and industrial applications, their advantages and disadvantages, and how they work. However, this chapter does not intend to compete with the existing comprehensive lighting engineering textbooks and lightning manuals, or to be added to the limited number of beautifully illustrated volumes containing finished projects. On the other hand, an understanding of lighting fundamentals, equipment, codes, and standards are essential for engineers and decision-makers, evaluating lighting upgrades, improvement or changes. After completing this chapter, the readers must have good understanding of the lighting system types, operation, characteristics and design, fundamental lighting terms, units, and definitions and to be able to apply photometric data, to analyze, compute, and design lighting systems, to select the proper lighting system components in order to provide the most adequate lighting levels and the most efficient and cost-effective lighting. The readers also have a basic understanding of the lighting system control, the importance of lighting system in the building energy system and to apply the lighting system retrofits to improve energy efficiency, as well as the requirements of the lighting standards and codes.

7.1 Introduction, lighting basics

Light is that part of the electromagnetic spectrum that is perceived by our eyes, which makes things visible. It is defined as electromagnetic radiation or energy transmitted through space or any medium in the form of electromagnetic waves, being defined as visually evaluated radiant energy. Light is that part of the electromagnetic spectrum visible by the human eye, in agreement with the illuminating engineering definition. The physical difference between radio waves, infrared, visible light, ultraviolet, and x-rays is their wavelength. A spectral color is a specific wavelength light, exhibiting deep chromatic saturation. Hue is the attribute of color perception denoted by what we call red, orange, yellow, green, blue, and violet. Lighting in conformity with standards is decisive for ensuring that a visual task can be identified and the related activities can be carried out. Consideration of the traditional lighting quality characteristics has a major impact on the visual task performances. Illuminating spaces, both interior and exterior, is a fundamental engineering function. To perform this function properly, a working knowledge of the characteristics and behavior of light, and a thorough understanding of the use of the space are necessary. Various activities that will take place within the space will have different illumination requirements.

Light represents a form of electromagnetic radiation generated for the physical processes from the Sun or converted from other energy forms in natural or man-made systems. Lighting represents the natural or converted light energy utilization to provide a desired or required visual environment for working and living. For the most part of our history, from the humans' origins to the most of the eighteenth century, there were basically two sources of light available. The first one is the daylight, the natural light originating from the Sun, whose properties the eye has adapted over millions of years. The second one is the flame discovered during the stone-age era, with its development of cultural techniques and tools, as an artificial light source. The lighting conditions remained almost same for a considerable period of time. Lighting was limited to daylight and flame and it was for this very reason that humans have continued to perfect the applications of these two light sources for tens of thousands of years. In the case of daylight this meant adapting the architecture to the natural lighting requirements. Entire buildings and individual rooms were aligned to the incidence of the sunrays. The room size was determined by the natural lighting availability and by ventilation. Light, not only serve to render the spatial bodies three-dimensional but is also excellent means for perception control on a psychological level. Similar processes took place in the realm artificial lighting, a development clearly confined by the luminous power provided by the available light sources. The story began when the flame was separated from fire, the source of warmth likely burning branches were removed from the fire and used for specific purposes. It soon became obvious that it was an advantage to select pieces of wood that combust and emit light well, and the branch was replaced by especially resinous pine wood. The next step involves of applying flammable materials to torches to produce more light. The development of the oil lamp and the candle meant that man then had compact, relatively safe light sources to use, economically fuel selection, while the torch holder was reduced to the wick as a means of wax or oil transport. The oil lamp, which was actually developed in prehistoric times, represented the highest form of lighting engineering progress for a very long time. The lamp itself, later to be joined by the candlestick, continued to be developed over course of history.

In contrast to the oil lamp and gas lighting, which both started as weak light sources and were developed to become ever more efficient, the electric lamp embarked on its journey in its brightest forms. From the beginning of the nineteenth century, it was a known fact that by creating a voltage between two carbon electrodes, an extremely bright arc is produced. Incandescent gas light was doomed to go the way of most lighting discoveries that were fated to be overtaken by new light sources just as they are nearing perfection. This also applies to the candle, which only received an optimized wick in 1824 to prevent it from smoking too much. Similarly, the Argand lamp was piped by the gas lighting development, using incandescent mantles, which in turn had to compete with the newly developed electric lighting. However, the electric light at that time requires manual adjustment, making difficult for it to gain acceptance, plus the fact that arc lamps first had to be operated on batteries, a costly business at that time. Following the arc lamp and the incandescent lamp, discharge lamps took their place as the forms of electric lighting. Again physical findings were available long before the lamp was put to any practical use. The three main functions of the lighting are (1) ensure the people safety, (2) facilitate the performance of visual tasks, and (3) help the creation of an appropriate visual environment (appearance and character). About mid-century, self-adjusting lamps were developed, thereby eliminating the problem of manual adjustment.

At earlier stage of its use, the incandescent lamp failed to establish itself as a new light source for technical reasons, much the same as the arc lamp. There are only a few materials, having a melting point high enough to create incandescence before melting, while the high level of resistance required thin filaments, difficult to produce, broke easily, and burnt up quickly in the air. First experiments made with platinum wires or carbon filaments did not produce much more than minimum service life. The lifetime was extended when the filament predominantly made of carbon or graphite was prevented from burning up by a glass bulb, which was either vacuum or filled with inert gas. The breakthrough was made by Edison, in 1879, which succeeded to develop industrial mass product out of the experimental constructions. This product corresponded to the today's incandescent lamp in many ways, right down to the construction of the screw cap. The filament was the only element that needed improvements. Following the arc and the incandescent lamps, the discharge lamps took their place as the third electric lamp form. Again physical findings were available long before the lamp was put to any practical use. For over one hundred years after the scientific research of new light sources began, all the standard electric lamps that we know today had been created in their basic forms. However, to this point, the only sufficient light available was during daylight hours. From now on, artificial light dramatically changed the humans' life. It was no longer a temporary daylight replacement but a form of lighting ranking with natural light. Illuminance levels similar to those of daylight are technically now produced in interior living and working spaces, or in exterior areas, such as street lighting, public spaces, or for the floodlighting of buildings, etc.

7.2 Lighting in engineering, architecture, industrial process, and building operation

In a new lighting design, the inadequate light sources had been often the main problem, while the lighting specialists also faced with the challenge of controlling excessive amounts of light, efficiency, and cost. Specialist engineers started to think about how much light is required in which situations and what lighting forms need to be applied. Task lighting in particular was examined in detail to establish how great an influence illuminance and the kind of lighting applied had on productivity. The result of these physiological investigations was a comprehensive reference work that contained the illuminance levels required for certain visual tasks plus minimum color rendering qualities and glare limitation requirements. Although these standards was designed predominantly as an aid for the lighting planning for workplaces, it soon became a guideline for lighting in general, and today lighting design in practice. The fact that the perception of an object is more than a mere visual task and that, in addition to a physiological process, vision is also a psychological process, which usually is not considered in lighting design. Quantitative lighting design, intended to provide uniform ambient lighting that meets the requirements of the most difficult visual task to be performed in the given space, while at the same time adhering to the standards regarding the glare limitation and color distortion. How one is seeing the architecture, under a given light, whether its structure is clearly legible, and its aesthetic quality has been enhanced by the lighting, goes beyond the realm of code rules. Lighting engineers still tend to practice a quantitative lighting philosophy. The architects were started to develop new concepts for lighting design, with daylight as defining agent. The light and shadow significances and the way the light can structure a building, is something every architect is familiar with. With the development of more efficient artificial light sources, the knowledge that has been gained of daylight technology was used for artificial lighting design.

Increasing demand for higher quality lighting design was accompanied by high quality lighting equipment demand. Differentiated lighting required specialized luminaires, designed to cope with specific lighting tasks. Different luminaires are required to achieve uniform light over wall areas, for accentuating individual objects, for the permanent lighting in theatre foyers, or for the variable lighting required in multipurpose halls or exhibition spaces. Technical developments and lighting applications led to the productive correlations; industry had to meet new luminaires designers' demands, and further developments in the lamp technologies and luminaire design were promoted to suit particular applications required by designers. New lighting developments served to allow spatial differentiation and more flexible lighting, followed by the new developments that offered time-related differentiation in lighting control systems. With the use of advanced control systems, it possible to plan lighting installations that not only offer one fixed application but are able to define a range of light scenes. Lighting control systems are a consequence of spatial differentiation, allowing a lighting installation to be utilized to the full, a seamless transition between individual scenes, which is simply not feasible via manual switching.

General lighting is the main source of illumination in a living or working space. Most of our information received about the world around us is through our eyes. Light is not only an essential prerequisite and the medium by which we are able to see but its properties, intensity, and distribution throughout a space create specific conditions that are influencing our environmental perception. Lighting design is, in fact, a visual environmental planning. Good lighting design aims to create perceptual conditions allowing working effectively and orienting safely, while promoting a well-being feeling in a particular environment and at the same time enhancing that same environment in an aesthetic sense. The physical qualities of a lighting situation can be calculated and measured. Ultimately, the actual lighting effects have on the space user, his subjective perception that decides whether a lighting design is successful or not. Lighting design must not be restricted to the technical concepts only. Human perception must be a key consideration in the lighting design. For example, the uniform, base level of lighting can easily become the focus of energy reduction, as the light levels from other fixtures can be lowered, by using, for example, LED and metal halide lamps. Recommended light levels for general lighting are 30–50 foot-candles, providing the area with overall illumination, specifically for orientation, general tasks, and control contrast ratios. Diffused general lighting is ensuring a well-being sense, making the employees to feel comfortable. A simple way to achieve it is by arranging recessed fixtures using reflectors, baffles, and lensed trims. Lighting is one of the best and easiest ways to improve living and office environments. The challenge is that office lighting plans must be cohesive and effectively illuminate different space types, coexisting under one roof: the reception area, open office space, and private offices of varying sizes. There also are energy codes to follow, concerns about energy costs and efficiency of the lighting system and the need to incorporate flexibility for easy future adjustments and changes in lighting needs. Scopes of lighting design are:

- Create an environment, enhancing the personnel well-being and productivity.
- Create flexible lighting settings enabling personnel to perform tasks and operations comfortably, effectively, and safely.
- Integrate and balance ambient, task, accent, and decorative lighting into any facility areas, allowing a comfortable transition from space to space.
- Design lighting systems for long-term employee comfort, higher lighting quality, with a proper energy use balance, savings and conservation, and future lighting system upgrades.
- Integrate, monitor, and control the daylight to improve performances, while reducing energy use.
- Address energy efficiency, conservation, while complying with energy codes and standards.

Designing a facility lighting plan involves more than calculations, lamps, equipment, and luminaire selection. Lighting solutions affect the facility ambiance, employee psychological well-being, interests, and enthusiasm, is affecting and enhancing the productivity, so consideration must be given to interior design to create stimulating work places. Employees need to perform tasks comfortably and effectively in the work place environment where they spend about one-third of their lives. Lighting is one of the best and easiest ways to improve office and living environments. Lighting design and plans must be cohesive and effectively illuminate different spaces that coexist in the same structure, such as reception area, open office space, home spaces, and private offices of varying sizes, at desired or required lighting levels, while representing and reinforcing the corporate or organization image. There are also energy codes to follow, concerns about energy costs and lighting system efficiency, the needs to incorporate flexibility for easy adjustments as the company and lighting needs change. Light has a triple effect on the human behavior:

1. Light is used for the visual functions, the area illumination in conformity with standards, codes and regulations, glare-free, and convenient;
2. Light systems are creating biological effects, stimulating, or relaxing; and
3. Light is used for emotional perceptions, enhancing architecture, creating scenes, and effects.

7.3 Lighting theory and illumination calculation methods

Light is a form of electromagnetic radiation (energy), emitted by natural or artificial sources. Human eyes are sensitive to only a very narrow band of electromagnetic spectrum, the visible spectrum, which extends from 380 to 780 nm in wavelengths. The visible spectrum portion covers a narrow band of wavelength from approximately 380 to 770 nm (1 nm $= 10^{-9}$ m). Wavelengths shorter or longer than these do not stimulate the human eye receptors. The electromagnetic radiation is transmitted to the free space at a speed of 3×10^8 m/s (186,282 mi/s). Light and other electromagnetic

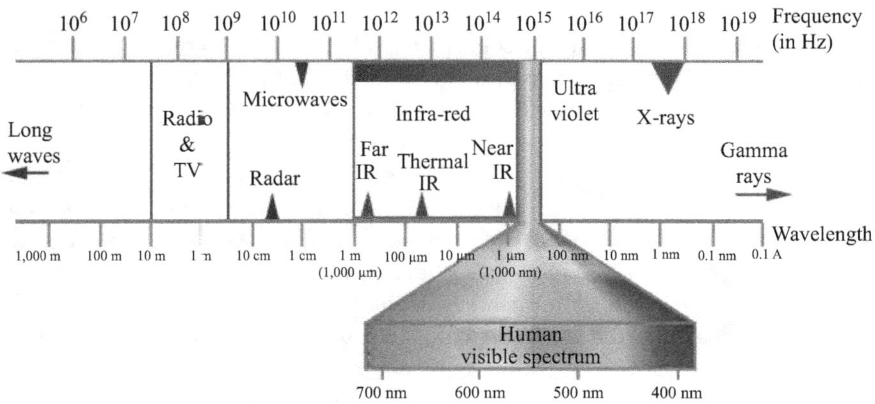

Figure 7.1 Electromagnetic spectrum, with the visible range included

radiation travel at lower speeds in other media, such as air or water. All electromagnetic radiation forms differ in wavelengths and frequencies, expressed by:

$$c = \lambda \times f \tag{7.1}$$

where c is the electromagnetic radiation (light) speed in m/s, f is the frequency in Hz, and λ is the wavelength in m. Electromagnetic radiation can be converted from other energy forms, via chemical, thermal or electrical processes, the last one being the most efficient. The electrical energy is also part of the electromagnetic spectrum, at range of lower frequencies. Light generated inside the buildings, is nearly all converted from the electricity. With the advent of the modern technologies, the efficiency of electrical energy conversion to lighting has advanced significantly, and all modern light sources used indoor and outdoor applications are electrical light sources. The electromagnetic spectrum is shown in the Figure 7.1.

Each light source is characterized by a spectral power distribution (SPD) curve or spectrum. The SPD curve (spectrum) of a light source shows the radiant power that is emitted by the source at each wavelength, over the electromagnetic spectrum (primarily in the visible region). With color temperature and color rendering index (CRI) ratings, defined in the next sections of this chapter, the spectrum can provide a complete picture of the color composition of a lamp's light output. Incandescent lamps and natural light produce a smooth, continuous spectrum, while other types of lamps may produce spectra with discrete lines or bands. For example, fluorescent lamps produce spectra with a continuous curve and superimposed discrete bands. The continuous spectrum results from the halo-phosphor and rare earth phosphor coating, while the discrete band or line spectrum results from the mercury discharge.

7.3.1 Basic parameters used in lighting physics

In lighting technology, a number of technical terms and units are used to describe the properties of light sources and the effects that are produced. Radiometry is the study of optical radiation, which includes light, ultraviolet radiation, and infrared radiation. Photometry, on the other hand, is concerned with humans' visual response to light. Radiometry is concerned with the total energy content of the radiation, while photometry examines only the radiation that humans can see. The common unit in radiometry is the watt (W), measuring the radiant flux (power), while the common photometric unit is the lumen (lm), which measures the luminous flux. For monochromatic light of 555 nm, 1 W is equal to 683 lumens. For light at other wavelengths, the conversion between watts and lumens is slightly different, the eye is responding differently to each wavelength. Similarly, the radiant intensity is measured in watts/steradian (W/sr.), while luminous intensity is measured in candelas (cd, or lm/sr.). Luminous flux describes the total amount of light emitted by a light source. This radiation can be measured or expressed in watt. However, this does not describe the light source optical effects adequately, since the varying eye spectral sensitivity is not taken into account. In order to include the eye spectral sensitivity to the luminous flux is measured in lumen. Radiant flux of 1 W emitted at the peak of the spectral sensitivity (in the photonic range at 555 nm) produces a luminous flux of 683 lm. The radiant flux produced is frequency dependent.

Luminous efficacy describes the luminous flux of a lamp in relation to its power consumption and is therefore expressed in lumen per watt (lm/W). The maximum value theoretically attainable when the total radiant power is transformed into visible light is 683 lm/W. The luminous efficacy varies from light source to light source but always remains well below this optimum value. The quantity of light, or luminous energy (US), is a product of the luminous flux emitted multiplied by time; luminous energy is generally expressed in klm·h.

Luminous flux, Φ is used to describe the quantity of light emitted by a light source. Its unit is lumen (lm). The luminous efficiency is the ratio of the luminous flux to the electrical power consumed (lm/W). It represents a measure of a light source's efficiency, being directly related to the light energy. Light energy degrades very fast into heat and cannot be stored. In order to maintain certain light level in any space, electrical energy must be supplied continuously. Light energy, W_{Light} measured in lm·s (or lm·h), and is defined by:

$$W_{Light} = \int \Phi \cdot dt \tag{7.2}$$

In physics, power is the rate of energy change, while in lighting light power is the luminous flux emitted by a light source in unit of time, given by:

$$\Phi = F \text{ (Luminous power)} = \frac{dW_{Light}}{dt} \text{ (lm)} \tag{7.3}$$

In the case of constant light power, the light energy is expressed as:

$$W_{Light} = F \times t$$

The luminous intensity describes the quantity of light that is radiated in a particular direction. An ideal point-source lamp radiates luminous flux uniformly into the space in all directions; its luminous intensity is the same in all directions. In practice, however, luminous flux is not distributed uniformly. This results partly from the design of the light source, and partly on the way the light is intentionally directed. It makes sense, therefore, to have a way of presenting the spatial distribution of luminous flux, i.e., the luminous intensity distribution of the light source. The unit for measuring luminous intensity is candela (cd). The candela is the primary basic unit in lighting engineering from which all other units are derived. The candela was originally defined by the luminous intensity of a standardized candle. In other words, the light intensity defines as the light flux density in a given direction, or the light source ability to produce illumination (or illuminance) in a given direction. Candela (cd) is defined as one lumen per steradian of solid angle, a useful measurement for directive lighting elements, such as reflectors. The light source luminous intensity distribution throughout a space produces a three-dimensional graph. A section through this graph results in a luminous intensity distribution curve (LDC); describing the luminous intensity on one plane, usually in a polar coordinate system as the function of the beam angle. To allow comparison between different light sources to be made, the light distribution curves are based on a 1,000 lm output. In the case of symmetrical luminaires, one light distribution curve is sufficient to describe them, while axially symmetrical luminaires require two curves, which are usually depicted in one diagram. The polar coordinate diagram is not sufficiently accurate for narrow-beam luminaires, e.g., stage projectors. In this case, it is usual to provide a Cartesian coordinate system. The luminous intensity is expressed as:

$$I = \frac{d\Phi}{d\Omega} \tag{7.4}$$

Here, Ω is the solid angle (measured steradian, or sr.) into which luminous flux is emitted, given by:

$$\Omega = \frac{dA}{r^2} \text{ (here, } r \text{ is the radius of an imaginary sphere)}$$

Radiant intensity, denoted by the letter I, is the amount of power radiated per unit solid angle, measured in W/sr. Luminous intensity is the amount of visible power per unit solid angle, measured in candelas (cd, or lm/sr.). Luminous intensity (I_v) is the fundamental SI quantity for photometry. The candela is the fundamental unit from which all other photometric units are derived.

Illuminance describes the quantity of luminous flux falling on a surface and its unit is Lux (lx). The light quantity is expressed, in engineering calculations in terms of average number of lumens per unit of area. In British system, the quantity of light is expressed in foot-candles (fc). One foot-candle being equal to 1 lumen per square foot, while in SI system, the light quantity is measured in lux (lx), where 1 lux being equal to 1 lm/m^2. Note that the conversion relation between foot-candle and lux is 1 foot-candle = 10.764 lux.

Example 7.1: Determine the illuminance of a $12 \times 12\ \text{ft}^2$ room, illuminated by a 6,480 lumens light source.

Solution: Illuminance $= \dfrac{6{,}480\ \text{lumens}}{12 \times 12\ \text{ft}^2} = 45\ \text{fc}.$

In physical term, the illuminance is expressed by the inverse square law, stating that it decreases by the square of the distance. Relevant standards specify the required illuminance. Illuminance is the means of evaluating the density of luminous flux, indicating the luminous flux amount from a light source falling on a given area. Illuminance need not necessarily be related to a real surface. It can be measured at any point within a space. Illuminance is determined from the light source luminous intensity, decreasing with the square of the distance from the light source. Luminance is the only basic lighting parameter that is perceived by the eye, specifying the surface brightness and is essentially dependent on its reflectance.

$$E = \frac{d\Phi}{dA} = I\frac{d\Omega}{r^2 \cdot d\Omega} = \frac{I}{r^2} \tag{7.5}$$

where A is the area hit by the luminous flux, measured in m^2. This is also known as the *inverse square law*, stating that the illuminance on a surface is direct proportional to the intensity, I, and inversely proportional to the square of the source-surface distance, the surface being normal to the direction of the light source. When the light receiving surface is not perpendicular to the light beams, the luminous flux will cover a larger area, and (7.5) is modified as:

$$E = \frac{I \times \cos(\theta)}{r^2} \tag{7.6}$$

Here, θ is the angle between luminous flux direction and the surface normal.

Example 7.2: A lighting fixture has an intensity of 8,500 cd, calculate the illuminance on a table of 2.5 m below.

Solution: From (7.6), the illuminance is:

$$E = \frac{8{,}500 \cdot \cos(0°)}{2.5^2} = 1{,}360\ \text{lx}$$

If the source is aimed at an angle toward a target on a vertical surface, the reduction in illuminance at the target is equal to the sine of the angle of incidence or tilt, θ, and is given by:

$$E = \frac{I \times \sin(\theta)}{r^2} \tag{7.7}$$

where again I is the intensity of the source (in candelas) in the direction of the light ray, θ is the angle of tilt between nadir and the direction of the target, and r is the distance from the source to the target.

Example 7.3: A lamp has an intensity of 18,000 cd calculate the illuminance, if the lamp is tilted at a 30° from nadir to cast light on a vertical surface 6 ft. away.

Solution: From (7.7), the illuminance is:

$$E = \frac{I \times \sin(\theta)}{r^2} = \frac{18{,}000 \times \sin(30°)}{(6)^2} = 250 \text{ fc}$$

When, the surface receiving the luminous lux is not perpendicular to the light source, the flux will cover a larger area, in relation to the cosine of the angle, and for a distance, h between the light source and the surface, at the normal to the surface, so the (7.5) became:

$$E = \frac{I \cdot \cos^3(\theta)}{h^2} \tag{7.8}$$

Example 7.4: If a lamp with intensity of 6,000 cd is focused at a painting from the light on the wall 6 ft. from the light at an angle of 45°, calculate the illuminance.

Solution: From (7.8):

$$E = \frac{I \cdot \cos^3(\theta)}{h^2} = \frac{6{,}000 \times (0.707)^3}{6^2} = 58.9 \text{ fc or } 10.76 \times 58.9 = 633.8 \text{ lx}$$

Following the practical notation, as an illuminated surface by a light source moves away from the light source, it appears dimmer. In fact, it becomes dimmer faster than it moves away from the source. The inverse square law, quantifying this effect (Figure 7.2), relates the illuminance (E_v) and intensity (I_v) as:

$$E_v = \frac{I_v}{d^2} \tag{7.9}$$

where d is the distance from the light source. For example, if the illuminance on a surface is 40 lux (lm/m^2) at a distance of 0.5 m from the light source, the illuminance decreases to 10 lux at a distance of 1 m. Note that, the inverse square law can only be used in cases where the light source approximates a point source. If the light is reaching the surface under an angle of incidence, θ other than 90°, the illuminance is due to the perpendicular component and is expressed as:

$$E_v = \frac{I_v}{d^2} \cos\theta \tag{7.10}$$

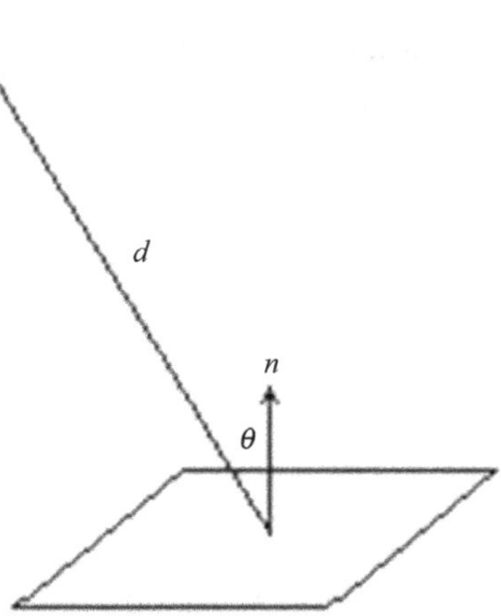

Figure 7.2 Illumination diagram

Example 7.5: A point source generates 3,000 cd in the direction of interest at an angle of 36°, determine the illuminance at a point of 13.5 ft. from the source.

Solution: The illuminance is given by:

$$E_v = \frac{I_v}{d^2}\cos\theta = \frac{3{,}000}{(13.5)^2}\cos 36^\circ = 13.3\ \text{fc}$$

Notice that the inverse-square method yields only a rough idea of what is perceived. Its main use is for comparisons, as when establishing the illuminance ratio between an object and its surrounding. Even here, the inverse-square method fails to account for any inter-reflections within the space. And more significantly, perceived brightness depends on the surface reflectance and the observer position. For the so-called Lambertian light sources, a useful guideline for the illuminance measurements is the *five times rule*: the measurement point to the light source distance should be greater than five times the largest source dimension for an accurate measurement. However, this rule does not work for directional light sources. When the case of a constant luminous flux, F, the illuminance is computed by using:

$$E = \frac{F}{A} \tag{7.11}$$

Example 7.6: A room $5 \times 8\ \text{m}^2$ is illuminated by 12 lighting fixtures, each of 3,500 lumens. If 75% of the light energy can be used at the desk level, what is the average desk illuminance?

Solution: From (7.11), the illuminance is:

$$E = \frac{12 \times 0.75 \times 3{,}500}{5 \times 8} = 787.5\ \text{lx}$$

When the light is reaching the surface at other incidence angles than 90°, then the illuminance is given by the normal component of the light ray reaching the surface. For light ray reaching the surface at an angle, θ, then (7.11) becomes:

$$E = E_0 = \frac{F}{A}\cos(\theta) \tag{7.12}$$

Example 7.7: A lamp produces 3,000 lumens in the direction of interest. If the angle of incidence with respect to the vertical is 30°, what is the average desk illuminance for the case, the geometry and 100% light energy used, and in Example 7.4?

Solution: From (7.12), the illuminance is:

$$E = \frac{F}{A}\cos(\theta) = \frac{3{,}500}{40}\cos(30°) = 75.8\ \text{lx}$$

The terms *illumination level* and *illumination* are still used in the lighting industry, either on lesser extent than in the past. These terms are more distinguishable than the *luminance*, which means brightness to a layperson. ***Exposure*** is described as the product of the illuminance and the exposure time. Exposure is an important issue, for example, regarding the calculating of light exposure on exhibits in museums.

A Lambertian surface reflects or emits equal (isotropic) flux in every direction. For example, an evenly illuminated diffuse flat surface, such as a piece of paper is approximately Lambertian, the reflected light being the same in every direction from which you can see the surface of the paper. However, it does not have isotropic intensity, because it varies according to the cosine law. Figure 7.3 shows a Lambertian reflection from a surface, the reflection follows the cosine law, and the

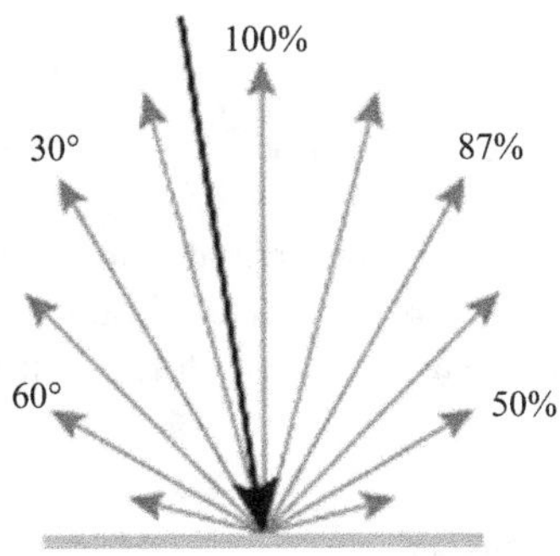

Figure 7.3 A Lambertian surface

amount of reflected energy in a particular direction (the intensity) is proportional to the cosine of the reflected angle. Remember that luminance is intensity per unit area. Because both intensity and apparent area follow the cosine law, they remain proportional to each other as the viewing angle changes. Therefore, luminance remains constant while luminous intensity does not. To compare illuminance and luminance on a Lambertian surface, consider the following example; a surface with a luminance of 1 lm/m^2/sr. radiates a total of πA lumens, where A is the area of the surface, into a hemisphere (which is 2π steradians). The illuminance of the surface is equal to the total luminous flux divided by the total area π lx/m^2. In other words, if you were to illuminate a perfectly diffuse reflecting surface with 3.1416 lm/m^2, its luminance would be 1 lm/m^2/sr.

Luminance, L, is the only basic lighting parameter that is perceived by the eye. Luminance is apparent brightness, how bright an object appears to the human eye. The illuminance indicates the lighting amount falling on a given surface, while the luminance describes the brightness of an illuminated or luminous surface. Luminance is defined as the ratio of luminous intensity of a surface (cd) to the projected area of this surface (m^2). In the illumination case, the light can be reflected by the surface or transmitted through the surface. In the case of diffuse reflecting (matt) and diffuse transmitting (opaque) materials luminance is calculated from the illuminance and the reflectance or transmittance. Luminance is the basis for describing perceived brightness; the actual brightness is, however, still influenced by the eye adaptation state, the surrounding contrast ratios and the information content of the perceived surface. So when we look at the world, we are seeing varying luminance patterns (ignoring the color component). It specifies the surface brightness, being essentially dependent on its reflectance (finish and color). In other words, it is directly related to the perceived brightness of a real or imaginary surface, the surface visual appearance, by the illuminance of a surface. The brightness is a surface subjective evaluation, while the illuminance is an objective surface characteristic. In fact, we are not seeing the surface illuminance but rather the surface brightness or the surface luminance differences (contrast), measured in cd/m^2, and defined as the luminous intensity, I in a given direction, θ per unit of projected area, A_θ from that direction:

$$L = \frac{dI_\theta}{dA_\theta} \tag{7.13}$$

Exitance, M, measured in lm/m^2 or lm/ft.^{2} represents the total luminous flux leaving a surface regardless the direction. It is the ability of a surface to emit light expressed as the luminous flux per unit area at a specified point on the surface M. The luminous exitance or luminous emittance of a radiator is the total flux emitted in all directions from a unit area of the radiator. For a sphere with radius, R the exitance is expressed as:

$$M = \frac{d\Phi}{dA} = \frac{1}{R^2}\int L\cos(\theta)\cdot dA \tag{7.14}$$

If the illuminance, E (in lx or fc) is known, the exitance in the case of surface reflectance and in the case of surface transmittance may be computed as:

$$\begin{aligned} M &= \rho \times E \\ M &= \tau \times E \end{aligned} \tag{7.15}$$

Here, ρ and τ are the per-unit reflectance and transmittance of a surface, respectively.

Example 7.8: The illuminance of a surface is measured to be 750 lx and surface reflectance is 0.75, what is the approximate exitance?

Solution: From (7.11), the exitance is:

$$M = 0.75 \times 750 = 562.50 \text{ lm/m}^2$$

***Contrast,* C**, is not a physical property, but rather our visual perception and the response to the light energy emitted and transmitted from object surfaces. The human eye to see the details must be either a difference in color (wavelength) or in luminous intensity (brightness) or both. Luminance is either generated by a light source or may be reflected or transmitted from an object. For example, to differentiate two objects of the same color situated side by side one must be more reflective than the other. This is called *contrast*, and is defined as:

$$C = \frac{|L_o - L_b|}{L_b} \tag{7.16}$$

Here, C the contrast is unitless, L_o, and L_b are the object and the background luminances, in candela per unit area.

Example 7.9: If the illuminance of an object is 2,000 cd/m^2 and the one of the background is 100 cd/m^2, what is the luminance contrast?

Solution: From (7.12), the luminance contrast is:

$$C = \frac{|2{,}000 - 100|}{100} = 19$$

7.3.2 The visible spectrum and color

The visible spectrum is the portion of the electromagnetic spectrum that is visible to the human eye. Electromagnetic radiation in this range of wavelengths is called visible light or simply light. A typical human eye will respond to wavelengths from about 390 to 700 nm. In terms of frequency, this corresponds to a band in the vicinity of 430–770 THz. The spectrum does not, however, contain all the colors that the human eyes can distinguish. Unsaturated colors, such as pink, or purple variations, such as magenta, are absent, because they can be made only by a mix of multiple wavelengths. Colors containing only one wavelength are also called pure colors or spectral colors. Visible wavelengths are passing largely unattenuated through the Earth's atmosphere. An example of this phenomenon is that clean air scatters blue light more than red wavelengths, and so the midday sky appears blue. The optical window is also referred to as the "visible window" because it overlaps the human visible response spectrum. The near infrared (IR) window lies just out of the human vision, as well as the medium wavelength IR window, and the far infrared (FIR) window, although other animals may experience them.

The human eye is our primary source of information about the outside world. It is first and foremost an optical system, described often as a camera, creating images on the retina, the surface on which the image occurs, where the pattern of luminances is translated into nervous impulses. The retina poses light sensitive receptors that are sufficient for high resolution of the visual images. One of the most remarkable eye properties is its ability to adapt to different lighting conditions, from moonlight to sunlight, although there is a difference by a factor of 10^5 in the illuminance. The extent of eye tasks is capable of performing is extremely wide, by perceiving a faintly glowing star in the night sky, producing only an illuminance of about 10-lux, by regulating incident light in a 1:16 ratio. Adaptation is performed to a large extent by the retina. Although vision is possible over an extremely wide luminance range, there are strict limits with regard to contrast perception in each individual lighting situation, due to the fact that the eye cannot cover the entire range of possible luminances at once and the same time but adapts to cover one narrow range in which differentiated perception is possible. Objects possessing too high luminance for a particular adaptation level cause glare, appearing to be extremely bright, while low luminance objects are often appearing to be too dark. The eye is able to adjust to new luminance conditions, by simply selecting a different but restricted range. The adaptation process needs time; however, the adaptation from dark to light situations occurs relatively rapidly, whereas the one from light to darkness requires longer time. Although vision is possible over an extremely wide luminance range there are clearly strict limits regarding to contrast perception in each individual lighting situation, due to the fact that the eye cannot cover the entire range of possible luminances at the same time, it adapts to cover one narrow range in which differentiated perception is possible. The facts that luminance contrast is processed by the eye within a certain range and that takes time to adapt to a new light level, or brightness have impacts on lighting design, or when adjusting lighting levels in adjacent spaces.

Objects possessing luminance too high for a particular adaptation level can cause glare, they appear extremely bright, while the ones with low luminance level appear to be too dark. Attempts to describe visual perception effectively must take into account the criteria by which the perceived information is affected. Any particular information is related to the observer current activity, work, movement-related or any other activity, requiring visual information. The specific information received depends on the activity type, with most of the required information from the need to feel safe, to be able to evaluate danger or to comprehend the environment, which applies both to orientation, the route, destination, and knowledge about the environment qualities and peculiarities. This information or lack of it determines the way a person feels and its behavior, which may lead tension and unrest in unknown or potentially dangerous situations, relaxation and tranquility in a familiar and safe environment. The information about the world around us is required to allow us to adapt our behavior to specific situations, including knowledge of weather conditions and the time of day as well as information relating to other activities occurring in the given environment. Areas that promise significant information be it in their own right, or through accentuation with the aid of light, which are perceived first, attracting our attention. The information content of a given object is responsible for its being selected as an object of perception. Moreover, the information content also has an influence on the way in which an object is perceived and evaluated.

The human eye that uses the part of the spectrum of electromagnetic waves to gather information about the world around us, perceiving the light amount and distribution that is radiated or reflected from objects, to gain information about their characteristics and perceiving the light color to acquire additional information about these objects. The seeing is performed in two ways: the color differences and luminance contrasts (brightness), through eye to brain transmitting information complex process, and transmitting the information back to the eye. For eyes to see there must be color differences and luminance contrasts, while high luminance contrast tending to improve the visual activities. However, higher luminance contrast for prolonged time may cause discomfort and must be taken into account in lighting applications and design. If the light source or object luminance is high enough to interfere with vision, excessive contrast or luminance being distracting and annoying, the brightness negative side, the glare. In the extreme, glare cripples vision by reducing or destroying the ability to see accurately. Glare is often misunderstood as "too much light," while in fact, the light coming from the wrong direction results in extreme luminance within the normal view field. An essential feature of good lighting is the extent to which glare is limited. If the glare is strong enough to cause physiological discomfort is called *discomfort glare*, while when it affects the ability to see is called *disability glare*. There are two aspects of glare: the objective visual performance depreciation and the subjective disturbance felt through excessive luminance levels or stark contrasts in luminance levels within the vision field. *Direct glare* is originating from a light source, while the indirect (reflected) glare is the one reflected from a surface. Direct glare can be avoided changing the light source position, while the indirect glare is minimized by replacing the reflecting surface with matte (nonglaring) or low-reflectance (dark) surfaces. The evaluation of luminances and luminance contrasts, leading to unwanted glare is predominantly dependent on the environment and the requirement that the lighting aims to fulfill. In the case of direct glare, there is the quantitative method of evaluating luminance limiting curves. Reflected glare is excessive uncontrolled luminance reflected from objects or surfaces in the field of view, which includes the reflected luminance from interior surfaces and/or the luminance of the lighting system. Specular surfaces have reflecting properties similar to those of a mirror. The luminance reflected is the mirrored image of the light source and/or of another lighted surface within the reflected field of view. Reflected glare can only be evaluated according to qualitative criteria, however.

In the case of objective depreciation in visual performance, the term physiological glare is applied. In this case, the light from a glare source superposes the luminance pattern of the visual task, thereby reducing visibility. The reason for this superimposition of the luminous intensities of visual task and glare source may be the direct overlay of the images on the retina. Super-imposition of scattered or disturbing light, which arises through the dispersion of the light from the glare source within the eye, is often enough to reduce the visual performance. The degree of light scattering depends primarily on the opacity of the inner eye. The latter increases with age and is the reason why older people are considerably more sensitive to glare. The most extreme case of physiological glare is disability glare. This arises for luminance levels higher than 10^4 cd/m^2 and is evident in the field of vision, e.g., when we look directly at artificial light sources or at the sun. Disability glare does not depend upon the luminance contrast in the environment. Suitable glare limitation can be achieved by the correct choice of the luminaires. Especially developed reflectors can guarantee that luminaires positioned above the critical angle do not produce any unacceptable luminances. By installing luminaires that emit only minimal direct light downward can also make a substantial contribution toward limiting reflected glare.

7.3.3 Color specifications and characteristics

Different wavelengths in the visible spectrum (the band from 380 to 780 nm) are producing different color sensations. In order to quantify color, the spectrum or wavelength composition of light must be known. An SPD, defined as the radiant power at each wavelength or band of wavelengths in the visible region, is typically used to characterize light. Depending on how light is generated by the source, the light SPD can vary from continuous across the visible spectrum to discrete across the spectrum to a narrow band at a particular wavelength. The common visible spectrum colors are blue, cyan (blue-green), green, yellow (green-red), red, and violet (magenta or red-purple). White light is a mixture of all visible spectrum colors, with a perfect white (an idealization) being a mixture of all colors of equal energy level. Daylight at noon is the closest match to the perfect white. Identical colors are produced not only by identical SPDs but also by many different SPDs that produce the same visual response. Physically different SPDs appearing to have the same color are called metamers. Wavelengths shorter than the violet, in the band between 200 and 380 nm are called ultraviolet, while the region with wavelengths higher than 780 nm is called infrared. Of all colors, red, green, and blue are the dominant, being the basis for forming all others, the *primary colors*, while the other colors are the *secondary colors*. Secondary colors are the result of mixing two or three primary colors. Mixing all three primary lights at equal energy levels produces the white light. Light is reflected from opaque or translucent materials, we are seeing most objects through the reflected light. Colors are a matter of visual perception's subjective interpretation. The color of a light or an object can be described by the following characteristics: (a) the basic colors and the mixture of these colors (the hue), (b) the color shade (the value), and (c) the intensity or degree of color saturation (the chroma). There several systems for color specification, with Munsell and CIE (Commission Internationale de l'Eclairage) the most common used. Brightness is referred to in this case as the reflecting coefficient of an object color, hue is the color tone, and the term saturation refers to the degree of color strength from pure color to the noncolored gray scale.

The color of illuminated objects is the result of the spectral composition of the light falling on a body and its ability to absorb or transmit certain components of this light and only reflect or absorb the remaining frequency ranges. In addition to the resulting, objective calculable or measurable color stimulus the eye color adaptation, also playing a role in the actual perception of things. The eye is able to gradually adapt to the predominant luminous color, similar to its adaptation to luminance levels, meaning that in the case of a lighting situation that comprises different luminous colors virtually constant perception of the scale of object colors is guaranteed. The degree of deviation is referred to as the light source color rendering, defined as the degree of change, which occurs in the color effect of objects through lighting with a specific light source in contrast to the lighting with a comparative reference light source, being a comparison of the similarity of color effects under two lighting types. Since the eye can adapt to different color temperatures, color rendering is determined in relation to luminous color. In order to determine the color rendering of a light source, the color effects of a scale of eight object colors are calculated under the lighting types to be evaluated, and under comparison standard lighting, then the two are compared. The quality of the color rendering established is expressed as CRI, which can be related to both general color rendering, and the rendering of individual colors. The maximum index of 100 represents optimum color rendering, and lower values correspondingly less adequate color rendering.

Human eyes are sensitive to the visible range of the electromagnetic spectrum, while the colors are matter of visual perception, being subjective to each person. Color is a perception caused by the unbalanced mixture of light wavelengths. An object can be described in terms of color (plain, deep, bright, etc.) being inadequate to describe the color precisely. The object and light color can be described by the following three terms. *Hue* is the basic color, such as red, yellow, green or blue, and the mixture of them. *Value* represents the color shade, such as light or dark blue or red. It has more significance in painting or architecture rather than in lighting. *Chroma*, *purity*, or *saturation* is the color saturation degree or intensity, meaning if the color is dull or vivid. It describes the freedom of the color from white or gray. *Tints*, *tones*, or *shades* are colors diluted white, gray, and black. Lighting represents the dilution of the spectrum saturated light with white lights. The colors are specified using the above-mentioned systems. A colored lamp or lighting system has most of its energy concentrated around a specific wavelength, while a colored object reflects most of incident light in a narrow wavelength band. Correlated color temperature (CCT), measured in °K, describes the light appearance generated by a hot object (a heated object produces light), such as an incandescent filament. The generated light is correlated to the black body curve. At lower temperatures, reddish light is generated, as the temperature increases, the light appears to shift from red to reddish-yellow to yellowish-white to white to bluish-white at high temperatures. The light source color can be quantified using absolute temperature (K) to describe the light emitted by a blackbody radiator. For blackbody temperature of 800 K, red light is emitted, yellow corresponds to 3,000 K, white to 5,000 K, pale blue to 8,000 K, and bright blue to 60,000 K. Lamp color temperature rating is a convenient way to describe the lamp's color characteristics. However, this has nothing with the lamp operating temperature or with lamp color rendering. The illuminated object color is the result of the spectral composition of the light falling on a body and the body ability to absorb or transmit certain light components, while only reflecting or absorbing the remaining frequency ranges. In addition to the result, the objective measurable color stimulus and the eye color adaptation also plays a role in the actual perception. The eye is able to gradually adapt to the predominant luminous color, in a similar way that it adapts to a luminance level, means that in the case of lighting situation, comprising different luminous colors virtually constant perception of the scale of object colors is guaranteed.

The degree of deviation is referred to as the color rendering of the light source. Color rendering is defined as the degree of change, which occurs in the color effect of objects through lighting with a specific light source in contrast to the lighting with a comparative reference light source. It is therefore, a comparison of the similarity of color effects under two types of lighting. Since the eye can adapt to different color temperatures, color rendering must be determined in relation to luminous color. One single light source cannot therefore serve as the reference source the comparison standard is rather a comparable light source with a continuous spectrum, a thermal radiator of comparable color temperature, or daylight. Other useful color measures can be derived from calorimetry. The most commonly used are CRI and CCT. CRI is a measure of how colors of surfaces will appear when illuminated by a light source. Light that has an even spectral distribution across the visible spectrum, such as daylight or incandescent light, has a high CRI (the maximum is 100). Light that has gaps in its spectral distribution has a lower CRI. To determine the light source color rendering, the color effects of a scale of eight object colors are calculated under lighting types to be evaluated as well as under comparison standard lighting, and the two then are compared. CRI is one of the ways to quantify the color rendering quality of a light source, being a measure of the shifts when standard color samples are illuminated by the light source, compared to the reference (standard) light source. The established color rendering quality, expressed as CRI, relates to both general color rendering (Ra) and the rendering of individual colors. The maximum index of 100 represents optimum color rendering, and lower values correspondingly less adequate color rendering. The color rendering quality is an important criterion when choosing light sources. The degree of color fidelity represents, therefore, by which illuminated objects are rendered in comparison to reference lighting.

7.3.4 Light control and basic concepts in optics

Light is traveling in clean air or similar transparent medium without bending or notable loss until it is intercepted by another medium, which is reflecting, absorbing, transmitting, refracting, diffusing, or polarizing the light. These material characteristics are used as methods to control the light and to achieve better and more appropriate lighting. When light encounters a surface, it can be either reflected away from the surface or refracted through the surface to the material beneath. Once in the material, the light can be transmitted, absorbed, or diffused (or combinations) by the material. These properties apply not only to the light, but to all other forms of electromagnetic radiation. However, to simplify this discussion, it will be limited to light. There are three general types of reflection: *specular*, *spread*, and *diffuse*. A specular reflection, such as what you see in a mirror or a polished surface, occurs when light is reflected away from the surface at the same angle as the incoming light's angle. A spread reflection occurs when an uneven surface reflects light at more than one angle, but the reflected angles are all more or less the same as the incident angle. A diffuse reflection, called also *Lambertian scattering* or *diffusion*, occurs when a rough or matte surface reflects the light at many different angles. Specular reflections demonstrate the law of reflection, which states that the angle between the incident ray and a line that is normal (perpendicular) to the surface is equal to the angle between the reflected ray and the normal, as shown in Figure 7.4(a) and (b). The angle between an incident ray and the normal is called the incident angle, denoted by the symbol θ. The angle between a reflected ray and the normal is called the reflected angle, denoted by the symbol θ'.

When light travels from one material to another (such as from air to glass), it refracts, bends, and changes velocity. Refraction depends on two factors: the incident angle (θ) and the refractive index of the material, denoted by the letter n.

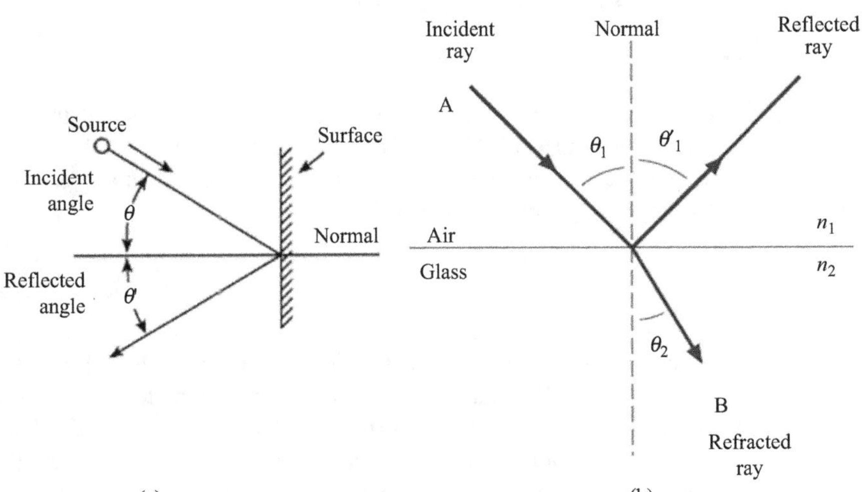

Figure 7.4 Diagrams of (a) law of reflection, and (b) refraction and Snell's law

The index of refraction for a particular material is the ratio of the speed of light in a vacuum to the speed of light in that material:

$$n = \frac{\text{light speed in vaccum}}{\text{light speed in material}} = \frac{c}{v} \tag{7.17}$$

The speed of light in air is almost identical to the speed of light in a vacuum, so the index of refraction for air is considered to be 1 ($n_{air} = 1.000293$). The index of refraction for almost all other substances is greater than 1, because the speed of light is lower as it passes through them. As shown in Figure 7.4(b), Snell's law of refraction shows the relationship between the incident angle and the refractive index:

$$n_1 \cdot \sin(\theta_1) = n_2 \cdot \sin(\theta_2) \tag{7.18}$$

where n_1 is the refractive index of medium 1, n_2 is the refractive index of medium 2, θ_1 is the incident angle of the light ray (with respect to the normal), θ_1' is the reflected angle (with respect to the normal), and θ_2 is the refracted angle (with respect to the normal). By using Snell's law (for sin 0° = 0) means that light with a normal incident angle does not bend at a boundary. Snell's law also shows that light traveling from a medium with a low index to one with a high index ($n_1 < n_2$) bends toward the normal, while light traveling from a medium with a high index to one with a low index ($n_1 > n_2$) bends away from the normal.

Example 7.10: Determine the angle of refraction, for light ray entering a piece of crown glass ($n = 1.52$) from the air ($n = 1$) at an incident angle of 45°.

Solution: From (7.14), the refraction angle is:

$$1 \times \sin(45^\circ) = 1.52 \times \sin(\theta_2) \Rightarrow \theta_2 = 28^\circ$$

A transparent substance transmits almost all light, reflecting only a small fraction of light at each of its two surfaces. This reflection occurs whenever light travels through a change in the refractive index. At normal incidence (incident angle = 0°), Fresnel's law of reflection quantifies the effect:

$$r_\lambda = \left(\frac{n_2 - n_1}{n_2 + n_1}\right)^2 \tag{7.19}$$

where r_λ is the reflection loss, n_1 is the refractive index of medium 1, and n_2 is the refractive index of medium 2. For example, when light strikes a material that has a refractive index of 1.5 (such as glass) at a normal incident angle, and each of the two material boundaries with air reflects less than 5% of the incident light. As the angle of incidence increases, so does the reflected light. As Snell's law shows for light traveling from a material with a higher index of refraction to one with a lower index of refraction (e.g., light moving through a piece of glass toward air), the refracted light bends away from the normal, leading to the phenomenon called total internal reflection. If a light ray is incident on the interface at an angle greater than the critical angle, and is totally reflected into the same medium from which it came. If a beam of light's angle of incidence increases far from normal, it reaches an angle (the critical angle, θ_c) at which the light is refracted along the boundary between the materials instead of being reflected or passing through the boundary. The total internal reflection is exploited when designing light propagation in fibers by trapping the light in the fiber through successive internal reflections along the fiber or in lighting control systems. At even higher angles of incidence, all the light is reflected back into the medium, which allows fiber optics to transport light along their length with little or no loss except for absorption. Critical angle calculation (θ_c), based on the Snell's law, and for n_r and n_i the refractive indexes of the medium with lower refractive index and the incidence medium, respectively is then:

$$\theta_c = \sin^{-1}\left(\frac{n_r}{n_i}\right) \tag{7.20}$$

Example 7.11: Find the critical angle θ_c in the core; at the core-cladding interface of a fiber optics having a core index of 1.53 and a cladding index of 1.39.

Solution: From (7.16) the critical angle is:

$$\theta_c = \sin^{-1}\left(\frac{n_r}{n_i}\right) = \sin^{-1}\left(\frac{1.39}{1.53}\right) = 65.3^\circ$$

The index of refraction depends on the incident light wavelength. Materials typically have higher refraction indexes for shorter wavelengths, so the blue light bends more than red light. This phenomenon is called *dispersion*. White light passes through the nonparallel faces of a prism is spreading into its spectral components. When light passes through an object, it is called *transmission*. Absorption, reflection, refraction, and diffusion (explained in the following sections) affect light transmission. Instead of completely transmitting light, objects and materials can absorb part or the entire incident light, usually converting it into heat. Many materials absorb some wavelengths, transmitting others, the so-called selective absorption. Lambert's law of absorption states that equal thicknesses of a given homogenous material absorb the same fraction of light, being given by this relationship:

$$I = I_0 \exp(-\alpha \cdot x) \tag{7.21}$$

where I is the intensity of transmitted light, I_0 is the intensity of incident light, entering the material (excluding surface reflection), α is the absorption coefficient in inverse length units, and x is the sample thickness (measured in the same unit as α). Beer's law further breaks down the absorption coefficient α in two variables: β, the absorption per unit concentration coefficient, and c, the material concentration.

Beer's law states that equal absorbing material amounts absorb equal light fractions. As with Lambert's law, each wavelength should be considered separately for Beer's law. The two laws can be combined into a single equation, including both the material thickness and the concentration, in the Beer–Lambert law:

$$I = I_0 \exp(-\beta \cdot c \cdot x) \tag{7.22}$$

Here, β is the absorption per concentration coefficient (inverse length per inverse grams or moles per liter), c is the concentration of the absorbing material, and x is the path length (length). When light strikes a perfectly smooth surface, the reflection is specular, while when striking a rough surface, the light is reflected or transmitted in many different directions at once, the so-called *diffusion* or *scattering*. The amount of diffuse transmission or reflection that occurs when light moves through one material to strike another material depends on two factors: the difference in refractive index between the two materials, and the size and shape of the particles in the diffusing material compared to the wavelength of the light. For example, the molecules in air happen to be in the right size to scatter light with shorter wavelengths, giving us blue sky. A transmissive filter is a material that absorbs some wavelengths and transmits others, while a reflective filter absorbs some wavelengths and reflects others. For example, a red filter absorbs all but the longest wavelengths of visible light; a reflective red filter reflects the longest wavelengths, and a transmissive red filter transmits the longest wavelengths. The amount of light absorbed by a filter depends also on the filter's thickness.

7.4 Lighting equipment and systems

Light may originate in many ways, from solar energy, the so-called daylight, from chemical reactions and combustion or from electricity conversion. Daylight is apart from all other light sources plentiful and free of charge, with only drawbacks being variable (fluctuating), available only during the day and sometimes being too bright for the visual comfort. However, when is properly controlled is the most economical of all light sources. Commercial, industrial, and retail facilities use several different light sources. Each light source (lamp) type has particular advantages; selecting the appropriate source depends on installation requirements, life-cycle cost, color qualities, dimming capability, and the effect wanted. Representation of different kinds of light sources according to the means of their light production and specific characteristics are important factors and criteria for a properly light source selection for a specific application. The important factors of a light source selection include: efficiency, illumination characteristics, lifetime, maintenance requirements, auxiliary equipment needed and the light requirements for the premise.

7.4.1 Light sources and systems

Light, the basis for all vision, is an element of our lives that we take for granted. We are so familiar with brightness, darkness, and the spectrum of visible colors that another form of perception in a different frequency range and with different color sensitivity is difficult for us to imagine. Visible light is in fact just a small part of an essentially broader spectrum of electromagnetic waves, which range from cosmic rays to radio waves. The lighting industry makes millions of electric light sources, called lamps. Those used for providing illumination can be divided into three general classes: incandescent, discharge, and solid-state lamps. Incandescent lamps produce light by heating a filament until it glows. Discharge lamps produce light by ionizing the gas through electric discharge inside the lamp. Solid-state lamps use a phenomenon called electroluminescence to convert electrical energy directly to light. In addition to manufactured light sources, daylight (sunlight received on the Earth, either directly from the Sun, scattered and reflected by the atmosphere, or reflected by the moon), provides illumination. The prime characteristic of daylight is its variability. Daylight varies in magnitude, spectral content, and distribution with different meteorological conditions, at different times of the day and year, and at different latitudes. The illuminances on the Earth's surface produced by daylight can cover a large range, from 150,000 lx on a sunny summer's day to 1,000 lx on a heavily overcast day in winter. The spectral composition of daylight also varies with the nature of the atmosphere and the path length through it. In the case of technical lamps, the main distinction is between thermal radiators and discharge lamps. Discharge lamps are further subdivided into high-pressure and low-pressure lamps. Current developments show a marked trend toward the development of compact light sources, such as low voltage halogen lamps, compact fluorescent lamps (CFL), metal halide lamps, fiber optic sources, and LED systems.

Incandescent lamps: The incandescent lamp is a thermal radiator with most common shape as shown in Figure 7.5. The filament wire begins to glow when it is heated to a sufficiently high temperature by an electric current. With increases of the temperature, the spectrum of the radiated light shifts toward the shorter wavelength range, the red heat of the filament shifts to the warm white light of the incandescent lamp. Either simple in operation, there are a substantial number of practical problems involved in the construction of an incandescent lamp. There are only a few conducting materials, that have a sufficiently high melting point and at the same time a sufficiently low evaporation rate below melting point that render them suitable for filament wire use. Depending on lamp type and wattage the temperature of the filament can reach up to 3,000 K, in the case of halogen lamps over 3,000 K. Maximum radiation at these temperatures still lies in the infrared range, with the result that in comparison to the visible spectrum there is a high degree of thermal radiation and very little UV radiation. However, the lack of a suitable material for the filament means that it is not possible to increase the temperature further, which would increase the luminous efficacy and produce a cool white luminous color.

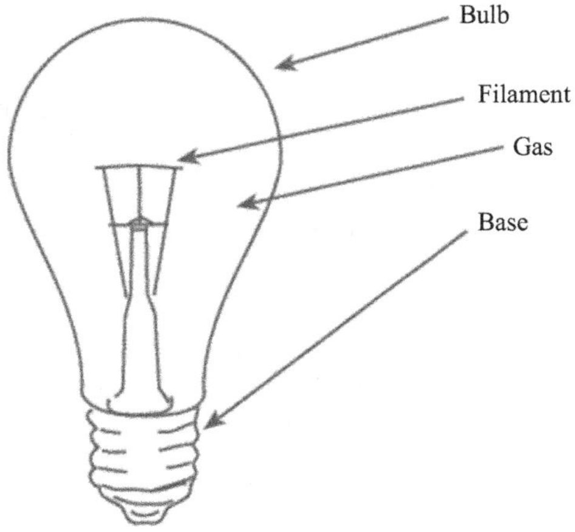

Figure 7.5 Pictorial diagram of a typical incandescent lamp

Nowadays, practically only tungsten is used for the manufacture of filament wires, because it only melts at a temperature of 3,653 K and has a low evaporation rate. The tungsten is made into fine wires, wounded to make single or double coiled filaments, the filament is located inside a soft glass bulb, which is relatively large in order to keep light loss, due to deposits of evaporated tungsten (blackening), to a minimum. To prevent the filament from oxidizing, the outer envelope is evacuated for low wattages and filled with nitrogen or a nitrogen-based inert gas mixture for higher wattages. The thermal insulation properties of the gas used to fill the bulb increases the temperature of the wire filament but at the same time reduces the evaporation rate of the tungsten, which in turn leads to increased luminous efficacy and a longer lamp life. The inert gases predominantly used are argon and krypton. The krypton permits a higher operating temperature, and greater luminous efficacy, however, being expensive krypton is only used in special applications. A characteristic feature of these lamps is their low color temperature, so their light is warm in comparison to daylight. The continuous color spectrum of the incandescent lamp provides excellent color rendition. As a point source with a high luminance, sparkling effects can be produced on shiny surfaces and the light easily controlled using optical equipment.

Incandescent lamps can be easily dimmed. No additional control gear is required for their operation and the lamps can be operated in any burning position. In spite of these advantages, there are a number of disadvantages: low luminous efficacy, for example, and a relatively short lamp life, while the lamp life relates significantly to the operating voltage. Special incandescent lamps are available with a dichroic coating inside the bulb that reflects the infrared component back to the wire filament, which increases the luminous efficacy by up to 40%. General-purpose service lamps (A lamps) are available in a variety of shapes and sizes. The glass bulbs are of clear, matt, or opal type. Special forms are available for critical applications (e.g., rooms subject to the danger of explosion, or lamps exposed to mechanical loads), as well as a wide range of special models available for decorative purposes. A second basic model is the reflector lamp (R lamp). The bulbs of these lamps are also from soft glass, and in contrast with the A lamps, which radiate light in all directions, the R lamps control the light via their form and a partly silvered area inside the lamp. Another range of incandescent lamps are the PAR (parabolic reflector) lamps. The PAR lamp is made of pressed glass to provide a higher resistance to changes in temperature and a more exact form; the parabolic reflector produces a well-defined beam spread. In the case of cool-beam lamps, a subgroup of the PAR lamps, a dichroic, i.e., selectively reflective coating is applied. Dichroic reflectors reflect visible light, but allow a large part of the IR radiation to pass the reflector. The thermal load on illuminated objects can therefore be reduced by half. Incandescent lamps are strongly affected by input voltage. For example, reducing input voltage from the normal 110 to 104.5 V (95%) can double the life of a standard incandescent lamp, while increasing voltage to just 115.5 V (105% of normal) can half its life. In summary, the main advantages of incandescent lamps include: inexpensive, easy to use, small and does not need auxiliary equipment, easy to dim by changing the voltage, excellent color rendering properties, directly work at power supplies with fixed voltage, no toxic components, and instant switching. Disadvantages of incandescent lamps include: short lamp life (about 1,000 h), low luminous efficacy, high heat generation, lamp life, and other characteristics that are strongly dependent on the supply voltage, and the total costs are high due to high operation costs. The traditional incandescent lamps are progressively replaced with more efficient light sources.

Halogen lamps: Unlike incandescent lamps, halogen lamps use halogen gas fill (typically iodine or bromine), to produce what is called a "halogen cycle" inside the lamp. Halogen lamps are available in a wide range of models, shapes (from small capsules to linear double ended lamps), with or without reflectors. There are reflectors designed to redirect forward only the visible light, allowing infrared radiation to escape from the lamp back. There are halogen lamps available for mains voltages or low voltages (6–24 V), the latter needing a step-down transformer. Low-voltage lamps have better luminous efficacy and longer lamp life than the high-voltage lamps but the transformer implicates energy losses in itself. The latest progress in halogen lamps has been reached by introducing selective-IR-mirror-coatings in the bulb. The infrared coating redirects infrared radiations back to the filament. This increases the luminous efficacy by 40%–60% compared to other designs and lamp life is up to 4,000 h. In the halogen cycle, halogen gas combines with the tungsten that evaporates from the lamp filament, and eventually redepositing the tungsten on the filament instead of allowing it to accumulate on the bulb wall as it does in the standard incandescent lamps. However, it is not so much the melting point of the tungsten (which, at 3,653 K, is still a long way from about 2,800 K of the incandescent lamp operating temperatures), hindering the construction of more efficient incandescent lamps but rather increasing the filament evaporation rate accompanying the temperature increase. This initially leads to lower performance due to the blackening of the surrounding glass bulb until finally the filament burns through. The increase in luminous efficiency is at a cost of shorter lamp life. One technical way of preventing the blackening of the glass is the adding of halogens to the gas mixture inside the lamp. The evaporated tungsten combines with the halogen to form a metal halide, which takes on the form of a gas at the temperature in the outer section of the lamp and can therefore leave no deposits on the glass bulb. The metal halide is split into tungsten and halogen once again at the considerably hotter filament and the tungsten is then returned to the coil. The temperature of the outer glass envelope has to be over 250 °C to allow the development of the halogen cycle to take place. To achieve this, a quartz glass compact bulb is fitted tightly over the filament, forming not only means an increase in temperature and in gas pressure, which in turn reduces the evaporation rate of the tungsten. Compared with the conventional incandescent the halogen lamp gives a whiter light, a result of its higher operating temperature of 3,000–3,300 K, and its luminous color is still in the warm white range. The tungsten-halogen lamp has several differences from incandescent lamps:

1. The lamps have a longer life (2,000–3,500 h).
2. The bulb wall remains cleaner because the evaporated tungsten is constantly redeposited on the filament by the halogen cycle, allowing maintaining its lumen output throughout its life.
3. The higher operating temperature of the filament improves luminous efficacy.
4. The lamp produces a "whiter" or "cooler" light, which has a higher CCT than standard incandescent lamps.
5. The bulbs are more compact, offering opportunities for better optical control.

Halogen lamps are sometimes called "quartz" lamps because their higher temperature requires quartz envelopes instead of the glass used for the incandescent lamps. Their main advantages include small size, directional light (narrow beams for some models), low-voltage options, easy to dim, instant switching and full light output, and excellent color rendering properties, while the disadvantages of tungsten halogen lamps are low luminous efficacy, surface temperature is high, and lamp life and other characteristics are strongly dependent on the supply voltage. The recommended design practice is the use of halogen lamps, in applications requiring instant switch on and instant full light, excellent color rendering, easy dimming, frequent switching and, or short on-period, directional light, and compact size of the light source.

LED: Light emitting diode sources are made from semiconductor material that emits light when energized. An LED is an electronic semiconductor component that emits light when a current flows through it. The wavelength of the light depends on the semiconductor material and its doping. The spectrum of LEDs offers a major benefit only light (eradiation in the visible range) and no ultraviolet or infrared radiation is emitted. LEDs provide instant on/off capability and can be dimmed. LEDs can be extremely small and durable; some LEDs can provide much longer lamp life than other sources. The plastic encapsulate and the lead frame is occupying most of the volume. The light-generating chip is quite small (typically a cuboid with one side equal to 0.25 mm). Light is generated inside the chip, a solid crystal material, when current flows across the junctions of different materials.

The composition of the materials determines the wavelength and therefore the color of light. Major LED features include: long service life (e.g., 50,000 h at 70% luminous flux), emitted light is only in the visible range, i.e., no UV or infrared radiation, compact size, high luminous efficiency (lm/W), good to excellent CRI (Ra), luminous flux, and service life highly temperature-sensitive, no or little environmentally harmful materials (e.g., mercury), resistant to vibrations and impact, saturated color, about 100% luminous flux after switching on, no ignition, boosting or cooling time, high-precision digital dimming via pulse-width modulation, and no shifting of color locations during dimming. Following are the three major types of LED:

1. Standard through-hole LED, often used as indicator light source, although with low light output. Due to their shorter service life, higher probability of failure and sensitivity to UV radiation, they are not used in lighting technology.
2. Surface mounted device LED, in which an LED that is reflow-soldered to the surface of a printed circuit board (using a reflow oven). Basically, it consists of an LED chip protected by silicon coating mounted in or on housing or a ceramic plate with contacts.
3. Chip on board LED, where the LED chip is mounted directly on the printed circuit board. This allows a dense arrangement of chips close to each other.

Heat management is one of the main concerns for LED luminaires. Proper heat management will allow for long LED life at or above 50,000 h and is necessary to maintain proper light output. The efficacies for LEDs luminaires are as good as fluorescent and HID sources and are continuing to be improved. The more efficacious LEDs look whiter with typically high CCT in the range of 5,000–8,000 K. Warmer color temperatures are available but their efficacy will be reduced. LEDs can generate red, yellow, green, blue or white light, have a life up to 100,000 h, and are widely used in traffic signals and for decorative purposes. White light LEDs are a recent advance and may have a great potential market for some general lighting applications. However, a critical issue when comparing LED luminaires of various suppliers is the indication of luminous flux levels. In catalogues you will find details regarding the luminous flux and efficiency of individual LEDs at a junction temperature of 25 °C in the LED chip, details regarding the luminous flux levels of the LED boards used, or details regarding the luminous flux levels of luminaires and luminaire efficiency levels, including the power loss of ballasts and any potential loss of efficiency through lighting optics, such as lenses, reflectors or mixing chambers. Future lighting systems are expected to require intelligent features. In this regard LED-based lighting systems have an important advantage due to their easy controllability. Intelligent features combined with the inherent high energy-saving potential of LEDs will be an unbeatable combination in a wide range of applications. Advantages of LEDs are small size (heat sink can be large), physically very robust, long lifetime expectancy (with proper thermal management), switching has no effect on life, very short rise time, contains no mercury or other potentially dangerous materials, excellent low ambient temperature operation, and high luminous efficacy (LEDs are developing fast and their range of luminous efficacies is wide), new luminaire design possibilities, possibility to change colors, and no optical heat on radiation. Major disadvantages of LEDs: high price, however, keep declining, low luminous flux/package, often low CRI, risk of glare due to high output with small lamp size, need for thermal management, and lack of standardization.

Discharge lamps: Discharge lamps produce light by passing an electric current through a gas that emits light when ionized by the current. In contrast to incandescent lamps, light from discharge lamps is not produced by heating a filament but by exciting gases or metal vapors. This is effectuated by applying voltage between two electrodes located in a discharge tube filled with inert gases or metal vapors. Through the voltage current is produced between the two electrodes. On their way through, the discharge tube the electrons collide with gas atoms, which are in turn excited to radiate light, when the electrons are traveling at a sufficiently high speed. For every type of gas there is a certain wavelength combination; radiation, i.e., light is produced from one or several narrow frequency ranges. An auxiliary device known as a ballast supplies voltage to the lamp's electrodes, which have been coated with a mixture of alkaline earth oxides to enhance electron emission. Two general categories of discharge lamps used to provide illumination are *high-intensity discharge* and *fluorescent lamps*.

It soon becomes evident that discharge lamps have different properties than incandescent lamps. Whereas incandescent lamps have a continuous spectrum dependent on the temperature of the filament, discharge lamps produce a narrow band spectrum, typical for the respective gases or metal vapors. The spectral lines can occur in all regions of the spectrum, from infrared through the visible region to ultraviolet. The number and distribution of the spectral lines results in light of different colors. These can be determined by the choice of gas or metal vapor in the discharge tube, and as a result white light of various color temperatures can be produced. Moreover, it is possible to exceed the given limit for thermal radiators of 3,650 K and produce daylight-quality light of higher color temperatures. Another method for the effective production of luminous colors is through the application of fluorescent coatings on the interior surfaces of the discharge tube. Ultraviolet radiation in particular, which occurs during certain gas discharge processes, is transformed into visible light by means of these fluorescent substances, through which specific luminous colors can be produced by the appropriate selection and mixing of the fluorescent material. The quality of the discharge lamp can also be influenced by changing the pressure inside the discharge tube. The spectral lines spread out as the pressure increases, approaching continuous spectral distribution. This results in enhanced color rendering and luminous efficacy.

To ignite a discharge lamp, there must be sufficient electron current in the discharge tube. As the gas that is to be excited is not ionized before ignition, these electrons must be made available via a special starting device. Once the discharge lamp has been ignited there is an avalanche-like ionization of the excited gases, which in turn leads to a continuously increasing operating current, which would increase and destroy the lamp in a relatively short time. To prevent this from happening, the operating current must be controlled by means of ballast. Additional equipment is necessary for both the ignition and operation. In some cases, this equipment is integrated into the lamp; but it is normally installed separate from the lamp, in the luminaire. Discharge lamps can be divided into two main groups depending on the operating pressure. Each of these groups has different properties. One group comprises low-pressure discharge lamps. These lamps contain inert gases or a mixture of inert gas and metal vapor at a pressure well below 1 bar. Due to the low pressure inside the discharge tube there is hardly any interaction between the gas molecules. The result is a pure line spectrum. The luminous efficacy of low-pressure discharge lamps is mainly dependent on lamp volume. To attain adequate luminous power the lamps must have large discharge tubes. High-pressure discharge lamps, on the other hand, are operated at a pressure well above 1 bar. Due to the high pressure and the resulting high temperatures there is a great deal of interaction in the discharge gas. Light is no longer radiated in narrow spectral lines but in broader frequency ranges. In general, radiation shifts with increasing pressure into the long-wave region of the spectrum.

High intensity discharge (HID): High intensity discharge sources include mercury vapor, metal halide, and high pressure sodium (HPS) lamps. Light is produced in HID and low pressure sodium (LPS) sources through a gaseous arc discharge using a variety of elements. Each HID lamp consists of an arc tube which contains certain elements or mixtures of elements which, when an arc is created between the electrodes at each end, gasify and generate visible radiation. The major advantages of HID sources are their high efficacy in lumens per watt, long lamp life and point-source characteristic for good light control. Disadvantages include the need for a ballast to regulate lamp current and voltage as well as a starting aid for HPS and some MH and the delay in re-striking after a momentary power interruption.

High Pressure Sodium (HPS): In the 1970s, as increasing energy costs placed more emphasis on the efficiency of lighting, HPS lamps (developed in the 1960s) gained widespread usage. With efficacies ranging from 80 to 140 lm/W, these lamps provide about 7 times as much light per watt as incandescent and about twice as much as some mercury or fluorescent. The efficacy of this source is not its only advantage. An HPS lamp also offers the longest life (over 24,000 h) and the best lumen maintenance characteristics of all HID sources. The major objection to the use of HPS is its yellowish color and low color rendition. It is ideal mainly for some warehouse and outdoor applications. The most common application today is in street and road lighting. Their characteristics include high luminous efficacy (80–100 lm/W), and lamp life is 12,000 h (16,000 h), and the CCT is 2,000 K. Improvement of the CRI is possible by pulse operation or elevated pressure but this reduces the luminous efficacy. Color improved HPS lamps have CRI of about 65 and white HPS lamps of more than 80. Their CCT is 2,200 and 2,700, respectively. Among advantages of HPS lamps are very good luminous efficacy, long lamp life (12,000 h or 16,000 h), high luminous flux from one unit for street and area lighting, while disadvantages include low CCT, about 2,200 K, low CRI, about 20 (color improved 65, white 80), and long starting and restarting time, 2–5 min.

Metal halide (MH): Metal halide lamps are similar in construction to mercury lamps with the addition of various other metallic elements in the arc tube. The major benefits of this change are an increase in efficacy to 60–100 lm/W and an improvement in color rendition to the degree that this source is suitable for commercial areas. Light control of a metal halide lamp is also more precise than that of a deluxe mercury lamp since light emanates from the small arc tube, not the total outer bulb of the coated lamp. Pulse-start metal halide lamps have several advantages over standard (probe-start) metal halide: higher efficacy (110 lm/W), longer life, and better lumen maintenance. A disadvantage of the metal halide lamp is its shorter life (7,500–20,000 h) as compared to induction, LEDs and HPS. Starting time of the metal halide lamp is approximately 4–7 min depending on the ambient temperatures. Restriking after a voltage dip has extinguished the lamp, however, can take substantially longer, depending on the time required for the lamp to cool. Among advantages of metal halide lamps are good luminous efficacy, alternatives with good color rendering available, and different color temperatures available, while their major disadvantages are expensive, long starting, and restarting time (2–5 min), differences in CCT between individual lamps and changes of CCT during burning hours. These differences are much reduced with new ceramic metal halide lamps.

Low pressure sodium (LPS): LPS offers the highest initial efficacy of all lamps on the market today, ranging from 100 to 180 lm/W. However, because all of the LPS output is in the yellow portion of the visible spectrum, it produces extremely poor and unattractive color rendition, being more difficult than with HID sources because of the large size of the arc tube. The average life of LPS lamps is 18,000 h. While lumen maintenance through life is good with LPS, there is an offsetting increase in lamp watts, reducing the efficacy of this lamp type with use.

Fluorescent lamps: Fluorescent lighting accounts for two-thirds of all electric light in the United States. Fluorescent lamps are the most commonly used commercial light source in North America. In fact, fluorescent lamps illuminate 71% of the commercial space in the United States. Their popularity can be attributed to their relatively high efficacy, diffuse light distribution characteristics, and long operating life. Fluorescent lamp construction consists of a glass tube with the following features: filled with argon or argon-krypton gas and a small amount of mercury, coated on the inside with phosphors; and equipped with an electrode at both ends, as shown in Figure 7.6. The fluorescent lamp is a low-pressure discharge lamp using mercury vapor. It has an elongated discharge tube with an electrode at each end. The gas used to fill the tube comprises inert gas, which ignites easily and controls the discharge, plus a small amount of mercury, the vapor of which

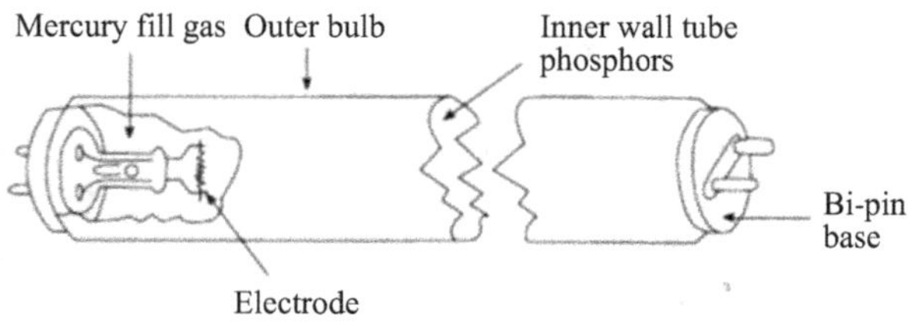

Figure 7.6 Construction of a linear fluorescent lamp

produces ultraviolet radiation when excited. The inner surface of the discharge tube is coated with a fluorescent substance that transforms the ultraviolet radiation produced by the lamp into visible light by means of fluorescence. The fluorescent lamp is a gas discharge source that contains mercury vapor at low pressure, with a small amount of inert gas for starting. Once an arc is established, the mercury vapor emits ultraviolet radiation. Fluorescent powders (phosphors) coating the inner walls of the glass bulb respond to this ultraviolet radiation by emitting wavelengths in the visible spectrum. Ballasts, which are required by both fluorescent and HID lamps, provide the necessary circuit conditions (voltage, current, and wave form) to start and operate the lamps. Two general types of ballasts are available for fluorescent lamps: magnetic and electronic. Electronic ballasts are often more expensive but lighter, quieter, and eliminate the lamp flicker associated with magnetic ballasts. Fluorescent lamps are described in terms of the diameter of the lamp tube, with the diameter is given in eighths of an inch.

Full-size fluorescent lamps are available in several shapes, straight, U-shaped, and circular configurations, with diameters from 1″ to 2.5″. The most common lamp type is the four-foot (F40), 1.5″ diameter (T12) straight fluorescent lamp. More efficient fluorescent lamps are now available in smaller diameters, including the T10 (1.25″) and T8 (1″). Fluorescent lamps are available in color temperatures ranging from warm (2,700 K) "incandescent-like" colors to very cool (6,500 K) "daylight" colors. "Cool white" (4,100 K) is the most common fluorescent lamp color. Neutral white (3,500 K) is becoming popular for office and retail lighting. Improvements in the fluorescent lamp phosphor coating have improved color rendering and made fluorescent lamps acceptable in applications previously dominated by incandescent lamps. Linear fluorescent lamps range in length from 6 in. to 8 ft., and in diameter from 2/8 in. (T2) to 2-1/8 in. (T17), with power ranges from 14 to 215 W. Figure 7.6 shows the construction of a linear fluorescent lamp. CFLs produce light in the same manner as linear fluorescent lamps. Their tube diameter is usually 5/8 in. (T5) or smaller. CFL power ranges from 5 to 55 W. Their advantages include inexpensive, good luminous efficacy, long lamp life (10,000–16,000 h), large variety of CCT and CRI, while main disadvantages of fluorescent lamps are ambient temperature affects the switch-on and light output, need of auxiliary ballast and starter or electronic ballast, light output depreciates with age, contain mercury, and short burning cycles shorten lamp life. Fluorescent lamps are ideal for general lighting in most working places (shops, hospitals, open spaces, etc.), but also in some residential applications. The choice of the lamp is always related to the application. Always consider the CCT and the CRI.

7.4.2 Lamp efficiencies, control, and electrical requirements

Electric lighting consumes about 19% of the world total electricity use, so any improvement in energy efficient lighting, has also significant impacts on the energy conservation and savings. Every change in technologies, in customers' consumption behavior, even in lifestyle, has influences on global energy consumption and indirectly, on environment. Therefore, lighting energy saving and the saving

Table 7.1 Efficacies of common light sources. Adapted from IESNA Lighting Handbook

Light source	Power (W)	Efficiency (lm/W)	CRI	Typical life (h)
Standard incandescent	100	18	Excellent	1,000
Linear tungsten-halogen	300	20	Excellent	2,000–4,000
Fluorescent lamps (T-5)	28	50	Good	5,000
Compact fluorescent lamps (CFL)	100	60	Very Good	8,000–10,000
Metal-halide (low voltage)	100	80		
Metal-halide (high voltage)	400	90		
High-pressure mercury lamp	1000	50	Fair	5,000
Xenon sort-arc lamp	1000	30		
High-pressure sodium (HPS)	70	90	Fair	6,000–12,000
Low-pressure sodium (LPS)	180	180	Poor	1,200–10,000

methods should be considered at different levels company and corporation, town, city, state, country, region, and by the international organizations, too. Artificial lighting is based on systems: lamps, ballasts, starters, luminaires and controls, and each component must be considered in the energy management, saving, use and conservation. IESNA defines lamp efficacy as "the quotient of the total luminous flux emitted divided by the total lamp power input." It is expressed in lumens per watt (lm/W). For fluorescent and HID lamps, you must also include both the ballast wattage and any reduction in lumen output associated with the lamp-ballast combination to determine the system efficacy. Table 7.1 compares efficacies of some common lamp types. To summarize, energy savings and efficiency and economics are dependent on improvement of lighting technologies, better use of available cost-effective and energy efficient lighting technologies, optimum, and good lighting design (identify needs, avoid misuses, proper interaction of technologies, automatic controls, daylight integration), proper building design, through daylight integration and architecture, knowledge dissemination to final users and operators (designers, sellers, decision makers), reduction of resources by recycling and proper disposal, size reduction, by using less aluminum, mercury, etc., and finally the life cycle cost assessment (LCCA).

Auxiliary equipment for lighting consists of two major categories: transformers and ballasts. All discharge-type lighting systems (all lighting technologies discussed in this chapter except incandescent and LED) require a ballast to supply electricity to the lamp. This auxiliary equipment usually consumes a small amount of electrical power, adding to the total amount of lighting system wattage. Low-voltage light sources require the use of a transformer to step down the standard building service of 120 V or 220–6 V, 12 V or 24 V. Transformers are placed either within (integral to) the luminaire or in a remote location. The smaller size of low-voltage light sources allows for the design of smaller luminaires. In the case of recessed luminaires the transformer is hidden above the ceiling and out of view. Surface- or pendant-mounted luminaires usually have their transformers enclosed within the housing, however, the luminaire volume increases. Where ceiling conditions permit, surface and pendant-mounted luminaires can be designed with the transformer recessed in the ceiling and out of view. Track-mounted luminaires usually contain their transformers. It is also possible to provide low-voltage service to a length of track, locating the transformer in the ceiling or in an ancillary space. The high amperage of low-voltage lamps strictly limits the number of track luminaires per transformer. If remote transformers are used to maintain the compactness of the lighting element, the increased distance between the source and its transformer requires larger wire sizes to prevent a voltage drop from occurring over the longer wiring run. In modern lighting systems, the electromagnetic conventional transformers are replaced with power electronics converters, which are cheaper, more compact, smaller in size, and more efficient. Rectangular shape transformers are relatively large and heavy. However, properly sized for the lamp load, they have a long life expectancy. They sometimes cause a noise problem by producing an audible 60 Hz "hum." Toroidal (doughnut-shaped) magnetic transformers are quieter but they also hum when controlled by some kind of electronic dimmers. The hum grows with the number of luminaires in a room, and the luminaires, if improperly designed, will resonate with their transformers.

Lamps, with the exception of incandescent lamps and LEDs, require ballast to operate properly and safely. Every discharge source has negative resistance characteristics. If the arc discharge is placed directly across a nonregulated voltage supply, it will draw large currents that almost instantly and the lamp is quickly destroyed. Therefore, a current-limiting device called, ballast is inserted between the discharge lamp and the power supply to limit the electric current through the arc discharge. Besides limiting the current flow, the ballast also provides the correct voltage to start the arc discharge, by adjusting the available line voltage to that required by the lamp. Most of the old types of ballasts are not interchangeable, being designed to provide the proper operating characteristics for only one kind of lamp. However, modern electronic ballasts are designed to operate more than one connected load. Lamp wattage is controlled by the ballast, not by the lamp. Unlike incandescent lamps, the rated wattage of a discharge lamp is the wattage at which it is designed to operate, not the wattage at which it operates. Therefore, in order to reduce discharge system energy use must change not only the lamp's wattage, but also the ballast must be changed. Notice that the lamp wattage is controlled by the ballast, not by the lamp. If a 100 W HPS lamp is operated by 400-W ballast operates at the ballast power, 400 W, to the detriment of the lamp's performance, which may lead to premature ballast failure.

Ballasts are devices, electromagnetic or electronic type, designed for: (a) provides sufficient ignition voltage to start the lamp, (b) acts as a constant current source when the lamp is starting, and (c) acts as a constant power source when the lamp is operating. These functions are implemented using an iron-core reactor (inductor) for current limitation in combination with a capacitor (starter) to provide ignition voltage, or by a variety of solid-state circuits. Solid-state ballasts are used on the new systems, however, the core-and-coil ballasts, used for many years, are still in service. During the ignition phase, the ballast provides sufficient voltage across the lamp electrodes to initiate and maintain discharge, and must also provide sufficient current at discharge voltage to force a transition from glow to arc, are specific to the lamp type, so ballasts must be correctly matched to the lamps they are supplying. During the warm-up phase, the resistance of the lamp increases, and the ballast must provide a constant current to the lamp, which linearly increases power to the lamp. In the operating phase, the lamp resistance takes on a value close to the arc impedance, a low resistance requiring current limitation to prevent lamp damage. As the voltage supplied to the lighting system changes, the lamp reacts differently depending on the change rate. If the voltage change over several seconds, a corresponding lamp current change tends to occur, and a constant current supply is needed to keeps the lamp operating properly. Sudden changes in voltage could cause the arc extinction or a sudden current increase. The ballast constant current characteristic allows proper and safe lamp operation for most voltage transients. If the magnitude or duration of the transient exceeds the ballast capabilities, the lamp is shutting down, and needs to cool before arc restrike. High-pressure sodium lamps experience a rise in lamp voltage over their lifetime. This voltage rise is substantial, often as much as 170% of the lamp voltage experienced when the lamp is new. Therefore, the ballast must keep the lamp power within an acceptable power range over the life of the lamp.

The ballast *power factor* shows how effective the ballast converts the supplied power by the electrical distribution system to the power delivered by the ballast to the lamp. Perfect phase relationship would result in a power factor of 100%. The power factor of an inductive circuit is lagging, while the one of a capacitive circuit is leading. When discharge lamps are operated in conjunction with simple inductive ballasts, the overall power factor is 60% or less. With a capacitor, the leading current drawn compensates for the lagging current in the remainder of the circuit, improving the power factor. Ballasts are classified according to one of the following three categories: (1) high power factor: 90% or greater, (2) power factor corrected: 80%–89% and (3) low (normal) power factor: 79% or less. High-power-factor ballasts use the lowest level of current for the specific amount of power needed, reducing wiring costs by permitting more luminaires on branch circuits. Low power factor ballasts use higher current levels, about twice the line current needed by high-power-factor ballasts, so fewer luminaires can be connected per branch circuit, increasing wiring costs. Power factor is not an indication of the lamp-ballast system ability to produce light, only the ballast ability to use the power that is supplied. The initial lumen and mean-lumen ratings published by lamp manufacturers are based on the operation of the rated lamps by *reference ballasts*. In practice, when a lamp is operated by commercially available ballasts, it provides fewer lumens than the rated value. Because of the electrical resistance created by the passage of a current through the core-and-coil of electromagnetic ballast, some power is converted to heat, the ballast loss, is not used to produce light from the lamp. The disparity between light provided by the reference ballast and the commercially available ballast is called the ballast factor (BF), defined as the ratio of light output produced by lamps operated by commercially available ballasts to the theoretically one supplied by lamps powered by laboratory-reference ballasts. The

Table 7.2 Luminances of common light sources. Adapted from the IESNA Lighting Handbook

Light source	Approximate average luminance (cd/m^2)
Sun	1.6×10^9
Moon	6×10^6
Clear sky	8×10^3
Overcast	2×10^3
60-W incandescent lamp	1.2×10^5
Tungsten-halogen lamp	1.3×10^7
T-5 Fluorescent lamp	2×10^4
T-8 Fluorescent lamp	1×10^4
High-pressure mercury lamp	2×10^8

ballast efficacy factor is a ratio of the BF to the ballast power input. This is used to compare the efficiency of various lamp-ballast systems. Ballast efficacy factors are meaningful only for comparing different ballasts when operating the same quantity and kind of lamp.

7.4.3 Common lamp luminances and luminaires

Light sources are generating a wide range of luminances. The light direction is based on three principles: reflection, refraction, and diffraction, as discussed earlier. They are applied to define the photometric properties of luminaires in terms of lighting patterns. Table 7.2 shows the approximate luminances of common light sources. Luminaires are devices that produce, distribute, control, filter, and transform the light emitted from their lamps. The luminaire includes all the parts needed for fixing and protecting the lamps, except the lamps themselves. Luminaires may also include the necessary circuit auxiliaries, together with the means for connecting them to the electric supply. A wide variety of luminaire designs are available to meet virtually all lighting applications. In general, considerations when selecting a luminaire for a specific application include construction and installation codes and standards, physical and environmental conditions, electrical and mechanical requirements, thermal characteristics, cost, and most importantly, safety. The basic physical principles used in optical luminaire are reflection, absorption, transmission, and refraction. Most luminaires are fitted with reflectors, refractors, and/or diffusers, in order to specifically control the distribution of light. Photometric data, including plots of candela distribution and isoilluminance, can be obtained from the manufacturers. The performance of any luminaire system depends on how well its components work together. With fluorescent lamp-ballast systems, light output, input watts, and efficacy are sensitive to changes in the ambient temperature. When the ambient temperature around the lamp is significantly above or below 25 °C (77 F), the performance of the system can change. Exhibit 6 shows this relationship for two common lamp-ballast systems: the F40T12 lamp with magnetic ballast and the F32T8 lamp with electronic ballast.

Table 7.3 NEMA beam types

Beam type	Beam spread (degrees)	Projection distance (ft.)
1	10–18	>240
2	18–29	200–240
3	29–46	175–200
4	46–70	145–175
5	70–100	105–145
6	100–130	80–105
7	>130	<80

Luminaires are usually classified both by applications, used by the lighting manufacturers to present their products, and by their photometric characteristics. Applications include residential, commercial, and industrial with subclassification by lighting technology, mounting method, and luminaire construction types. Photometric classifications are developed by engineering and professional organizations, such as the International Commission on Illumination (CIE), NEMA, and IESNA. The CIE classification system is based on upward-directed light to downward-directed light ratio, being applied to indoor luminaires. Categories include direct lighting (90%–100% downward light), semi-direct lighting (60%–90% downward light), general diffuse lighting (approximately equal upward and downward components), semi-indirect lighting (with 60%–90% upward light), and indirect lighting (90%–100% upward light).

Outdoor luminaires are usually described by cutoff characteristics. Three methods are used in photometric reports: physical cutoff, optical cutoff, and shielding angle. Physical cutoff is the angle measured from the downward-directed vertical axis, or nadir, to the point where the lamp is fully occluded. Optical cutoff measures the angle from the nadir to the point where reflection of the lamp in the luminaire's reflector is fully occluded. The shielding angle is the angle measured from the horizontal at which the lamp is just visible. The NEMA classification system is based on the distribution of luminous flux, used primarily for sport's lighting and flood-lighting, considering the spread of the light beam in degrees and the projection distance in feet, separated in seven beam types, as is given in Table 7.3. The IESNA classification system applies to outdoor luminaires, and is based on the shape of the area illuminated, and is commonly used in lighting for roadways, parking lots, and streets lighting. Six intensity distribution types are defined by IESNA, separated into four cutoff classifications, as shown in Table 7.4.

The luminaire efficiency is the percentage of lamp lumens produced that actually exits the fixture. The use of louvers can improve visual comfort; however, they are reducing the lumen output of the fixture, so the efficiency is reduced. In general, the most efficient fixtures have the poorest visual comfort (e.g., bare strip industrial fixtures), while the fixtures providing the highest visual comfort level, and are the least efficient. A lighting designer must determine the best compromise between efficiency and visual comfort, when selecting luminaires. In the last years,

Table 7.4 IESNA cutoff classification

Classification	Intensity distribution description
Full cutoff	At 90° and greater above nadir, zero candela intensity, not exceeding 10% of the maximum candela intensity at 80° above nadir at all lateral angles around luminaire.
Cutoff	At 90° above nadir, candela intensity, not exceeding 2.5% of the maximum candela intensity, and not exceeding 10% at 80° above nadir, and at all lateral angles around luminaire.
Semi-cutoff	At 90° above nadir, candela intensity, not exceeding 5% of the maximum candela intensity, and not exceeding 20% at 80° above nadir, and at all lateral angles around luminaire.
Noncutoff	No candela limitations in the zone above maximum candela intensity

manufacturers started to offer fixtures with excellent visual comfort and efficiency. These so-called "super fixtures" combine state-of-the-art lens or louver designs to provide the best of both worlds. Surface deterioration and accumulated dirt in older, poorly maintained fixtures can also cause reductions in luminaire efficiency. Each of the luminaires consisted of several components, designed to work together to produce and direct light in the best way possible for a specific application. Reflectors are designed to redirect the light emitted from a lamp in order to achieve a desired distribution of light intensity outside of the luminaire. In most incandescent spot and flood lights, highly specular reflectors are usually built into the lamps. An energy-efficient upgrade option is to install custom-designed reflectors to enhance the light control and the fixture efficiency, allowing partial delamping. Retrofit reflectors are useful for upgrading the efficiency of older, deteriorated luminaire surfaces. A variety of reflector materials are available: highly reflective white paint, silver film laminate, and two grades of anodized aluminum sheet (standard or enhanced reflectivity). Silver film laminate is generally considered to have the highest reflectance, but is considered less durable. Proper design and installation of reflectors can have more effect on performance than the reflector materials. In combination with de-lamping, the use of reflectors may result in reduced light output and may redistribute the light, which may not be acceptable for a specific space or application. To ensure acceptable performance from reflectors, trial installation, calculations and measure "before" and "after" light levels are required.

Most indoor commercial fluorescent fixtures use either a lens or a louver to prevent direct viewing of the lamps. Light that is emitted in the so-called "glare zone" (angles above 45° from the fixture's vertical axis) can cause visual discomfort and reflections, reducing contrast on work surfaces or computer screens. By using lenses and louvers these issues are controlled. Lenses made from clear ultraviolet-stabilized acrylic plastic deliver the most light output and uniformity of all shielding media, providing fewer glares than louvered fixtures. Clear lens types include prismatic, batwing, and polarized lenses. Lenses are usually less expensive than louvers. White translucent diffusers are much less efficient than clear lenses, and they result in relatively low visual comfort probability. New low-glare lens materials are available for retrofit and provide high visual comfort (over 80) and high efficiency. Louvers provide superior glare control and high visual comfort compared with lens diffuser systems. The most common application of louvers is to eliminate the fixture glare reflected on computer screens. The deep-cell parabolic louvers (with 5″–7″ cell apertures and depths of 2″–4″) provide a good balance between visual comfort and luminaire efficiency. Although small-cell parabolic louvers provide the highest visual comfort level, tend to reduce luminaire efficiency to about 35%–45%. For retrofit applications, both deep-cell and small-cell louvers are available for use with existing fixtures. Note that the deep-cell louver retrofit adds 2″–4″ to the overall depth of a troffer; verify that sufficient plenum depth is available before specifying the deep-cell retrofit.

Distribution is one of the primary functions of a luminaire, meaning to direct the light to where it is needed, being characterized by the Illuminating Engineering Society as follows:

1. Direct, 90%–100% of the light is directed downward for maximum use.
2. Indirect, 90%–100% of the light is directed to the ceilings and upper walls and is reflected to all parts of a room.
3. Semi-direct, 60%–90% of the light is directed downward with the rest directed upward.
4. General diffuse (direct-indirect), in which equal light portions are directed upward and downward.
5. Highlighting, the beam projection distance and focusing ability characterize this luminaire.

The lighting distribution that is characteristic of a given luminaire is described using the candela distribution, intensity distribution curve, or candlepower curve, provided by the luminaire and lamp manufacturers. It represents the amount of luminous intensity (cd) generated in each direction by a light source in a plane through the center of the source, giving a picture of the total light pattern produced by a source. LDCs are available from luminaire manufacturers and are often found on the back of the manufacturer's product data sheet. Polar graphs are used to represent the distributional intensity of a luminaire, and rectilinear or Cartesian graphs to represent the intensity of directional lamps. The candlepower distribution is represented by polar graphs showing the relative luminous intensity 360° around the fixture, looking at a cross-section of the fixture. In the candlepower polar graph, the luminaire or the lamp is located at the center of the radiating lines, which are representing specific degrees of angular rotation from the 0° axis of the luminaire (nadir). The concentric circles represent intensity (cd), with values entered along the vertical scale. For luminaires with symmetrical light distributions, a single curve fully describes the luminaire light intensity distribution. Often only one side of the polar graph is shown; the other side is a symmetrical (mirror) image. A luminaire with an asymmetrical distribution, such as a linear fluorescent down-light, requires curves in several planes to adequately represent its distribution, one curve is parallel to the luminaire, another is perpendicular to the luminaire, and either a third plane at 45° or other three planes at 22½° intervals. Reflector lamp sources or luminaires with

Table 7.5 Candlepower data

Angle	0° plane	45° plane	90° plane
0	3,670	3,670	3,670
10	3,600	3,630	3,680
20	3,410	3,520	3,600
30	3,080	3,270	3,410
40	2,480	2,750	2,800
50	1,600	1,700	1,850
60	1,010	800	1,000
70	500	370	480
80	230	290	240
90	0	0	0

directional distributions and abrupt cutoffs, having abrupt light intensity changes within a small angular area, have values that are difficult to read on a polar graph. Consequently, a Cartesian graph is substituted to portray the candlepower distribution or the date is tabulated. On this graph, the horizontal scale represents degrees from the beam axis and the vertical scale represents the light intensity. In general, for the non-symmetrical lighting distributions the candlepower data, representing various vertical through the fixture are presented in tabular format, as Table 7.5. From the fixture top the plan intersecting the fixture long dimension is designated 0°, while the direction across the fixture is designated, 90°. When required the data must be interpolated for intermediate angles.

When selecting luminaires for a lighting application, the proposed luminaire *and its source* must be precisely those shown in the manufacturer's photometric test data. It is inaccurate to extrapolate from one source or reflector finish to another unless the photometric report includes multipliers for various tested sources and reflector finishes. The candlepower distribution information is useful because it shows how much light is emitted in each direction and the relative proportions of down-lighting and up-lighting. The cut-off angle is the angle, measured from straight down, where the fixture begins to shield the light source and no direct light from the source is visible. The **shielding angle** is the angle, measured from horizontal, through which the fixture provides shielding to prevent direct viewing of the light source. The shielding and cut-off angles add up to 90°. The candlepower distribution provides the light intensity at the point source and at various points away from the fixture. The candlepower distribution curve and data are specified for fixture type, number of lamps, ballast, and lamp lumen output.

Example 7.12: Using the data from Table 7.5 and appropriate interpolation, determine the candlepower along the fixture, at an incidence angle of 30° and 45°, and the candlepower in perpendicular plan and at an incidence angle of 30°.

Solution: From the tabulated data, at 30°, $I = 3{,}080$ cd. For 45°, by using linear interpolation $I = 2{,}100$ cd, while for perpendicular plan at 30°, $I = 3410$ cd.

7.5 Indoor and outdoor lighting design

Humans depend on light, a vital for existence and for all activities. Light is a natural phenomenon, taken often for granted, and in fact, life involves day night cycles beginning with sunrise and ending with sunset. Pre-historic humans had activities limited only to day time. Artificial light enables extended activity period employing in a planned optimized manner, while minimizing the resources. The primary function of lighting in workspaces is to support work and to enhance personnel performances. Vision is the most important sense accounting for 80% of the human information acquisition. Information may be acquired through sun or moon light (direct/reflected) or by using artificial light. It is well accepted that the lighting is good, when our eyes can clearly and pleasantly perceive the things around us, often achieved by employing multiple light sources. However, the light sources need to be economic and energy efficient. All light sources today employ electrical energy. Lighting system affects the ambiance in the office, affecting company's view and image by employees, clients or customers, having profound effects on the feeling of well-being and productivity of the staff, being essential that lighting to be included as a vital consideration to the successful operation of any business. A designer must consider a variety of characteristics when developing a lighting plan, including lamp life, system efficiency, lumen maintenance, color rendering and appearance, daylight integration and control, light distribution, system cost, control, and flexibility. The prime objectives behind a lighting system design are as follows:

1. ***The safety and comfort of occupants***—The nature of a task or process performed in a space will dictate the illuminance level which must be provided by the lighting system (lx or lm/m^2). Tasks involving high degrees of visual acuity will require higher lighting levels.
2. ***The minimization of energy consumption***—Energy consumption reduction involves the development of the most energy efficient lighting systems which is suitable for the task that can be achieved by selecting high efficiency equipment and making use of available daylight.
3. ***Color rendering or the creation of a specific atmosphere***—The color characteristics of a lighting scheme will affect tasks performed when the lighting system is on. For example, tasks which require the accurate color representation need spectral characteristics of daylight. Alternatively, to create a "warm atmosphere" in a restaurant requires the selection of lights from the red end of the spectrum.

Achieving the required illuminance level does not necessarily ensure good lighting quality. The quality as well as the illuminance quantity is important in producing a comfortable, productive, aesthetically pleasing lighted environment. The lighting system quality includes aspects of lighting, such as proper color, good uniformity, proper room surface luminances, adequate brightness control, and minimal glare. Lighting system can affect impressions of visual clarity, spaciousness, and pleasantness because they occur in spaces that are uniformly lighted with emphasis on higher luminances on room surfaces. A lighting system design has several important design steps, such as:

1. Identification of the requirements for the lighting system, illuminance levels, color requirements, available space, etc.
2. Selection of equipment, lamps, and luminaires: lighting systems consist of numerous components, the two most important of which are: ***lamps,*** which influence the lighting level, color characteristics and efficiency of the lighting system, and ***luminaires***, which affect the efficiency with which the light is distributed and so affect lighting efficiency and uniformity.
3. Design of the lighting system to achieve a reasonably uniform distribution of light on a particular plane (usually horizontal), avoidance of glare with a minimum expenditure of energy.
4. Include the optimum system control for the designed lighting system to make daylight maximum use, through selection of appropriate switching mechanisms and daylight responsive controls.

A designer must consider a variety of key characteristics when developing their lighting plan including lamp life, system efficiency, lumen maintenance, color rendering and appearance, daylight integration and control, light distribution, points of interest, cost, system control, and flexibility. Lighting requirements are primarily dictated by the function of a space or the tasks being performed within it, being usually specified by the required lighting level and the color rendering requirements. Selection of components follows from the identification of systems requirements. The luminaires are normally chosen first, particular types of luminaire for specific tasks. These also come with different types of reflectors, lens, etc. for different applications. Lamps are selected based on those which are compatible (lamp type, dimensions, frequency of operation, efficiency, etc.) with the selected luminaire and which have the appropriate color-rendering index. A lighting designer has four major objectives:

1. Provide the visibility required based on the task to be performed and the economic objectives.
2. Furnish high quality lighting by providing a uniform illuminance level, where required, and by minimizing the negative effects of direct and reflected glare.
3. Choose luminaires aesthetically complimentary to the installation with mechanical, electrical, and maintenance characteristics designed to minimize operational expense.
4. Choose sustainable products that minimize energy usage while achieving the visibility, quality, and aesthetic objectives.

7.5.1 Factors affecting the selection of the light sources and equipment

Light output is the most common measure of light output (or luminous flux) is the lumen. Light sources and light fixture's output are labeled with an output rating in lumens. For example, a 40-W fluorescent lamp may have a rating of 3,050 lm. As lamps and fixtures age and become dirty, their lumen output decreases, i.e., lumen depreciation occurs. Most lamp ratings are based on initial lumens (when the lamp is new). Light intensity measured on a plane at a specific location is called illuminance, being measured in foot-candles, which are work-plane lumens per square foot, measured by using a light meter located on the surface where tasks are performed. Using calculations and the manufacturer photometric data, the illuminance for a defined space can be estimated. Lux is the illuminance metric unit, measured in lumens per square meter, and to convert foot-candles to lux, foot-candles are multiplied by 10.76. Another light measurement, the luminance, sometimes called brightness, is measuring the light *leaving* a surface in a particular direction, considering the illuminance on the surface and the reflectance of the surface. The eye does not see illuminance but its luminance. Therefore, the amount of light delivered into the space and the surface reflectances in the space affects our ability to see. IESNA has developed a procedure for determining the appropriate average light level for a particular space, being used extensively by designers and engineers. It recommends a target light level by considering the following factors, the performed task(s) (contrast, size, etc.), the occupant ages, and the importance of speed and accuracy. Then, the appropriate type and quantity of lamps and light fixtures is selected based on the following: fixture efficiency, lamp lumen output, surrounding surface reflectances, the effects of light losses from lamp lumen depreciation (LLD) and dirt accumulation, the room size and shape, and the daylight availability.

When designing a new or upgraded lighting system, one must be careful to avoid over lighting a space. In the past, spaces were designed for as much as 200 fc in places where 50 fc may not only be adequate, but superior. This was partly due to the misconception that the more light in a space, the higher the quality. Not only does over-lighting waste energy but it can also reduce lighting quality. The light levels are specified by the IESNA codes. Within a listed range of illuminance, three factors dictate the proper level: age of the occupant(s), speed and accuracy requirements, and the background contrast. For example, to light a space that uses computers, the overhead light fixtures should provide up to 30 fc of ambient lighting. The task lights should provide the additional foot-candles needed to achieve a total illuminance of up to 50 fc for reading and writing. For illuminance recommendations for specific visual tasks, refer to the IESNA Lighting Handbook or to the IES Recommended Practice No. 24 (for VDT lighting). Improvements in lighting quality can yield high dividends for businesses. Worker productivity gains may come from providing corrected light levels with reduced glare. Although the cost of energy for lighting is substantial, it is small compared with the cost of labor. Therefore, these gains in productivity may be even more valuable than the energy savings associated with new lighting technologies. In retail spaces, attractive and comfortable lighting designs can attract clientele and enhance sales. Three quality issues are needed to be addressed in the design processes are: glare, uniformity of illuminance, and color rendition. You can reduce glare or luminance ratios by not exceeding suggested light levels and by using lighting equipment designed to reduce glare. A louver or lens is commonly used to block direct viewing of a light source. Indirect lighting, or up-lighting, can create a low glare environment by uniformly lighting the ceiling. Also, proper fixture placement can reduce reflected glare on work surfaces or computer screens. Standard data now provided with luminaire specifications include tables of its visual comfort probability (VCP) ratings for various room geometries. The VCP index provides an indication of the percentage of people in a given space that would find the glare from a fixture to be acceptable. A minimum VCP of 70 is recommended for commercial interiors, while luminaires with VCPs exceeding 80 are recommended in computer areas. The uniformity of illuminance is a quality issue that addresses how evenly light spreads over a task area. Although a room's average illuminance may be appropriate, two factors may compromise uniformity: improper fixture placement based on the luminaire's spacing criteria (ratio of maxim recommended fixture spacing distance to mounting height above task height), and the fixtures that are retrofit with reflectors that narrow the light distribution. Nonuniform illuminance causes several problems: inadequate light levels in some areas, visual discomfort when tasks require frequent shifting of view from underlit to overlit areas bright spots and patches of light on floors and walls that cause distraction and generate a low quality appearance.

Electric light sources have three characteristics: efficiency, color temperature, and CRI. Some lamp types are more efficient in converting energy into visible light than others. The efficacy of a lamp refers to the number of lumens leaving the lamp compared to the number of watts required by the lamp (and ballast). It is expressed in lumens per watt. Sources with higher efficacy require less electrical energy to light a space. Another characteristic of a light source is the color temperature. This is a measurement of "warmth" or "coolness" provided by the lamp. People usually prefer a warmer source in lower illuminance areas, such as dining areas and living rooms, and a cooler source in higher illuminance areas, such as grocery stores. Color temperature refers to the color of a blackbody radiator at a given absolute temperature, expressed in Kelvins. A blackbody radiator changes color as its temperature increases, first to red, then to orange, yellow, and finally bluish white at the highest temperature. A "warm" color light source has a lower color temperature. For example, a cool-white fluorescent lamp appears bluish in color with a color temperature of around 4,100 K. A warmer fluorescent lamp appears more yellowish with a color temperature around 3,000 K. The CRI is a relative scale (ranging from 0 to 100), indicating how perceived colors match actual colors. We have to remember that the higher the CRI, the less color shift or distortion occurs. The CRI number does not indicate which colors will shift or by how much, being an indication of the average shift of the standard colors. Two different light sources may have identical CRI values but their colors may appear quite different.

Selection of the right lamp or lighting equipment depends on what and where the lighting is required. Incandescent lamps were popular for private domestic use for many years. However, due to their poor efficiency and short service life, are being replaced by more environmentally compatible alternatives of higher quality, such as LED lamps. Discharge lamps are the perfect choice for professional applications thanks to their efficient operating mode. LED light sources are taking over in application areas because of their high luminous efficiency and long service life. They can legitimately be regarded as the light source of the future. Thus, it is part of the expertise of the lighting designer to find the most suitable lamp for a lighting task. The performance characteristics of lamps are essentially defined by the following concepts:

1. The electric power is the power consumed by a light source. The system power takes the power consumption of the control gear as well as that of the light source into account.
2. Luminous flux defines the total quantity of light emitted from a light source. The unit used is the lumen [lm]. The ratio of luminous flux to the required electric power gives the luminous efficiency [lm/W]. The system luminous efficiency also takes the ballast losses into account. Luminous efficiency, *i* describing the light source efficiency is now one of the most important performance characteristics of all.
3. The average service life is usually quoted, which is the time after half of the lamps are statistically still serviceable, in other words half of the lamps failed. This test is subject to standardized operating conditions. Lamp manufacturers display this failure rate by curves, and they are shown as maintenance factors (LSF). Special service-life data apply to some light sources, such as LEDs.
4. Drop in luminous flux, means that the initial luminous flux of a new lamp decreases over its time of operation (lumen maintenance), due to the ageing of its chemical and physical components. Lamp manufacturers display this drop in luminous flux by curves, shown as maintenance factors (LLWF).
5. The color code, a three-digit numerical value (e.g., 840) describes the lighting quality of a white light source. First digit denotes color rendering, the second and third denote color temperature (light color). The light color describes the color impression of a white light source as relatively warm (ww = warm) or relatively cool (nw = intermediate, tw = cool), affected by the spectrum red and blue color components.
6. The spectral components of the light determine how well various object colors can be reproduced. The higher the CRI (Ra or CRI), or the lower the color rendering group number, the better the color rendering in comparison with the optimum reference light. The maximum CRI value is 100. Values in excess of 80 are considered to be very good. Eight test color samples (R1 to R8) are used for the general CRI, and there are another six more vivid high-saturation colors (R9 to R14). The CRI is calculated for a light source relative to a "known" reference light source. Color fields can only convey an impression of the original reflection patterns.
7. Discharge lamps need between 30 s and several minutes to warm up to output the full luminous flux, while high-pressure discharge lamps need to cool down for several minutes before they can be started again.
8. Incandescent and halogen incandescent lamps and almost all fluorescent and compact fluorescent lamps can be dimmed as required. Most of the metal halide lamps are not suited for dimming, having uncontrolled effects on lighting quality

and lamp service life. The new series of special models for indoor and outdoor applications constitute exceptions. The sodium vapor lamps and high-pressure mercury lamps are restricted in stages. LED light sources can be switched and dimmed as required.

9. Manufacturers specify the permitted operating positions for their lamps. For some metal halide lamps, only certain operating positions are allowed so as to avoid unstable operating states. Compact fluorescent lamps may usually be used in any operating position; however, important properties such as the luminous flux vs. temperature curve may vary with the position.

7.5.2 Lighting design project structure and criteria

Achieving lighting energy savings is considered one of the fundamental energy efficiency measures with numerous opportunities and supporting benefits. Lighting design is mainly done in the framework of guidance, specifications, and recommendations, rather than fixed design rules. In general there is more than a single optimum solution for a lighting problem. Quite often there are requirements to which priorities need to be set before a satisfactory compromise can be made. General guidance involves: lighting requirements, design process decisions, and selected calculation procedures. Lighting projects and design executed properly and comprehensively can be easily justified for a number of reasons including:

1. Energy savings, often a 25% internal rate of return or even better.
2. Emission reductions, direct correlation between energy and emission reduction.
3. Maintenance cost savings by replacing inefficient and/or old systems.
4. Increasing light levels for occupants and employees comfort or improved safety and productivity considerations.
5. Improved CRI to enhance comfort or productivity.

The objective of *quality* lighting designs is to provide a safe and productive environment, whether for business or pleasure. A good lighting design must provide the proper amount of light in every room, to be built and constructed within the budget, code, guideline, and other constraints, environmentally responsible, responding to the architecture and interior design requirements and specifications, produce good color, while achieving the desired moods of each space and are able to control the lights. Last but not least it mast looks good. The lighting objectives can be considered in three broad categories: (1) safety and health, (2) performance, and (3) appearance and comfort. These are often accomplished by a redesign or upgrade to ensure that the appropriate light quality and quantity is provided for the space users, at the lowest operating and maintenance cost. A *quality* lighting design addresses more than the cost issues. Proper data evaluation, planning, and execution are essential for the successful implementation. Building systems are interrelated. For example, removing 10 kW of lighting energy from a commercial building has a significant impact on the heating, ventilation air conditioning system. It is necessary for the lighting designer to have a clear understanding of all the building systems and how they interrelate. The methodology used to evaluate the energy savings for a lighting project, either for a retrofit or a comparison for new projects, is critical to the success of installing a complete energy efficient solution. Too often the simple payback method is used, which undervalues the financial benefit to the organization.

Light level, or more correctly, illuminance level, is easily measured using an illuminance meter. Illuminance is the light energy striking a surface. The IESNA regularly publishes tables of recommended illuminance levels for all possible tasks. It is also important to realize that the illuminance level has no relevance to the lighting quality, being entirely possible to have the recommended illuminance in a space but with light sources, producing enough glares to affect work. This accounts for many of the complaints of either too much or not enough light. There are a number of methods for determining whether a lighting installation is efficient. One method is for the lighting designer to check with the current version of the ASHRAE/IESNA 90.1 lighting standard. This document, which is revised regularly, provides a recommendation for lighting power density. It is usually possible for a capable lighting designer to achieve better results than the ASHRAE/IESNA 90.1 recommendations.

7.5.3 Indoor lighting design methods

It is essential that the decision about the method of lighting is taken at an early stage in the design of the building and the architect should consult the lighting engineer and others concerned during the conceptual stage. The first step is to establish the general requirements for the artificial lighting in terms of the main visual tasks to be carried out in the building. The illuminance levels are usually the ones required for designated spaces and activities, specified by the codes and standards, publishing the general accepted values for the lighting design use, the desired values for specific applications. Table 7.6 summarizes the recommended

Table 7.6 Recommended and minimal illuminance levels

Application	**Illuminance (lx = lm/m^2)**
Emergency lighting	0.2
Suburban street lighting	5
Dwelling	50–150
Corridors	100
Rough tasks with large-detail storerooms	200
Elevators, public corridors	200
Lobbies, atria	200
Training rooms	300
General offices, retail shops	400
Kitchens	500
Child care centers	500
Conference rooms	500
Drawing office	600
Prolonged task with small detail	900

and minimal illuminance levels for the common space and activities. The next step is to determine the lighting requirements in terms of building designation and helping to create the right character of the interior (the building lighting). The architect and the lighting engineer should be able to consider lighting design detailed aspects under the following headings:

1. The extent to which artificial lighting is used alone, or to supplement the day lighting.
2. The illuminances required for lighting specific visual tasks.
3. The required luminance throughout the interior.
4. The evaluation of discomfort glares in terms of the whole visual environment.
5. The directional lighting characteristics required for the desired effects and to reveal form and texture.
6. The main features of the building interior color schemes in terms of type, chrome, and color rendering.

Illumination calculations involved in lighting design are based on the principle of luminous flux transfer from the light source to the designated surfaces (areas). In order to design a luminaire layout that best meets the illuminance and uniformity work place requirements, two types of information are generally needed: average illuminance level and illuminance level at a given point. Calculation of illuminance at specific points is often done to help the designer to evaluate the lighting uniformity, especially when using luminaires, and the spacing recommendations are not supplied, or where task lighting levels must be checked against ambient. Most important quantities, involved in these calculations are the lighting power density (illuminance), and to a lesser degree the luminance (brightness) and the contrast (luminance ratio of the surfaces). The most used methods to calculate the illuminance are:

1. *Lumens method* that is based on the definition of illuminance (light power density), E on a surface, $E = F/A$, expressed in foot-candles or in lux, for the area in square feet, or in square meters, respectively.
2. *Point method* using the fact that the square law of illuminance, (7.9) and (7.10), with illuminance expressed in foot-candles or in lux, depending on the units used for the distance.

When these illuminance definitions (equations) are applied in actual lighting design and analysis, they are modified to include correction factors (coefficients) because not the all luminous flux is coming from the light source(s) to the surface. Furthermore, the light flux degrades over the lifetime of the light source(s) and the depreciation must be accounted. Depending on the receiving surface and the light source(s), either one or both methods are applicable. If the light source is small or very small in connection with surface of interest, then the point method can yield to more accurate results, while for large light source, such as the fluorescent lamps or diffused ones, such as luminaires with defusing lens the Lumen method is the choice. Most of the computer-based lighting calculation packages are based on the point method and ray-tracing algorithms. The point method is also used frequently in outdoor lighting calculations and design.

The lumen method or *zonal cavity method* is a widely used approach to the systematic design of electric lighting, especially for indoor lighting design. It is the most used hand calculation method to estimate the average illuminance levels for indoor areas, unless the light distribution is radically asymmetric, being an accurate method for indoor applications because it takes into consideration the effect that inter-reflectance has on the level of illuminance. Although it takes into account several variables, the basic premise that foot-candles are equal to luminous flux over an area is not violated. The Zonal Cavity Method basis is that a room is divided (made up) of three spaces or cavities. The space between the ceiling and the fixtures, if they are suspended, is defined as the *ceiling cavity*, the space between the work plane and the floor, the *floor cavity*, and the space between the fixtures and the work plane, the *room cavity*. Once the concept of these cavities is understood, it is possible to calculate numerical relationships, the *cavity ratios*, which can be used to determine the effective reflectance of the ceiling and floor cavities and then to find the coefficient of utilization (CU). The method depends essentially on the accuracy of the utilization factor. For example the estimation of the ratio of the lumens which are received on the working plane to the total output of the lamps in the room. The aim of the lumen method is to give a reasonably even spread of light over the horizontal working plane. How this spread of light is achieved depends upon the way the light is distributed from the fittings, not only in relation to fittings lay be related to the height at which the fittings are mounted over the working plane. The ratio mounting height to the spacing of the fittings varies with the choice of fitting: the greater the concentration of light distribution from the fitting, the closer must be the spacing relative to the mounting height. There are four basic steps of the calculation of the illuminance level (1) determine the cavity ratios; (2) determine the effective cavity reflectances; (3) select the CU; and (4) compute the average illuminance level. In the case of rectangular surfaces, the cavity ratios may be computed using the following relationships:

$$CCR = \frac{5h_{cl}(L+W)}{L \times W} \tag{7.23a}$$

$$RCR = \frac{5h_{lw}(L+W)}{L \times W} \tag{7.23b}$$

$$CCR = \frac{5h_{wf}(L+W)}{L \times W} \tag{7.23c}$$

Here *CCR* is the ceiling cavity ratio, *RCR* is the room cavity ratio, *FCR* is the floor cavity ratio, *CR* is the cavity ratio, h_{cl} is the distance from luminaire to ceiling, h_{lw} is the distance from luminaire to work plane, h_{wf} is the distance from work plane to floor, *L* is the room length, and *W* is the room width, all distances, length, and width are in feet or meters (Figure 7.7). Effective cavity reflectances are then determined for the ceiling cavity and for the floor cavity, the values are given in Table B5 (Appendix B) under the applicable combination of cavity ratio and actual reflectance of ceiling, walls, and floor. The effective reflectance values found is then ρ_{cc} (effective ceiling cavity reflectance) and ρ_{fc} (effective floor cavity

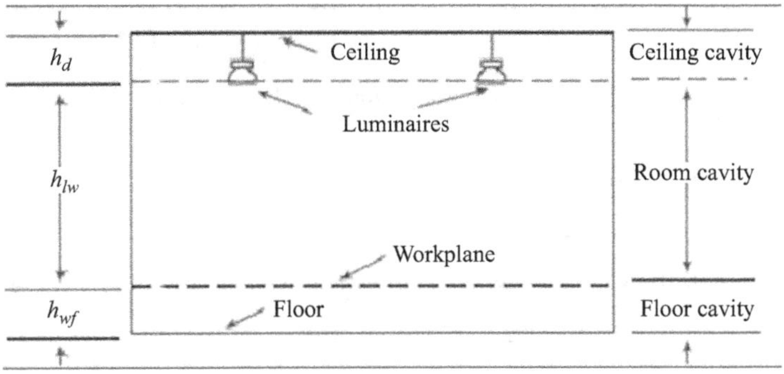

Figure 7.7 Zonal cavity method parameters

reflectance). If the luminaire is recessed or surface mounted, or if the floor is the work plane, the CCR or FCR is 0 and then the actual reflectance of the ceiling or floor will also be the effective reflectance. With the values of ρ_{cc}, ρ_{fc}, and ρ_w (wall reflectance), and the calculated RCR, the CU is found in the luminaire CU Table B6 (Appendix B), linear interpolations may be used for exact cavity ratios and reflectance combinations. The CU found is for a 20% effective floor cavity reflectance. Thus, it is necessary to correct for the determined ρ_{fc}. This is done by multiplying the previously determined *CU* by the factor from Table B6, as:

$$CU_{final} = CU(20\%\ \text{floor}) \times \text{Multiplier}\ (\text{Actual}\ \rho_{fc}) \tag{7.24}$$

If it is other than 10% or 30%, then it is needed to interpolate or to extrapolate and multiply by this factor. Next, computation of the illuminance level is performed with the lumen method, using (7.23). Alternate relationship for calculating any cavity ratio (*CR*) is given by:

$$CR = 2.5h_c \times \frac{P_{cav}}{A_{cav}} = 2.5h_c \times PAR_{cav} \tag{7.25}$$

where p_{cav} is the cavity perimeter (ft. or m), A_{cav} is the area of the cavity base (ft^2 or m^2), and PAR_{cav} represents the ratio of the perimeter to the floor area ($ft.^{-1}$ or m^{-1}), determined by the room geometry, and expressed by:

$$PAR_{cav} = \begin{cases} \dfrac{2(L+W)}{L \times W}, & \text{for rectangular rooms} \\ \dfrac{4}{D}, & \text{for circular rooms} \\ \dfrac{3.27}{D}, & \text{for semi-circular rooms} \end{cases} \tag{7.26}$$

Here, *L* and *W*, as previous defined, and *D* is the diameter of the circular room (m or ft.).

Example 7.13: A room is 25 × 32 ft., and h_c is equal to 7.5 ft. Find the RCR.

Solution: For (7.20), the ratio of the perimeter to the floor area, for a rectangular room is

$$PAR_{cav} = \frac{2 \times (25 + 32)}{25 \times 32} = 0.143 \text{ ft}^{-1}$$

And from (7.20), the cavity ratio for this room is:

$$CR = 2.5 \times 7.5 \times 0.143 = 2.67$$

Once the cavity ratios are determined, the next step consists of finding the effective cavity reflectances determined for ceiling and floor cavities. The surface ability to reflect incident light is given by its luminance factor. The common reflectance values are given in Table B5 (Appendix B) under the applicable combination of cavity ratio and actual reflectance of ceiling, walls, and floor. Sample of the luminance factor, determined by the reflectance values are included in Table 7.7. Note that if the luminaire is recessed or surface mounted, or if the floor is the work plane, the CCR or FCR are 0 and then the actual reflectance of the ceiling or floor is the effective reflectance. The recommendations are to use the actual ceiling reflectance value, when fixtures are surface mounted or recessed; and to use the actual floor reflectance, if the floor is the work plane. The values of ρ_{cc}, ρ_{fc}, and ρ_w (wall reflectance), knowing the RCR previously calculated, are used to find the CU in the luminaire CU table. Note that the linear interpolations can be used for exact cavity ratios and reflectance combinations. The coefficient or factor of utilization (CU) found is for a 20% effective floor cavity reflectance. Thus, it is necessary to correct for the previously determined ρ_{fc}. This is done by multiplying the previously determined CU by the factor from Table B5 (see Appendix B), CU final is equal to the CU, at 20% floor time the multiplier for actual ρ_{fc}. If it is other than 10% or 30%, interpolate or extrapolate and multiply by this factor, to get the utilization coefficient. Computation of the

Table 7.7 Luminance factors for painted surfaces

Surface	Typical color	Luminance factor range (%)
Ceiling	White, cream	70–80
Ceiling	Sky blue	50–60
Ceiling	Light brown	20–30
Walls	Light stone	50–60
Walls	Dark gray	20–30
Walls	Black	10
Floor	—	10

illuminance level is performed using the standard lumen method relationship, used for lighting design:

$$E = \frac{F \times N \times CU \times LLF}{A} \tag{7.27}$$

where E is the average horizontal illumination at the working plane in ***lx or fc,*** F is the rated lamp ***lumens*** (as published by each lamp manufacturer), N is the number of lamps (number of luminaires times number of lamps per luminaire), CU is the utilization factor or coefficient, LLF is the light loss factor, and A is the area of the working plane (ft^2 or m^2). CU is determined as discussed in the earlier paragraph, based on the room size, configuration, room surface reflectance, and the performance characteristics of each of the luminaires, per unit. The total LLF consists of three basic factors: LLD, luminaire dirt depreciation (LDD) and BF. If initial levels are to be found, a multiplier of 1 is used. LLFs, along with the total lamp lumen output, vary with manufacturer, lamp or luminaire type are determined by consulting the manufacturer data. BF is defined as the ratio between the published lamp lumens and the lumens delivered by the lamp on the ballast used. Typical HID BFs vary between 0.9 and 0.95. Occasionally, other LLFs may need to be applied when they are applicable. Some of these are luminaire ambient temperature, voltage factor, and room surface dirt depreciation. When the initial illuminance level required (specified in codes or standards) is known and the number of fixtures (luminaires), N_{lm} needed to obtain the desired or required illuminance level, a variation of the standard lumen formula may be used:

$$N_{lm} = \frac{E_{desired/required}(\text{fc or lx}) \times A\left(\text{ft}^2 \text{ or m}^2\right)}{\#\frac{lamp}{fixture} \times F \times CU \times LLF} \tag{7.28}$$

Example 7.14: A university amphitheater is 72′ long and 40′ wide with a 15′ ceiling *height. Reflectances are ceiling 80%, walls 30%, floor 10%. Four-lamp, Prisma-wrap type* per each luminaire are used, having 3,000 lm per lamp, on 6′ stems and the work plane is 2′ above the floor. Find the illuminance level if there are 24 luminaires in the room.

Solution: From (7.19a)–(7.19c), the CCR, RCR, and FCR are calculated as:

$$CCR = \frac{5h_{cl}(L + W)}{L \times W} = \frac{5 \times 4 \times (72 + 40)}{72 \times 40} = 0.78$$

$$RCR = \frac{5h_{lw}(L + W)}{L \times W} = \frac{5 \times (15 - 4 - 2)(72 + 40)}{72 \times 40} = 1.75$$

$$FCR = \frac{5h_{wf}(L + W)}{L \times W} = \frac{5 \times 2(72 + 40)}{72 \times 40} = 0.39$$

In Table B5 (see Appendix B), for the above ceiling and floor cavities, the effective reflectance is determined as 65, while ρ_{fc} for the floor cavity is 10%. The effective cavity reflectances for the ceiling and floor cavities are 65% and 10%, respectively. Using the RCR value, 1.75 and the effective reflectances values, the CU is calculate, using Table B5 (Appendix B), via interpolation as 0.63. However, this CU is for an effective reflectance of 20% while the actual effective reflectance of the floor ρ_{fc} is 10%. To correct for this, locate the appropriate multiplier in Table B6 for the $RCR = 1.75$, which is 0.952. The final CU is computed: $CU_{final} = 0.63 \times 0.952 = 0.60$. Assuming $LLF = 1$ in (7.22), the initial (desired) illumination level is:

$$E = \frac{24 \times 4 \times 3{,}000 \times 0.60 \times 1}{72 \times 40} = 60\ \mathrm{lm/ft^2}$$

Notice that once the CU coefficient is found, we can determine the required total lumens (F_{tot}), number of luminaires (N), and area per luminaire (A_{lm}), by using the relationships given below. An approximation but quite accurate one for LLF is the product of LLD and LDD, only.

$$F_{tot} = \frac{E \cdot A}{CU \cdot LLF} \tag{7.29a}$$

$$N = \frac{F_{tot}}{F} \tag{7.29b}$$

$$A_{lm} = \frac{A}{N} \tag{7.29c}$$

The notations, here are the ones defined above and with the specified units. The LDD factor values are determined from the luminaire category provided by the manufacturers or found in the codes or standards, and is depending of the lamp and the replacement schedule. The CU is basically the ratio of the lumens received on the working plane to the lamp total flux outputs.

Example 7.15: For a room of $20' \times 30'$, the adjusted utilization factor was found 0.47, and the LLD and LDD are 0.84 and 0.85, respectively. Find the total illuminance level, number of luminaires, and the area per luminaire, if the desired horizontal average illuminance is 50 lm/ft^2. Each luminaire has two lumps of 3,150 lm.

Solution: From (7.24a) the total desired lumens for this room are:

$$F_{tot} = \frac{E \cdot A}{CU \cdot LLF} = \frac{50 \cdot (30 \times 20)}{0.47 \times 0.84 \times 0.85} = 89{,}397\ \mathrm{lm}$$

The total number of luminaires is given, by (7.24b) as:

$$N = \frac{F_{tot}}{F} = \frac{89{,}397}{2 \times 3{,}150} = 14.2$$

The closest integer is 14, which is the selected number of luminaires. Area per luminaire is computed by using (7.24c) as:

$$A_{lm} = \frac{A}{N} = \frac{20 \times 30}{14} = 42.86 = 43 \text{ ft}^2$$

The aim of the lumen method is to give a reasonably even spread of light over the horizontal working plane. How this spread of light is achieved depends upon the way the light is distributed from the fittings, not only in relation to fittings but also be related to the height at which the fittings are mounted over the working plane. The ratio mounting height to the spacing of the fittings will vary with the choice of fitting: the greater the concentration of light distribution from the fitting, the closer must be the spacing relative to the mounting height. A slightly modified version of (7.26) is given below, in which the LLF is replaced by the maintenance factor or coefficient (M). M represents the ratio taking into account the light lost due to an average expectation of dirtiness of light fittings and the room surfaces. The maintenance factor is often determined individually, and takes the installation reduction in luminous flux caused by soiling and ageing of lamps, luminaires, and room surfaces into account. For normal conditions, a factor of 0.8 may be used, for air-conditioned rooms a factor of 0.9 may be used, while for an industrial atmosphere, where cleaning is difficult, a factor as low as 0.5 may sometimes be used. In order to have accurate M estimates, the maintenance schedule (the cleaning and maintenance intervals for the lamps and installation) must also be documented. With previous notation, with units specified above for each of the parameters involved the E, the average horizontal illumination at working plane is expressed as:

$$E = \frac{F \times N \times CU \times M}{A} \tag{7.30}$$

Another design factor that is usually considered in lighting design is the spacing-to-height ratio (SHR) is the center-to-center (S) distance between adjacent luminaires to their mounting height (H) above the working plane. Manufacturer catalogs can be consulted to determine maximum SHR's, e.g., for a luminaire with trough reflector is about 1.65 and for an enclosed diffuser about 1.4. Must keep in mind, that there is a large range of room dimensions but it has been found that the light behavior in rooms is a function not of room dimensions, but the room index (RI), which is the ratio of the area of the horizontal surfaces to that of the vertical surfaces in the room. For the lumen method of design the vertical surfaces are measured from the working plane to the center of the fitting. This is expressed by the equation:

$$RI = \frac{\mathrm{A}}{H(L+W)} = \frac{L \times W}{H(L+W)} \tag{7.31}$$

Here, H is the room height (ft. or m). Equation (7.31) is for rectangular room, only. However, adaptations for other room geometries are easy to be made.

Example 7.16: A general office measuring $15 \times 9 \times 3$ m^3 high is to be illuminated to a design level of 400 lux using 85 W fluorescent fittings. The fittings are to be flush with the ceiling and the working plane is to be 0.85 m above the floor. Design the lighting system for the office when the installed flux is 8,000 lumens per fitting. Assume that the maintenance factor is 0.8, and the height of fitting above the working place, H is 1.65. For the room index of this space of 3.4, calculated by using (7.24), utilization factor from the tables is found to be 0.56.

Solution: From (7.23), the number of luminaires is:

$$N = \frac{E \times A}{F \times CU \times M} = \frac{400 \times (15 \times 9)}{8{,}000 \times 0.56 \times 0.80} = 15$$

In terms of illumination, 15 fittings would provide about 398 lx and would probably be satisfactory. In terms of spacing arrangement, however, 16 fittings are required, which would provide the following illumination level:

$$E = \frac{F \times N \times CU \times M}{A} = \frac{8{,}000 \times 16 \times 0.56 \times 0.8}{15 \times 9} = 425 \text{ lx}$$

The mounting height for fittings is: $3 - 0.85 = 2.15$ m.

7.5.4 Outdoor lighting design

One of the main lighting design applications involves the lighting of open outdoor areas, such as parking lots, sidewalks, building entrance areas, loading docks, etc. In such applications, lighting is necessary for vehicular and pedestrian safety, protection against crime, and the user convenience. The outdoor lighting design objectives need to conform to those specified in the codes, standards, and technical manuals, as well as to the local and federal regulations and requirements and other requesting organizations. The aim for road and public space lighting schemes can include any or all of the following:

1. Facilitation of safe movement of vehicles and people
2. Discouragement of illegal acts
3. Contributing to the prestige and amenity of an area through increased aesthetic appeal
4. Minimum light spill and glare
5. Cost and efficient use of the energy.

This lighting design type is based on maintaining specific illuminance levels and keeping the ratio of maximum illuminance to minimum illuminance below a predetermined value. The values for maintained illuminance and *uniformity ratio*, the ratio of maximum illuminance to minimum illuminance levels, listed in the *IESNA Lighting Handbook* for various applications, as shown in Table 7.8. Horizontal illuminance is defined as the quantity of luminous flux falling on a horizontal plane. Similarly, vertical

Table 7.8 Recommended illuminance values for parking lots and loading docks

Perimeter	Basic use (fc)	Enhanced security (fc)
Minimum horizontal illuminance	0.2	0.5
Uniformity ratio, maximum-to-minimum	20:01	15:01
Minimum vertical illuminance	0.1	0.25

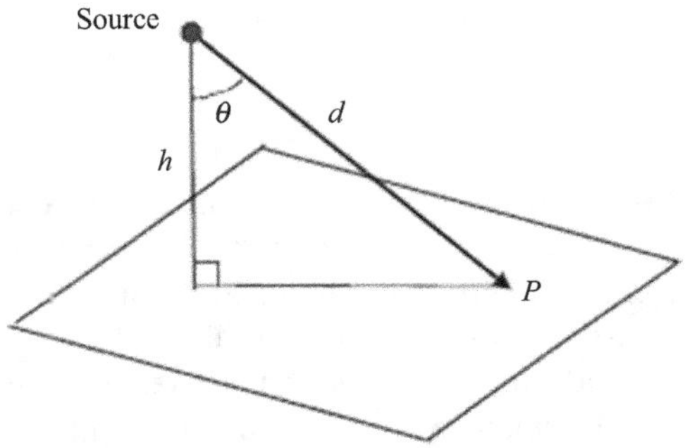

Figure 7.8 The illumination plan and geometry

illuminance is the amount of luminous flux striking a vertical plane. These illuminances are minimum values whereas the uniformity ratio is a maximum value. Actual requirements for a particular application may surpass these values. Parking lot, sidewalk and loading dock lighting are typically designed based on illuminance criteria. The IESNA recommends a minimum horizontal illuminance level of 0.2 ft-cd and a minimum vertical illuminance level of 0.1 ft-cd for low-activity open parking lots and loading docks to help assure safety and to deter crimes. A uniformity ratio, compared to the area with highest illuminance to the area with lowest illuminance, of 20:1 or lower is also specified for parking lots. A suitable lighting technology must be selected in the design process, as well as the luminaire style and lamp type.

For outdoor lighting, it is reasonable to treat the light source as a point source, considering the luminaire and area of interest dimensions. In this case the inverse square, as shown in Figure 7.8, can be applied to calculate the illuminance for a point P as:

$$E = \frac{I(\theta, \phi)}{d^2} \cos(\theta) \tag{7.32}$$

Or

$$E = \frac{I(\theta, \phi)}{h^2} \cos^3(\theta) \tag{7.33}$$

where E is the illuminance (lx), d is the distance from the source to the point (m), θ is the angle of the light from the normal, $I(\theta,\phi)$ is the intensity of the source in the direction of θ *(cd)* as specified by manufacturer, h is the perpendicular distance from the source to the plane (m), ϕ is the lateral angle. The values are given in tabular format, as the one found in Appendix B for parking areas and other outdoor applications. In a tridimensional space, with x, y, the horizontal coordinates, and z, the vertical coordinate, then d and the lateral and vertical angles are calculated as:

$$d = \sqrt{x^2 + y^2 + z^2} \tag{7.34}$$

$$\theta = \tan^{-1}\left(\frac{\sqrt{x^2 + y^2}}{z}\right) \tag{7.35}$$

And

$$\phi = \tan^{-1}\left(\frac{y}{x}\right) \tag{7.36}$$

A variation and (7.33) is used for roadway lighting and sometimes for other outdoor applications, calculates how far apart the fixtures must be spaced to produce the necessary average illuminance. A utilization curve, as the ones included in Appendix B, shows the percent of light which falls onto an area having a designated width and an infinite length. This width is expressed on the utilization curve in terms of a ratio of the width of the area to the luminaire mounting height. Separate CUs are given for the area to the street side and area to the house side of the fixture and are used to find illumination on the roadway or sidewalk areas or added to find the total light on the street in the case of median mounted luminaires. A relationship for street lamp separation or spacing, taking into account the lamp characteristics, CU and LLF values and the desired foot-candle level is:

$$Spacing = \frac{I_{lamp} \times CU \times LLF}{\text{Avg MTD fc} \times W_{road}} \tag{7.37}$$

Here, I_{lamp} is the lamp intensity, W_{road} is the road width, and Avg MTD is the desired foot-candle level as required in codes and/or guides.

Example 7.17: A roadway 24 ft. wide is lightened to maintain illumination level of 1.0 fc. HPS lamps (35,000 cd), mounted on 30 ft. poles which are setback 30 ft. from the road, are used. Assuming a $CU = 0.23$, found from the curve included in Appendix B, and $LLF = 0.82$. Calculate the recommended spacing for the desired foot-candle level.

Solution: Spacing is then calculated as:

$$Spacing = \frac{35{,}000 \times 0.23 \times 0.82}{1.0 \times 24} = 275 \text{ ft.}$$

7.6 Chapter summary

Lightning represents the utilization of the natural or artificial lighting energy to provide the desired visual environment of working and living. Lighting is an essential service in all the industries. Artificial illumination for both functional and decorative purposes is a major consumer of primary energy, and developed civilizations have become used to very high illumination standards with consequently high electricity consumption. The power consumption by the industrial lighting varies between 2% and 10% of the total power depending on the type of industry. Innovation and continuous improvement in the field of lighting, has given rise to tremendous energy saving opportunities in this area. Lighting is an area, which provides a major scope to achieve energy efficiency at the design stage, by incorporation of modern energy efficient lamps, luminaires, and gears, apart from good operational practices. Luminous intensity, measured in cd, is one of the seven basic SI measurement units. The quantity of light emitted by a source is the luminous flux, and is measured in lumens, where one lumen equals one candela-steradian. The total luminous flux incident on a surface per unit area is called illuminance, and is measured in foot-candles, where one foot-candle equals one lumen per square foot. Luminous efficacy quantifies the conversion efficiency from electricity to light. Various lighting technologies and systems have been developed and evolved over time, including incandescent, low-pressure discharge types, such as fluorescent and low-pressure sodium, and HID types, such as mercury vapor, high-pressure sodium, and metal halide. LEDs produce photons when electrons change energy states while propagating through semiconductor material. Luminaires are fixtures designed to enclose lamps and provide specific dispersion patterns for the light. Lighting systems are usually designed using illumination criteria determined by IESNA, with criteria are based on the anticipated use of the space being illuminated. This chapter also discusses the methods and tools needed to design and analysis to produce lighting applications, using modern and advanced light sources, luminaires, and control techniques. An overall presentation and discussion of lighting design methods and processes, including lighting quality issues, guidelines, and recommendations for advanced lighting design are presented in details. Several examples, questions and problems, as well as further readings and essential references are included here.

References and further readings

[1] R.R. Janis and W.K.Y. Tao, Mechanical and Electrical Systems in Buildings (5th ed.), Pearson, Upper-Saddle River, NJ, 2014.

[2] T.R. Bosela, Electrical Systems Design, Prentice Hall, Upper-Saddle River, NJ, 2002.

[3] IESNA Lighting Handbook (9th ed.), Illuminating Engineering Society of North America, New York, 2000.

[4] R.N. Helms. Illumination Engineering for Energy Efficient Luminous Environments, Prentice Hall, Upper-Saddle River, NJ, 1980.

[5] J.L. Lindsey, Applied Illumination Engineering, Prentice Hall, Upper-Saddle River, NJ, 1991.

[6] J.L. Lindsey, Applied Illumination Engineering, Fairmont Press, Lilburn, GA, 1996.

[7] Design of R. Fehr, Industrial Power Distribution (2nd ed.), Wiley, Hoboken, NJ, 2015.
[8] F.J. Trost and I. Choudhury, Mechanical and Electrical Systems in Buildings, Pearson-Prentice Hall, Upper-Saddle River, NJ, 2004.
[9] R.R. Janis and W.K.Y. Tao, Mechanical and Electrical Systems in Buildings (6th ed.), Pearson, Upper-Saddle River, NJ, 2019.
[10] R. Fehr, Industrial Power Distribution (2nd ed.), Wiley, Hoboken, NJ, 2015.
[11] G. Gordon, Interior Lighting for Designers (5th ed.), Wiley and Sons, New York, 2015.
[12] F.J. Trost and I. Choudhury, Design of Mechanical and Electrical Systems in Buildings, Pearson, Upper-Saddle River, NJ, 2004.
[13] F. Kreith and D.Y. Goswami (eds.), Energy Management and Conservation Handbook, CRC Press, Boca Raton, FL, 2008.
[14] H.W. Beaty, Handbook of Electrical Power Calculations, McGraw Hill, New York, 2001.
[15] A. Sumper and A. Baggini, Electrical Energy Efficiency – Technology and Applications, Wiley, Chichester, 2012.
[16] G. Petrecca, Energy Conversion and Management - Principles and Applications, Springer, Cham, 2014.

Questions and problems

1. What is the visible spectrum range?
2. What is the relationship between wavelength and frequency for EM waves?
3. What are the three primary color of the light?
4. Which waves from the EM spectrum does the Sun emit?
5. What are the illuminance and its measure unit?
6. Define the glare, direct and reflected glare, discomfort, and disability glare.
7. What are the basic characteristics of color?
8. Define the luminous intensity.
9. What is the standard unit of luminous intensity?
10. What is the standard procedure to measure luminosity?
11. What is the wavelength range of the visible spectrum?
12. What is the refracted angle inside a glass ($n = 1.52$) if the light enters from air at a 45° from the normal axis?
13. Define glare.
14. What are the types of commonly used lamps?
15. Which is illumination intensity?
    (a) Luminosity
    (b) Light level
    (c) Lumens
    (d) Lux
    (e) Lumen per watt

16. What are the main characteristics in selecting a lighting system?
17. What are the primary benefits of LED lighting system?
18. What do the following terms mean?
    (a) Illuminance
    (b) Luminous efficacy
    (c) Luminaire
    (d) Control gear
    (e) CRI
19. What is a lumen?
    (a) Unit of lighting not presently in use.
    (b) 1,000 candela/m^2
    (c) SI unit of light output or received
    (d) A directional measurement of light
    (e) Lighting power of a source
20. What is the color temperature of florescent lamps?
21. What means the lamp efficiency?
22. Define the lamp glare.
23. Describe how an incandescent lamp, a fluorescent lamp, and LED systems are producing light.
24. What is the function of ballast in a lighting system?
25. Define the terms used in lighting system design.
26. What are the three primary colors?
27. What does luminaire mean?
28. What is the purpose of lighting system ballast?
29. List some of the applications of point method.
30. What are the most important criteria in developing an outdoor lighting design strategy?
31. Determine the illuminance of a 12 × 9 ft^2 office area, if a total 6,300 lumens are directed from a light source.
32. A point light source has 2,800 cd intensity in the direction of interest. Determine the illuminance at a distance of 9 ft. and 6 m.
33. Compare the energy efficiency and color-rendering of different lamp types, stating suitable applications for each.
34. If a point source has lighting intensity of 2,800 cd in the direction of interest, determine the illuminance at a point 15 ft. from the source, and the angle of incidence with respect to the vertical is 36°.
35. By using data from Table 7.4, determine (a) the candlepower along the fixture at an incidence angle of 35°; and (b) the candlepower in a vertical plan, at an incidence angle of 45°.
36. Calculate the room index for an office 24 × 12 m^2 in plan, 3.2 m high, where the working plane is 0.85 m above floor level.
37. Find the utilization factor for a bare fluorescent tube light fitting having two 58 W, 1,500 mm lamps in a room 5 m by 3.5 m in plan and 2.5 m high. The working plane is 0.85 m above floor level. Walls and ceiling are light stone and white, respectively.

38. On what factor does the arrangement of luminaires depend?
39. Ten incandescent lamps of 500 W and 10,800 lm are used for an area of 60 m^2. If the utilization coefficient and LLF are 0.65 and 0.80, respectively, calculate the illuminance and the lamp efficiency.
40. Determine the RCR for a space 60-ft. long, 30-ft. wide, and 15-ft. high. If the room described in has a work surface 2.5 ft. above the floor, and the luminaire are suspended 2.5 ft. below, calculate also the floor cavity ratio and the ceiling cavity ratio.
41. For the room of problem 40, determine the 20,000 lm luminaires are needed to maintain an illuminance of 50 fc. Assume a ceiling reflectivity of 80%, wall reflectivity of 50%, a floor reflectivity of 20%, and a LLF of 0.85.
42. If an illuminance of 40 fc must be maintained at the work surface of the room described in problem 23, how many 20,000-lm luminaires are needed assuming a ceiling reflectivity of 77%, a wall reflectivity of 50%, a floor reflectivity of 20%, and an LLF of 0.88.
43. An office 12×8 m^2 long requires an illumination level of 400 lux on the working plane. It is proposed to use 80 W fluorescent light fittings having a rated output of 7,350 lumen each. Assuming a utilization factor of 0.6 and a maintenance factor of 0.85, calculate the number of light fittings required.
44. Determine the RCR for a space 90 ft. long, 50 ft. wide, and 13.5 ft. high. If the room has a work surface of 2.85 ft. above the floor, and the luminaires are suspended 2 ft. below the ceiling, calculate the ceiling cavity ration and the floor cavity ratio.
45. Find CU of a luminaire installed in a 15×20 m room with 3.5 m ceiling height, 0.75 work place, and luminaires are mounted 0.65 m from below of the ceiling. The reflectance coefficients are $\rho_{cc} = 0.80$, $\rho_w = 0.60$, and $\rho_{fc} = 0.20$. Assuming LLF = 0.85, lamp intensity of 3,200 lm, and 2 lamps per fixture, how many fixtures are required to maintain 540 lx?

Chapter 8

Motor control and protection, drives, and applications

Objectives and abstract

Electric motors are used in a vast range of applications, types, shapes, sizes, or constructions. In our power systems, the generators in power plants are connected to a three-phase network while most of the industrial equipment and large- or medium-size electric motors, pumps in heating, water and air conditioning systems, refrigerator, dryers, vacuum cleaner or most of the appliances are connected to a single phase AC, switched on or off by simple contactors. In cars, a DC battery is providing power to the starter motor, windshield wiper motors, and other car subsystems. The DC car motors are usually activated by a relay switch without any control. However, many of other electric motor applications often require advanced control, depending on the application and the load requirements. Motor protection and control are essential functions for proper operation and safeguard electric motors and their connection cables from the effects and/or damages caused by overheating or improper motor operation. For example, overload, stalling, and single-phasing result in overheating, and the motor protection must detect these conditions and prevent their effects. Electric motors are the major prime-mover in industrial and commercial facilities and in building electrical, mechanical, and thermal systems. Most of the electric losses occur in the end user, and electric machines and drives are a large contributor. Electric motors and drives are important electrical system components, being the interface between the electrical and mechanical systems in an industrial process, a building, industrial, or commercial facility. These are creating unique challenges for motor control and protection which, in turn, led to the solutions that are critical in all electrical motor applications. By completing this chapter, the readers must have a good understanding of the electric motor control, starting, stopping, speed changes, breaking, and motor protection methods. It is very important to understand motor characteristics, in order to choose the right one for the application requirements. The learning objectives for this chapter include understanding the basic principles of operation of AC and DC motors, understand their operation and basic characteristics, control and protection methods and schemes, compute their electrical and mechanical parameters using the equivalent circuit, and to be able to select the most appropriate electric motor for a specific application. Readers must also understand and learn the structure, configurations, characteristics, and the operation of electric drives and their major applications. The chapter also includes appropriate references to the electric motor specifications of the codes and standards.

8.1 Introduction

The knowledge and understanding of the electrical motors, characteristics, and performances are critical for motor selection, operation, control, and protection, since electrical motors form an integral part of all of the industrial processes, building energy, and electromechanical systems. One of the main advantages of electrical energy is that it can easily be converted into mechanical energy. Over 60% of the electrical energy generated in the US, with similar percentages in most of the developed countries is used by electric motors. Most of the electric loads in today's power systems (70% or even higher in the industrial sectors) are electrical motors, with the remaining portion consisting primarily of heating, lighting, and electronic system loads. Electric motors are a part of everyday life that we seldom give a second thought, being found in applications from computer disk drives, appliances, cars, industrial processes, pumps, conveyors, weaving machines, air conditioning and heating systems, and many more. Modern building services and industrial facilities are heavily reliant upon electric motors and drives, found at the heart of air handling and heating systems, industrial processes, industrial equipment, transportation subsystems, and even in the theatre and film industries. If something is moving or has parts in motion, it is likely to assume that electric motors are responsible for the motion. It is estimated that electric motors consume about or over 50% of the world's electricity. With the energy cost steadily increasing and to reduce the overall costs, the industry is also focused on replacing less efficient constant-speed electric motors and drives with microprocessor-based, variable-speed drives. This motor control technology can reduce the energy consumption up to 30% or more. While these variable-speed control schemes add cost to the electric motors, the energy savings and increased motor functionality offset the initial investments within a few years. Apart from lighting, electric motors represent the largest loads in the industry and commercial facilities. Their function, to convert electrical energy into mechanical energy, making them significant in economic terms, and hence, they cannot be ignored by installation or machinery designers and/or users.

There are several electric motor types in use, both DC and AC, however, the three-phase asynchronous (induction) motors, and in particular squirrel-cage induction motors, are the most used in the industry, commercial facilities and large buildings above certain power levels, usually mid- and high-range power levels. Moreover, although they are ideal for many applications when controlled by specialized devices, with the increasing electronic equipment use, is significantly widening their application field. This is the case for motor control with soft

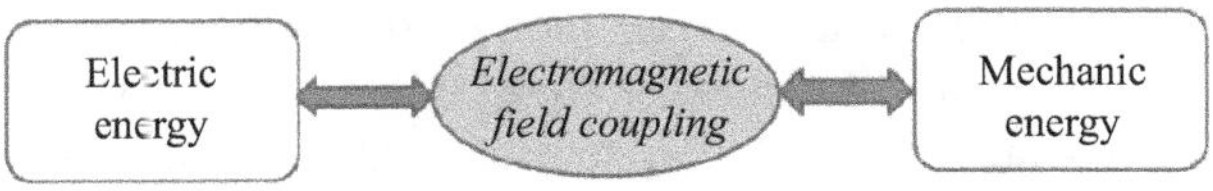

Figure 8.1 Block diagram of an electromechanical energy conversion system

start-stop units, and when precise speed adjustment is also necessary with variable speed drives/regulators. However, slip-ring asynchronous motors are used for certain high power applications in industry, and single-phase motors remain suitable for lower power applications, mainly for building applications, tools, low-power equipment, and devices. The use of synchronous motors, mostly brushless or permanent magnet type motors, combined with power electronics converters is becoming increasingly common in applications requiring high performance levels, in particular in terms of dynamic torque (on starting or on a duty change), precision and speed range. Electrical motors are electromechanical energy conversion devices, involving the interchange of energy between an electrical system and a mechanical one through the electromagnetic fields. The primary quantities involved in electrical side are voltage (E) and current (I) while the analogous quantities in the mechanical system are torque (τ) and angular speed (ω), respectively. Figure 8.1 shows a block diagram of an energy conversion device.

Electrical motors are designed with speed-torque characteristics to match the requirements of common applications. The four standard NEMA induction motor (IM) designs, A, B, C, and D, have different characteristics, as described in standards and/or codes. Because motor torque varies with speed, the relationship between speed and torque, speed-torque curve, or diagram is very important in motor selection, control, and protection. *Starting torque*, also referred to as *locked rotor torque*, is the torque that the motor develops each time it is started at rated voltage and frequency. When voltage is initially applied to the motor stator, there is an instant before the rotor turns. *Starting current*, also referred to as *locked rotor current*, is the current supplied to the motor when the rated voltage is initially applied with the rotor at rest. *Full-load current* is the current supplied to the motor with the rated voltage, frequency, and load applied and the rotor up to speed. For a NEMA B induction motors, starting current is typically 600%–650% of full-load current. Knowledge of the current requirements for a motor is critical for the proper application of overcurrent protection devices. Protection of the electric motors against over-currents or short circuits, undervoltage operation, or unbalanced voltages is critical for their proper operation, and for the motor life time. For example, the insulation lifetime decreases by half, if motor operating temperature exceeds thermal limit by 10 °C for any period of time. The chapter contains an in-depth presentation of the AC and DC electrical motor control and protection methods and techniques, including motor starting, stopping, breaking, speed change control, and the motor protection methods and schemes, since they form an integral part of industrial processes, equipment, and in building electromechanical systems. The aim of this chapter is to gather knowledge about the following topics of the types and principle of operation of three-phase motors, their equivalent circuits, performance calculation by means of finding torque, slip and efficiency, different types of starters like auto-transformer starter, star-delta starter, and various methods of speed control schemes. A comprehensive discussion of electric drives is also included in this chapter. The chapter ends with a discussion of the electrical motor applications in compressed air systems; also heating and air compressors are discussed since they are major energy consumers in many industrial processes and part of the electromechanical systems of most of the modern buildings. Then, some sample case studies of the process energy management and conservation opportunities are also provided.

8.2 Poly-phase induction motor control schemes and methods

Many applications driven by electric motors require more or less advanced control. All motors must have a device to start and stop the motor, a motor controller device or system. A motor controller is the actual device that energizes and deenergizes the motor circuits to start and stop it. A controller is defined as *a group of devices or a device that serves to govern, in some predetermined manner, the electric power delivered to the apparatus to which it is connected*. Motor starting contactors are available as integral units with externally operable switches, defined as a combination controller. Electrical motor controllers may include some or all of the following control functions: starting, stopping, overcurrent and overload protection, reversing, speed changing, jogging, plugging, sequence control, and pilot light indication. Controllers range from simple to complex systems and may provide control for one motor, several motors grouped together, or auxiliary equipment, such as brakes, clutches, heaters, or other signal controlling devices and equipment. For example, lowering the speed of a fan or a pump is relatively simple control process, while the dynamic positioning of a tug in a wafer-stepper with nanometer accuracy when accelerating at several g's or the controlled drive of an electric crane that must be able to move a hook at high speed, navigate heavy loads up and down at moderate speeds, and making a soft touchdown as close as possible to its intended final position are incomparable more difficult. Other applications, such as assembly robots, electric elevators, or control of hybrid electric vehicles, trains, streetcars, or CD-players can, with regard to complexity, be situated in between of the previous two examples.

Motor starting contactors are available as integral units with externally operable switching means is defined a combination controller. A starter is defined as a form of electric motor controller that includes the switching means it is necessary to start and stop a motor in combination with suitable overload protection, a combination starter, which includes the motor switching contactor as well as overload protection and an integral disconnecting device, is a type of combination controller. Low voltage manual and magnetic controllers are classified as Class A, B, or V according to their interrupting medium and their ability to interrupt currents, motor classes described in details in the next chapter section. Design and analysis of all electric drive systems requires not only knowledge of dynamic properties of different motor types but also a good understanding of the way these motors interact with power electronic converters and their loads. These power converters are used to control motor currents or voltages in various ways. Once the proper motor is selected, understanding the various control devices available and their uses and limitations becomes an important part related to the reliable operation and protection of the motor and the personnel using the motor. On the other hand, motor control circuits are an effective way to reduce cost by using smaller wire and reduced-amperage devices to control a motor. There are four major motor control topics or categories to consider. Each of these has several subcategories and sometimes the subcategories overlap to some extent. Certain pieces of motor control equipment can accomplish multiple functions from each of the topics or categories. The four motor control categories include (1) methods for starting the motor, (2) motor protection, (3) stopping and breaking the motor, and (4) motor operation control. A good and complete understanding of each of these areas is necessary to effectively apply motor control principles and equipment to effectively operate and protect a motor.

Motor control circuits are an effective way to reduce the overall motor operation costs by using smaller wires and reduced-amperage devices. Smaller size motors often use the same size conductors for both control and power circuits; however, as the power increases, this becomes impractical. Motor control circuits are often connected to lower voltages than the motors they control to make it safer for operators and maintenance personnel. A motor control circuit, for the most part, is simply a switch (or group of switches). Many large- and medium-size motors are controlled by computerized control systems, solid-state logic controls, or programmable logic controllers (PLCs). However, the fundamental principles of control systems are still applying. The PLC controls an external output based on the logic of a control program, and that output controls the motor or groups of motors by using a magnetic starter, and in some cases, additional relays. PLCs and other solid-state control devices were originally invented to provide less expensive replacements for older automated systems that used large numbers of relays and mechanical timers. In some cases, a single PLC can replace thousands of relays resulting in less expensive wiring systems that offer greater flexibility in control designs and operation.

There are several methods for starting electric motors, and the most common are outlined here. In addition, introduction in-depth discussions of the motor-starting and control devices and schemes are also included. The most common motor starting device is the low voltage motor-starting contactor. A contactor is defined as a *two-state ON-OFF device* for repeatedly setting or interrupting an electric power circuit. Contactors are designed for lifetime and for optimum performance when switching loads, and not for interrupting short-circuits and therefore, motors require separate short-circuit protection. Contactors are usually closed magnetically via their control coils and are typically referred to as magnetic control. For small motors, typically fractional-horsepower, manual control switches are also available. Motor starting contactors and switches in the United States are typically designed and manufactured per NEMA ICS-1, NEMA ICS-2, and UL 508. A controller is defined in literature as *a device or group of devices that serves to govern, in a predetermined manner, the electric power delivered to the apparatus or equipment to which it is connected.* Electric motor starting contactors are available as integral units with externally operable switching means, as a combination controller. A motor starter is defined as *a form of electric motor controller that includes the switching means necessary to start and stop an electric motor in combination with suitable overload protection.* This combination starter, which includes the motor switching contactor as well as overload protection and an integral disconnecting device, is a type of combination controller. Low voltage manual controller and magnetic type controllers are classified as Class A, B, or V according to their interrupting medium and their ability to interrupt electric currents:

1. Class A controllers are AC air-break, vacuum break, or oil-immersed manual or magnetic controllers for service on 600 V or less, capable of interrupting operating overloads but not short circuits or faults beyond operating overloads.
2. Class B controllers are DC air-break manual or magnetic controllers for service on 600 V or less, capable of interrupting operating overloads but not short circuits or faults beyond operating overloads.
3. Class V controllers, AC vacuum-break magnetic controllers for service on 1,500 V or less, capable of interrupting operating overloads but not short circuit or faults beyond operating overloads. Low-voltage NEMA-rated contactors are designated in sizes 00 (smallest) to 9 (largest) for various duty applications.

Control of contactors using maintained-contact devices is referred to as two-wire control. Use of the momentary contact devices in the contactor controls is referred to as three-wire control. Three-wire control has the advantage of allowing the contactor to open and remain open, if the line voltage fail. This arrangement is typical providing the undervoltage protection for motors and prevents inadvertent reenergization after a power failure. Medium-voltage contactors typically are using vacuum for interrupting. Unlike a circuit breaker, a medium-voltage vacuum contactor is specifically designed for long life in load-interrupting duty rather than for short-circuit interrupting duty. However, unlike their low-voltage counterparts, the medium-voltage contactors are able to interrupt short-circuit currents beyond operating overloads. Medium voltage air-break, vacuum, or oil-immersed controllers are classified as class E, further divided into class E1, employing their contacts for starting and stopping the motor and interrupting short circuits or faults exceeding operating overloads and class E2 controllers employing their contacts for starting and stopping the motor and employ fuses for short circuits or faults exceeding operating overloads. It is worth to notice that above 7.2 kV, motor control is accomplished using circuit breakers.

8.2.1 Induction motor starting methods

The induction motor (IM) starting, besides the switching alone, involves the control of the starting inrush current, starting torque, or both, and its overload and short-circuits protection. The two important factors to be considered in the induction motor starting are: the starting current drawn from the supply, and the starting torque. The starting current must be kept low to avoid motor overheating and excessive voltage drops in the supply network. The starting torque must be over 50% higher than the rated load torque to ensure that the motor runs up in a reasonable time. There are three major problems in starting and connecting the electrical energy supply to the induction motors: if under low voltage conditions at starting, the motor will successfully run up at rated speed, the voltage drop effects on the other equipment, when the motors are starting from the standstill, and the problems caused by the currents supplied back into the electric network by the motors when the supply voltage is suddenly dropped because of the network faults. Wound-rotor induction motors can be connected directly to the electric supply with the rotor winding terminals in open-circuit or through high resistances connected to the slip-ring brushes. At full operating speed, the brushes are short-circuited in each phase. In the case of squirrel-cage induction motors, there are several starting methods, depending on the motor size and type. To understand the starting and control methods for the induction motors it is worse to briefly review the construction, structure and motor equivalent circuits. The IM stator is laminated iron core with slots, while the coils are placed in these slots to form a three or single phase winding(s). Rotor is of two different types (1) squirrel-cage rotor, and (2) wound-rotor type. In the squirrel-cage rotor, the rotor winding consists of copper or aluminum bars placed in the slots and short-circuited by end-rings on both sides of the rotor. Most of single-phase induction motors have squirrel-cage rotor. In the wound-rotor induction motors, an insulated three-phase winding similar to the stator winding wound for the same number of poles as the stator is placed in the rotor slots. The ends of the star-connected rotor winding are brought to three slip rings on the shaft so that a connection can be made to it for starting or speed control, being usually for large three-phase induction motors. The stator is supplied by a balanced three-phase voltage that drives a three-phase current through the winding. This current induces a voltage in the rotor, which has the following simplified equivalent circuit (as shown in Figure 8.2).

At speeds close to the synchronous speed, the slip s is very small so R_2/s is very large, the rotor and the stator currents are very small, while at low speeds or stand still R_2/s is quite low and the rotor current became very large and in consequence

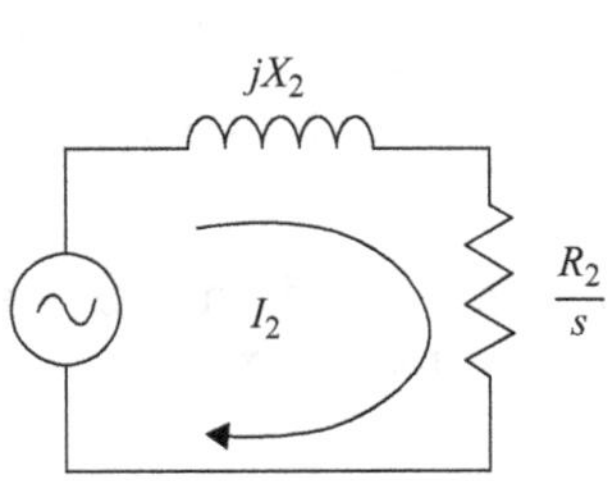

Figure 8.2 Simplified equivalent circuit of an induction motor

the stator current (the current drawn by the motor from the power supply) are also very large. Typical induction motor starting currents are between 5 and 8 times the normal running currents. Hence, the starting motor currents should be reduced. The most usual methods of starting three-phase induction motors are: for slip ring motors, the common starting method is the *rotor resistance starting method,* while the most common starting methods for squirrel-cage induction motors include the *direct-on-line starting,* the *reduced-voltage staring,* and the *current limiting* methods. Most of induction motors are rugged enough that they can start across the electric line without any damages of the windings during the starting even the starting current can be 5–7 times than the rated current flowing through the stator windings at the standstill voltage. In many cases, the direct-on-line (DOL) starting method is used. In this method, the full voltage is applied across the stator windings. This is an ideal starting method, is the most simple and economical, being an inexpensive starting method of an induction motor. The motor is switched on directly for full supply voltage. The initial starting current is large, up to 8 times the rated current but the starting torque is likely to be 0.75–2 times the full load torque. To avoid excessive supply voltage drops because of large starting currents, the method is restricted to small motors only. To decrease the starting current, the squirrel-cage motors (medium and larger sizes) are started at a reduced supply voltage. The torque developed by the motor is at a maximum value, while the acceleration is fast and the heat of starting is low. For heavy rotating masses, with large moments of inertia, this is an ideal switching method. The only limitation is the initial heavy inrush current, which may cause severe voltage disturbances to nearby feeders (due to a large $I_{st} \cdot Z$ voltage drop). With this in mind, even local electricity authorities sometimes restrict the use of DOL starting beyond a certain rating, often, 10 HP, for small installations. However, DOL starting techniques are not recommended for large power induction motors, because the line capacity may not be enough to carry the starting currents and the large starting currents may cause large voltage dips resulting in large voltage drops across the motors. The squirrel-cage induction motor current is estimated from its nameplate code letter, horsepower, and rated voltage. The motor starting apparent power is estimated from rated power (in HP) and the code letter factor from NEMA motor and generator standards, and then the starting current is:

$$I_{Start} = \frac{S_{Start}}{\sqrt{3}V_t}, \quad S_{Start} = (\text{Rated horsepower})(\text{code letter factor}) \tag{8.1}$$

Example 8.1: Estimate the locked-rotor (starting) current of 1.5 HP, 277 V motor with an H motor code.

Solution: From Table 8.1, an H motor code implies a range of 6.3 kVA per HP up to 7.10 kVA per HP so the locked-rotor current is:

$$I_{LR} = \frac{1}{277}(6{,}300 \div 7{,}100) \times 1.5\ \text{HP} = 34.1\ \text{A} \div 38.5\ \text{A}$$

Table 8.1 Motor code letters. Adapted from NEMA Generators and Motors Standards

Code letter	Locked-rotor current (kVA per HP)	Locked-rotor current (kVA per kW)
A	0.00–3.14	0.00–4.25
B	3.15–3.54	4.25–4.80
C	3.55–3.99	4.80–5.40
D	4.00–4.49	5.40–6.00
E	4.50–4.99	6.00–6.70
F	5.00–5.59	6.70–7.50
G	5.60–6.29	7.50–8.40
H	6.30–7.09	8.40–9.50
J	7.10–7.99	9.50–10.70
K	8.00–8.99	10.70–12.10
L	9.00–9.99	12.10–13.40
M	10.00–11.19	13.40–15.00
N	11.20–12.49	15.00–16.80
P	12.50–13.99	16.80–18.80
V	22.40 and up	

Reduced voltage starting methods for induction motors consist of a reduced voltage and is supplied to the motor stator and gradually increased to the rated voltage when the motor speed is about 25% of its rated speed or higher. The most common reduced voltage starting methods are *wye-delta (Y-Δ) starting method*, *auto-transformer starting*, *starting using solid-state voltage controller*, and *reduced voltage impedance (reactor or resistor) starting methods*. Motors that have windings in which both ends of each stator windings are brought to terminals, being configurable in either wye or delta connections are candidates for the wye-delta starting method. This is applicable to motors designed for delta connection in normal running conditions. The each phase ends of the stator winding are brought out and connected to a three-phase change-over switch. Wye-delta starting method starts the motor in a wye configuration, which supplies 57.7% of the line-to-line voltage to each winding. During the starting process, the motor is reconnected back into delta, supplying 100% of the line-to-line voltage to each winding, as the motor is approaching the full rated speed, the stator windings are reconnected in delta, via a switch. A disadvantage of this method is that the starting torque (which is proportional to the square of the applied voltage) is also reduced to one-third of its delta value. Both normally open and normally closed transition schemes are available, typically by using three contactors. Motors which have stator windings in two parts with at least six terminal leads may be started with part-winding starting. Part-winding starting energizes part of the transformer windings, typically half or two-third of the entire winding per phase, to allow a lower inrush and smoother acceleration. This scheme typically uses two contactors and is a closed-transition scheme. Separate overload relays are provided for each part of each winding.

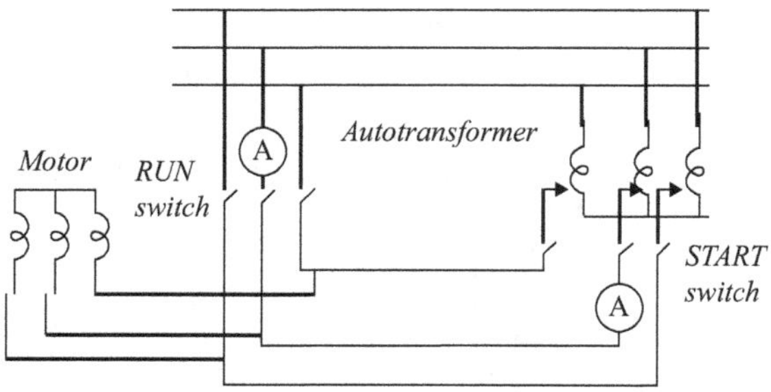

Figure 8.3 Autotransformer starting method diagram

Reduced-voltage autotransformer starting consists of initially starting the motor with an autotransformer, then removing the autotransformer from the circuit as the motor accelerates, as shown in Figure 8.3. The principle is similar to star/delta starting and has similar limitations. Notice that a wye-delta starting is equivalent to an autotransformer with a ratio $n = 1/\sqrt{3}$. The advantage of the method is that the current and torque can be adjusted to the required value, by taking the correct tapping on the autotransformer. The method is expensive because of the additional autotransformer, while resulting in a lower inrush current as the motor starts; it also results in less available output torque when the autotransformer is in the circuit. Autotransformer windings typically are tapped at 80%, 65%, and 50% voltage levels, while the available output torque is related to the output torque when at full voltage by the equation:

$$T_{RV} = T_{FV}\left(\frac{\%\ \text{Autotransformer tap}}{100}\right) \tag{8.2}$$

where T_{RV} is the motor output torque at the autotransformer reduced voltage, and T_{FV} is the motor output torque with full voltage applied. The motor output torque at the 80% autotransformer tap is 64% of the full-voltage torque value, at 65% tap the torque is 42.25% of the full-voltage torque, and at 50% tap the torque is 25% of the full-voltage value. However, care must be taken that the starting at the selected tap value, the autotransformer thermal duty capabilities (NEMA ICS-9) are taken into account, limiting the lowest tap to which the motor may be connected without damage to the autotransformer during starting.

Example 8.2: For an autotransformer starting, determine the voltage and current of a 440-V induction motor as a function of autotransformer taps of 50%, 65%, and 80% of line voltage, and starting current is 655 A at rated voltage.

Solution: From (8.2), the voltages and the currents, computed from the transformer current relationship are:

Tap (%)/Voltage	N_P/N_S ratio	$I_{ST}(N_P/N_S)$ (A)
100–440	1.00	655.0
80–352	0.80	524.0
65–286	0.65	425.8
50–220	0.50	327.5

Reduced-voltage reactor or *resistor starting method* consists of adding a resistor or impedance in series with the motor at the starting cycle, later short-circuited as the motor accelerates. A slip-ring induction motor cannot be started direct on-line with its rotor windings short-circuited, due to unacceptable current peaks. Moreover, the torque of the induction machine is dependent on the rotor resistance, while its maximum value is independent of the rotor resistance. The slip at maximum torque is dependent on the rotor resistance, therefore, it is expected that if the rotor resistance is changed, the maximum torque point shifts to higher slip values, while retaining a constant torque. However, while the maximum torque and synchronous speed remain constant, the maximum torque' slip increases with the rotor resistance, as well as the motor starting torque. Further, rotor resistance control could also be used as a means of generating high starting torque. For all its advantages, the scheme has two serious drawbacks. First, in order to vary the rotor resistance, it is necessary to connect external variable resistors (winding resistance itself cannot be changed), therefore, necessitates a slip-ring machine, since only in that case rotor terminals are available outside. For squirrel-cage rotors, there are no rotor terminals. Second, the method is not efficient since the additional resistance and operation at high slips entails losses. Resistors must therefore be inserted in the rotor circuit and then gradually short-circuited, while the stator is powered at full mains voltage. The resistors connected to the slip-ring brushes should have good power dissipation capability. The resistance inserted in each phase is calculated to ascertain the torque-speed curve with strict accuracy. The result is that it has to be fully inserted on starting and that full speed is reached when it is completely short-circuited. The current absorbed is more or less proportional to the torque supplied at the most only a little greater than the theoretical value. By adding external resistance to the rotor circuits, any starting torque up to the maximum torque can be achieved, and by gradually cutting out the resistance, a high torque can be maintained throughout the starting period. The added resistance also reduces the starting current, so the starting torque is about 2.5 times the full load torque, at a starting current up to 1.5 times the full load current. Notice that if a series reactance is used in the starting, the current power factor is poor and may produce significant disturbances in the line, while if a resistance is used, the power factor is better but the losses are higher. However, even this starting method is a less expensive method than the reduced-voltage autotransformer method suffers the same limitations due to the reactor or resistor thermal limits. This starting method does not alter the connection of the motor windings so the ends of each winding do not need outputs on a terminal board. The resistance value is calculated using the peak current on starting or the minimum starting torque required for the load torque. During the acceleration stage with the resistors, the voltage applied to the motor terminals is not constant but equals the mains voltage minus the voltage drop in the starting resistance. The voltage drop is proportional to the current absorbed by the motor. As the current reduces with the motor acceleration, the same happens to the resistance voltage drops. The voltage applied to the motor terminals is therefore at its lowest value at starting and then gradually increases. As the torque is proportional to the squared motor terminal voltage, it increases faster than in star-delta starting where the voltage remains constant throughout the star connection. This starting system is well suited to the machines with a resistive torque that increases with the speed (e.g., fans and centrifugal pumps). It has the drawback of a rather high current peak on starting, that can be lowered by increasing the resistance value but that would cause the voltage to drop further at the motor terminals and thus a steep drop in the starting torque. Standard reactors are available to limit the motor starting voltages to 50%, 75%, and 90% of the terminal voltage. When starting with series impedance, the apparent power (kVA) drawn from the electric supply is reduced directly proportional to the applied voltage and the motor torque is reduced proportional to the voltage squared. If x is the reduced voltage fraction by the series impedance, then the starting current and torque are given by:

$$I_{ST} = xI_{LR(SC)} \tag{8.3a}$$

And

$$\frac{T_{ST}}{T_{fl}} = \left(\frac{I_{ST}}{I_{fl}}\right)^2 s_f = x^2\left(\frac{I_{LR}}{I_{fl}}\right)^2 s_f \tag{8.3b}$$

By using (8.3a) and (8.3b), the starting current, I_{ST}, and the starting torque T_{ST} can be evaluated if the full-load current I_{fl}, short-circuit current $I_{SC(LR)}$, the slip at rated (full) load, s_f, the full-load torque, T_{fl} and the fraction of applied voltage are known.

Example 8.3: A 40-HP, 480 V, class letter C, 78 A three-phase induction motors has 0.67 reduced voltage fraction. If the full load torque is 980 Nm and full slip is 0.064, calculate the starting current and starting torque for this motor.

Solution: From Table 8.1, the maximum starting kVA and maximum starting current, at rated conditions of this motor are:

$$S_{ST} = 40 \times 3.99 = 159.60 \text{ kVA}$$

$$I_{ST} = \frac{159{,}600}{\sqrt{3} \times 480} = 192.0 \text{ A}$$

From (8.3a) and (8.3b) the reduced voltage starting current and torque are:

$$I_{ST} = xI_{LR(SC)} = 0.67 \times 192.0 = 128.6 \text{ A}$$

And

$$T_{ST} = x^2\left(\frac{I_{LR}}{I_{fl}}\right)^2 s_f \times T_{fl} = (0.67)^2\left(\frac{192}{78}\right)^2 \times 0.064 \times 980 = 170.6 \text{ Nm}$$

Figure 8.4 Block diagram of a solid-state controller stating method of an induction motor

Solid-state soft-starting ramps the voltage at the motor terminals linearly, producing smooth acceleration. Recent innovations include motor output torque-control models which linearly ramp the motor output torque, resulting in even smoother, almost linear, acceleration. Central to the operation of a solid-state soft-starter is the silicon-controlled rectifier (SCR, thyristor), a device which conducts current in one direction when current is injected into its gate terminal, and blocks current in the other direction. A typical implementation of a solid-state soft-start controller is shown in Figure 8.4. The control scheme firing circuit causes the SCR's to vary the motor voltage as required by the starting parameters. Once the motor has accelerated to full speed, the motor is connected directly to full voltage. The soft-start controller can also decelerate the motor in the same manner. Because the SCR's dissipate heat, the equipment heat dissipation and the ambient temperature are of the major concerns when applying soft-start controllers and must be considered carefully. The motor power factor correction capacitors must be switched out during motor starting when adjustable-speed drives are employed, to reduce the harmonic voltage interactions. Surge capacitors should not be used at motors which are soft-started for the same reason. Because soft-starters are often microprocessor-based devices, are typically supplied with communications and internal diagnostic capabilities, making them a cutting-edge motor control method. For example, in many cases single-phase motors over 5 or 7.5 HP need to use an electronic soft starter. Soft starting a motor limits the starting current up to 2 times the full-load current. However, the reduced starting current also reduces the available starting torque to 50%–90% of full-load torque. Therefore, this type of motor starter is matched to easy starting loads, such as drier fans, forage blowers, and irrigation pumps.

Another method of reducing high inrush currents when starting large motors is a capacitor starting system. This maintains acceptable voltage levels throughout the system, while the high motor inductive component of the reactive starting current is offset by the addition, during the starting period only, of capacitors to the motor bus. This differs from the practice of applying capacitors for running motor power factor correction. A motor-starting study can provide information to allow optimum sizing of the starting capacitors and determination of the time the capacitor must be energized. This method also maintains acceptable voltage levels throughout the system.

8.2.2 Voltage drop during the start-up of induction motors

One of the classical electrical engineering problems is the induction motor starting, having multiple aspects, such as the motor stator must withstand the starting voltage surge, the electric supply evaluated in connection with the motor characteristics must guarantee adequate starting capabilities, and the protection devices are evaluated against the staring currents. One of the most studied effects of motor starting is the voltage dip experienced throughout an industrial power system as a direct result of starting large motors. Available accelerating torque drops appreciably at the motor bus due to large voltage dips, extending the starting interval and affecting adversely, the overall motor-starting performances. Acceptable voltage for motor starting depends on the motor and load torque characteristics. Requirements for minimum starting voltage can vary over a wide range. Voltages can range from 80% or lower to 95% or higher. Moreover, from the mechanical point of view, the motor must develop enough torque to reach the operating speed in a reasonable time range. During the initial electrical voltage surge, in the first few cycles of the IM motor starting interval, the highest electrical currents occur. At locked-rotor conditions, the induction motor is a two-port constant-impedance device connected to the electrical system, contrasting strongly with the steady-state motor operation. Usually the starting (locked-rotor) motor current is about six times higher than the rated load (nameplate) current, and the voltage drop across the system equivalent circuit is causing a momentary voltage drop at the large motor terminals. A relationship to estimate the percentage in voltage variation by motor starting is:

$$\Delta V\% = \frac{\sqrt{3} I_{Start(LR)}}{V_T} (R \cos\phi + X \sin\phi) \times 100\% \tag{8.4}$$

Here, $I_{Start(LR)}$ is the starting (locked-rotor) motor current, V_T is the system phase voltage, R and X are the resistive and reactive system components, and ϕ is the motor starting phase angle. Care must be taken to use R and X appropriate values, including the feed and return path. If the locked-rotor current is large and the system equivalent impedance is high, the IM terminal voltage sags at levels, affecting other equipment connected to the motor electrical supply. During motor starting, the terminal voltage should be maintained at about 80% of the rated voltage level. This value results from the motor speed-torque characteristics (at about 150% of the starting torque at full-rated voltage) and the need to successfully accelerate fully loaded motors at reduced voltage (the torque varies with the voltage square). If other induction motors, connected to the same bus are drawing higher stator currents in an effort to maintain the power outputs may result in even higher voltage drops, lowering the bus voltage, and if these voltage drops are really severe, the induction motors can pull-out and stall, and the undervoltage relays can drop-out casing even larger problems. The calculations in (8.4) is performed by using the actual values, with ϕ positive for lagging power factor and negative for leading power factor or in per-unit values. Notice that, the power factor of the motor starting current is very low, 0.2–0.3 lagging. If $R << X$, for such low power factors the voltage drop (ΔV) is almost in phase with the supply voltage, therefore, the motor starting voltage drop, under the motor starting inrush current in a cable with impedance Z_{cable} is simply given by:

$$\Delta V = I_{st} \times Z_{cable} \tag{8.5}$$

Here, Z_{cable} in this relationship is the cable effective impedance, often found in the manufacturer or vendor data sheets; however, the actual cable impedance can also be computed using the impedance relationship:

$$Z = R + jX, \quad \text{and } |Z| = \sqrt{R^2 + X^2} \tag{8.6}$$

The voltage drop from (8.5) can further be approximated as $I_{st} \cdot X$ in longer (large) cable where $X >> R$ and $I \cdot R$ in the case of short length cable where $R >> X$.

Example 8.4: A three-phase 35 mm^2 copper cable 50-m-long supplies a 400 V, 50 Hz motor taking: a current of 100 A at a $cos\ \varphi = 0.8$ on normal (rated condition) permanent load, and a starting current of 500 A (5 times the rated current) at a $cos\ \varphi = 0.35$ during the motor start-up. Calculate the cable voltage drop during the motor staring and normal operation.

Solution: The resistivity of copper conductor ρ is 0.0225 $\Omega \cdot$ mm^2/m, while the linear cable reactance (Table B2 of Appendix B) at 50 Hz is 0.081 Ω/km the 50 m cable resistance and reactance are:

$$R = \frac{0.0225 \times 50}{35} = 0.0357\ \Omega$$

$$X = 0.081 \times 50/1{,}000 = 0.00405\ \Omega$$

The cable reactance can be neglected, as for most of the short-length cables. The voltage drops at motor starting and rated (normal load) conditions are computed form (8.5) as:

$$\Delta V_{st} = R_{cable} \cdot I_{st} = 0.0357 \times 500 = 17.85\ \text{V}$$

$$\Delta V_{rd} = R_{cable} \cdot I_{rd} = 0.0357 \times 100 = 3.57\ \text{V}$$

8.2.3 *Induction motor speed control*

The speed torque characteristic of an induction machine is determine is nonlinear and has the following characteristics: for small values of the slip s, the torque is directly proportional to s, while for large values of the slip s, the torque is inversely proportional to s. In the stable region of operation in the motoring mode, the curve is rather steep and goes from zero torque at synchronous speed to the stall torque at a value of the slip s_{pl}, at the maximum torque. Normally s_{pl} may be such that stall torque is about three times that of the rated motor operating torque, and hence may be about 0.3 or less. This means that in the entire loading machine range, the speed change is quite small. The machine speed is quite stiff with respect to load changes. The entire speed variation is only in the range N_S to $(1-s) \cdot N_S$, where the synchronous speed N_S is set by the supply frequency and number of poles. In the case of an induction motor torque-speed characteristic (Figure 8.5), there exists stable and

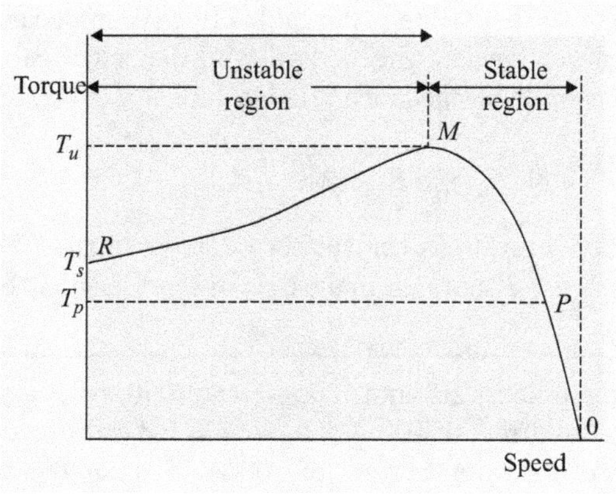

Figure 8.5 Torque-speed diagram of an induction motor

unstable ranges in this curve. Since it is impossible to reliably operate in the unstable range, simple voltage control (open loop control) is limited to controlling the speed in a narrow range of values. The torque versus speed curve is affected by the value of rotor resistance, as is shown in the expression above that higher values of resistance result in higher slip at rated conditions, meaning lower values of the slope of the torque–speed curve at the zero slip intercept. Clearly, high efficiency motors should have lower rotor resistance. There is, of course, a tradeoff here, for lower rotor resistances result in lower values for starting torque. Note that peak, or pull-out torque, is the same for all values of rotor resistance. This is because torque is proportional to air-gap power, maximized when the quantity R_2/s is equal to the source impedance.

The output power of an induction, from the equivalent circuit is given by the difference between the air-gap power and the rotor copper, stray (miscellaneous), windage, and friction losses. The output torque is given by the output power divided by the shaft angular speed of rotation (in rad/s), expressed as:

$$\tau_{Out(Load)} = \frac{P_{Out}}{\omega_m} = 3R_{rot}I_{rot}^2\left(\frac{1-s}{s}\right) = 3\left(\frac{P}{2}\right)I_{rot}^2\left(\frac{R_{rot}}{s \times \omega_{el}}\right) \approx 3\left(\frac{P}{2}\right)\frac{V_{Supply}^2}{R_{rot} \times \omega_{el}^2} \tag{8.7}$$

Here, $\tau_{Out(Load)}$ is the induction motor output (load) torque (Nm), R_{rot}, and I_{rot} are the rotor resistance and current, P is the motor number of poles, ω_{el} equal to $2\pi f_{el}$ is the supply (electrical)angular frequency, while the f_{el} is the electrical frequency, (usually 60 Hz or 50 Hz). Remember that the synchronous speed (RPM) depends on the motor number of poles and the supply frequency, as expressed by:

$$N_S = \frac{120f}{P}, \text{ RPM} \tag{8.8}$$

Example 8.5: A 460 V three-phase, 60 Hz, four-pole, 10 HP, and delta-connected induction motor operates at full load speed of 1,725 RPM, and the rotor current is 10 A. If the supply voltage fluctuates in the rage of the ±10%, calculate the current and torque ranges.

Solution: The motor synchronous speed, the actual motor angular velocity, and the motor output (load) torque are:

$$N_S = \frac{120 \times 60}{4} = 1{,}800 \text{ RPM}$$

$$\omega_m = \frac{2\pi \times 1725}{60} = 180.64 \text{ rad/s}$$

$$\tau_{Out} = \frac{P_{Out}}{\omega_m} = \frac{10 \times 746}{180.64} = 41.3 \text{ Nm}$$

From (8.7), the output torque of an induction motor is proportional to the square of the supply (line) voltage and the minimum and maximum torque when the voltage fluctuates in the ±10% range are:

$$\tau_{Out-low} = \tau_{Out}\left(\frac{0.9 \times 460}{460}\right)^2 = 41.3 \times 0.81 = 33.5 \text{ Nm}$$

$$\tau_{Out-high} = \tau_{Out}\left(\frac{1.1 \times 460}{460}\right)^2 = 41.3 \times 1.21 = 50.0 \text{ Nm}$$

The corresponding motor currents are proportional to the voltage fluctuations:

$$I_{2-low} = I_2\left(\frac{0.9 \times 460}{460}\right) = 10 \times 0.9 = 9 \text{ A}$$

$$I_{2-high} = I_2\left(\frac{1.1 \times 460}{460}\right) = 10 \times 1.1 = 11 \text{ A}$$

The electric motors, when operating is essential to control the machine speed; however, many industrial drives, fan, or pump applications, have typically constant speed requirements and hence the induction machine is ideally suited for these. On the other hand, the induction machine, especially the squirrel cage type, is rugged and has a simple construction, being also a good candidate for variable speed applications, if this can be achieved. The induction motor is operating within a few percent (typically 3%–5%) of its synchronous speed. There are several methods of controlling the speed of the induction motors, including changing the number of motor poles, power supply voltage changes, the resistance changes for wound-rotor motors, and the power supply frequency changes. The speed control of induction motor is performed by varying the slip, s, the number of poles, P or the electric supply frequency, f. The IM slip increases proportionally to the decrease in voltage squared. However, this simultaneously decreases the breakdown torque and can cause the motor to stall. This method of induction motor speed control is not recommended. Slip can also be varied by changing the rotor resistance for wound-rotor motors. This method of speed control uses external resistors in series with the rotor circuit to limit the rotor current. This method causes the speed to vary with the load and also is inefficient due to the power lost by the resistors. The ability of varying any one of the above three qualities, provides the speed control methods of an induction motor. Constant V/F method is commonly used for constant and variable speed control of induction motor. The different methods of speed control of IM can be broadly classified into scalar and vector control methods. In this paper, scalar control methods are used. The scalar methods of speed control can be classified as:

(A) *Stator (supply) voltage control*: This speed control method, a very simple and economical, consists of varying the stator voltage at constant frequency. From (8.7), by varying the stator voltage leads to the changes of the developed torque, proportional to the square of the stator voltage, meaning that the variations of the supply voltage is producing a variation of the shaft speed. Therefore, a continuous speed control may be achieved by the adjustment of the supply voltage without any alteration of the supply frequency. This method is generally used for small squirrel-cage motors where cost is an important criterion and efficiency is not. Main applications of this speed control methods are for loads, such as fans, centrifugal pumps, blowers, etc. However this method has rather limited range of speed control. However, the salient features of the voltage speed control method are not very suitable for constant torque-speed loads, poor power factor, and for low slip motors, the speed range is very limited.

(B) *Rotor resistance control*: This method is applicable only to the wound rotor induction motors. Speed variation is obtained by inserting external resistance in the rotor circuit, placed in series with the rotor windings during starting, limiting the starting current, which is several times the rated current. Depending on the size of the machine, it can draw 300% to over 900% of full-load current. Once the motor is started, the external resistance can be cut out to obtain high torque throughout the accelerating range.

(C) *Supply frequency control*: By changing the supply frequency, the synchronous speed is changed and thus the torque-speed of a three-phase induction motor can be controlled. The synchronous speed can be varied by changing the stator supply frequency. This can be achieved by using power electronic circuits called inverters which convert DC to AC of desired frequency. Depending on the type of control scheme of the inverter, the AC generated may be of variable-frequency-fixed amplitude or variable-frequency variable-amplitude type. Power electronic control achieves smooth variation of voltage and frequency of the AC output. This when fed to the machine is capable of running at a controlled speed. However, consider the equation for the induced emf in the induction machine, as expressed by:

$$E_b = 4.44 Z \Phi_m f \tag{8.9}$$

where Z is the number of the turns per phase, Φ_m is the peak flux in the air gap and f is the electric frequency. In order to change the motor speed, the frequency has to be varied. If the frequency is reduced while the voltage is kept constant, thereby requiring the amplitude of induced EMF to remain the same, flux has to increase. This is not advisable since the machine likely to enter into saturation. To avoid the saturation, the flux level must be maintained constant, implying that voltage must be reduced along with frequency. The *V/f* ratio is held constant to maintain the flux level for maximum torque capability. Actually, it is the voltage across the magnetizing branch of the exact equivalent circuit that must be maintained constant, being one determines the induced EMF. Under conditions where the stator voltage drop is negligible compared to the applied voltage, the above equation is valid. In this operation mode, the voltage across the magnetizing inductance in the equivalent circuit reduces in amplitude with reduction in frequency, so the inductive reactance, and the current through the inductance and the motor flux remains constant. The speed-torque characteristics at any frequency may be estimated as before. There is one curve for every excitation frequency considered corresponding to value of synchronous speed. The synchronous speed of the motor at rated conditions is known as the base speed. By using variable frequency control, it is possible to adjust the motor speed around the base speed. Increase in frequency increases the torque-speed relation and a decrease in frequency decreases the torque-speed relation of the motor. Notice that with this kind of control, it is possible to get a good starting torque and steady-state performance. However, under dynamic conditions, this control is insufficient. Advanced control techniques, such as field-oriented control (vector control) or direct torque control are necessary.

(D) Stator *voltage and frequency control, i.e., Volts per Hertz control*: By varying ratio of the stator voltage to frequency, the induction motor speed and torque can be controlled. If a low voltage and a low frequency are applied to the motor, the maximum torque available decreases at reduced speeds. If the ratio of voltage to frequency is kept constant, this method allows the induction motor to deliver its rated torque at speed up to its rated speed, because the motor air gap flux is reduced due to the drop in the stator impedance while motor operates at a low frequency. In short, the speed control is achieved by varying the stator voltage in such a way that the flux remains constant by simultaneously varying the supply frequency such that the ratio *V/f* remains constant.

(E) *Pole changing control*: One of factors in determining the motor synchronous speed is the number of poles, given by (8.8). The speed control method, by changing the number of poles, is an option for some of the induction motors with pole changing capabilities, the so-called Dahlander motors or pole changing motors. By using several sets of windings, several speeds can be achieved. If the induction machines have a special stator winding capable of being externally connected to form different number of pole numbers, so the synchronous speed is changed. With the slip now corresponding to the new synchronous speed, the operating motor speed is changed. This method of speed control is a stepped variation and generally restricted to two steps. If the changes in stator winding connections are made so that the air gap flux remains constant, then at any winding connection, the same maximum torque is achievable. Such winding arrangements are therefore referred to as constant-torque connections. If however such connection changes result in air gap flux changes that are inversely proportional to the synchronous speeds, then such connections are called constant-horsepower type. The maximum number of speeds available by changing the pole number is usually limited to four; the two speed motors are the most common pole changing motors. If the changes in stator winding connections are made so that the air gap flux remains constant, then at any winding connection, the same maximum torque is achievable. Such winding arrangements are therefore referred to as constant-torque connections. If such connection changes result in air gap flux changes that are inversely proportional to the synchronous speeds, then such connections are called constant-horsepower type. The number of stator poles can be changed by: multiple stator winding, method of consequent poles, and pole amplitude modulation.

Example 8.6: A four pole 60 Hz induction motor has the pole changing capabilities to six and eight poles. What are the motor synchronous speeds?

Solution: From (8.8), this motor has three synchronous speeds, one for each pole configuration:

$$N_{S-4} = \frac{120 \times 60}{4} = 1{,}800 \text{ RPM}$$

$$N_{S-6} = \frac{120 \times 60}{6} = 1{,}200 \text{ RPM}$$

$$N_{S-8} = \frac{120 \times 60}{8} = 900 \text{ RPM}$$

8.3 Starting and speed control of synchronous motors

Synchronous motors are usually employed in very large power applications, are constant-speed motors with no starting torque, so they are not self-starting machines. During one rotation cycle, the net torque of a synchronous motor is zero; therefore, the motor is not able to develop starting torque, a major drawback of the synchronous motors compared with the induction motors. Two methods are common for synchronous motor starting (a) the use of a variable-frequency power supply, and (b) start the machine as an induction motor. Starting synchronous motors as an induction is a two-step process. The first step is to accelerate the motor to near synchronous speed. In order to do that, some types of the synchronous motors are equipped with squirrel-cage type windings on the rotor for starting purposes, differing from an induction squirrel-cage motor, not being rated to carry load but only to assist in starting. The synchronous motor is started as a squirrel-cage induction motor, using any of the IM starting methods: *full voltage*, *reduced impedance*, *autotransformer*, etc. Protection, required for the stator and this stage of the starting sequence is the same as required for similar size induction motors. Once the motor reaches near synchronous speed, usually defined as about 95% speed, the DC voltage is applied to the rotor field windings, pulling the rotor into step at 100% of synchronous speed. This is the second step of the starting process. During the acceleration stage, the field winding is shorted through a field discharge resistor, which is removed from the circuit before the DC voltage is applied. A synchronous motor must be brought up to the speed closer to its synchronous speed, in order that the rotor locks with the rotating magnetic field. In fact their most important characteristic is that their output speed is very exact. Most synchronous motors used in industry are quite very large horse-power units and are running at relatively low speed ranges. For example, a synchronous motor rated at 750 HP and 360 RPM is lower in cost to install and operate than a similar squirrel-cage induction motor. These large synchronous motors are usually starting like three-phase induction motors. Then, when they approach their synchronous speed, a DC voltage is applied to the armature (rotor) to create poles of fixed polarity. These poles are attracted to the rotating magnetic poles of the stator and "lock" into the synchronous speed of the stator field. Synchronous motor speed can be controlled by both design and the AC line frequency. The second starting approach is the use of a variable-frequency power supply to bring a synchronous motor from standstill to the desired speed. The motor is started with the power supply set at very low frequency in order that the stator magnetic field is rotating at very low speed and the rotor poles can follow the stator poles, and then the power supply frequency is gradually increased and the motor is brought at the desired speed. However, the frequency converter is an expensive power conditioning unit, therefore this method is expensive.

Basically there are quite a few methods that are used to start a synchronous motor, and three of the most common starting synchronous motor methods are summarized here:

1. A very effective starting method is to reduce the speed of the stator rotating magnetic field to a low enough value that the rotor can easily accelerate and lock in with it during one half-cycle of the rotating magnetic field rotation. This is done by reducing the frequency of the applied electric power, through variable frequency power sources, by using a cyclo-converter or an inverter. This method is usually followed in the case of inverter-fed synchronous motor operating under variable speed drive applications. These methods have the advantage that they can also be used to control the synchronous motors in normal operating conditions in applications that close speed tracking is required.
2. Starting by using an external prime mover to accelerate the rotor of the unloaded synchronous motor near to its synchronous speed and then supply the rotor as well as stator. The stator is paralleled to the busbars as a synchronous generator. Of course care should be taken to ensure that the direction of rotor

rotation and that of the rotating magnetic field of the stator are the same. Since the motor is unloaded, the external prime mover (a DC motor or an induction motor) rating must be large enough to overcome the inertia and the friction of the synchronous motor. Then, the power supply to the prime mover is disconnected so that the synchronous machine will continue to operate as a motor.

3. The most common starting method of synchronous motors is to use *damper* or *amortisseur* windings if these are provided in the synchronous machine. The damper or amortisseur windings are provided in most of the large synchronous motors in order to nullify the oscillation of the rotor whenever the synchronous machine is subjected to a periodically varying load. These windings are not capable of carrying the rated motor load but are capable of starting the synchronous motor as a poly-phase induction motor. Notice that in the case of a large synchronous motor, it may be necessary to reduce the inrush current during starting by reducing voltage starting methods, as described above, for the induction motors. During the starting phase the rotor windings are short-circuited to prevent insulation damages due to high induced voltages. A side beneficial effect of this method is that the resulting current creates a magnetic field that assists the damping windings. The pull-in torque of a synchronous motor, defined as the maximum constant load that the motor is pulled into the synchronism at rated voltage and frequency when the excitation filed is applied. NEM defines standardized values of the of the pull-in torque as 100% for high-speed general purpose motors in the power range 0.746–149.2 kW (1–200 HP), 60% for large high-speed motors from 186.5 to 932.5 kW (250–1250 HP) and 30% for all low-speed synchronous motors. The speed-torque characteristics of a synchronous motor using damping windings during starting can be modified by controlling the resistance of the squirrel-cage windings. Also notice that the load and motor inertia momentum values are very important for the motor starting. High-inertia loads can damp out the rotor oscillations at the synchronization point preventing locking of the rotor into synchronization with the stator field. Also notice that for high-speed motor larger than 373 kW (500 HP), the inrush starting currents are 3–4 times higher than the rated full-load current, while for high-speed motors less than 373 kW the inrush currents are 4.5–6 times higher the rated ones. Such large inrush currents may exceed the utility specifications. However, in order to fully understand the synchronous motor operation, starting, and speed control, the developed power and torque relationships are needed:

$$P_{dvp} = \frac{3E_g V_\phi \sin\delta}{X_S} \text{ (W)} \tag{8.10a}$$

And

$$\tau_{dvp} = \frac{3E_g V_\phi \sin\delta}{\omega_m X_S} \text{ (N} \cdot \text{m)} \tag{8.10b}$$

Here, E_g is the excitation voltage per-phase voltage, V_ϕ is the per-phase terminal voltage, δ is the torque angle, ω_m is the shaft angular velocity, and X_S is the motor synchronous reactance. Maximum motor (pull-out) power and torque occur at a torque angle of 90°, assuming that the other parameters remain constant. As applied voltage and power supply frequency are basically constant, once a synchronous motor is brought to the synchronous speed, its speed remains constant for all load torques up to the pull-out torque. The synchronous motors have flat torque-speed characteristics from no-load to pull-out torque. The condition for constant power operation is set based on (8.10a) and (8.10b) as:

$$E_g \cdot \sin\delta = \text{constant}$$

If the synchronous motor field current is kept constant the internal generated voltage is proportion to the speed, which is directly determined by the power supply frequency:

$$E_g = K_1 \cdot f \tag{8.11}$$

Here K_1 is a constant, determined by the machine construction, and f is the power supply frequency. The speed of a synchronous motor can be controlled by changing the frequency of the power supply. From (8.10) and (8.11), the motor speed remains constant at any given frequency even for the load changing conditions, unless the motor loses its synchronization. The most common speed control methods for synchronous motors are the *speed control method* by *changing the frequency and voltage* with an AC–AC power converter, and the *self-controlled motor* in which the frequency is automatically adjusted by the motor speed. Former speed control method is based on the fact that the motor torque is controlled by the terminal voltage and frequency, as expressed by following relationship:

$$\tau_{dvp} = K_m \frac{V_\phi}{f} \sin\delta \tag{8.12}$$

Here, K_m is a machine constant. From this equation, a base speed is defined from the motor rated voltage and frequency. If the ratio V_ϕ/f corresponding to this base speed is maintained at lower motor speeds by changing voltage with frequency, while the maximum torque is maintained at its value of the base speed. Care must be taken if the frequency is suddenly changed or is changed at higher rates, because the motor can lose the synchronization, so this method is not suitable for motors running loads that may change suddenly. To avoid this problem, the second speed control method is used, in which the rotor position is sensed and the rotor speed is adjusted to the frequency, preventing the above issue. This scheme consists of the two controlled rectifiers, and control and feedback circuits, being quite suitable for large power synchronous motors, while the rectifiers are less expensive than the controlled inverters.

Example 8.7: A three-phase, 460 V, 105 HP, 0.85 power factor leading, 60 Hz, wye-connected synchronous motor has a negligible armature resistance and a synchronous reactance of 10 Ω. The motor speed is controlled over the range of 600–1,800 RPM. Determine the range of power supply frequency variation, and the internal generated voltage at rated condition.

Solution: The synchronous speed is 1,800 RPM, the developed power, the torque per phase, and the excitation voltage at rated conditions, in the case of wye-connected machine are:

$$P_{\max} = \frac{105 \times 746}{0.85} = 92{,}153 \text{ W}$$

$$\tau_{\max} = \frac{P_{3\phi}}{\omega_{syn}} = \frac{92{,}153}{2\pi \times 60} = 244.6 \text{ Nm}$$

$$E_g = \frac{10 \times 92{,}153}{3 \times \left(460/\sqrt{3}\right)} = 1{,}158 \text{ V}$$

Assuming the torque remains the same as at the rated based voltage and frequency, the maximum speed at is 1,800 RPM, and the minimum speed is 600 RPM, for which the frequency is:

$$f_{\min} = \frac{600 \times 4}{120} = 20 \text{ Hz}$$

8.4 AC motor protection methods

For many years, the motor protection is considered as an essential activity to safeguard the motors and their cables or connecting wiring system for damages caused by overheating. Overload, stalling, or single-phasing can result in overheating, however, most of the motor protection schemes can detect these abnormal conditions and take protective measures. Motor failure rate is conservatively estimated at a rate up to 5% per year. However, in mining, pulp and paper industries, the motor failure rate is as high as 12%. Motor failures are divided in three categories: failure due to electrical causes, about 33% of the motor failures, due to mechanical causes, about 31%, and the rest, about 36% due to adverse and harsh environmental conditions, improper maintenance, and other reasons. Every electric motor has operating limits. Overshooting these limits can damage or even destroy the motor and the systems it drives, the immediate effect being operating shutdown and losses. The electric motors can encounter electrical or mechanical incidents that may damage them. Electrical incidents include power surges, voltage drops, unbalance, and phase losses causing variations in the absorbed current, short circuits where the current can reach levels that can destroy the motor. Mechanical incidents, on the other hand, include rotor stalling, momentary or prolonged overloads increasing the current absorbed by the motor and dangerously heating its windings. The cost of these incidents can be high, including among others production loss, loss of raw materials, the cost of the repairs of the production equipment, nonquality production and delivery delays. The businesses economic necessity to be more competitive implies reducing the costs of discontinuous output and lower quality. These incidents can also have a serious impact on the people safety in direct or indirect contact with the motor.

Protection is necessary to overcome these incidents, or at least mitigate their impact and prevent them from causing damage to equipment and disturbing the power supply. It isolates the equipment from the mains power by means of a breaking device, which detects and measures electrical variations (voltage, current, etc.). Every starter motor unit or scheme should include protection against short circuits, to detect and break abnormal currents, several times greater than the rated current, as fast as possible, and protection against overloads to detect current increase up to about 10 rated current and open the power circuit before the motor heats up, damaging the insulation. These protection types are ensured by fuses, circuit breakers and overload relays or by integral devices with a range of protections. Motor failure cost is a major part of the overall repair or replacement, removal and installation costs, and to the production losses. Thermal stress potentially can cause the failure of all the major motor parts *Stator*, *Rotor*, *Bearings*, *Shaft*, and *Frame*. Insulation lifetime decreases by half if motor operating temperature exceeds thermal limit by 10 °C for any period of time. Motor protection involves protection of any abnormal conditions. The most common abnormal condition is an overload, which can produce damaging through heating effects in the motor. For this reason, overload relays are the primary means of motor protection. However, short-circuit protection is also required to minimize damage to the motor from an internal short-circuit. Other protective devices are also available, their use depending upon the size of the motor and the cost of protection vs. the cost of the motor.

Motor protection involves thermal overload, caused due to the excessive load, high ambient conditions, such as hot, blocked ventilation, power supply issues, such as voltage or current unbalance, harmonics, phase and/or ground fault, abnormal operating conditions, overvoltage and undervoltage, under-frequency operating condition, voltage and current unbalance, load loss, jamming, and jogging. A motor can run overloaded without a fault in motor or supply. A primary element of the motor protection relay is the thermal overload element, usually designed through motor thermal image modeling. This model accounts for the thermal processes, while motor is starting and running at normal load, running overloaded, and stopped. Algorithm of the thermal model integrates both stator and rotor heating into a single model. The overall result of an overvoltage condition is a decrease in load current and poor power factor. Although old motors had robust design, new motors are designed close to saturation point for better use of the core materials and increasing the *V*/*f* ratio may cause air gap flux saturation, leading to the motor heating. The overvoltage element is set at about 110% of the motors nameplate. Over the last two decades have been significant changes in motor protection devices from thermal-based protection toward electronic and microprocessor-based motor protection relays. Today, all motors with the exceptions of the smallest ones are protected by the new devices. A ground fault is a fault that creates a path for current to flow from one of the phases directly to the neutral through the earth bypassing the load. Ground faults in a motor occur (a) When its phase conductor's insulation is damaged, for example, due to voltage stress, moisture, or internal fault occurs between the conductor and ground; and (b) To limit the level of the ground fault current, impedance is connected between the supplies neutral and ground. This impedance can be in the form of a resistor or grounding transformer sized to ensure maximum ground fault current is limited.

8.4.1 Unbalanced phase motor protection

A three-phase system can be balanced or unbalanced, the last one is when the three voltages are of unequal amplitude and/or are not phase-shifted by 120° in relation to each other. Unbalance can be due to phase opening (dissymmetry fault), single-phase loads in the motor's immediate vicinity or the source itself. The result of unbalance in the voltage power supply is an increase of current for the same torque, invert component, thereby overheating the motor. The IEC 60034-26 standard has a derating chart for voltage unbalance which should be applied when the phenomenon is detected or likely in the motor power supply. The derating factor is used to oversize a motor to take into account the unbalance or to lower the motor operating current in relation to its rated current. A three-phase system is unbalanced when the three voltages are of unequal amplitude and/or are not phase-shifted by 120° in relation to each other. Unbalance can be due to phase opening (dissymmetry fault), single-phase loads in the motor's immediate vicinity or the source itself. If the three-phase voltages are unbalanced (not equal in magnitudes or phase shifts), it is causing not only unequal phase currents in the rotor and stator circuits but also the current unbalance can be up to 10 times larger than the percentage voltage unbalance. The current unbalance conditions are resulting in high I^2R losses, overheating the motor insulations, shortening its life and decreasing in the motor locked-rotor torque and break-down torque leading to improper operation and performance of the motor. Voltage unbalance can be calculated by using the following relationship:

$$\Delta V_{unbalanced} = Max\left(\frac{V_{\max} - V_{avrg}}{V_{avrg}}, \frac{V_{avrg} - V_{\min}}{V_{avrg}}\right) \tag{8.13}$$

Here, V_{max} and V_{min} are the maximum and the minimum values of the three-phase voltages, and the mean (average) voltage value is calculated from the individual phase voltages, as:

$$V_{avrg} = \frac{V_a + V_b + V_c}{3}$$

The result of unbalance in the voltage power supply is an increase of current for the same torque, invert component, thereby overheating the induction motor. The IEC 60034-26 standard has a derating chart for voltage unbalance, which should be applied when the phenomenon is detected or likely in the motor power supply. This derating factor is used to oversize a motor to take into account the unbalance or to lower the operating current of a motor in relation to its rated current. The net effect of the temperature rise due the voltage unbalance (imbalance) is that for any 10 °C temperature increase the motor life is reduced by half, the so-called *ten-degree rule*, while the temperature reduction has opposite effect. Empirical data from laboratory tests showed that the percentage motor temperature increase (%ΔT) due to the voltage unbalance is approximately 2 times the square of the percentage voltage unbalance:

$$\Delta T\% = 2 \times (\Delta V_{unbalance}\%)^2 \tag{8.14}$$

Assuming the rated load operation, the expected temperature rise, T_{UBV} due to the three-phase voltage unbalance is given by:

$$T_{UBV} = T_{rated} \cdot \left(1 + \frac{\%\Delta T}{100}\right) \tag{8.15}$$

Example 8.8: A three-phase, 460 V, 60 Hz four pole induction motor is operated at rated power form an unbalanced three-phase supply, having line-to-line voltages, 460 V, 455 V, and 445 V. The expected life time of the motor is 18 years. The approximate expected motor temperature at rated load is 40 °C. Determine the voltage unbalance, the temperature rise, and the approximate reduction of the motor life due to the voltage unbalance temperature rise.

Solution: The average voltage and the maximum voltage deviation are:

$$V_{avrg} = \frac{460 + 455 + 445}{3} = 453.33 \text{ V}$$

$$\%\Delta V = (8.33/453.33) \cdot 100 = 1.84\%$$

The temperature rise, for the operating conditions, (8.12) and (8.13) and the data assuming a service factor of one are giving:

$$\%\Delta T = 2 \times (1.84)^2 = 6.75\%$$

$$T_{UBV} = T_{rated} \cdot \left(1 + \frac{\%\Delta T}{100}\right) = 110 \cdot \left(1 + \frac{6.75}{100}\right) = 117.4\ °\text{C}$$

The temperature rise is 7.4 °C, the life reduction is then given by:

$$\text{Life reduction ratio} \approx \frac{1}{2^{\delta T/10}} = \frac{1}{2^{7.4/10}} = 0.60 \text{ or } 60\%$$

$$\text{Operation life} = 18 \text{ years} \times 0.6 = 10.75 = 11 \text{ years}$$

8.4.2 Low voltage, undervoltage, voltage drop, and break motor protection

Low-voltage motor protection typically involves overload and short-circuit protection, first one being the protection from the thermal effects of overloads. The motor input current is always larger than would normally be dictated by the output power due to losses and the motor power factor. NEC Article 430 gives typical full-load currents where a machine's actual full-load current is not known. However, for overload protection purposes the motor nameplate full-load current rating must be used, as recommended in Article 430.32 (A). NEC requirement for the overload protection is 125% of the nameplate rating for motors with service factors of 1.15 or greater and with a marked temperature rise of 40 °C or less, and 115% for all other motors. This requirement takes into account the maximum long-time setting of the overload relay but for low-voltage motors fine-tuning of the relay selection is made according to the manufacturer recommendations. Overload relays for low-voltage motors are classified as *melting alloy*, *bimetallic*, or *solid-state*. Melting alloy relays are usually hand-reset devices, whereas bimetallic relays can be either self-resetting or hand resetting. Voltage drops or breaks (a sudden voltage loss at a point in the power supply) are limited by standards to 90% of nominal voltage for half a cycle, i.e., 10 ms to 1 min, while a short break is when the voltage falls below 90% of nominal voltage for less than 3 min. A long brake is when the duration exceeds 3 min. A microvoltage drop or brake is one that lasts about a millisecond or less. Voltage variations are caused by random external phenomena (faults in the mains supply or an accidental short circuit) or phenomena related to the connection of heavy loads, such as big motors or transformers, may have strong effects on motor. When the voltage drops, the torque in an asynchronous motor (proportional to the square of the voltage) drops suddenly and causes speed reduction, which depends on the amplitude and duration of the drop, the inertia of rotating masses and the torque-speed characteristic of the driven load. If the developed motor torque drops below the resistant torque, the motor stops (stalls). Moreover, after a voltage break, the restoration causes a reacceleration inrush current, closed to the motor starting current (up to 10 times higher than the rated current). When a facility has several motors, simultaneous reacceleration can cause a voltage drop in the upstream power supply network. This prolongs the drop and can hamper reacceleration (lengthy restarting with overheating) or prevent it (driving torque below the load torque). Bimetallic relays are available as temperature-compensated or noncompensated types, and the last one has an advantage when the relay and motor are at the same ambient temperature since the relay opening time changes with the temperature in the same manner to the motor. Temperature compensated relays are designed for operation where the motor is at a constant ambient temperature but the relay is at a varying ambient temperature. While melting alloy and bimetallic overload relays are selected to suit the motor. In the case of solid-state relays the same physical relay may be used for different motor types, with the relay settings adjusted to match the protected motor. Solid-state relays can also provide phase-loss protection. Note that NEMA ICS 2-2000 classifies motor overload relays into three classes, Class 10, 20, and 30, depending upon the time delay to trip on locked-rotor current. NEMA Class 10 overload relays trip in 10 s at 6 times the relay overload rating, Class 20 relays trip in 20 s at 6 times the overload rating, while Class 30 ones trip in 30 s.

Short-circuit protection usually involves fuses or circuit breakers (motor circuit protectors). Short-circuit protection, as required by the NEC article 430-52 sets the limits for various types of motor and protective device combinations. In addition to protecting the motor, the short-circuit protection also protects the motor power supply circuit and the contactor. Motor circuit conductors, as specified in NEC

Article 430.22 (A), must be sized to have an ampacity not less than the 125% of the motor full-load current as determined from the tables in Article 430, not from the full-load current nameplate rating. The purpose is to avoid undersized cables should the motor be replaced in the future with a different make and model of the same HP rating, since the power rating does not clearly define the full-load motor current. Unlike conventional branch circuits, overcurrent protective devices in motor branch circuits are not dictating the conductor sizes. For this reason, care must be taken to insure that the motor branch circuit protection device protects the motor conductors for short-circuits. To show how the overload and short-circuit protective devices coordinate with the motor and motor cable damage curves, consider a 480 V, 300 HP squirrel-cage induction motor with a full-load current nameplate rating of 355 A. The NEC Table 430.250 full-load current rating for sizing the motor branch-circuit conductors is 361A. From NEC table 310-16, 500 kcmil cable per phase, with an ampacity of 380 A, is selected to supply the motor. A Class 10 melting-alloy overload relay, sized for this motor as manufacturer's recommendations, is selected for overload protection. A magnetic-only circuit breaker, sized at 800 A, which is in accordance with the 800% of motor full-load current from NEC Table 430.52, is used for short-circuit protection. A three-phase motor may also be damaged due to the single-phasing of the power supply. There are two basic types of single-phasing conditions. The most destructive is the opening of a transformer primary of delta-wye or wye-delta transformer or transformer bank, leading to a 115% increase in the current in two phases, while in the third one the current rises to 230% of its rated value. The second type, less severe is the secondary single-phasing. In this case, the current in one phase is zero but the currents in other two legs rise at 173% of normal. In either case the motor protection must be set based on these current levels. Keep in mind that no overcurrent device can prevent single-phasing form occurring; however, properly sized overload relays and fuses can protect motors from single-phasing damaging effects.

The overall result of an undervoltage condition is an increase in current and motor heating and a reduction in overall motor performance. The undervoltage protection element can be thought of as backup protection for the thermal overload element. In some cases, if an undervoltage condition exists, it may be desirable to trip the motor faster than thermal overload element, with the undervoltage trip usually set to 80%–90% of nameplate unless otherwise stated on the motor data sheets. Motors that are connected to the same source/bus may experience a temporary undervoltage, when one of motors starts. To override this temporary voltage sags, a time delay set-point should be set greater than the motor starting time. The short circuit element provides protection for excessively high overcurrent faults, noticing that phase-to-phase and phase-to-ground faults are common types of short circuits. When a motor starts, the starting current, which is typically 6 times the full-load (rated) current and has asymmetrical components. These asymmetrical currents may cause one phase to see as much as 1.7 times the RMS starting current. In order to avoid nuisance tripping during starting, set the short circuit protection pick up to a value at least 1.7 times the maximum expected symmetrical starting current of motor. The breaker or contactor must have an interrupting capacity equal to or greater than the maximum available fault current or let an upstream protective device interrupt fault current. A thermal protector, automatic or manual, mounted in the end frame or on a winding, is designed to prevent a motor from getting too hot, causing possible fire or damage to the motor. Protectors are generally current and temperature-sensitive. Some motors have no inherent protector but they should have protection provided in the overall system's design for safety. Never bypass a protector because of nuisance tripping. This is generally an indication of some other problem, such as overloading or lack of proper ventilation.

8.5 Single-phase motor control

There are probably more single-phase AC induction motors in use today than the total of all the other types put together. It is logical that the least expensive, lowest maintenance type of AC motor should be used most often. The single-phase AC induction motor fits that description. Unlike poly-phase induction motors, the stator field in the single-phase motor does not rotate. Instead, it simply alternates polarity between poles as the AC voltage polarity. Voltage is induced in the rotor as a result of magnetic induction, and a magnetic field is produced around the rotor. This field will always be in opposition to the stator field. The interaction between the rotor and stator fields is not producing rotation. There are several types of single-phase induction motors; however, they are identical except for the starting methods. Once they are up to operating speed, all single-phase induction motors operate the same. One type of induction motor, which incorporates a starting device, is called a split-phase induction motor. Split-phase motors are designed to use inductance, capacitance, or resistance to develop a starting torque, based on the behavior such elements in alternating current. The first type of split-phase induction motor is the capacitor start type. The stator consists of the main winding and a starting winding (auxiliary). The starting winding is connected in parallel with the main winding and is placed physically at right angles to it. A 90° electrical phase difference between the two windings is obtained by connecting the auxiliary winding in series with a capacitor and starting switch. When the motor is first energized, the starting switch is closed. This places the capacitor in series with the auxiliary winding. The capacitor is of such value that the auxiliary circuit is effectively a resistive-capacitive circuit, through the capacitive reactance, X_C. In this circuit, the current leads the line voltage by about 45° (because X_C is about equals R). The main winding has enough resistance-inductance, X_L to cause the current to lag the line voltage by about 45° (because X_L is about equals R). The currents in each winding are therefore 90° out of phase, so the magnetic fields that are generated. The effect is that the two windings act like a two-phase stator and produce the rotating field required to start the motor. When nearly full speed is reached, the centrifugal device, the starting switch is cutting out the starting winding. The motor then runs as a plain single-phase induction motor. Since the auxiliary winding is only a light winding, the motor does not develop sufficient torque to start heavy loads. Split-phase motors, therefore, come only in low power range and small sizes.

Another type of split-phase induction motor is the resistance-start motor. This motor also has a starting winding figure in addition to the main winding. It is switched in and out of the circuit just as it was in the capacitor-start motor. The starting winding is positioned at right angles to the main winding. The electrical phase-shift between the currents in the two windings is obtained by making the impedance of the windings unequal. The main winding has a high inductance and a low resistance. The current, therefore, lags the voltage by a large angle. The starting winding is designed to have a fairly low inductance and a high resistance. Here the current lags the voltage by a smaller angle. For example, suppose the current in the main winding lags the voltage by about 70° or so. The current in the auxiliary winding lags the voltage by 40°. The currents are, therefore, out of phase by 30°. The magnetic fields are out of phase by the same amount. Although the ideal angular phase difference is 90° for maximum starting torque, the 30° phase difference still generates a rotating field. This supplies enough torque to start the motor. When the motor comes up to speed, a speed-controlled switch disconnects the starting winding from the line, and the motor continues to run as an induction motor. The starting torque is not as great as it is in the capacitor-start.

The shaded-pole induction motor is another single-phase motor. It uses a unique method to start the rotor turning. The effect of a moving magnetic field is produced by constructing the stator in a special way. This motor has projecting pole pieces just like some DC motors. In addition, portions of the pole piece surfaces are surrounded by a copper strap called a shading coil. A pole piece with the strap in place is shown in figure below. The strap causes the field to move back and forth across the face of the pole piece.

Note the numbered sequence and points on the magnetization curve in the figure. As the alternating stator field starts increasing from zero, the lines of force expand across the face of the pole piece and cut through the strap, so a voltage is induced in the strap. The current that results generates a field that opposes the cutting action (and decreases the strength) of the main field. This produces the following actions: As the field increases from zero to a maximum at 90°, a large portion of the magnetic lines of force are concentrated in the unshaded portion of the first pole. At 90° the field reaches its maximum value. Since the lines of force have stopped expanding, no EMF is induced in the strap, and no opposing magnetic field is generated. As a result, the main field is uniformly distributed across the second pole. From 90° to 180°, the main field starts decreasing or collapsing inward. The field generated in the strap opposes the collapsing field. The effect is to concentrate the lines of force in the shaded portion of the third pole face. You can see that from 0° to 180°, the main field has shifted across the pole face from the unshaded to the shaded portion. From 180° to 360°, the main field goes through the same change as it did from 0° to 180°; however, it is now in the opposite direction. The direction of the field does not affect the way the shaded pole works. The motion of the field is the same during the second half-cycle as it was during the first half of the cycle. The motion of the field back and forth between shaded and unshaded portions produces a weak torque to start the motor. Because of the weak starting torque, shaded-pole motors are built only in small sizes. They drive such devices as fans, clocks, blowers, and electric razors.

8.6 DC motor protection and control methods

There are different types of DC motors, based on the internal structure and configurations, the way that field and armature circuits are connected or not, or placement of the armature and field windings on the stator and the rotor but they all work on the same principles. It is interesting to note that the same DC machine can be used either as a motor or as a generator, only by reversing the terminal connections and input power. The torque developed in a DC motor is determined by the armature current and the total magnetic flux (the one of the field circuit and the armature reaction) and is expressed as:

$$\tau_m = K \cdot \phi \cdot I_A \tag{8.16}$$

where ϕ is the magnetic flux per pole (Wb), K is a constant depending on coil geometry and the DC motor internal structure, the so-called machine constant, and I_A is the armature current. The mechanical power generated, by any motor is the product of the machine torque and the angular velocity (mechanical velocity of rotation), ω_m, and in the case of a DC motor is given by:

$$P_{Out} = \tau_{Out} \times \omega_m = K\phi I_A \times \omega_m \tag{8.17}$$

In most applications, DC motors are used for driving mechanical loads. Some applications require that the speed remain constant as the load on the motor changes. In some applications, the speed is required to be controlled over a wide range. It is therefore important to study the relationship between torque and speed of the motor. In order to effectively use a DC motor for an application, it is necessary to understand its characteristic curves. For every motor, there is a specific Torque/Speed curve and Power curve. The relation between torque and speed is important in choosing a DC motor for a particular application or to select the optimum motor control and protection scheme or method. Many applications require the speed of a motor to be varied over a wide range. One of the most attractive features of DC motors in comparison with AC motors is the ease with which their speed can be varied. The relationship between the back EMF for a separately excited DC motor can be rearranged as:

$$\omega_m = \frac{V_T - E_b}{K \cdot \phi} \tag{8.18}$$

From this equation, it is evident that the speed of a separately DC motor can be varied by using any of the following methods: *armature voltage control*, by *varying the terminal voltage*, V_T; *magnetic field control*, by *varying the magnetic field* through *the control (variation) of the field current*, in the field circuit; and through the *armature resistance control*, by *varying the armature resistance*. The armature voltage control method is usually applicable to separately excited DC motors. In this method of speed control, the motor armature resistance, R_a and magnetic flux, ϕ are kept constant. In normal operation, the drop across the armature resistance is very small compared to the back (internal generated) voltage, E_b. Therefore, the

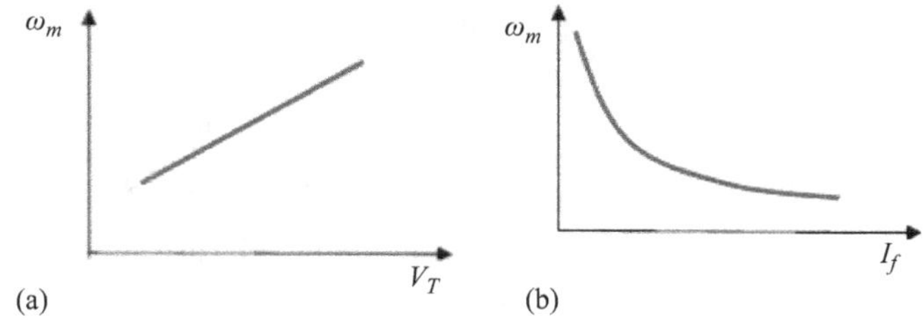

Figure 8.6 (a) Variation of speed with applied voltage for separately excited DC motor; and (b) variation of speed with field current

back EMF value is very close to the terminal voltage value $E_b \approx V_T$, so the motor angular speed is:

$$\omega_m \cong \frac{V_T}{K\phi} \tag{8.19}$$

From this equation, if the magnetic flux is kept constant, the separately DC motor speed changes linearly with the terminal (supply) voltage V_T, as the terminal voltage is increased, the speed increases and vice versa. The relationship between speed and applied voltage is shown in Figure 8.6(a). This method provides smooth variation of speed control for separately excited DC machines.

For the shunt series and compound DC motors is basically proportional with the back (internal generated) electromotive force and invers proportional to the magnetic flux, expressed by:

$$N_m(RPM) = K\frac{V_T - R \cdot I_A}{\phi} \tag{8.20a}$$

Or

$$\omega_m(\text{rad/s}) = k\frac{V_T - R \cdot I_A}{\phi} \tag{8.20b}$$

Here R is equal to the armature reaction, R_a in the case of a shunt DC machine and is equal to the $R_a + R_{SE}$, in the case of a series DC motor, K and k are machine constant, adjusted for this speed type.

Example 8.9: A 400 V DC shunt motor has an armature current of 20 A when running at 1,000 RPM against full load torque. The armature resistance is 0.35 Ω. What is the value of the control resistance that must be inserted in series with the armature resistance to reduce the motor speed to 500 RPM at the same torque? Assume the flux to remain constant throughout and neglect brush contact drop.

Solution: The motor back EMF is:

$$E_{b-initial} = V_T - R_a \cdot I_a = 400 - 0.5 \times 40 = 380 \text{ V}$$

Assuming the magnetic flux and torque remain the same, the armature current is remaining at the same value and from (8.15), the new back EMF is then:

$$\frac{N_{m-initial}}{N_{m-new}} = \frac{1000}{500} = \frac{V_T - R_a \cdot I_A}{V_T - (R_{Control} + R_a)I_A} = \frac{400 - 0.35 \times 20}{400 - (R_{Control} + 0.35) \cdot 20}$$

And solving for the control resistance in series with the armature resistance, we are getting:

$$R_{Control} = \frac{2 \times 400 - 40 - 393}{80} = \frac{367}{80} = 4.588\ \Omega$$

8.7 Electric drives

Today, industry applications make high demands of drive technology with regard to dynamic performances, speed and positioning accuracy, control range, torque stability, and overload capacity. Drives are employed in any system requiring motion control, such as transportation equipment and systems, fans, robots, pumps, tools, packaging equipment, etc. Prime movers are required in drives to provide motion and energy, coming from various sources, such as diesel engines, internal combustion engines, hydraulic motors, or electrical motors. Drives that use electrical motors as the prime movers are known as electrical drives. It is estimated that about half of electricity generated is converted to mechanical energy by the electric drives. Design and analysis of all electric drive systems requires not only knowledge of dynamic properties of electrical motors but also a good understanding of the way these motors interact with power electronic converters and loads. These power converters are used to control motor currents or voltages. Control of electrical motors always was in the highlight of designers of mechanisms, machines, industrial, and transport equipment. Any mechanism is usually complex, and quite often, its behavior is vague, and its reaction on influences and disturbances is unforeseen. To a considerable degree, this concerns the electric drives too. Advances in electric motor technology and variable frequency drive (VFD) hardware have enabled more sophisticated control schemes and higher power and speed electric motors. Due to these advances and other economic or environmental factors, electric drives are becoming more common for centrifugal and reciprocating compressors. As with any new technology, the electric motor drive system (which includes the adjustable speed drive, gearbox, bearings, and frame) offers some advantages at the cost of some limitations compared to conventional technology.

The appropriate drive technology and the selection of the primary components is a function of the load, such as compressor, pump, or equipment operation. An electric drive is the electromechanical system that converts electrical energy to mechanical energy of the driven machine. Figure 8.7 gives the functional diagram of the electric drive, which includes a motor (or several ones), a mechanical transmission (gear, gearbox), an optional power converter, and a control system

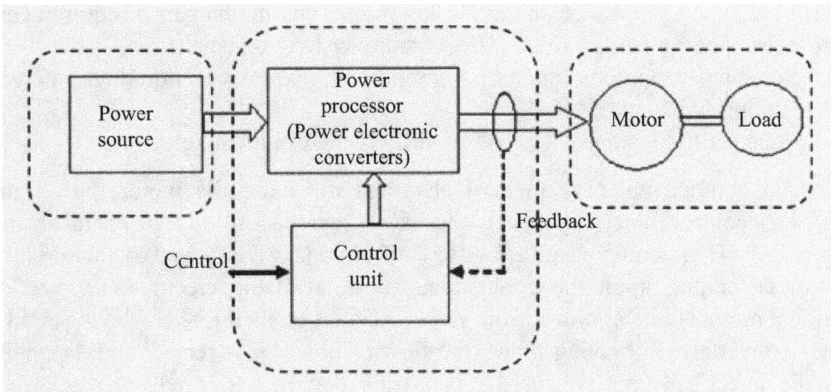

Figure 8.7 Functional diagram of electric drive

(controller). The power converter transforms electrical energy of the grid to motor supply energy in response to the set-point control. The motor is an electromechanical converter, which initially converts input electrical energy into mechanical work on the motor shaft. The gear transforms mechanical energy to the load in accordance with the requirements of the driven machine. The controller (regulator) compares the set-point with outputs and disturbance, and generates the references on its inputs. The part of electric drive, which involves the mechanical transmission and the motor rotor, is called the mechanical system.

Most of the applications are exploiting the general-purpose electric drives of low and mean accuracy which constitutes approximately 80% of the word driving complexes. They are usually presented by the mains-operated open-end mechanisms consisting of the motor, transmission, and control system, which also provide protection. They have neither the power converter nor the feedbacks. The accurate variable-speed electric drives that comprise the rest drive are the converter-operated closed-loop systems built on the microcontrollers. They present the high performance drives of the very broad speed range and positioning requirements. Developments in power electronics and microelectronics resulted in an unprecedented growth of adjustable speed drives offering a wide range of advantages from process performance improvement to comfort and power savings. Nowadays, electric drives can be found nearly everywhere, in heating, ventilation and air conditioning, compressors, washing machines, elevators, cranes, water pumping stations and wastewater processing plants, conveyors and monorails, centrifuges, agitators, and this list could continue on and on. Electric drives use approximately 70% of generated electrical energy. There are several advantages of electrical drives:

- Flexible control characteristics, when power electronic converters are used and the dynamic and steady-state motor characteristics are controlled by the applied voltage or current.
- Available in wide range of speed, torque, and power; they are available in mW up to MW range.

- High efficiency, cleaner operation, low noise, and maintenance requirements.
- Electric energy is easy to be transported over long distances.
- Adaptable to various operating conditions: explosive, liquid submerged, mounting types, etc.
- Can be started instantly and can be fully loaded immediately.

The main components of a modern electrical drive are the motor, power unit processor, control unit (regulator), electrical source, and the mechanical load (Figure 8.7). These components are different from one drive system to another drive system, depending upon the applications, cost, available electrical source, etc. Electrical motors transfer power from electrical sources, therefore can be regarded as energy converters. In braking mode, the flow of power is reversed, and depending upon the power converter type used, it is possible that the power to be fed back to the sources in regenerative braking, than dissipated as heat in dynamic braking. There are several types of motors used in electric drives, while the choice of the type to be used depends on applications, cost, environmental factors, and on the type of power sources available. Power processor units are used because, typically, electrical sources are uncontrollable. For instance, if it is an AC source, the frequency and magnitude are fixed (from the utility company), or maybe both are varying randomly (e.g., wind generator). It is therefore necessary to provide an interface between the electric sources, controlling the power flow, and hence the motor speed and/or torque can be systematically regulated and controlled. With controllable sources, the motor can be reversed, brake or can even be operated with variable speed. Conventional power processor (nonpower electronics ones) used, for example, variable impedance or relays, to shape the supplied voltage or current to the motor are not flexible, inefficient, and have limited control capability. In modern electric drive systems, power electronic converters are used to shape the desired voltage or current that is supplied to the motor. The power converters are commonly used to convert one form of electrical power to another (e.g., AC to DC, DC to AC). The main advantage of using power electronic converters is their higher efficiency. With power electronic converters, characteristic of the motors can be changed and adapted to the load requirements. Power electronic converters have several advantages over classical methods, such as: higher efficiencies (since ideally no losses occur in power electronic converters); flexible voltage and current can be shaped by simply controlling the switching functions of the power converter; compact (smaller, compact, and higher ratings solid-state power electronic devices are continuously being developed); and the prices are getting cheaper and cheaper.

Power electronic converters are typically consists of power semiconductor devices and passive elements, such as inductors and/or capacitors. The losses in power semiconductor devices are minimized since they are operated in switching modes. The conversion of electrical power from one form to another can be performed with either single-stage conversion or multiple-stage conversion, depending on the application requirements, such as control bandwidth, output voltage or current ripples, cost, etc. Control unit is used to generate the switching signals to the power switches of the power converters. Most of the time, the control unit is electrically isolated from the power converters for the safety, malfunction in power circuit that may damage control circuit, and power quality reasons. Electrical sources or power supplies provide the energy to the electrical motors. Power sources can be of AC or DC and usually are uncontrollable, i.e., their magnitudes and frequencies are either fixed or varying, depending on the sources of energy, such as battery, power utility, fuel cell, etc. Fixed frequency and fixed magnitude AC source is normally obtained from power utility and can be either three-phase or single-phase; three-phase sources are normally for high power applications. In order to efficiently control the motor, regardless of whether it is a DC or AC source (and depending on the motor), and it is regulated using power electronics converters before being fed to the motor. Power electronics converters typically have poor input power factor and it is sometimes necessary for the power converters to be operated with high power factor; if this is needed, power factor correction circuit has to be introduced. Sensors for voltage, current, speed, or torque are required for closed-loop operation and protections in electrical drive systems. For high-performance drive system, the speed is obtained from high-resolution speed encoders or resolvers. The terms "sensorless drive" is normally referred to a drive system that does not need a mechanical speed sensor but rather the speed is estimated using motor terminal variables, i.e., voltages and current.

8.8 Summary

Induction and synchronous electrical motors are valuable assets to today's industrial and large commercial facilities. Induction motors are now the preferred choice for industrial motors due to their rugged construction, absence of brushes (which are required in most DC motors) and the ability to control the speed of the motor. We have discussed problems and issues related to the starting of induction motors, synchronous, and DC machines. We also examined various schemes enabling to control the speed of electric motors. A chapter section is dedicated to the motor protection due to the phase imbalance, single-phasing, overloading, or undervoltage conditions and their effect on the motor components and life. The temperature rise of motor dictates its life and operation characteristics. When applied, thermal protection can prevent loss of motor life. Additional protection elements, such as overvoltage, undervoltage, unbalance, ground fault, differential, short circuit, and stator, the thermal model protection and provide complete motor protection. Harsh conformal coating of motor protection relays should be considered to avoid the environmental effects of harsh gaseous chemicals. Most of the motor failure contributors and failed motor components are related to motor overheating. Thermal stress potentially can cause the failure of all the major motor parts: Stator, Rotor, Bearings, Shaft, and Frame. A motor can run overloaded without a fault in motor or supply. A primary motor protective element of the motor protection relay is the thermal overload element and this is accomplished through motor thermal image modeling. This model must account for thermal process in the motor while motor is starting and running at normal load (rated conditions), running overloaded and stopped. Algorithm of the thermal model integrates both stator and rotor heating into a single model.

References and further readings

[1] G. McPherson, An Introduction to Electrical Machines, Wiley, NY, 1981.
[2] R.H. Engelman, Static and Rotating Electromagnetic Devices, M. Dekker, NY, 1982.
[3] D.R. Patrick and S.W. Fardo, Rotating Electrical machines and Power Systems, Prentice-Hall, NJ, 1985.
[4] P. Schavemaker and L. van der Sluis, Electrical Power System Essentials, John Wiley and Sons, Chichester, UK, 2008.
[5] M.A. Pai, Power Circuits and Electro-mechanics, Stipes Publishing, LLC, Champaign, IL, 2003.
[6] Safety Standard and Guide for Selection, Installation, and Use of Electric Motors and Generators, NEMA Standards Publication MG 2-2001.
[7] Industrial Controls and System: General Requirements, NEMA Standards Publication ICS 1-2000.
[8] Industrial Control and Systems: Controllers, Contactors, and Overload Relays Rated 600 Volts, NEMA Standards Publication ICS 2-2000.
[9] The National Electrical Code, NFPA 70, The National Fire Protection Association, Inc., 2005 Edition. Industrial Control and Systems: Medium Voltage Controllers Rated 2001 to 7200V AC.
[10] P.C. Sen, Principles of Electric Machines and Power Electronics, John Wiley & Sons, New York, 1989.
[11] R.A. Pearman, Electrical Machinery and Transformer Technology, Harcourt College, Publisher, San Diego, CA, 1994.
[12] J.R. Cogdell, Foundations of Electrical Power, Prentice Hall, Upper-Saddle River, NJ, 1999.
[13] C.I. Hubert, Electric Machines—Theory, Operation, Applications, Adjustment and Control (2nd ed.), Prentice Hall, New York, 2002.
[14] B.S. Guru and H.R. Hiziroglu, Electric Machinery and Transformers, Oxford University Press, New York, 2001.
[15] S.J. Chapman, Electrical Machinery and Power System Fundamentals (2nd ed.), McGraw Hill, New York, 2002.
[16] M.E. el-Hawary, Principles of Electric Machines with Power Electronic Applications, Wiley-IEEE Press, NJ, 2002.
[17] A.E. Fitzgerald and C. Kingsley, Electric Machinery (6th ed.), McGraw Hill, USA, 2003.
[18] M.E. el-Hawary, Introduction to Power Systems, Wiley-IEEE Press, NJ, 2008.
[19] M.A. El-Sharkawi, Electric Energy: An Introduction (2nd ed.), CRC Press, Boca Raton, FL, 2009.
[20] J.L. Kirtley, Electric Power Principles: Sources, Conversion, Distribution and Use, Wiley, Chichester, UK, 2010.
[21] T. Gonen, Electric Machines with MATLAB (2nd ed.), CRC Press, Boca Raton, FL, 2012.

[22] P. Sen, Principles of Electric Machines and Power Electronics (3rd ed.), Wiley, USA, 2014.
[23] S.L. Herman, Electrical Transformers and Rotating Machines (4th ed.), Cengage Learning, UK, 2017.
[24] M.A. El-Sharkawi, Fundamentals of Electric Drives, Cengage Learning, Pacific Grove, CA, 2000.

Questions and problems

1. How does the developed induction motor torque vary with the applied voltage?
2. Why do we need a starter for starting a three-phase induction motor?
3. What is the inrush current and what causes it?
4. Why are three-phase induction motors, very popular as drives for industrial applications?
5. What are the various types of three-phase induction motors as per the rotor construction?
6. Connection diagrams are also called wiring diagrams. It is true or false? Explain.
7. Explain the connections and types of the DC motors.
8. Why it is needed to use reduced-voltage starting with induction motors? Draw a neat diagram showing the connections of three-phase induction motor with star (wye)-delta starter. Explain how the above starter reduces the starting current.
9. Draw the diagram of an auto-transformer starter used for three-phase induction motor and explain its operation.
10. Why single-phase induction motors are not self-starting?
11. What are the various types of single-phase induction motors?
12. What factors determine if the full-voltage (DOL) can be employed?
13. Explain the various techniques used for speed control of three-phase induction motor.
14. What are the unbalanced line voltages adverse effects on induction motors?
15. What is an electric drive? Draw the block diagram and explain the roles of its main components.
16. What are the reasons for the expansion of the adjustable-speed drive market?
17. In electrical drives, the component which is used to control, process, and modulate the power from source to the motor?
18. Visit your local machine-tool shop or small industrial facility and make a list of various electric drive types and applications.
19. Explain the various techniques used for speed control of three-phase induction motor.
20. Explain rotor resistance speed control of three-phase induction motor.
21. Explain the starting methods of poly-phase induction motors.

22. Many types of overload motor protection circuits or schemes operate on the relationship between heat and current. It is true or false? Explain.
23. A three-phase 440 V, six pole, 50 Hz squirrel-cage induction motor is running at a slip of 5%. Calculate the speed of stator magnetic field to rotor magnetic field and speed of rotor with respect of stator magnetic field.
24. If an auto transformer is used for reduced voltage starting of an induction motor to provide 1.5 per unit starting torque, estimate the auto transformer ratio (%).
25. What is the starting current of a 20 HP, three-phase induction motor has a nameplate code letter H?
26. For a three-phase, 75 kW, 50 Hz, six pole induction motor having two possible connections, a power factor of 0.87, a rated efficiency of 0.93, a shaft speed of 975 RPM and rotor resistance of 0.035 Ω stator winding (per-phase) resistance, calculate the nominal stator current (line) for star and delta connections of stator winding, apparent nominal power (power drawn by the stator from the line), active and reactive power absorbed from the mains for nominal load, nominal torque and nominal slip, and iron core losses.
27. For the induction motor of the previous problem calculate the values of the starting torque and currents for direct switch-on starting (at nominal conditions of the power supply), and the starting torque and current for Y/Δ starting.
28. A 2.35 kV three-phase, four-pole, wye-connected synchronous motor has a negligible armature resistance and asynchronous reactance of 2.65 Ω. The motor developed power is 890 kW at rated line voltage and the internal generated induced voltage is 2.50 kV. Calculate the torque angle, the line current, the synchronous speed, and the motor power factor.
29. A 220 V DC shunt motor has a rotation speed of 1,200 RPM and an armature current of 7.5 A at no load. Given that the armature resistance is 0.5 Ω, determine the motor speed when the armature current is 40 A at full load.
30. A 250 V, DC series motor runs at 500 RPM when taking a line current of 30 A. The armature resistance is 0.25 Ω and the of the series field resistance is 0.8 Ω. At what speed this motor run when developing the same torque when armature diverter of 10 Ω is used? Assume a straight line magnetization curve.
31. A three-phase, 6 HP, 208 V, 60 Hz, four-pole synchronous motor has a negligible synchronous resistance and a 7.54 Ω synchronous reactance. Determine the based speed in RPM and the ratio V_{ϕ}/f.
32. A DC shunt motor takes an armature current of 30 A from a 260 V supply. If the armature circuit resistance is 0.75 Ω. For reducing the speed by 50%, calculate the resistance required in the series, with the armature resistance, if (a) the load torque is constant, and (b) the load torque is proportional to the square of the shaft speed.

Chapter 9

Wind and solar energy

Objectives and abstract

In this chapter, we are focusing on the understanding of the basic characteristics of the Sun and the solar radiation, solar energy conversion, wind velocity, wind power, and wind energy conversion systems, the methods to estimate, analyze, and assess the solar or wind energy resource potential. The solar radiation has directional characteristics that are defined by a set of angles that determine the angle of incidence of the radiation on a surface. After completing this chapter, the readers are able to compute these angles and to estimate the available solar radiation incident on horizontal and tilted surfaces. Wind regime and wind characteristics are influenced by synoptic circulation, mesoscale dynamics, being strongly shaped by the local circulation, topography, and conditions. The most important characteristics of wind are its variability and intermittency on a broad range of spatio-temporal scales. The assessment of wind energy potential, design, or operation of wind energy conversion systems requires in-depth knowledge of wind regime and characteristics. In this chapter, we have also included those topics that are based on the extraterrestrial radiation and the geometry of the Earth and Sun. Knowledge about the effects of the atmosphere on the solar radiation, measurement techniques, direct, diffuse, and global radiation are also presented and discussed. Similar topics, such as wind velocity statistics, wind velocity measurements are included and discussed in this chapter. After successfully completing this chapter, the readers or students have a good understating, and become familiar with solar and wind energy system parameters, characteristics, principles of operation, performances, and estimation methods. They also are able to analyze and perform basic calculations and design of wind energy and/or solar energy conversion systems, estimates and assess wind or solar energy potential, select appropriate systems and/or components for a specific application.

9.1 Introduction

Sun, our only primary source of energy, emits continuous energy as electromagnetic radiation at an extremely large and relatively constant rate. The rate at which this energy is emitted is equal to the energy emitted from a blackbody at a temperature of about 6,000 K (10,340 °F). If we are able to harvest the energy coming from just

10 hectares (25 acres) of the Sun surface, it will be enough to supply the world current energy demands. However, this cannot be done due to three reasons. First, the Earth is displaced from the Sun, the solar energy decreases with inverse of the distance squared, only a small fraction of the energy leaving the Sun surface reaches an equal area on the Earth. Second, the Earth rotates about its polar axis, so any collection device located on the Earth receives the Sun's radiant energy for only about one-half of each day. The third and least predictable factor is the atmosphere conditions, accounting for about 30% reductions in the Sun's energy reaching the ground. However, the weather conditions can stop all but a minimal amount of solar radiation from reaching the Earth's surface for many days in a row. The rate at which solar energy reaches a unit area at the Earth is called the *solar irradiance* or *insolation* (W/m^2). Solar irradiance is in fact a power density of the solar radiation and is varying over time. The maximum solar irradiance value is used in the solar energy system design to determine the peak rate of energy input into the system. If storage is included in a system design, the designer also needs to know the variation of solar irradiance over time in order to optimize the system design. The use of solar and wind energy for lighting, heating, or as mechanical power has a long history.

The use of wind energy has its roots in antiquity, and for long time, it was the major source of power for pumping water, grinding grain, or long distance transportation (sailing ships). The farm's windmills were instrumental in the Great Plains settlement, during the nineteenth century. Among the wind energy advantages are renewable, ubiquitous, and does not require water for the generation of electricity. The disadvantages are variable and low power density, which means high initial investment costs. Wind energy is a converted form of solar energy, produced by the nuclear fusion of hydrogen (H) into helium (He) into the Sun's core. The (H $\rightarrow$ He) fusion process creates heat and electromagnetic radiation streams out into the space from the Sun. Even if only a small portion of the solar radiation is intercepted by the Earth, it can provide all our energy needs. Wind turbines convert the wind kinetic energy into mechanical and eventually into electrical energy that can be used for a variety of tasks. Regardless the task, wind offers an inexpensive, clean and reliable form of mechanical power. It represents an important energy source of new power generation trends and an important player in the energy market. As a leading renewable energy technology, wind power's technical maturity and speed of deployment is acknowledged, along with the fact that there is no practical upper limit to the percentage of wind that can be integrated into the electricity system. It is estimated that the total solar power received by the Earth is approximately 1.8×10^{11} MW. Of this solar input, only 2% (i.e., 3.6×10^9 MW) is converted into wind energy and about 35% of wind energy is dissipated within 1,000 m of the Earth's surface. Therefore, the available wind power that can be converted into other forms of energy is approximately 1.26×10^9 MW, value that represents 20 times the rate of the present global energy consumption, meaning that wind energy in principle could meet entire energy needs of the world. Worldwide development of wind energy expanded rapidly starting in the early 1990s. The average annual growth rate from 1994 to 2015 of the world installed capacity of wind power has been over 35%, making the wind industry one of the fastest growing in the field. Unlike the last surge in wind power development during 1970s which was due mainly to the oil embargo of the OPEC countries, the current wind energy development is driven by many forces that make it favorable. These include its tremendous environmental, social and economic benefits as well as its technological maturity, the deregulation of electricity markets, public support and government incentives. Even among other applications of renewable energy technologies, power generation through wind has an edge because of its technological maturity, good infrastructure, and relative generated energy cost competitiveness.

The designer of solar energy collection systems is also interested in knowing how much solar energy has fallen on a collector over a period of time, such as a day, week, or year. This summation is called *solar radiation or irradiation.* The common solar radiation units are joules per square meter (J/m^2), more often in watt-hours per square meter (Wh/m^2). *Solar radiation* is simply the integration or summation of *solar irradiance* over a time period. For system design optimization studies, it is considered better to use actual recorded weather databases. The system designer must know how much solar irradiance is available in order to predict the rate of energy that will be incident on a solar collector aperture. To do this, the position of the Sun relative to a collector that is not parallel to the surface of the Earth must be found. Combining the amount of solar irradiance falling on the collector, with the orientation of the collector relative to the Sun, the designer then knows the rate of solar energy being input into that collector.

There are different solar energy system types that are converting the solar resource into a useful form of energy. A block diagram showing three of the most basic renewable system types is shown as Figure 9.1. In the top diagram, the solar

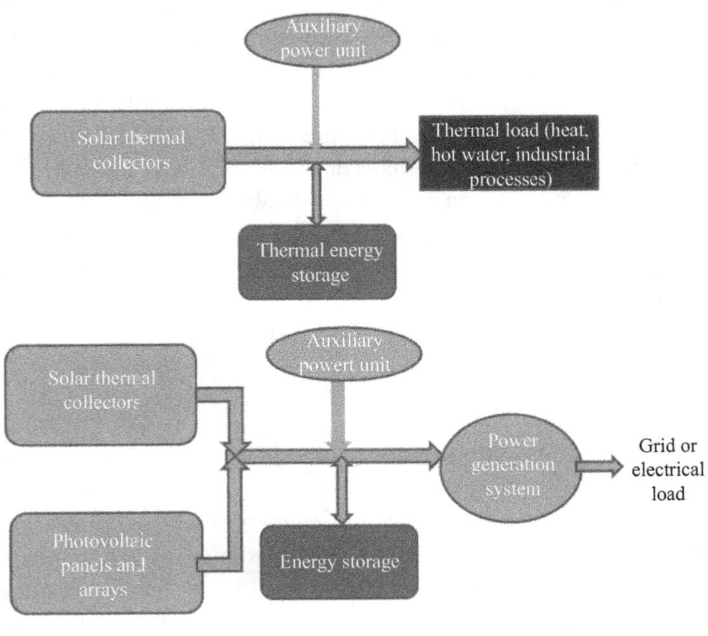

Figure 9.1 Diagram of basic solar energy conversion systems

radiation is converted into heat, supplied directly to a demand for thermal energy (thermal load), such as house heating, hot water heating, or heat for industrial processes. Such system may not include thermal storage, and usually include an auxiliary energy source, in order to meet the demand during long periods with no sunshine. If the demand (load) to be met is an electrical load, there are two common methods of converting solar energy into electricity. One method consist of collecting solar energy as heat and converting it into electricity using a typical power plant or engine, while the other method consists of using photovoltaics to convert solar energy directly into electricity, the bottom diagram of Figure 9.1, while in the middle there is a wind energy conversion system. If the solar energy conversion systems are connected to a large electrical grid, no storage or auxiliary energy supply is needed. If the solar energy conversion system is operating in standalone mode, storage and auxiliary energy supply are usually incorporated. If the thermal route is chosen, storage of heat rather than electricity is used to extend the system operating time. Auxiliary energy may either be supplied as heat before the power conversion system or as electricity after it. If the photovoltaic option is chosen, extra electricity may be stored, usually in batteries, thereby extending the operating time of the system. For auxiliary power, an external electricity source is the only choice for photovoltaic systems.

Both wind and solar energy are highly intermittent electricity generation sources. Time intervals within which fluctuations occur span multiple temporal scales, from seconds to years. These fluctuations can be subdivided into periodic fluctuations (diurnal or annual fluctuations) and nonperiodic fluctuations related to the weather change. Wind and solar energy are complementary to each other in time sequence and regions. In the summer, sunlight is intensive and the sunshine duration is long but there is less wind. In the winter, when less sunlight is available, wind becomes strong. During a day, the sunshine is strong while wind is weak. After sunset, the wind is strengthened due to large temperature changes near the ground. It has been reported that the effects of complementarity are more dramatic in certain periods and locations in many areas of the globe. Because the major operating time for wind and solar systems occurs at different periods of time, wind–solar hybrid power systems can ensure the reliability of electricity supply. The applications of wind–solar hybrid systems ranges extensively from residential houses to municipal and industrial facilities, either grid-connected or standalone configurations.

9.2 Wind energy

Wind energy is a special form of kinetic energy in air as it flows. Wind energy can be either converted into electrical energy by power converting machines or directly used for pumping water, sailing ships, or grinding grains. Wind energy is not a constant source of energy. It varies continuously and gives energy in sudden bursts. About 50% of the entire energy is given out in just 15% of the operating time. Wind strengths vary and thus cannot guarantee continuous power. It is best used in the context of a system that has significant reserve capacity, such as hydro-power, compressed air storage, or reserve load, such as a microgrid, desalination plant, to mitigate the economic effects of the resource variability. Electricity generation from wind can be economically achieved only where a significant wind resource exists. Wind power is one of the most potential and techno-economically viable renewable energy sources for power generation. However, the technical know-how is not yet fully adequate to develop reliable wind energy conversion systems for all wind speed regimes. For securing maximum power output, wind energy resource assessment at a prospective site is critical. Due to wind speed and available wind energy relationship accurate knowledge of the wind characteristics is critical to all wind energy exploitation aspects, from the site identification and predictions of the economic viability of wind energy projects through the wind turbine design and understanding the effects on electricity networks. The most striking characteristic of the wind is its spatio-temporal variability, persisting over a very wide range of spatio-time scales. Because of the relationship between wind velocity and output energy, sites with small differences in average wind speeds can have substantial differences in available energy. Therefore, accurate and thorough monitoring of wind resource at potential sites is essential in the wind turbine siting. Accurate measurements of wind speed frequency spectrum are an important factor in wind energy potential analysis and assessment. For that, investigators have used wind speed distributions that are parameterized solely by the wind speed arithmetic mean. However, an assessment of wind turbine power output is accurate, only if the wind speeds and directions are measured at the turbine hub height.

Knowledge of the local wind capacity remains vital to the industry, yet commercially viable renewable-related geospatial products that meet the wind industry needs are often suspect. There are three stage involved with wind power project planning and operations during which accurate wind characterization plays a critical role (1) prospecting: uses historical data, retrospective forecasts, and statistical methods to identify potential sites for wind power projects; (2) site assessment: determines the optimum placement of a wind power project; and, (3) operations: uses wind forecasting and prediction to determine the available power output for hour-ahead and day-ahead time frames. The most critical is identifying and characterizing the wind resources. Appropriate statistical and modeling methods to compute the wind speed probability density function (PDF) are critical in wind resource analysis. In addition, although there has been an increasing awareness of wind energy as a viable energy source, there has not been a concomitant increase in the awareness of the impacts that any spatial and temporal trends in the wind resource may have on long-term production, use, and implementation. Despite environmental benefits and technological maturation, penetration of wind-generated power represents a challenge for reliability and stability of the power grids due to the highly variability and intermittent nature of winds. Moreover, for the purposes of wind energy use and wind turbine design, the wind vector is considered to be composed of a steady wind plus fluctuations about the steady wind. Whereas for designing wind turbines, the steady wind and the wind fluctuations have to be considered, the power and energy obtained from wind can be based only on the steady wind speed. The power available in the wind varies with the cube of the wind speed, and depends also on the air density. From kinetic energy of the wind by taking it time derivative, the available power in the wind can be expressed as:

$$P_{wind} = 0.5\ \rho A v^3 \tag{9.1}$$

Here ρ is the air density (kg/m^3), A is the cross-sectional area (m^2), and v is the wind speed (m/s). Examining (9.1) reveals that in order to obtain a higher wind power, it requires a higher wind speed, a longer length of blades for gaining a larger swept area, and a higher air density. Because the wind power output is proportional to the cubic power of the mean wind speed, a small variation in wind speed can result in a large change in wind power. A common unit of measurement is the wind power density, or the power per unit of area normal to the wind direction from the wind is blowing:

$$p_w = \frac{P_{wind}}{A} = 0.5\ \rho v^3 \tag{9.2}$$

Here p_w is wind power density (W/m^2). The ultimate wind energy project goal is to extract the wind energy and not just producing power, an important parameter in site selection is the mean wind power density, expressed as if the wind frequency distribution $f_{PDF}(v)$ is known as:

$$\bar{p}_w = 0.5\ \rho v^3 f_{PDF}(v) \tag{9.3}$$

Example 9.1: For an average wind speed of 10 mph, a small wind turbine produces 100 W/m^2. What is the power density for a speed of 40 mph?

Solution: From (9.2), the power density is proportional to cube of the wind speed so:

$$\text{Speed ratio} = \frac{40}{10} = 4$$

$$P_{40} = 4^3 P_{10} = 64 \times 100 = 6.4\ \text{kW}$$

Weibull or Rayleigh probability distribution, are mostly used in the wind energy assessment, characterization, and analysis. Instead of integration, the mathematical Weibull function, the mean value of the third power of the wind speeds in appropriate time intervals can also be used. Wind resource maps often estimate the potential of the wind resources in terms of wind power classes referring to the annual wind power density.

Example 9.2: The diameter of a large offshore wind turbine is 120 m, assuming the air density is compute the available wind power for wind speed of 5, 10, and 15 m/s. What the available wind power density for wind speed of 8.5 m/s?

Solution: By using (9.1), the available power densities are:

$$P_5 = 0.5 \cdot 1.2\left(\pi\frac{120^2}{4}\right)(5)^3 = 847.8\ \text{kW}$$

$$P_{10} = 0.5 \cdot 1\ 2\left(\pi\frac{120^2}{4}\right)(10)^3 = 6.7824\ \text{MW}$$

$$P_5 = 0.5 \cdot 1.2\left(\pi\frac{120^2}{4}\right)(5)^3 = 22.8906\ \text{MW}$$

The power density for a wind speed of 8.5 m/s is then calculated as:

$$p_w = 0.5 \cdot 1.2(7.5)^3 = 253.12\ \text{W/m}^2$$

9.2.1 Wind energy resources

Wind results from the movement of air due to atmospheric pressure gradients. Wind flows from regions of higher pressure to regions of lower pressure. The larger the atmospheric pressure gradient, the higher the wind speed and thus, the greater the wind power that can be captured from the wind by means of wind energy-converting machinery. The wind regime is determined due to a number of factors, the most important factors being uneven solar heating, the Coriolis force, due to the Earth's rotation, and local geographical conditions. For example, the surface roughness is a result of both natural geography and manmade structures. Frictional drag and obstructions near the Earth's surface generally retard with wind speed and induce a phenomenon known as wind shear. The rate at which wind speed increases with height varies on the basis of local conditions of the topography, terrain, and climate, with the greatest rates of increases observed over the roughest terrain. A reliable approximation is that wind speed increases about 10% with each doubling of height. In addition, some special geographic structures can strongly enhance the wind intensity. Air masses move because of the different thermal conditions, and the motion of air masses can be a global phenomenon (i.e., the jet stream), or a regional or local phenomenon. The regional phenomenon is determined by orography (e.g., the surface structure of the area) and also by global phenomena. Wind turbines utilize the wind energy close to the ground. The wind conditions in this area, known as boundary layer, are influenced by the energy transferred from the undisturbed high-energy stream of the geostrophic wind to the layers below and by regional conditions. Owing to the roughness of the ground, the local wind stream near the ground is turbulent, and wind speed varies continuously as a function of time and height.

Wind energy resources rely on the incident wind speed and direction, both of which vary in time and space due to changes in large-scale and small-scale circulations, surface energy fluxes, and topography. Since the wind power density is proportional to the cube of the wind speed, any small errors in forecasted wind speeds can result in significant differences between forecasted and actual wind energy output. Consequently, accurate assessment and forecasting of spatial and temporal characteristics of the winds and turbulence remains the most significant challenge in wind energy production. The wind energy production viability is governed by factors, such as the potential for large scale energy production, the predictability of the power to be supplied to the grid, and the expected return on investment. The various wind energy uncertainties impact the reliable determination of these viability factors. Currently the worldwide capacity of wind-powered generators is approximately 2.5% of the electricity consumption, with a steady annual growth. For wind to play a more prominent role in the future energy market, improvements in the generation technology are needed, while some advancement can be realized in part through appropriate uncertainties quantification and explicit consideration of their influences in the optimal design. A robust and optimal wind farm planning include (a) site optimal selection based on the quality of the local wind resources, (b) maximization of the annual energy production and/or minimization of the cost of generated energy, and (c) maximization of the reliability of the predicted energy output. The most important activity in a site selection is to determine the wind resource potential, consisting in the estimated local wind PDF. Another important activity is to determine the levels of turbulence and the resulting wind loads at the concerned site, promoting better decision making, in selecting the most suitable wind turbines for that site and in optimum life cycle cost prediction; higher wind loads generally result in higher costs. Among others site selection criteria include (1) local topography, (2) distance to electric grid, (3) vegetation, (4) land acquisition issues, and (5) site accessibility for turbine transport and maintenance. For a better estimate of the wind power potential for any extended time period, you would need to know the frequency distribution of the wind speeds; the amount of time for each wind speed value, or a wind speed histogram; and the number of observations within each wind speed range. It is also important to have information about wind regime characteristics, air density, and turbulence intensity (TI).

Wind power density is a comprehensive index in evaluating the wind resource at a particular site. It is the available wind power in airflow through a perpendicular cross-sectional unit area in a unit time period. The classes of wind power density at two standard wind measurement heights are listed in Table 9.1. Some of the wind resource assessments utilize 50 m towers with sensors installed at intermediate levels (10 m, 20 m, etc.). For large-scale wind plants, class rating of 4 or higher are preferred. The use of wind power classes to describe the magnitude of the wind resource was first defined in conjunction with the preparation of the 1987 US Department of Energy Wind Energy Resource Atlas. The atlas is currently available through the American Wind Energy Association and is an excellent source of regional wind resource estimates for the United States and its territories. The atlas wind resource magnitude is expressed in terms of the seven wind power classes, as well as the wind velocity. The wind power classes range from class 1, for winds containing the least energy to class 7, for winds containing the greatest energy

Table 9.1 Classes of wind power density at 10 m and 50 m

Wind power class	10 m—Wind power density (W/m^2)	50 m—Wind power density (W/m^2)
1	<100	<200
2	100–150	200–300
3	150–200	300–400
4	200–250	400–500
5	250–300	500–600
6	300–350	600–700
7	➢400	➢8,000

(Table 9.1). Mean wind speed estimates hare are based on Rayleigh wind speed probability distribution of equivalent mean wind power density, for standard sea-level conditions, and to maintain the same power density, speeds are increased by 3%/1,000 m (5%/5,000 ft.) elevation.

Wind resource assessment is the most important step in planning a wind project because it is the basis for determining initial feasibility and cash flow projections, being vital for financing. Assessment and project progress through several stages (1) initial assessment; (2) detailed site characterization; (3) long-term data validation; and (4) detailed cash flow projection and financing. Prediction of wind energy resources is crucial in the development of a commercial (large-scale) wind energy installation. The single most important characteristic to any wind development is the wind velocity. The performance and wind farm power output is very sensitive to uncertainties and errors in wind velocity estimates, so the wind resource assessment must be extremely accurate in order to procure funding and accurately estimate the project economics. Commercial wind resource assessment performed by wind developers is using both numerical and meteorological data. Wind speed and direction measurements are collected by permanent or semi-permanent meteorological towers designed to measure wind velocity using a variety of wind sensors, (sonic, LIDAR, cup and vane, sonic, etc.), at different hub heights. An important aspect is to gain an understanding of the wind profile both spatially across the location of interest and in elevation above terrain level. The main factors impacting the wind flows are: orography, surface roughness and the atmospheric stability. The latter represents the thermal effects on the wind flow, being rarely taken into account for the wind power assessments, since the wind statistics are averaged over a long period, and the atmosphere is generally considered as neutral. However, this assumption presents some limitations (1) sites where the average wind speed is low (<5 m/s), the thermal effects are starting to be significant; (2) on offshore site, where the atmospheric stability is predominant over the orography and the roughness; and (3) for short-period simulations (hours or days), for short-term prediction, supervision of operation, and power curves measurement with site calibration, as defined in the IEC 61400-12 standard.

9.2.2 Air density, temperature, turbulence, and atmospheric stability effects

Since wind speed generally increases with height, higher elevation sites potentially offer greater wind resources than comparable lower ones, being advantageous to site wind turbines at higher elevations, and taking advantage of higher wind speeds. However, the decrease of air density with height can make an impact on the output power, wind power density being proportional to air density, so a given wind speed therefore produces less power from a particular turbine at higher elevations, because the air density is less. Output power and the power curve depend on the air density. For example, the air density values encountered at measurement sites in western Nevada are mostly between 0.936 kg/m^3 and 1.025 kg/m^3 with a multi-annual mean value of 0.982 kg/m^3, significantly lower than the standard air density of 1.225 kg/m^3. Power curves for various air density effects must be accounted for to improve the power output estimate accuracy. Air density is usually computed from temperature and pressure data, as expressed by:

$$\rho = \rho_0 \left(\frac{T}{T_0} \right)^{-(g/cR+1)} \quad \text{or} \quad \rho = \rho_0 \left(1 + \frac{c \cdot z}{T_0} \right)^{-(g/cR+1)} \tag{9.4}$$

where T is the local air temperature (K), T_0 is the air temperature at the ground (K), z is the elevation in m, $c = dT/dz$, is the atmosphere thermal gradient (~9.80 °C/km), R is the gas constant (287 J/kg-K for air). Alternate relationships to estimate the air density dependence on the elevation are:

$$\rho = 1.229 \frac{P - VP}{760} \frac{273}{T} \ \text{kg/m}^3 \tag{9.5a}$$

$$\rho = \frac{353.049}{T} \exp\left(-\frac{0.034 \cdot z}{T} \right) \tag{9.5b}$$

Here, the atmospheric pressure, P is expressed in mm Hg, VP is the vapor pressure in mm Hg, and T is the local absolute temperature in Kelvin degrees. This relationship yields to a value of 1.225 kg/m^3 for dry atmosphere in standard atmospheric conditions. The vapor pressure represents a small correction, around 1%, and can be neglected. High temperatures and low pressures reduce the air density, which reduces the wind power. A major factor for air density change is the pressure change with elevation. One km increase in elevation reduces the pressure by 10%, reducing the wind power by 10%. If only elevation is known, air density can be estimated by using:

$$\rho = 1.225 - 1.194 \times 10^{-4} z \tag{9.6}$$

Depending on the turbine's method of control, either the power or velocity is normalized for use in power density calculations, as here were the velocity is normalized with the reference air density ρ_0:

$$v_{norm} = \bar{v} \left(\frac{\bar{\rho}}{\rho_0} \right)^{1/3} \tag{9.7}$$

Example 9.3: For a pressure of 750 mm Hg, and local air temperature equal to 21.5 °C estimate the air density and the normalized value of a wind speed of 10 m/s. Assume VP = 0.

Solution: The air density, by using (9.5) is 1.124 kg/m^3, and by using (9.7) the normalized value of the wind speed (10 m/s) is equal to: 9.18 m/s.

At today's usual hub heights at 80 m or so, turbine rotors encounter large vertical gradients of wind speed and boundary layer turbulence. Rotors are susceptible to fatigue damage that results from turbulence. Wind turbulence represents the fluctuation in wind speed in short time scales, especially for the horizontal velocity component. The wind speed $v(t)$ at any instant time t can be in two components: the mean wind speed V_{mn} and the instantaneous speed fluctuation $v'(t)$, i.e.,

$$v(t) = V_{mn} + v'(t)$$

Wind turbulence has a strong impact on the wind turbine power output fluctuations. Heavy turbulence may generate large turbine dynamic fatigue loads, reducing the expected turbine lifetime, or resulting in turbine failure. In selection of wind farm sites, the knowledge of wind TI is crucial for the stability of wind power production, turbine control and wind turbine design. For example, understanding of the impact of turbulence on the blades helps in designing long-term operational and maintenance schedules for wind turbines, or can lead to the design of control schemes to mitigate such loads. Quantification of the turbulence effects on wind turbine is done by computing an equivalent fatigue load, as function of the wind fluctuation amplitudes within an averaging period, blade material properties, number of averaging bins, and a total number of samples. Turbulent fluctuations are the main source of the blade fatigue. The TI, a measure of the overall turbulence level, is defined as:

$$TI = \frac{\sigma_v}{v} \tag{9.8}$$

where σ_v is the wind speed standard deviation (m/s), usually at the nacelle height over a specified averaging period (e.g., 10 min). There also are differences in the output power standard deviations. In the wind speed range 4–15 m/s, the standard deviation of certain TI classes (4%–8% and 10%–15%) differs up to about 50% with the standard deviation for all turbulence intensities. TI is affected by atmospheric stability, so the theoretical wind turbine power curves. A TI correction factor can be expressed as:

$$v_{corr} = v_{norm}\left(1 + 3(TI)^2\right)^{1/3} \tag{9.9}$$

Example 9.4: If the standard deviations for the following wind speeds 6.5 m/s, 10 m/s, and 13.5 m/s are 0.90 m/s, 1.05 m/s, and 1.15 m/s, respectively. What are the turbulence intensities corrected wind speeds?

Solution: From (9.8), the TI levels for these data are:

$$TI_{6.5} = \frac{0.90}{6.5} = 0.1385$$

$$TI_{10.0} = \frac{1.05}{10.0} = 0.1050$$

$$TI_{13.5} = \frac{1.15}{13.5} = 0.0852$$

By using modified (9.9), the corrected wind speeds are:

$$v_{corr} = 6.5\left(1 + 3(0.1385)^2\right)^{1/3} = 6.624 \text{ m/s}$$

$$v_{corr} = 10.0\left(1 + 3(0.105)^2\right)^{1/3} = 10.11 \text{ m/s}$$

$$v_{corr} = 13.5\left(1 + 3(0.0852)^2\right)^{1/3} = 13.60 \text{ m/s}$$

Notice that these are corrected speed, while the wind turbine power output is depended of cube of the wind speed.

9.2.3 Wind shear, wind profile, wind gust, and other meteorological effects

Vertical wind shear is important as wind turbines become larger and larger. It is therefore questionable how well the hub height wind speed is representative. Various methods exist concerning the extrapolation of wind speed to the wind turbine hub height. There also are several theoretical expressions used for determining the wind speed profile. However, from practical point of view, the surveys must resort to simpler expressions and secure satisfactory results even when they are not theoretically accurate. Obstacles can cause the displacement of the boundary layer, affecting the wind velocity. The roughness length (z_0) is the height at which the wind is zero, meaning that surfaces with large roughness lengths have larger effects on the wind. It ranges from 0.0002 for open sea, 0.005–0.03 for open land, 0.03–0.1 for agricultural land, to 0.5–2 m for very rough terrain or urban areas. The increase of wind speed with height should be considered for the large wind turbine installations. Wind speed is usually recorded at 10 m, the standard meteorological height, while wind turbines have hub heights are 60 m or higher. In cases which lack elevated measurements hub height wind velocity is estimated by extrapolations of the surface measurements. The vertical extrapolation coefficients may contain errors and uncertainties due to terrain complexity, atmospheric stability and turbulence. There exist several wind speed extrapolation methods at the wind turbine hub height. The wind speed $v(z)$ at a height z can be calculated directly from the wind speed $v(z_{ref})$ at reference height z_{ref} (the standard measurement level) by using the logarithmic law (the Hellmann exponential law) expressed by:

$$\frac{v(z)}{v_0} = \left(\frac{z}{z_{ref}}\right)^{\alpha} \quad (9.10)$$

where, $v(z)$ is the wind speed at height z, v_0 is the speed at z_{ref} (usually 10 m height, the standard meteorological wind measurement level), and α is the friction coefficient or power low index. This coefficient is a function of the surface roughness at a specific site and the thermal stability of the Prandtl layer. It is frequently assumed to be 1/7 for open land. For 10 m and $z_0 = 0.01$ m, the parameter $\alpha = 1/7$, which is consistent with the value of 0.147 used in the wind turbine design standards (IEC standard, 61400-3, 2005) to represent the change of wind speeds in the lowest levels of the atmosphere. However, this parameter can vary diurnally and seasonally as well as spatially. It was found that a single power law is insufficient to adequately project the power available from the wind at a given site, especially during nighttime and also in presence of the low-level jets. However, there are significant discrepancies of values for α, especially for arid and dry regions, with ranging from 0.09 to 0.120, quite smaller comparing to the standard 0.147 value. Moreover, α can vary for one place to other, during the day and year. Another formula, known as the logarithmic wind profile law and widely used across Europe, is the following:

$$\frac{v}{v_0} = \frac{\ln\left(\frac{z}{z_0}\right)}{\ln\left(\frac{z_{ref}}{z_0}\right)} \quad (9.11)$$

where z_0 is called the roughness coefficient length and is expressed in meters; it depends basically on the land type, spacing, and height of the roughness factor (water, grass, etc.) and it ranges from 0.0002 up to 1.6 or more. These values can be found in the common literature.

Example 9.5: A meteorological tower is located close to a town limit. If the wind speed is 7 m/s at 10 m height (standard measurement level), what is the wind speed at 50 m, using (9.10)? If you are using (9.11) and select $z_0 = 1.2$, what is the wind speed at 50 m level?

Solution: By using (9.10) with $\alpha = 0.147$, the wind speed at 50 m is:

$$v = 7\left(\frac{50}{10}\right)^{0.147} = 8.7 \text{ m/s}$$

By using (9.11), the wind speed at 50 m is:

$$v = 7\frac{\ln\left(\frac{50}{1.2}\right)}{\ln\left(\frac{10}{1.2}\right)} = 12.3 \text{ m/s}$$

This compares to 8.7 m/s using the power law with a shear exponent $\alpha = 0.147$.

In addition to the land roughness, these values depend on several factors, varying during the day and at night, and even during the year. For instance, the reading or monitoring stations can be within farming land; it follows that the height/length of the crops will change. Once the speeds have been calculated at other heights, the relevant equations are used for calculating the useful energy potential via different methods. If the type of ground cover is known, the wind speed at other heights can be estimated. In some cases, we may have data from a reference location (e.g., meteorological station) at a certain height. These data are to be transformed to a different height at another location with similar wind profile but different roughness height (e.g., the wind turbine site). Under such conditions, it is safe to assume that the wind velocity is not significantly affected by the surface characteristics beyond a certain height. From (9.10), after a simple calculation, a relationship can be derived as:

$$v(z_{WT}) = v(z_{ref})\left(\frac{\ln\left(\frac{h_{WT}}{z_{0ref}}\right)}{\ln\left(\frac{h_{WT}}{z_0}\right)}\cdot\frac{\ln\left(\frac{z_{WT}}{z_0}\right)}{\ln\left(\frac{z_{ref}}{z_{0ref}}\right)}\right) \tag{9.12}$$

Here, z_{ref} are the wind measurement level at the reference station, z_{WT} is the wind turbine site level, h_{WT} is a reference height (e.g., 60 m), z_{0ref} and z_0 are the reference station and wind turbine sire roughness coefficient lengths, respectively. Aside from ground level to hub height shear, wind shear over the rotor disc area can also significant. The standard procedure for power curve measurements is given by the IEC standard (IEC Standard, 6-1400-12-1, 2005) where the wind speed at hub height is considered to be representative of the wind over the whole turbine rotor area. This assumption can lead to considerable wind power estimate inaccuracies, since inflow is often nonuniform and unsteady over the rotor-swept area. In most studies about the effect of wind shear on power performance, the wind speed shear is described by the shear exponent, obtained from the assumption of a power law profile. By integrating the wind profile over the rotor span, the corrected wind speed at the turbine nacelle can be obtained:

$$U_{avrg} = \frac{1}{2R}\int_{H-\frac{D}{2}}^{H+\frac{D}{2}} v(z)dz = v(H)\cdot\frac{1}{\alpha+1}\cdot\left(\left(\frac{3}{2}\right)^{\alpha+1} - \left(\frac{1}{2}\right)^{\alpha+1}\right) \tag{9.13}$$

where H is the nacelle height and D is the rotor diameter. From (9.13), it is obvious that the hub height wind speed $z(H)$ is α corrected based on the profile it is experiencing. It was noticed that these corrections have more or less the same effect. For wind speeds in the range 5–25 m/s (the useful wind turbine speed regime), the corrected power differs in general less than 5% from the uncorrected power, while often corrected power is larger than the uncorrected power.

Example 9.6: The wind velocity measured at 10 m height (standard level) at a meteorological station is 6.5 m/s. Find out the velocity at 30 m height at a wind turbine site having similar wind profile. The roughness coefficient lengths at the observatory and wind turbine location are 0.03 m and 0.1 m, respectively. Assume the h_{WT} equal to 60 m.

Solution: Using (9.12), in the problem assumptions, the velocity and the wind turbine height is then:

$$v(z_{WT}) = 7.5 \cdot \left(\frac{\ln\left(\frac{60}{0.03}\right)}{\ln\left(\frac{60}{0.1}\right)} \cdot \frac{\ln\left(\frac{30}{0.1}\right)}{\ln\left(\frac{10}{0.03}\right)} \right) = 8.45\ \text{m/s}$$

An additional wind property that can make the impact on wind turbine operations is wind gustiness. Wind gust refers to a phenomenon that a wind blasts with a sudden increase in wind speed in a relatively small interval of time. In case of sudden turbulent gusts, wind speed, turbulence, and wind shear may change drastically. In order to reduce rotor imbalance while maintaining the power output of wind turbine generator constant during such sudden turbulent gusts calls for relatively rapid changes of the rotor control parameters. Moreover, sudden turbulent gusts may also significantly increase tower fore-aft and side-to-side bending moments due to increase in the effect of wind shear. To ensure safe operation of wind farms, wind gust predictions are highly desired. Proper design and operation of a wind turbine for a specific wind climate requires knowledge of wind extremes and gustiness, often defined by a wind gust factor, being especially true in areas where wind climate is determined by inherently strong gusty winds (e.g., downslope windstorms). In sites with high ambient turbulent intensity and gusty winds, turbines are subject to extreme structural loading and fatigue. The gust factor (G) is defined as:

$$G = \frac{u_g}{U} - 1 \tag{9.14}$$

where u_g is the gust speed and U is the mean daily wind speed. One expects higher gusts to be associated with higher mean speeds; however, we may also expect that the normalized gust speed u_g/U and, consequently, the gust factor, G, decreases with the increasing mean speed. The following equation relates the gust factor to the mean daily wind speed:

$$G = AU^n \tag{9.15}$$

where the parameters A and n are obtained by using a least-square fit of the logarithm of G versus the logarithm of the mean daily wind speed. While gusts generally decrease as wind speed increases, in extreme cases the wind gusts can easily reach over twice the strongest wind speeds ($v > 20$ m/s) and damage a wind turbine. However, wind gusts over 25 m/s, the upper wind speed limit of a large wind turbine, are quite unlikely in many areas or regions. Gusts associated with stronger winds may cause considerable losses by reducing the energy production of the wind turbine which would otherwise operate at nominal output power. Another effect of wind gusts is the additional stress on the wind turbine structure, which may reduce its lifespan.

There some other factors that can affect the local wind regime with strong impacts on any wind development project, such as the low-level jet, channeling, breezes, etc. However, the discussions of such factors are beyond the scope of our textbook. The low-level jet is a mesoscale phenomenon associated with the night-time very stable boundary layer that can have a width of hundreds of kilometers and a length of a thousand kilometers. They have been observed worldwide. During nighttime and over land, ground surface cools at a faster rate than the adjacent air and stable stratification forms near the surface and propagates upward. Downward mixing of the winds is reduced and winds aloft become decoupled from the surface and accelerate. The maximum wind speeds are usually 10–20 m/s or more at elevations mainly at 100–300 m and occasionally as high as 900 m above ground. Consequently, it is not possible to accurately estimate winds aloft at hub and blade heights from routine surface measurements. Additionally, a strong wind shear and associated turbulence develop at the bottom and top of the jet layer and may have impacts of wind turbine power output.

9.2.4 Wind velocity statistics

Wind speed is the most critical data needed to assess the power potential of a candidate site, due to the cubic dependence of the wind power density. The wind is never steady at any site. It is influenced by the weather system, the local land terrain, and its height above the ground surface. Wind speed varies by the minute, hour, day, season, and even by the year. Therefore, the annual mean speed needs to be averaged over 10 years or more. In this subsection the most common wind speed probability distributions used by the wind energy community are discussed in some details. In many wind power studies the features and characteristics of such distributions are used for assessment and analysis of wind resources, wind power plant operation, grid integration as well as for wind turbine designs. Usually the time series of wind speeds and directions are rather large, differences among parameter estimation methods is not as important as differences among distributions. There are several PDF parameters' estimators, such as the Moment Method, Maximum Likelihood, Least-Square, and Percentile Estimators Methods. These estimators are unbiased, so there is no reason to give preference to any of them. Once the wind probability distribution function is obtained, the mean power available can be deduced. The goal of any wind energy assessment and analysis is to obtain expressions allowing in giving responses to questions about statistical distribution of the maximum power obtainable from the wind, regardless of the WT chosen. The temporal variation of the wind velocity leads to the use of statistical measures. The lowest (first) order statistic is the time average (mean), defined as:

$$V_m = \frac{1}{N}\sum_{i=1}^{N} v_i \tag{9.16}$$

Here N is the number of wind speed measurements. It is important to note that since the wind turbine power scales as V^3, the average available wind power is:

$$P_{avg} = \frac{1}{N}\sum_{i=1}^{N} v_i^3 \neq V_m^3 \tag{9.17}$$

Example 9.7: Consider the following set of time-varying velocity measurements (m/s): $V_i = (3.5, 4.3, 4.7, 8.3, 6.2, 5.9, 9.3, 10.2, 11.4)$. Compute the mean wind speed and the average available wind power.

Solution: The mean wind speed, computed using (9.16) is: 7.09 m/s. The available wind power computed by using the correct relationship, (9.17) is: 509.9 $(\text{m/s})^3$. However, if the wind power is computed as:

$$P_{av} = V_m^3 = (7.09)^3 = 356.4\ (\text{m/s})^3$$

As a result, the incorrect approach underestimates the power generation by approximately 30%. Based on this, a "power component" time-averaged wind speed is defined as:

$$V_{3\text{-pwr}} = \left[\frac{1}{N}\sum_{i=1}^{N} v_i^3\right]^{1/3} \tag{9.18}$$

In this case, the available wind power and the speed are related as $P = V_{3\text{pwr}}$.

Wind speed variations have a strong impact on a wind turbine's power generation, so it is needed to quantify the variations occurring in the wind speed over time. One such statistical measure is the standard deviation, s or second statistical moment which is defined as

$$s = \left[\frac{1}{N}\sum_{i=1}^{N}(v_i - V_m)^2\right]^{1/2} = \left[\frac{1}{N}\sum_{i=1}^{N}(v_i)^2 - \left(\frac{1}{N}\sum_{i=1}^{N} v_i\right)^2\right]^{1/2} \tag{9.19}$$

The second part of (9.19) is usually applied in practical computation of the standard deviation, being more convenient to compute since the sum of the v_i and $(v_i)^2$ can be accumulated together and subtracted at the end. Wind data are most often grouped in the form of a frequency distribution, such as shown in Table 9.2. This shows the number of hours per month in which the wind speed is within a specified range. In this case of the frequency data, the power-weighted time wind speed average is given by:

$$V_{3\text{-pwr}} = \left[\frac{\sum_{i=1}^{N} f_i v_i^3}{\sum_{i=1}^{N} f_i}\right]^{1/3} \tag{9.20}$$

Table 9.2 Sample frequency distribution of monthly wind velocity

Wind speed range (m/s)	Hours per month	Cumulative hours
0–1	13	13
1–2	37	50
2–3	50	100
3–4	62	162
4–5	78	240
5–6	87	327
6–7	90	417
7–8	77	494
8–9	65	559
9–10	54	613
10–11	40	653
11–12	31	684
12–13	21	705
13–14	14	719
14–15	9	728
15–16	6	734
16–17	5	739
17–18	4	743
18–19	2	745
19–20	1	746
>20	1	747

While the standard deviation using frequency data is given by:

$$s = \left[\frac{\sum_{i=1}^{N} f_i \left(v_i - V_{3\text{-pwr}} \right)^2}{\sum_{i=1}^{N} f_i} \right]^{1/2} \tag{9.21}$$

Example 9.8: For the frequency data of Table 9.2, compute the standard deviation and the power-weighted time average wind speed.

Solution: Applying the above relationships the values are: $V_{3\text{-pwr}}$ is 8.35 m/s and the standard deviation is 0.81 m/s.

We also note that $V_{3\text{-pwr}}$ is not the most probable wind velocity. It generally does not unless the "skewness" (3rd statistical moment) is zero. This occurs only if the distribution is Gaussian.

9.2.5 Wind statistical models

In order to predict the power generated on a yearly basis, statistical models of the wind velocity frequency of occurrence are needed. It has been found that Weibull and Rayleigh probability distributions can be used to describe wind variations with acceptable accuracy. The advantage of using well known analytic distributions like these is that the probability functions are already formulated plenty of information and papers are available. The Weibull density distribution is a commonly applied statistical distribution to model wind speed distributions. The Weibull curve is a PDF and indicates both the frequency and magnitude of a given wind speed over a period of time. It has been established that the Weibull distribution can be used to characterize wind speed regimes in terms of its probability density and cumulative distribution functions, and it is commonly used to estimate and to assess wind energy potential. Weibull distribution, well accepted and widely used for wind data analysis is given by:

$$f_{WB} = k\frac{v^{k-1}}{c^k}\exp\left(-\left(\frac{v}{c}\right)^k\right) \tag{9.22}$$

The Weibull distribution is a function of two parameters, k, the shape parameter, and c, the scale factor, defining the shape or steepness of the curve, and the mean value of the distribution. These coefficients are adjusted to match the wind data at a particular site. For wind analysis or modeling, typical k values range from 1 to 2.5 and can vary drastically form site to site, as well as during years and/or seasons. The scale parameter, c, corresponds to the average wind speed for the site. The main inaccuracy of the Weibull distribution is that it always has a zero probability of zero wind speed, which is not the case, since there are frequently times in which no wind is blowing. However, the fault is virtually without consequence because most turbines will not operate in speeds below 3 m/s and the distribution is more accurate, compared to measured data, within the zone most used by turbines: 5–25 m/s. The higher the k value, the sharper the increasing part of the curve is. The higher c values correspond to a shorter and fatter distribution, with a higher mean value. Ideally the mean value would correlate with the rated wind speed of the turbine, producing rated power for the greatest period of time annually. The cumulative probability function for Weibull distribution is given by:

$$F(v) = 1 - \exp\left[-\left(\frac{v}{c}\right)^k\right] \tag{9.23}$$

In all of these statistical representations, c and k are Weibull coefficients are dependent on the elevation and location. In general frequency data would be accumulated for a particular site and wind turbine hub height elevation being considered. The data would then be fitted to a Weibull distribution to find the best c and k. The availability of high quality fitted wind speed distributions is crucial to accurate forecasts of annual energy production for a wind turbine or to assess the site wind energy potential. Statistical distributions are suffice for early energy potential estimations and assessment, while the actual wind speed measurements are necessary for accurate predictions. Once the distribution is found, important parameters to characterize wind regime can be computed. The average wind speed, letting $x = (v/k)^2$, and by using Gamma function, is then:

$$V_m = c\int_0^{\infty} e^{-x}x^{1/k}dx = c\Gamma\left(1+\frac{1}{k}\right) \tag{9.24}$$

The cumulative distribution function, *F(v)*, can be used to estimate the time over which the wind speed is between some interval, V_1 and V_2, such as:

$$F(V_1 \le v \le V_2) = F(V_2) - F(V_1) = \exp\left(-\left(\frac{V_2}{c}\right)^k\right) - \exp\left(-\left(\frac{V_1}{c}\right)^k\right) \tag{9.25}$$

Example 9.9: A wind turbine with a cut-in velocity of 4 m/s and a cut-out velocity of 21 m/s is installed at a site where the Weibull coefficients are $k = 2.0$ and $c =$ 7.85 m/s. How many hours in a 24-h-period will the wind turbine generate power?

Solution: Applying (9.25), the probability that the wind speed is between cut-in and cut-off wind turbine speeds is:

$$F(4 \le v \le 21) = \exp\left(-\left(\frac{21}{7.85}\right)^{2.0}\right) - \exp\left(-\left(\frac{3}{7.85}\right)^{2.0}\right)$$

$$= 0.86959 - 0.02125 = 0.84834$$

Therefore, the number of hours in a 24-h-period (1 day) where the wind speed is between 4 and 21 m/s is: H = (24)(0.84834) = 20 h and 22 min. The Weibull shape and scale parameters are also height dependent. Suggested corrections to Weibull coefficients *k* and *c* to account for different altitudes, *z*, than the reference level and corresponding values are given as:

$$k = k_{ref}\frac{1 - 0.088 \cdot \ln\left(\frac{z_{ref}}{10}\right)}{1 - 0.088 \cdot \ln\left(\frac{z}{10}\right)}$$

$$c = c_{ref}\left(\frac{z}{z_{ref}}\right)^{\beta}$$

$$\beta = \frac{0.037 - 0.088 \cdot \ln(c_{ref})}{1 - 0.088 \cdot \ln\left(\frac{z_{ref}}{10}\right)}$$

9.2.5.1 Methods for Weibull model fits

The most used methods for estimating the best *k* and *c* for a Weibull distribution include:

1. Graphical method,
2. Standard deviation method,

3. Moment method,
4. Maximum likelihood method, and
5. Energy pattern factor method.

Graphical method is based on the use of Weibull cumulative distribution function, and consist of plotting $ln\ [-ln(1-F(v))]$ versus $ln(v_i)$ for the velocity samples v_i, for $i = 1, \dots, N$, the slope of the best fit straight line represents the Weibull coefficient, k, and the y-intercept represents $-k \cdot ln(c)$, from which the Weibull scale factor, c can be found. Alternatively, one can perform a least-square curve fit of the linear function to find the slope and intercept. Another method involves the wind speed sample mean and standard deviation. The shape and scale parameter, k and c can be found by using:

$$k = \left(\frac{s}{V_m}\right)^{-1.09} \tag{9.26a}$$

$$c = \frac{V_m}{\Gamma\left(1 + \frac{1}{k}\right)} \tag{9.26b}$$

Energy pattern method is based on the energy pattern factor, EPF, which is the ratio of the total power available in the wind and the power corresponding to the cube of the mean wind speed:

$$EPF = \frac{\frac{1}{N}\sum_{i=1}^{N} v_i^3}{\left[\frac{1}{N}\sum_{i=1}^{N} v_i\right]^3}$$

Having found the energy pattern from the wind velocity data at a given site, the approximate value of k is found from:

$$k = 3.957 \times (EPF)^{-0.898} \tag{9.27}$$

The value for c can be found using (9.26b), for example. For the sake of brevity, moment method, and maximum likelihood method are not discussed here. Interested readers are directed to the references at the end of chapter for details and implementation of these methods, or elsewhere in the literature.

9.2.6 Rayleigh probability distribution

Another common used probability distribution in wind energy analysis and assessment is the Rayleigh distribution, which is a special case of the Weibull distribution where $k = 2$. The Rayleigh distribution depends only on the mean wind speed, and is given by:

$$f_{RL}(v) = \frac{\pi}{2}\frac{v}{c^2} exp\left[-\frac{\pi}{4}\left(\frac{v}{c}\right)^2\right] \tag{9.28}$$

These two probability distribution functions are the most commonly used for wind energy analysis and assessment. The simpler of the two is the Rayleigh distribution which has a single parameter c. The Rayleigh distribution is actually a special case of the Weibull distribution with $k=2$. Setting $k=2$ in the Weibull distribution gives the Rayleigh distribution. For both distributions, $V_{min}=0$ and $V_{max}=\infty$. Setting $k=2$ in this result gives the cumulative Rayleigh distribution.

$$F(v) = 1 - \exp\left(-\left(\frac{v}{c}\right)^2\right) \tag{9.29}$$

For the Rayleigh distribution the single parameter, c, relates the following the property:

$$c = V_m\sqrt{2} = \frac{2\mu}{\sqrt{\pi}} = \sigma\sqrt{\frac{4}{8-\pi}} \tag{9.30}$$

The Rayleigh distribution can be written using V_m or the mean velocity, μ. The usual determination of the mean and standard deviation from experimental data for the normal distribution are well known. The minimum-least-squares-error (MLE) estimate of the mean of the normal distribution is the arithmetic mean. The MLE estimate of the variance is also familiar. The parameter c in the Rayleigh distribution is evaluated from a set of N wind velocity, v_i. When experimental data are used to determine parameters of probability distributions, the computed result is called an estimate of the true parameter. The symbol is used here to indicate that the equation below gives us only an estimate of the true distribution parameter, c.

$$\widehat{c} = \sqrt{\frac{1}{N}\sum_{i=1}^{N} V_i^2} \tag{9.31}$$

9.3 Wind direction

Wind direction is one of the main wind characteristics. Statistical data of wind directions over a long period of time are very important in the site selection of wind farm and the layout of wind turbines in the wind farm. Changes in wind direction are due to the general atmospheric circulation, on an annual basis (seasonal) to the mesoscale (up to 5 days) or even smaller scale, such hours. The seasonal changes of prevailing wind direction could be as little as 30° in trade wind regions to as high as 180° in temperate regions. In the US plains, the predominant directions of the winds are from the south to southwest in the spring and summer and from the north in the winter. Traditionally, wind direction changes are illustrated by a graph, which indicates percent of winds from that direction, the wind rose diagram (Figure 9.2). The wind rose diagram is a useful tool of analyzing wind data that are related to wind directions at a particular location over a specific time period (year, season, week, etc.). This circular diagram displays the wind direction relative

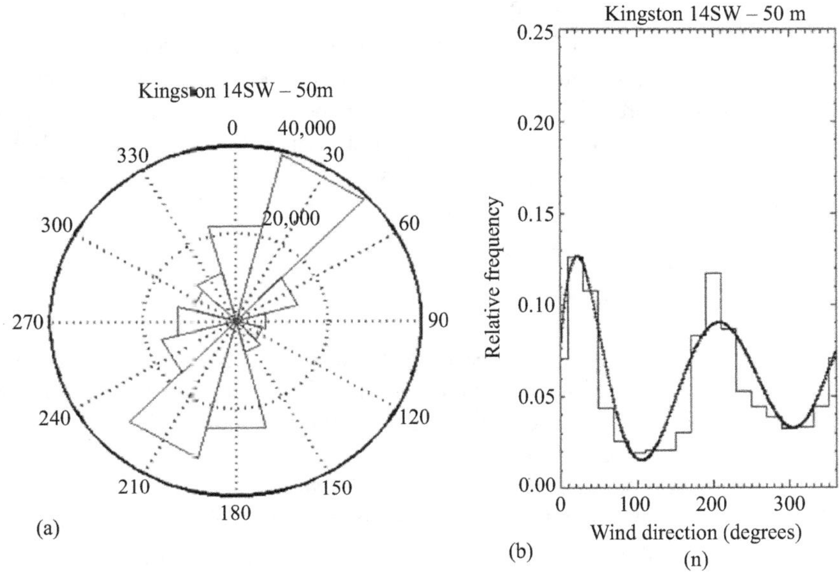

Figure 9.2 (a) A wind rose diagram, and (b) the wind direction histogram and the fitted von Mise wind direction probability distribution

frequencies in 8 or 16 principal directions. There can also be change in wind direction on a diurnal basis. However, a wind shear of change in wind direction with height is generally very small, except for very short time periods as weather fronts move through.

To ensure the most effective wind turbine use, it should be exposed to the most energetic wind. The wind may blow frequently from some predominant directions, so more wind energy may come from directions with stronger winds. It is very important to find out which directions have the best winds for electricity production. The wind direction distribution is crucially important for the evaluation of utilizing wind power, being given by wind roses or histograms of wind directions. A wind rose chart, which is generated from your wind resource assessment, is a helpful tool to determine wind direction and distribution. The wind rose diagrams and wind direction frequency histograms provide useful information on the prevailing wind direction and availability in different wind speed bin. Notice that a wind vane points toward the source of the wind. Wind direction is reported as the direction from which the wind blows, not the direction toward which the wind moves. A north wind blows from the north toward the south. The wind direction varies from station to station due to differential local features (topography, altitude, orientation, distance from the shore, vegetation, etc.). Notice that there can also be changes in the wind directions on diurnal, seasonal, or annual basis. The wind direction can also be analyzed using continuous variable probability models to represent distributions of directional wind speeds, such as von Mises circular statistics. The model usually comprised of a finite mixture of the von Mises distributions. The presentation of the wind circular statistics and probability distribution is beyond the scope of this chapter however interested readers are directed to use the end-of-chapter references. While obtaining the average wind speed at a given location is useful, it does not accurately portray the power of the wind resource. The statistical distribution of wind speeds vary depending on climate conditions, landscape, and surface roughness. Strong gale force winds are rare while consistent low speed velocities are most common.

9.4 Wind energy estimation

The ultimate estimate objective to be made in selecting a site for a wind turbine, wind power plant, or wind farm is the energy that is available in the wind at that specific site. This involves calculating the wind energy density, E_{den}, for a wind turbine unit rotor area and unit time, which is a function of the wind speed and its temporal distribution at the site. Three important wind speed parameters used for wind energy assessment or in turbine design are the most probable (frequent) wind speed, V_{mP}, the wind speed carrying the maximum energy, V_{MaxE}, and the most frequent speed, V_{MF}. The parameters are estimated, once the wind probability distribution (i.e., Weibull or Rayleigh) is determined. The most frequent wind velocity corresponds to the maximum of the probability distribution, $f(v)$. As a result, that the power generated scales as the wind velocity cube, the maximum energy usually corresponds to velocities that are higher than the most frequent. Horizontal wind turbines are usually designed to operate most efficiently at its design power wind speed, V_d. Therefore, it is advantageous if V_d and V_{MaxE} at the site are made to be as close as possible. Once V_{MaxE} is computed for the selected site, it is then possible to match the characteristics of the wind turbine to be most efficient at the site conditions.

The wind power density, p_w, available in a wind stream of velocity V, is given by (9.2). For a given velocity, V, the unit amount of time that velocity is present is $1xf(V)$, $f(v)$ is the wind speed probability distribution. The total energy for all possible wind velocities at a site is therefore computed as:

$$E_{den} = \int_0^{\infty} p_w f(v) dv \tag{9.32}$$

In the case of Weibull distribution function, after some mathematical manipulations, the energy density is expressed as:

$$E_{den} = \frac{\rho \cdot k}{2c^k} \int_0^{\infty} v^{k+2} \exp\left[-\left(\frac{v}{c}\right)^k\right] dv = \frac{\rho \cdot c^3}{2} \Gamma\left(\frac{3}{k} + 1\right) \tag{9.33}$$

Applying the general reduction formula for a Gamma function, the following form for the energy density is obtained:

$$E_{den} = \frac{3\rho \cdot c^3}{2k} \Gamma\left(\frac{3}{k}\right) \tag{9.34}$$

With E_{den} estimated for a site or location, the density energy that is available over a period of time, T, is then given by:

$$E_T = E_{den} \cdot T = \frac{3\rho \cdot T \cdot c^3}{2k} \Gamma\left(\frac{3}{k}\right) \tag{9.35}$$

The most probable wind speed (v_{MP}), denoting the most frequent wind speed for Weibull distribution, is given by:

$$v_{MP} = c \cdot \left(\frac{k-1}{k}\right)^{1/k} \tag{9.36}$$

The wind speed carrying the maximum energy (v_{MaxE}) is expressed by:

$$V_{MaxE} = c \cdot \left(\frac{k+2}{k}\right)^{1/k} \tag{9.37}$$

The most frequent wind speed (v_{MF}), is defined by the maximum relative frequency:

$$V_{MF} = c \cdot \left(1 - \frac{1}{k}\right)^{1/k} \tag{9.38}$$

When considering a Rayleigh wind speed probability distribution, the wind energy density, with the mean wind speed, V_m, by using (9.32) is given by:

$$E_{den} = \int_0^\infty \frac{\pi\rho}{4V_m^2} v^4 \exp\left(\frac{\pi}{4}\left(\frac{v}{V_m}\right)^2\right) dv = \frac{3}{\pi}\rho V_m^3 \tag{9.39}$$

The energy density available at a site, having Rayleigh distribution of the wind speed, over a period of time, T, is then expressed as:

$$E_T = E_{den} \cdot T = \frac{3}{\pi}\rho \cdot T \cdot V_m^3 \tag{9.40}$$

The most frequent wind velocity in the case of a Rayleigh wind speed probability distribution is given by:

$$V_{MF} = \sqrt{\frac{2}{\pi}} V_m \tag{9.41}$$

The velocity that maximizes the energy density for a Rayleigh wind velocity distribution, over a unit period of time is given by:

$$V_{MaxE} = \sqrt{\frac{8}{\pi}} V_m \tag{9.42}$$

Example 9.10: The following monthly mean wind velocity data (m/s) at a location is given in the following table. Calculate the wind energy density, the monthly energy density availability, the most frequent wind velocity, and the wind velocity corresponding to the maximum energy, assuming a Rayleigh wind speed distribution and the air density 1.20 kg/m^3.

January	April	July	October
9.25	7.30	11.50	6.85

Solution: The energy density and the monthly energy density available are calculated with (9.39) and (9.40), respectively, while the most frequent wind speed and the wind speed maximizing the energy density are computed using (9.41) and (9.42). The results are summarized in the table below:

Month	E_{den} (W/m^2)	E_T (MW/m^2/Month)	V_{MF} (m/s)	V_{MaxE} (m/s)
January	907.0	2.4292	7.38	14.76
April	445.8	1.4940	5.82	11.65
July	1742.9	4.4680	9.18	18.35
October	368.3	0.9865	5.47	10.93

9.5 Wind energy conversion systems

A wind turbine is a rotating mechanical machine which converts the wind kinetic energy into mechanical energy. If the mechanical energy is then converted to electricity, the machine is called a wind generator, wind turbine, wind power unit (WPU), wind energy converter (WEC), or aero-generator. Wind turbines can be separated into two types based by the axis in which the turbine rotates. Turbines that rotate around a horizontal axis are more common. Vertical-axis turbines are less frequently used. Horizontal-axis wind turbines (HAWT) have the main rotor shaft and electrical generator at the top of a tower, and must be pointed into the wind. Most have a gearbox, which turns the slow rotation of the blades into a quicker rotation that is more suitable to drive an electrical generator. Since a tower produces turbulence behind it, the turbine is usually pointed upwind of the tower. Turbine blades are made stiff to prevent the blades from being pushed into the tower by high winds. Additionally, the blades are placed a considerable distance in front of the tower and are sometimes tilted up a small amount. The horizontal-axis wind turbines are the most used today for electricity generation, operate on the aerodynamic forces, not on trust forces that develop when wind flows around a blade of aero-foil design. Actually, the windmills that work on thrust forces are less efficient. The wind stream at the top of the aero-foil has to traverse a longer path than that at the bottom, leading to a difference in velocities, giving rise to a difference in pressure (Bernoulli's principle), and a lift force is produced. There is also a drag force that tries to push the aero-foil back in the direction of the wind. The aggregate force is determined by the resultant of these forces.

Vertical-axis wind turbines (or VAWTs) have the main rotor shaft arranged vertically. Key advantages of this arrangement are that the turbine does not need to be pointed into the wind to be effective. This is an advantage on sites where the wind direction is highly variable. VAWTs can utilize winds from varying directions. With a vertical axis, the generator and gearbox can be placed near the ground, so the tower doesn't need to support it, and it is more accessible for maintenance. Drawbacks are that some designs produce pulsating torque. Drag may be created when the blade rotates into the wind. Savonius rotors are extremely simple vertical-axis devices that entirely because of the thrust force of the wind. The basic equipment is a drum cut into two halves vertically. The two parts are attached to the two opposite sides of a vertical shaft. The wind blowing into the assembly meets two different surfaces—convex and concave; and different forces are exerted on them, giving torque to the rotor. Providing a certain overlap between drums increases the torque because wind blowing on the concave side turns around and pushes the inner surface of the other drum, which partly cancels the wind thrust on the convex side. An overlap of one- third of the drum diameter gives best results. In a Darrieus wind turbine, two or more flexible blades are attached to a vertical shaft. The blades bow outward taking the shape of a parabola, and are of a symmetrical aero-foil section. When the rotor is stationary, no torque is produced. It has to be started by some external means as it has no starting torque.

HAWT has based on the wind rotor configuration with respect to the wind direction and are classified as upwind and downwind wind turbines. The majority of horizontal-axis wind turbines used today is upwind turbines, in which the wind rotors face the wind. The main advantage of upwind designs is to avoid the distortion of the flow field as the wind passes though the wind tower and nacelle. For a downwind turbine, wind blows first through the nacelle and tower and then the rotor blades. This configuration enables the rotor blades to be made more flexible without considering tower strike. However, because of the influence of the distorted unstable wakes behind the tower and nacelle, the wind power output generated from a downwind turbine fluctuates greatly. In addition, the unstable flow field may result in more aerodynamic losses and introduce more fatigue loads on the turbine. Furthermore, the blades in a downwind wind turbine may produce higher impulsive or thumping noise. Downwind machines have been built, despite the problem of turbulence, because these do need additional mechanism for keeping them in line with the wind, and because in high winds the blades can be allowed to bend, which reduces their swept area and thus their wind resistance. Since cyclic (that is repetitive) turbulence may lead to fatigue failures most HAWTs are upwind machines.

The conversion of wind energy into electricity or mechanical power involves two steps: first, the wind kinetic energy is converted by the turbine rotor, and second the drivetrain transfer the mechanical power to a load, such as an electric generator. The amount of power of any rotating mechanical device is the product of the torque and angular velocity, being the power available at its shaft. Most operations of transferring shaft power try to have a large angular velocity because of structural considerations. The power coefficient is the power delivered by the device divided by the power available in the wind. Since the area cancels out, the power coefficient, C_P, in the case of wind turbine, considering the input power as specified in (9.1), the power available in the wind is then expressed as:

$$C_p = \frac{\text{Power out}}{\text{Power in}} = \frac{\text{Power out}}{0.5 \cdot \rho \cdot A \cdot v^3} \tag{9.43}$$

And

$$P_{WT} = 0.5\, \rho A C_P v^3 \tag{9.44}$$

Here P_{WT} is the output power of the wind turbine. A maximum value of the power coefficient is defined by the Betz limit, stating that a wind turbine can extract up to 59.3% of the power from an air stream. However, wind turbine rotors have maximum C_p values in the range 25%–45%. Applying the fluid mechanic principals, conservation of energy and momentum equation can be determined that the output power of a wind turbine is expressed as:

$$P_{WT} = 0.5\, \rho A v^3 a(1 - a)^2 \tag{9.45}$$

Here a is the fractional decrease in the wind velocity once it has reached the rotor, due to a change in pressure (depending on how much energy the rotor captured to slow the wind). We can define the performance power coefficient, C_P, as the ratio of the power in the rotor to the power in the wind:

$$C_P = 4a(1 - a)^2 \tag{9.46}$$

The power coefficient indicates the efficiency of the turbine based solely on the stream tube concept, without accounting for nonideal conditions and inevitable losses from the blades, the mechanics, turbine generator, control, and wind conversion system electronics. Taking the derivative of the power coefficient, (9.46) with respect to a, setting it equal to zero yields the axial induction factor of 1/3 which maximizes the efficiency. At this value of a, the power coefficient equals to $16/27 \approx 0.59$, the maximum extractable, raw incoming wind kinetic energy (the Betz's Limit). The wind turbine power curve displays the power output as function of the mean wind speed. Common ways to characterize wind turbine performances is expressing them through nondimensional characteristics performance curve. Power curves are usually determined from field measurements. The tip-speed ratio, λ, is a variable relating the peripheral blade speed and wind speed, computed as:

$$\lambda = \frac{\omega R}{v} \tag{9.47}$$

The tip-speed ratio, λ, and the power coefficient, C_P, are dimensionless and so can be used to describe the performance of any size of wind turbine rotor. The wind turbine aerodynamic performances are characterized by the variation of the non-dimensional C_P *versus* λ curve. Equation (9.44), the power extracted by wind turbine, is expressed as:

$$P_{WT} = 0.5\, \rho \cdot A \cdot C_P(\lambda) \cdot v^3 \tag{9.48}$$

Figure 9.3(a) shows that the maximum power coefficient is only achieved at a single tip–speed ratio and for a fixed rotational speed of the wind turbine this only occurs at a single wind speed. Hence, one argument for wind turbine operating at variable rotational speeds is that it is possible to operate at maximum C_p over a wind speed range. The wind turbine power output at various wind speeds is usually described by a power curve. The power curve gives the steady-state power output, function of the wind speed at the hub height, generally the power output as a function of the wind speed at the hub height measured in 10 min average data. A power curve example is given in Figure 9.3(b). The power curve has three key points on the velocity scale:

- Cut-in wind speed—the minimum wind speed at which the machine will deliver useful power.
- Rated wind speed—the wind speed at which rated power is obtained (rated power is generally the maximum power output of the electrical generator).
- Cut-out wind speed—the maximum wind speed at which the turbine is allowed to deliver power (usually limited by engineering loads and safety constraints).

As mentioned earlier, the mechanical energy captured by the wind turbine blades is further converted in electrical energy via turbine electric generator. In this stage the converted efficiency is determined by several parameters, such as gearbox efficiency, η_{gear}, generator efficiency, η_{gen}, and electrical transmission efficiency η_{ele}, counting for all losses in power electronics, converter, switches, control, and

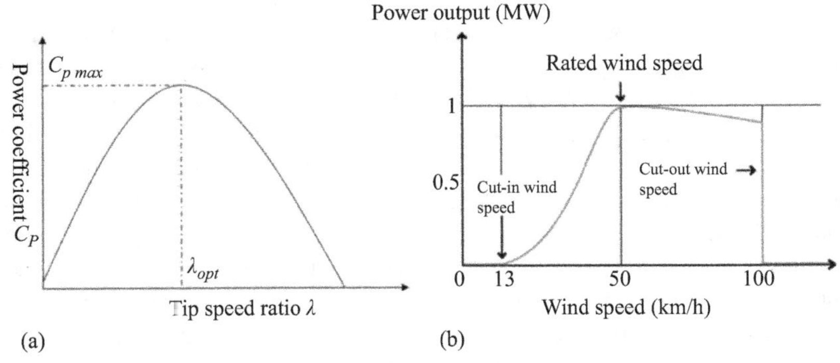

Figure 9.3 (a) Power coefficient versus TSR, and (b) typical large wind turbine power curve

cables. The total power conversion efficiency from wind to electricity, η_{tot} is then expressed as:

$$\eta_{tot} = C_P(\lambda) \cdot \eta_{gear} \cdot \eta_{gen} \cdot \eta_{ele}$$

The effective wind turbine power output becomes:

$$P_{eff} = \eta_{tot} \cdot 0.5\, \rho A v^3 \tag{9.49}$$

Example 9.11: An 8 m/s wind enters a wind turbine rotor with a diameter of 36 m. Assume the air density 1.2 kg/m^3. Calculate (a) the power of incoming wind, (b) the theoretical maximum power extracted by the wind turbine, and (c) if the gearbox, generator, and electrical transmission and processing efficiencies are 0.85, 0.96, and 0.95, respectively. What is the turbine effective converted power?

Solution:

(a) The incoming wind power is:

$$P_{\text{wind}} = 0.5\, \rho A v^3 = 0.5 \cdot 1.2 \cdot \left(\frac{\pi}{4} 36^2\right) \cdot 8^3 = 312,533 \text{ W}$$

(b) The maximum extracted power, the Betz limit is:

$$P_{\max} = 0.59 \cdot P_{\text{wind}} = 0.59 \times 312,533 = 184394.5 \text{ W}$$

(c) The effective power is then:

$$P_{eff} = 0.85 \cdot 0.96 \cdot 0.95 \cdot P_{\max} = 142942.6 \text{ W}$$

9.5.1 Wind energy conversion system components

Critical components of WTG system include the rotor, gearbox (not all WT has one but most of them), anemometer, generator, electric transmission, control system, tower, and foundation. The HWAT are either rotor-upwind, facing the wind or rotor-downwind types, enabling the wind to pass the tower and nacelle before entering the rotor. The rotor diameter, number and twist angle of blades, tower height, rated electrical power, and control strategy are the main factors considered in design. HWATs usually have two or three blades. In order to improve the power output performance, a selection of ratio between the rotor diameter and the hub height need to be considered. Two-blade rotors are faster and cheaper but three-blade ones operate more smoothly, with less flickers and higher efficiency. The tower height is also an important parameter regarding the WT performance. Higher towers offer usually more wind, larger rotors, higher power output on the trade-off of increased overall cost and installation. There are various types of towers for various classes of wind turbines. Foundation is important aspect of WT installation, requiring careful assessment of structural loads, materials, construction, geotechnical parameters, tower flange dimensions, and serviceability requirement.

The rotor is one the most critical element of a wind turbine. Downwind and upwind rotors are best suited for high capacity wind turbines, operating at higher tip speed ratio. TSR depends on the number of blades and both are affecting the rotor reliability. A rotor must have rotation speed according with its size, meaning correct diameter, and the wind speed. Rotor airfoils have an important effect on its performance, so require careful design. The rotor is designed to operate continuously, while its maintenance can be quite expensive. The most important rotor parameters in design are power at specific speed, power coefficient, TSR and the gearbox, generator, and electric transmission efficiencies. The WT performance is determined by the rotor blades, shape, number, and geometry with the most critical the rotor aerodynamics. Torque and area solidity increase when TSR is low, while efficiency and centrifugal stress increase when TSR is high. The most common rotor blade materials for high capacity turbine include wood laminates, glass-reinforced plastic, carbon fiber-reinforced plastics, steel, and aluminum. The material choice depends on the mechanical and structural properties of the fabrication. Hub height is the height from the ground to the rotor center and is a heavy turbine component The hub connects the blades to the main shaft and drivetrain and withstands all loads from the blades. The hub is rigid, hinged, or teetered and is usually made of ductile cast iron and is covered by a cone (made from the same material as the blades) in order to prevent environmental and visual effects.

In order to achieve dynamic stability, safe, and reliable operation under various wind conditions, a complex control system is required. Pitch control, is an expensive subsystem, located inside the hub where it rotates around radial axis as wind velocity changes. It changes the angle of attack by pitching the blades for almost optimal adjustment for every wind speed improving the dynamic stability and increasing the wind power capture. There are two types of pitch control hydraulic and electromechanical. A yaw control is also needed to maintain the system dynamic stability in turbulent environments. The nacelle is rotating with respect to the tower not with respect to the rotor, and this rotation is provided by the yaw system. This is necessary because the wind direction is not fixed. The yaw system directs the rotor in respect to the wind and consists of yaw bearing, motor, and drive. The way system is either active or passive, determined by the rotor type, upwind, or downwind.

Gearbox, another heavy turbine component is used to increase the rotational speed from a low-speed rotor to high-speed electric generator. Proper gearbox maintenance is required in order to reduce the operation cost, being expensive repair or replace it. Its lifetime is strongly affected by the wind regime. Wind turbines can run efficiently with minimum maintenance if the wind is smooth and there less turbulence effects. The nacelle is the housing located at the tower top and serves as protection for some wind energy system components, such as generator, gear box and control and must be strong enough to handle the loads. Moisture is the main source of damage for nacelles. Other hazard is fire (due to lightning or technical fault) nacelle caver is usually made of plastics. Large and medium size

WT may require heat exchanger to dissipate the heat generate by the electrical and mechanical subsystems. Other important component is the shaft, having main function to transfer torques. The low-speed and high-speed shafts are important components, being connected to the rotor and generator, respectively. Wind turbines must also be able to stop in case of failure of critical components or if the wind speed is higher than critical limit. There are two types of breaking systems: aerodynamic and mechanical breaking. Aerodynamic is the main breaking system, is turning the blade at 90° along their longitudinal axis and is stopping the system in a couple of rotation. The purpose of the mechanical breaking is to bring the rotor at rest and is used as backup for other breaking systems.

Electrical generators convert mechanical power into electrical power with the most common are induction and synchronous generators, first type being the most used in wind energy conversion. Induction generators are reliable and not very expensive, and having also mechanical properties that suit wind turbines. The most common type of induction generator rotor is squirrel-cage system. The generator speed changes according to the rotor speed, putting less stress on the tower, gearbox, and other components in the transmission lines or lower peak torque, and being important reasons for choosing induction generators rather than synchronous type. The synchronous generator is dependent on the speed of rotation, so the stator is connected to a DC-link converter system. This is increasing the system complexity and cost. It can operates at a wide range speed.

9.6 Solar energy

The potential solar resource reflects the ubiquitous nature of sunlight and yields a huge technical resource. Solar resource is exploited through solar photovoltaic and solar thermal panels and by heat pumps. While the potential solar resource is large solar energy is currently relatively costly to extract and deployment rates depend on economics. The practical resource for solar PV and thermal technologies is usually associated with buildings with suitable south facing roofs and is therefore a function of the built environment rather than the resource itself. The choice of whether solar PV or solar thermal technology is installed (or a combination thereof) is a decision taken building by building.

Solar radiation amount and intensity received by a given surface is controlled, at the global scale, *by the geometry of the Earth*, *atmospheric transmittance*, and *the relative Sun location.* At the local scale, solar radiation is controlled by surface slope, aspect and elevation. Clear sky solar radiation estimates for sloped surfaces are very important in renewable energy, civil engineering, and agricultural applications, which need accurate estimate of total energy striking a given surface. Figure 9.4 is showing Earth position relative to the sunrays at the winter solstice. At the winter solstice (around December 22), the North Pole is inclined 23.5° away from the Sun. All points on the Earth's surface north of 66.5° latitude are in total darkness while all regions within 23.5° of the South Pole receive continuous sunlight. At the time of the summer solstice (around June 22), the situation is reversed.

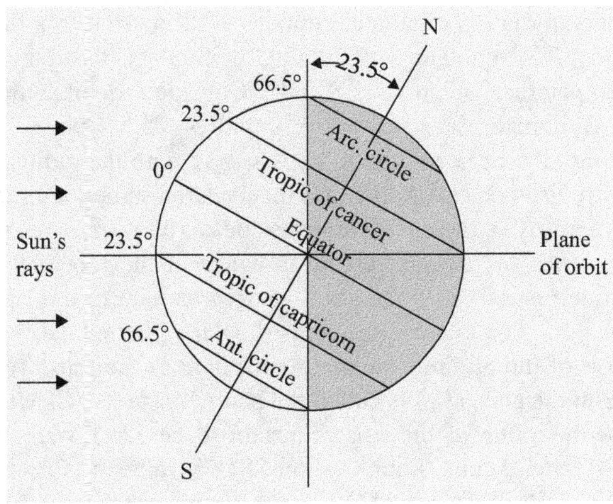

Figure 9.4 Position of the Earth in relation to the sunrays at time of winter solstice

Table 9.3 Day number of the first day of each month

Month	N	Month	N
January	1	July	182
February	32	August	213
March	60	September	244
April	91	October	274
May	121	November	305
June	152	December	335

At the two equinoxes (March 21 and September 21), both poles are equidistant from the Sun and all points on the Earth's surface have 12 h of daylight and 12 h of darkness. During this day a line from the center of the Sun to the center of the Earth passes through the equator and everywhere on Earth, so we have 12 h of daytime and night, hence the term equinox (equal day and night). Table 9.3 provides a list of day numbers for the first day of each month of the year, needed in solar radiation calculations.

Solar energy is in the form of electromagnetic radiation with the wavelengths ranging from approximately 0.3 μm (10−6 m) to over 3 μm, which correspond to ultraviolet (less than 0.4 μm), visible (0.4 and 0.7 μm), and infrared (over 0.7 μm). Most of this energy is concentrated in the visible and the near-infrared wavelength range. The Sun generated energy divided by the Sun surface area gives the Sun specific emission, 63.11 MW/m^2, the radiant power per square meter. One fifth of a square kilometer of the Sun surface emits every year an energy equal with total amount of the primary energy demand on Earth at present. Every object emits radiant energy in an amount that is a function of temperature. The usual way to describe how much radiation an object emits is to compare it to a theoretical body called a blackbody. To a good approximation, the Sun acts as a blackbody (perfect emitter) at a temperature of about 5,800 K. Considering the Sun a blackbody, applying Stefan-Boltzmann Law ($E_C = \sigma T^4$), with $\sigma = 5.67051 \cdot 10^{-8}$W/(m^{2}K^4), its surface temperature is estimate at 5,800 K. A sphere with the radius equal with the average Sun-Earth distance (1.5×10^8 km) receives the same total radiant power as the Sun surface. However, the energy density measured over 1 m^2, is much lower. This value determines the extraterrestrial radiance at the top of the Earth atmosphere. The extraterrestrial solar radiation also varies due to the Earth's elliptical orbit around the Sun. The average amount of solar radiation falling on a surface normal to the rays of the Sun outside the atmosphere of the Earth (extraterrestrial) at mean Earth–Sun distance (D_0) is called the solar constant, *SC*. Measurements by NASA indicated the value of the solar constant to be 1353 W/m^2 (±1.6%), 429 Btu/h·ft^2 or 1.94 cal/cm^2·min (Langleys/min). This value was revised upward to 1377 W/m^2 or 437.1 Btu/h·ft^2 or 1.974 Langleys/min, which was the value used in compiling SOLMET data in the United States. Recently, new measurements have found the value of the solar constant to be 1366.1 W/m^2. A value of 1,367 W/m^2 is also used by many references. However, the average distance variation causes a variation in radiant power density between 1,325 W/m^2 and 1,425 W/m^2, while the average solar constant is equal to:

$$SC = 1367 \pm 2 \frac{W}{m^2} \tag{9.50}$$

The seasonal variation in the solar radiation availability at the ground can be estimated from the relative movement geometry of the Earth around the Sun. Since the Earth's orbit is elliptical, the Earth-Sun distance varies during the year with about ± 1.7% form the average, expressed by the following relationship:

$$I_0 = SC \cdot \left(\frac{D_0}{D} \right)^2 \tag{9.51}$$

However, without much less accuracy the following approximate relationship is used in solar engineering applications:

$$I_0 = SC \cdot \left[1 + 0.034 \cos\left(\frac{360n}{365.25} \right)^{\circ} \right] \tag{9.52}$$

The parameter *n*, in both (9.52) is the day number, starting from January 1 as $n = 1$, as given in Table 9.3. The total energy flux incident on the Earth surface is obtained by multiplying *SC* by πR^2 (the area of the Earth disk), where *R* is the Earth radius. The average flux incident on a unit surface area is then obtained by dividing this number to the total surface area of the Earth ($4\pi R^2$), giving:

$$\frac{SC}{4} = 342 \text{ W}/m^2$$

9.6.1 Solar resources

The ultimate energy source for the vast majority of the energy systems is the Sun. Knowledge of the quantity and quality of solar energy available at a specific location is of prime importance for the design of any solar energy system. Although the solar radiation (*insolation*) is relatively constant outside the Earth's atmosphere, local climate influences can cause wide variations in available insolation on the Earth's surface from site to site. In addition, the relative motion of the Sun with respect to the Earth will allow surfaces with different orientations to intercept different amounts of solar energy. There are regions on the Earth of high or very high insolation where solar energy conversion systems are expected to produce the maximum amount of energy from a specific collector field size or type. However, solar energy is available over the entire globe, and only the size of the collector field needs to be increased to provide the same amount of heat or electricity as in the shaded areas. It is the primary task of the solar energy system designer to determine the amount, quality, and timing of the solar energy available at the site selected for installing a solar energy conversion system. Just outside the Earth's atmosphere, the Sun's energy is continuously available at the rate of 1,367 W, on every square meter facing the Sun. Due to the Earth's rotation, asymmetric orbit about the Sun, and the contents of its atmosphere, a large fraction of this energy does not reach the ground. In this section we will discuss the effects of the atmospheric processes that modify the incoming solar energy, how it is measured, and techniques used by designers to predict the amount of solar energy available at a particular location, both instantaneously and over a long term.

Earth's axis is tilted 23.5° with respect to the plan of its orbit around the Sun. This tilting results in longer days in the northern hemisphere from the spring equinox (approximately March 23) to the autumnal equinox (approximately September 22) and longer days in the southern hemisphere are during the other six months. On the equinoxes, the Sun is directly over the equator both poles are equidistant from the Sun, and the Earth experiences 12 h daylight and 12 h darkness. In the temperate latitude regions ranging from 23.450 to 66.50 north and south, variations in insolation are large. For example, at 40° N latitude, the average daily total extraterrestrial solar radiation varies from 3.94 kWh/m^2 in December to about 11.68 kWh/m^2 in June. To determine the extraterrestrial radiation at a specific location and time, it is necessary to know the Sun location in the sky. Sun position in the sky is a function of time and latitude, being defined by its solar altitude and solar azimuth angles. Sun's position relative to a location is determined by the location's latitude, L, a location's hour angle, W, and the Sun's declination angle. Latitude is the angular distance north or south of the Earth's equator, measured in degrees along a meridian. The hour angle is measured in the equatorial plane. It is the angle between the projection of a line drawn from the location to the Earth's center and the projection of a line drawn from the center of the Earth to the Sun's center. Thus, at solar noon, the hour angle is zero. At a specific location, the hour angle expresses the time of day with respect to solar noon, with one hour of time equal to 15° angle. By convention, the westward direction from solar noon is positive. The Sun's declination is the angle between projection of the line connecting the center of the Earth with the center of the Sun and the Earth's equatorial plane. Declination varies from −23.45° on the winter solstice (December 21), to +23.45° on the summer solstice (June 22), and is expressed by:

$$\delta = 0.006918 - 0.399912\cos\Gamma + 0.070257\sin\Gamma - 0.006758\cos 2\Gamma + 0.000907\sin 2\Gamma - 0.002697\cos 3\Gamma + 0.00148\sin 3\Gamma \tag{9.53}$$

where Γ is the daily angle, in radians given by:

$$\Gamma = 2\pi\frac{n-1}{365} \tag{9.54}$$

However, the above relationship is not usually employed in practical or engineering applications. For convenience, Table 9.3 provides a convenient list of day numbers for the first day of each month. Approximate estimates of declination angle, used in practical application are given by the following relationships:

$$\delta = 23.45 \cdot \sin\left[360 \cdot \frac{284+n}{365}\right] \tag{9.55}$$

And

$$\delta = \arcsin\left[0.4 \cdot \sin\left(\frac{360}{365}(n-81)\right)\right] \tag{9.56}$$

Or

$$\delta = 23.45 \cdot \sin\left[\frac{360}{365}(n-81)\right] \tag{9.57}$$

Example 9.12: Compute solar declination of the 21st day of each month of the year.

Solution: Using (9.55)–(9.57), the declination angles for the 21st day of each month of the year are:

Month	**Jan**	**Feb**	**Mar**	**Apr**	**May**	**Jun**	**Jul**	**Aug**	**Sep**	**Oct**	**Nov**	**Dec**
δ	−20.1	−11.2	0.0	11.6	20.1	23.4	20.4	11.8	0.0	−11.8	−20.4	−23.4

Solar declination as function of Julian day number is shown in Figure 9.5. Solar altitude angle (α) defines the elevation of the Sun above the location horizon. In the

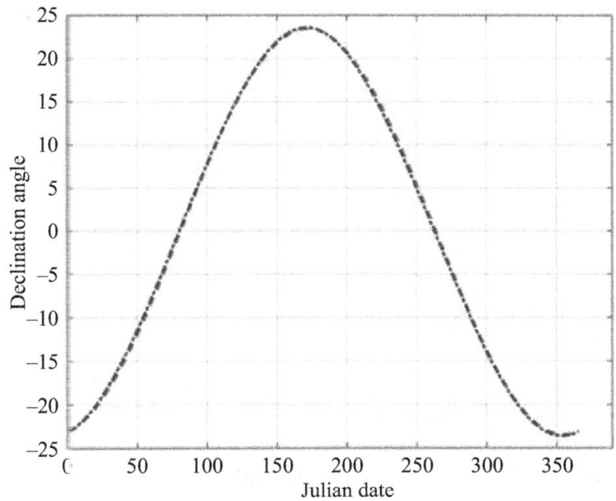

Figure 9.5 Declination angle versus Julian date

following, the term zenith refers to an axis drawn directly overhead at a site. The solar altitude is related to the solar zenith angle (θ_z), the angle between the Sun rays and the vertical, being calculated by using the following relationships:

$$\theta_z + \alpha = \frac{\pi}{2} = 90^\circ$$
$$\cos(\theta_z) = \sin(\alpha_S) = \sin(L) \cdot \sin\delta + \cos(L) \cdot \cos(H_S) \cdot \cos\delta \tag{9.58}$$

Here L is the local latitude, δ is the declination angle (Equations (9.55) or (9.57)), and H_S is the solar hour angle (i.e., the angular distance between the Sun and the local meridian line). In other words, this is the difference between the local meridian and the Sun meridian, with positive values occurring in the morning before the Sun is crossing the local meridian and negative values in the afternoon. Solar azimuth angle, the angle between the Sun and true north is then given by:

$$\sin\alpha_S = \frac{\cos(\delta) \cdot \sin(H_S)}{\cos(\alpha)} \tag{9.59}$$

For a given day, the calculation of declination, altitude, and azimuth angles is straightforward for latitude greater than the declination angle. Latitude angle L is the angle between the line from the center of the Earth to the location and the equatorial plane. The solar hour angle, H_S is equal to 15° times the number of hours from local solar noon. The values, east of due south, meaning the morning values are positive, while the values west, the afternoon ones are negative, and the numerical value of 15°/h is based upon the nominal time, 24-h required for the Sun to move 360° around the Earth, or 15° per hour. So H_S calculated by using a simple relationship:

$$H_S = \begin{cases} \dfrac{15^\circ}{\text{hour}} \cdot (\text{hour before solar noon}) \\ -\dfrac{15^\circ}{\text{hour}} \cdot (\text{hour after solar noon}) \end{cases} \tag{9.60}$$

Or

$$H_S = \frac{\text{Minutes from solar noon}}{4\ \text{min/degree}}$$

Example 9.13: Compute the solar altitude and solar azimuth angles at 10:00 a.m., 12:00 (noon) and 2:00 p.m. solar time Wilmington, Delaware ($L = 40°$ N) on April 10.

Solution: Using one of (9.55)–(9.57), the declination on April 10 ($n = 101$) is 7.9°. Then, by using (9.58) and (9.59), the following values of the solar altitude and the azimuth angles are computed:

10:00 a.m. 48.3°, 48.0°
12:00 p.m. 58.0°, 63.8°
2:00 p.m. 48.3°, –48.0°

In the early morning and late afternoon, during the spring or summer, the magnitude of the Sun's azimuth is liable to be more than 90° away from south, which never happens in the fall or winter. Since the inverse of a sine is ambiguous, sin $x = \sin(180 - x)$, we need to determine whether to conclude the azimuth is greater than or less than 90° away from south. In other words, (9.60) is correct, provided that:

$$\cos(H_S) < \frac{\tan(\delta)}{\tan(L)} \tag{9.61}$$

Otherwise, it means that the Sun is behind the E-W line, and the azimuth angle is:

$$\begin{cases} -\pi + |\alpha_S|, \text{for morning hours} \\ \pi - \alpha_S, \text{ for after noon hours} \end{cases} \tag{9.62}$$

In simpler words, if then we have:

$$\cos(H_S) \geq \frac{\tan(\delta)}{\tan(L)}, \text{ then } |\alpha_S| \leq 90^\circ, \text{ otherwise } |\alpha_S| > 90^\circ \tag{9.63}$$

Example 9.14: Find the altitude angle and azimuth angle for the Sun at 3:15 p.m. solar time in Detroit, Michigan (latitude 410.3° N) on the summer solstice.

Solution: The solar declination δ, for the summer solstice is +23.45°. The hour angle is computed as:

$$H = \left(\frac{15^\circ}{h}\right) \cdot (-3.25\text{h}) = -48.75^\circ$$

Using (9.58), the altitude angle is:

$$\begin{aligned} \sin\beta &= \cos(42.3^\circ) \cdot \cos(23.45^\circ)\cos(-48.75^\circ) + \sin(23.45^\circ) \cdot \sin(23.45^\circ) \\ &= 0.715 \\ \beta &= \sin^{-1}(0.715) = 45.7^\circ \end{aligned}$$

The azimuth angle from (9.59) is then:

$$\begin{aligned} \sin(\alpha_S) &= \frac{\cos(23.45^\circ) \cdot \sin(-48.75^\circ)}{\cos(45.7^\circ)} = -0.987 \\ \alpha_S &= -80.8^\circ \text{ and } \alpha_S = 180^\circ - (-80.8^\circ) = 260.8^\circ \end{aligned}$$

To resolve this ambiguity, and to decide which of these two options is correct, we apply relationships of (9.63).

$$\cos H = \cos(-48.75^\circ) = 0.660 \text{ and } \frac{\tan\delta}{\tan L} = \frac{\tan(23.45^\circ)}{\tan(42.3^\circ)} = 0.478$$

Since, cos (H) $\geq$ tan δ/tan (L), we are concluding that the azimuth angle is:

$$\alpha_S = -80.8^\circ (80^\circ \text{ west of south})$$

As, the solar noon, by definition, the Sun is exactly on the meridian (north-south line), and consequently, the azimuth angle is 0°. Therefore, the noon latitude angle (also known as the altitude angle), α_n is then given by:

$$\alpha_n = 90^\circ - L + \delta \tag{9.64}$$

Example 9.15: What are the maximum and minimum noon altitude angles for a location at 45° latitude?

Solution: Maximum declination angle is at summer solstice, $\delta = 23.5^\circ$, while the minimum is at winter solstice, when $\delta = -23.5^\circ$. The maximum and minimum noon altitude angles are then:

$$\begin{aligned} \alpha_n^{\max} &= 90^\circ - 45^\circ + 23.5^\circ = 68.5^\circ \\ \alpha_n^{\min} &= 90^\circ - 45^\circ - 23.5^\circ = 21.5^\circ \end{aligned}$$

During an equinox, at solar noon, the Sun is directly over the local meridian (line of longitude), the solar rays are striking a solar collector at the best possible angle, and they are perpendicular to the collector face. At other times of the year the Sun is a little high or a little low for normal incidence. However, on the average, it seems to be a good tilt angle. Solar noon is an important reference point for almost all solar calculations. In the Northern Hemisphere, at latitudes above the Tropic of Cancer, solar noon occurs when the Sun is due south of the observer. South of the Tropic of Capricorn, the opposite, it is when the Sun is due north, while in the tropics, the Sun may be either due north, due south, or directly overhead at solar noon. On the average, facing a collector toward the equator (in the Northern Hemisphere, meaning facing it south) and tilting it up at an angle equal to the local latitude is a good rule-of-thumb for better annual performances. Of course, if you want to emphasize winter collection, you might want a slightly higher angle, and vice versa for increased summer efficiency. The tilt angle that would make the Sun's rays perpendicular to the module at noon is given by:

$$Tilt = 90 - \alpha_n \tag{9.65}$$

Example 9.16: Find the optimum tilt angle for a south-facing photovoltaic module at latitude 32.3° at solar noon on May 1.

Solution: From Table 9.4 for May 1st, $n = 121$ and the declination angle (by using (9.56)), is then:

$$\delta = 23.45\sin\left[\frac{360}{365}(121 - 81)\right] = 14.9^\circ$$

Using (9.64) and (9.66), the tilt angle of the photovoltaic panel, facing south is:

$$\alpha_N = 90^\circ - 32.3^\circ + 14.9^\circ = 72.6^\circ$$
$$Tilt = 90^\circ - \alpha_N = 17.4^\circ$$

Table 9.4 US time zone meridians (West of Greenwich)

Time zone	Standard time meridian
Eastern time	75°
Central time	90°
Mountain time	120°
Pacific time	135°
Alaska and Hawaii	150°

There is strong relationship between the solar azimuth and hour angles, the first being the angle on the horizontal plane between the solar radiation projection and the line of the north-south direction, its positive values are indicating that the Sun is vest of the south, while the negative values indicate that the Sun is east of the south. The hour angle represents the angular distance between the Sun position at a particular time and its highest position for that day when crossing the local meridian at the solar noon. The length of the day varies for all latitudes during the year, so the solar altitude angle also changes hourly and daily (Equation (9.60)). To avoid the failures of (9.60) for negative numbers, it is advisable to implement the following set of equations. First one is coming from (9.60).

$$\alpha_S = \arctan\left(\frac{\sin(\alpha_S)}{\cos(\alpha_S)}\right) \tag{9.66}$$

When $\delta = -23.45°$, at the winter solstice, the locations in the Northern Hemisphere with latitude above 70° are in darkness, not illuminated at all during the day, and only negative values from (9.65) are obtained, while the South Pole is fully illuminated. For $L = 90°$ S, the solar altitude remains constant at 23.45° during the 24 h. The locations with latitude lower than 40° S are experiencing the greatest solar altitude during the day, and the solar altitude remains constant at 23.45° during 24 h. The opposite occurs during the summer solstice (June 22), $\delta = +23.45°$, the Southern Hemisphere locations with latitude lower than 70° S are not illuminated at all during the day, while the North Pole is fully illuminated 24 h. Locations with latitude lower than 40° N are experiencing the greatest solar altitude during the day, and the solar altitude remains constant at 23.45° during 24 h. Since the Earth rotates 15° per hour (4 min per degree), for every degree of longitude between one location and another, clocks showing solar time differ by 4 min. The only time two clocks would show the same time would be if they both were due north/south of each other. To deal with these longitude complications, the Earth is nominally divided into 24 1-h time zones, with each time zone spanning 15° of longitude. Of course, geopolitical boundaries invariably complicate the boundaries from one zone to another. The intent is for all clocks within the time zone to be set to the same time. Each time zone is defined by a Local Time Meridian located, ideally, in the middle of the zone, with the origin of this time system passing through Greenwich, England, at 0° longitude. Table 9.4 is showing the US time zone meridians. Since the Earth rotates 15° per hour (4 min per degree), for every degree of longitude between one location and another, clocks showing solar time would have to differ by 4 min. The Sun angles are calculated from the local solar time (ST), related to the local standard time (LST) through this relationship:

$$ST = LST + ET + (MR_{ST} - MR_{local}) \cdot 4\ \text{min/degree} \tag{9.67}$$

ET is the equation of time. A correction factor accounting for the irregularity of the Earth speed around the Sun, MR_{ST} is the standard time meridian, and MR_{local} is the local longitude. ET can be calculated using:

$$ET = 9.87 \cdot \sin 2B - 7.53 \cdot \cos B - 1.5 \cdot \sin B \tag{9.68}$$

The parameter B is equal to: 360 $(n-81)/365$. Together longitude correction and the equation of time give us the final relationship between local standard clock time (CT) and solar time (ST), expressed as:

$$ST = CT + \frac{4\ \text{min}}{\text{degree}}(MR_{ST} - MR_{local})^{\circ} + ET(\text{min}) \tag{9.69}$$

Example 9.17: Find the solar declination on October 5 in Cleveland, Ohio, the equation of time, ET, the solar time, ST at 2 p.m. on this day and location.

Solution: For October 5, $n = 279$ and by using (9.55)–(9.57) yields to $\delta = 6.02^{\circ}$. Negative declination makes sense the date is after the Fall Equinox ($n = 265$). On this day of the year, by using (9.68), ET, the equation of time is then:

$$B = \frac{360 \cdot (279 - 81)}{365} = 195.2^{\circ}$$

And

$$ET = 9.87 \cdot \sin(2 \cdot 195.82 \times 3.14/180) - 7.53 \cdot \cos(195.82 \times 3.14/180) \\ -1.5 \cdot \sin(195.82 \times 3.14/180) = 12.63\ (\text{min})$$

Cleveland longitude is 81.7° W, and by using (9.67), the solar time is:

$$ST = 2{:}00 + \frac{4\ \text{min}}{\text{degree}}(75 - 81.7)^{\circ} + 12.63\ \ (\text{min}) = 2{:}00 - 14.17\ (\text{min}) \\ = 1{:}45.43\ (\text{h})$$

For $L > \delta$, the solar time is due east (TE) or due west (TW) are calculated by:

$$TE = 12{:}00\ \text{Noon} - \frac{\left(\cos^{-1}\left[\frac{\tan\delta}{\tan L}\right]\text{degrees}\right)}{15^{\circ}/\text{hour}} \\ TW = 12{:}00\ \text{Noon} + \frac{\left(\cos^{-1}\left[\frac{\tan\delta}{\tan L}\right]\text{degrees}\right)}{15^{\circ}/\text{hour}} \tag{9.70}$$

Taking in account corrections to solar time (time zone and equation of time), the equation for hour angle in degrees, very useful in engineering and practical applications is given by:

$$H_S = \frac{15^{\circ}}{\text{hour}}(TS - 12\ \text{hour} + \text{ET}) + (MR_{ST} - MR_{local}) \tag{9.71}$$

For solar times earlier than TE or later than TW the Sun is north (south in the Southern Hemisphere) of east-west line and the absolute value of solar azimuth angle is greater than 90°. In this case, the correct value of solar azimuth angle is $a_s = 180^{\circ} - |a_s|$. For $L \leq \delta$, the Sun remains north, respectively; south in the austral hemisphere of the east-west line and the true values of a_s is greater than 90°. For $\alpha = 0$ in (9.71), the ***sunrise*** (H_{SR}) and ***sunset*** (H_{SS}) angles can be determined:

$$H_{SR(SS)} = \pm\cos^{-1}[-\tan L \cdot \tan \delta] \tag{9.72}$$

Since Earth rotates 15°/h, the hour angle can be converted to time of sunrise and sunset by:

$$\text{Sunrise/Sunset Time } (geometric) = 12{:}00 - \frac{H_{SR(SS)}}{15^\circ/\text{hour}} \tag{9.73}$$

Equations (9.72) and (9.73) are geometric relationships based on angles measured to the SUN center, hence the designation geometric sunrise in (9.73), being adequate for most of the solar engineering calculations. However, there is a difference between weather service sunrise and the geometric sunrise, due to two factors. The first deviation is caused by the atmospheric refraction, which bends the sunrays, making the Sun appearing to rise about 2.4 min sooner than geometrical calculation and then set 2.4 min later. The second factor is that the weather service definition of sunrise and sunset is the time at which the Sun upper limb (the top) is crossing the horizon, while the geometrical one is based on the center crossing the horizon. This effect is complicated by the fact that at sunrise or sunset the Sun pops up, or sinks, much quicker around the equinoxes when it moves more vertically than at the solstices when its motion includes much more of a sideward component. An adjustment factor Q that accounts for these complications is given by the following (US Department of Energy, 1978), relationship:

$$Q = \frac{3.467}{\cos(L) \cdot \cos(\delta) \cdot \sin(H_{SR})} \text{ (min)} \tag{9.74}$$

Since sunrise is earlier when it is based on the Sun top rather than the middle, Q is subtracted from geometric sunrise, while since the upper limb sinks below the horizon later than the Sun middle, Q is added to the geometric sunset. For mid-latitudes, the correction is typically in the range of about 4–6 min, which can be included or not depending on the applications.

Example 9.18: Determine the sunrise (geometric and conventional) that are occurring in Detroit, Michigan (latitude 42.3°) on July 15 ($n = 197$). Also find conventional sunset.

Solution: From (9.55), (9.56), or (9.57), the solar declination for this location and day is:

$$\delta = 23.45 \cdot \sin\left[\frac{360}{365}(187 - 81)\right] = 22.7^\circ$$

The hour angle at sunrise, by using (9.72) is then:

$$H_{SR(SS)} = \pm\cos^{-1}[-\tan(42.3°)\cdot\tan(22.7°)] = \pm 112.4°$$

From (9.73), the solar time of the geometrical sunrise is:

$$\begin{aligned}\text{Sunrise time}(\textit{geometric}) &= 12{:}00 - \frac{112.4}{15°/h} = 12{:}00 - 7.493h\\ &= 4{:}30.4 \text{ (a.m.)} - \text{solar time}\end{aligned}$$

From (9.74), the adjustment factor is then computed as:

$$Q = \frac{3.467}{\cos(42.3°)\cdot\cos(22.7°)\cdot\sin(112.7°)} = 5.49 \text{ (min)}$$

The upper limb will appear 5.5 min earlier than our original geometric calculation, so:

$$\textit{Sunrise} = 4{:}30.4 - 5.5 \text{ min} = 4{:}24.9 \text{ (a.m.)}$$

For Detroit, Michigan, the longitude is 83.04° W in the Eastern Time Zone with local time meridian 75°, and the estimate of the clock time follows.

$$B = \frac{360(187-81)}{365} = 104.2°$$

And the ET is the equation of time (Equation (9.68)) gives us:

$$\begin{aligned}ET &= 9.87\cdot\sin(2\cdot 104.2°) - 7.53\cdot\cos(104.2°) - 1.5\cdot\sin(104.2°)\\ &= -4.4 \text{ min.}\end{aligned}$$

The clock time is then computed using (9.69) as:

$$CT = 12{:}00 - 4(\text{min}/°)(75° - 83.04°) - (-4.4) = 12{:}00 + 27.76 \text{ min}$$

$$= 12{:}27.76$$

The local clock time is 27.76 min later than solar time, so sunrise will be at:

$$\text{Sunrise (upper limb)} = 4{:}24.9 + 27.76 = 4{:}52{:}66 \text{ (a.m.)}$$

Similarly, the geometric sunset is 7.493 h after solar noon, or 7:30.4 p.m. (solar time), while the upper limb will drop below the horizon 5.5 min later. Then adjusting for the 27.76 min difference between Detroit time and solar time gives:

$$\text{Sunset (upper limb)} = 7{:}30.4 + (5.5 + 27.76) = 8{:}03.26 \text{ (a.m.)}$$

Example 9.19: Determine the solar azimuth angle on May 1st for latitude 45° at 11:00 a.m.

Solution: On May 1st, $n = 121$ and using one of (9.55)–(9.57), the declination is: $\delta = 14.9°$. At 11:00 a.m., the hour angle is $w = -15°$. The solar azimuth angle is computed using (9.11), yielding to:

$$\cos(\theta_z) = \sin(45°)\sin(14.9°) + \cos(45°)\cos(14.9°)\cos(-15°)$$
$$\theta_z = 32.6°$$

9.7 Photovoltaics

Photovoltaic (PV) cells or solar cells are used to directly convert the solar energy (radiation) into electricity through the photoelectric effect. Solar electric energy conversion systems, or PV systems, are cost-effective and viable solutions to supply electricity for locations not connected to the conventional electrical grid or for special applications. PV systems are utilized almost everywhere, for terrestrial and space applications, and from Tropical to Polar Regions. However, the still higher PV capital cost means it is most economical to employ them for remote sites or applications where other, more conventional power generation options are not competitive. PV systems are used from very small low power applications to large power installations for electric utilities. Small power applications next to the utility grid can also be cost effective because the cost for a transformer is more than that of the PV system. An example is the flashing light for school lane crossings, warning signs, or street lighting. The photovoltaic term describes a process that produces direct electrical current from the radiant energy of the Sun. The PV effect can take place in solid, liquid, or gaseous materials. However, it is in solids, especially semiconductor materials, that acceptable conversion efficiencies have been found.

Solar cells are made from a variety of semiconductor materials and coated with special additives. The most widely used material for the various types of fabrication is crystalline silicon, representing over 90% of global commercial PV module production in its various forms. A typical silicon cell, with a diameter of 4 in., can produce more than 1 W of direct current (DC) electrical power in full Sun (about 1,000 W/m^2 solar radiation intensity). Individual solar cells can be connected in series and parallel to obtain desired voltages and currents. These groups of cells are packaged into standard modules to protect the cells from the environment while providing needed levels of voltage and current. PV modules are extremely reliable because they are solid state and have no moving parts. Silicon PV cells manufactured today can provide over 40 years of useful service life, with the average lifecycle of PV modules of about 25 years. Large scale PV applications for power generation, either on the house rooftops or in large fields connected to the utility grid are promising electricity generation option, clean, reliable, safe and strategically sound alternatives to current methods of electricity generation.

Solar cell is the component responsible for converting solar radiation into electricity. Some materials, silicon being the most common can produce a PV effect, consisting of freeing electrons, when sunlight is striking the cell material. The freed electrons cannot return to the positively charged sites (holes) without flowing through an external circuit, thus generating a current. Solar cells are designed to absorb as much light as possible and are interconnected in series and parallel electrical connections to produce desired voltages and currents. A PV module is composed of interconnected solar cells, encapsulated between a glass cover and weatherproof backing. The modules are typically framed in aluminum frames suitable for mounting and protection. PV modules are connected in series and parallel to form a PV arrays, thus increasing total available power output to the needed voltage and current for a particular application. PV modules are rated by their total power output (W). A peak Watt is the amount of power output a PV module produces at standard test condition (STC): 25 °C operating temperature and full noon-time sunshine (irradiance) of 1,000 W/m^2. However, PV modules often operate at temperatures higher than 25 °C in all but cold climates, thus reducing crystalline module operating voltage and power by about 0.5% for every 1 °C above STC. Therefore, a 100 W module operating at 45 °C (20° hotter than STC, yielding a 10% power drop), producing about 90 W. Amorphous PV modules do not have this effect. PV cells have been made with silicon (Si), gallium arsenide (GaAs), copper indium diselenide (CIS), cadmium telluride (CdTe), and a few other materials.

The common denominator of PV cells is that a *p–n* junction, or the equivalent, is needed to enable the photovoltaic effect. Understanding the *p–n* junction is thus critical for understanding how a PV cell converts sunlight into electricity and how a PV system operates. The main parameters, used to characterize the performance of solar cells are the peak power P_{max}, the short-circuit current density J_{SC}, or short-circuit current I_{SC}, the open-circuit voltage V_{oc}, and the fill factor *FF*. These parameters are determined from the illuminated *I–V* characteristic, as shown in Figure 9.6. The conversion efficiency η_{pv} is determined from these parameters. Short-circuit current I_{SC} is the current that flows through the external circuit when the electrodes of the solar cell are short circuited. The short-circuit current of solar cells depends on the photon flux density incident on the solar cell, determined by the spectrum of the incident light. For standard solar cell measurements, the solar spectrum is standardized to the AM1.5 spectrum. The I_{SC} depends on the area of the solar cell. In order to remove the dependence of the solar cell area on I_{SC}, often the short-circuit current density is used to describe the maximum current delivered by a solar cell. The maximum current that the solar cell can deliver strongly depends on the optical properties of the solar cell, such as absorption in the absorber layer and reflection. The gross current generated by a solar cell, I_L (the light current), since it occurs when the cell is illuminated in calculated taking into account the losses occurring in the cell. When a solar cell is connected to an external circuit, the photo-generated current then flows from the *p*-type semiconductor-metal contact,

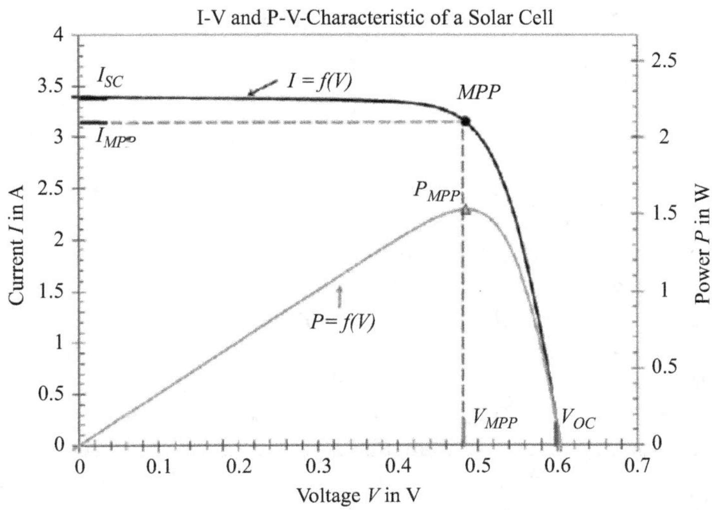

Figure 9.6 Solar cell I–V characteristics and output power versus voltage

through the conductor loop, powers the load, until reaches the n-type semiconductor-metal contact. Under a certain sunlight illumination, the current passed to the load from a solar cell depends on the external voltage applied to the solar cell normally through a power electronic converter for a grid-connected PV system. If the applied external voltage is low, only a low photo-generated voltage is needed to make the current flow from the solar cell to the external system. Nevertheless, if the external voltage is high, a high photo-generated voltage must be built up to push the current flowing from the solar cell to the external system. This high voltage also increases the diffusion current, so that the net output current of the solar cell is reduced. The I–V characteristic curve for the p–n junction diode is described by the (Shockley) diode equation, diode current, I_D, expressed as:

$$I_D = I_0\left(\exp\left(\frac{qV_D}{mkT}\right) - 1\right) \tag{9.75}$$

Here, I_0 is the reverse saturation current (A) and V_D is the forward diode voltage, T is the junction temperature (K), $k = 1.381 \times 10^{-23}$ J/K, Boltzmann constant, $q = 1.605 \times 10^{-19}$, C the elementary (electron) charge, and m, the diode factor depends on the voltage at which the cell is operating. The diode factor m is equal to 1 for an ideal diode; however, a diode factor between 1 and 5 allows a better description of PV cell characteristics. The so-called thermal voltage $V_T = kT/q$ has a value of 25.7 mV at 25 °C (STC) and the magnitude of the saturation current I_0 is of the order of 10^{-10} to 10^{-5} A. For standard temperature of the junction, 25 °C (Equation (9.75)), and ideal diode ($m = 1$) has the form:

$$I_D = I_0[\exp(38.9 \cdot V_D) - 1] \tag{9.76}$$

The dark current, I_D is flowing in the opposite direction of the photovoltaic (light) current, I_L, so the net diode current is computed as:

$$I = I_L - I_D = I_L - \left[\exp\left(\frac{qV_D}{mkT}\right) - 1\right] \tag{9.77}$$

Plotting current I versus voltage V, using (9.77) for the representative cell parameters and the insulation level the I–V diagram (Figure 9.6) is obtained. I–V curve typically passes through the two end points: the short-circuit current, I_{SC}, and the open-circuit voltage, V_{oc}. I_{SC} is the current produced with the positive and negative terminals of the cell shorted, the voltage between the terminals is zero, corresponding to zero load resistance. The V_{oc} is the voltage across the positive and negative terminals under open-circuit conditions with no current, corresponding to infinite load resistance, and the peak power point is located on the farthest upper right corner of where the rectangular area is greatest under the curve. At zero voltage, the amount of current produced by the PV cell is the *short-circuit current*, I_{SC}, which is equal to the light current since the dark current is making no contribution, the exponential term in (9.77) is equal to 1. The *open-circuit voltage* V_{OC}, the voltage at which no current flows (due to the exponential nature of I_D) through the external circuit is the maximum voltage that a solar cell can deliver, depending on the photo-generated current density. V_{OC} is calculated, assuming that the net current is zero, Equation (9.77):

$$V_{OC} = \frac{mkT}{q} \ln\left(\frac{I_L}{I_0} + 1\right) \approx \frac{mkT}{q} \ln\left(\frac{I_L}{I_0}\right) \tag{9.78}$$

For 25 °C, the standard junction temperature, the above equation becomes:

$$V_{OC} = 0.0257 \ln\left(\frac{I_L}{I_0} + 1\right) \tag{9.79}$$

Equation (9.78) shows that V_{oc} depends on the saturation (reverse diode) current of the solar cell, I_0, and the photo-generated current, I_L. The photo-generated current density J_{ph}, typically has a small variation, key effect being the saturation current, since it may vary by orders of magnitude. The saturation current density, J_0, depends on the recombination processes in the solar cell, so V_{oc} is a measure of the amount of recombination in the device. Laboratory crystalline silicon solar cells have a V_{oc} of up to 720 mV under the standard *AM1.5* conditions, while commercial solar cells typically have V_{oc} exceeding 0.6 V. depends on the recombination in the solar cell.

Example 9.20: An ideal PV cell has a saturation current of 10^{-8} A and is operation at 35 °C. Find the open-circuit voltage, assuming that light current is 600 mA.

Solution: The short-circuit current is equal to the light current at zero voltage:

$$I_{SC} = I_L = 600 \text{ mA}$$

Assuming the ideal diode ($m = 1$) in (9.78), the open-circuit voltage is:

$$V_{OC} = \frac{1.38 \times 10^{-23}(35 + 273.15)}{1.602 \times 10^{-19}} \ln\left(\frac{600 \times 10^{-3}}{10^{-8}} + 1\right) = 0.4755 \text{ V}$$

The net current can be written in terms of the fixed cell parameters I_{SC} and V_{OC} and independent variable, V, the diode voltage. We are now able to determine the values V_{mp} and I_{mp} (the so-called Maximum Power Point, MPP on the I–V curve, as shown in Figure 9.6) that maximize the solar cell power output, P_{max}. The PV cell may be operated over a wide range of voltages and currents, by varying the load resistance from zero (a short circuit) to infinity (an open circuit), it is possible to determine the highest efficiency as the point where the cell delivers maximum power. Because power is the product of voltage and current, the maximum-power point (P_{max}) occurs on the I–V curve, where the product of current (I_{mp}) and voltage (V_{mp}) is a maximum. No power is produced at the short-circuit current with no voltage or at open-circuit voltage with no current, so maximum power generation is expected to be between these points. Note that maximum power is generated at only one point on the power curve; this occurs at the knee of the curve. This point represents the maximum efficiency of the device in converting sunlight into electricity. Recall that the power is the product of voltage and current. The values of I_{mp} and V_{mp} are determined by taking the derivative of power and setting equal to zero.

$$\frac{dP}{dV} = V\frac{dI}{dV} - I = 0$$

The derivative of the current versus voltage is computed from (9.77), using the thermal voltage parameter $V_T = q/kT$, as:

$$\frac{dI}{dV} = -\frac{I_0}{V_T}\exp\left(\frac{V}{V_T}\right)$$

Combing these two equations yields to:

$$\frac{dP}{dV} = (I_L + I_0) - I_0\left(1 + \frac{V}{V_T}\right) \cdot \exp\left(\frac{V}{V_T}\right)$$

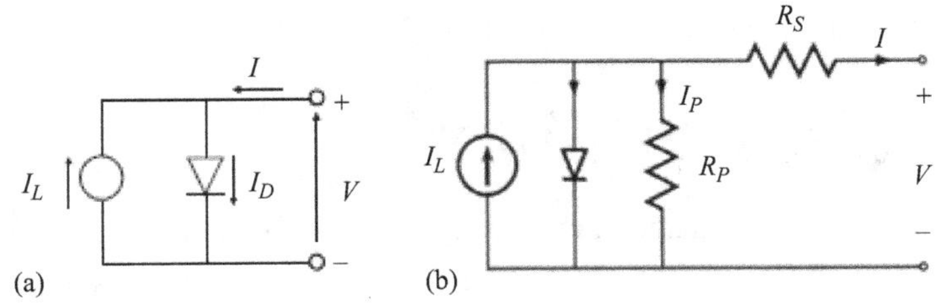

Figure 9.7 Solar cell equivalent circuit: (a) simple equivalent circuit; (b) extended one diode equivalent circuit model

Setting the above power derivative equal to zero, yields to a relationship that can be used to determine V_{mp}.

$$\left(1+\frac{V_{mp}}{V_T}\right)\cdot \exp\left(\frac{V_{mp}}{V_T}\right)=\frac{I_L+I_0}{I_0} \tag{9.80}$$

Equation (9.80) is nonlinear and has no analytic solution. However, it can be solved by using numerical methods. Once V_{mp} is determined, substituting it in (9.77) yields to I_{mp}.

$$I_{mp}=I_L-I_0\left[\exp\left(\frac{V_{mp}}{V_T}\right)-1\right] \tag{9.81}$$

Keep in mind that the power of a PV cell delivers to a load depends also on the load resistance. The optimum operating point occurs at V_{mp} and I_{mp}. The relationship between the maximum power $P_{max}=V_{mp}{\cdot}I_{mp}$ and the product of open-circuit voltage and short-circuit current is referred as the fill factor, *FF*. The fill factor is the ratio between the maximum power (P_{max}) generated by a solar cell and the product of V_{oc} with I_{sc}, expressed as:

$$FF=\frac{P_{\max}}{V_{OC}\cdot I_{SC}} \tag{9.82}$$

Example 9.21: An ideal solar cell is operating at 25 °C has a saturation current of 2 nA and photocurrent of 0.8 A. Compute the voltage and current at the maximum power point and the fill factor.

Solution: The thermal voltage at 25 °C, $T=273.15+25$ is:

$$V_T=\frac{kT}{q}=0.025683\ \text{V}$$

The open-circuit voltage, assuming ideal diode ($m=1$) is:

$$V_{OC}\approx\frac{1.38\times10^{-23}(273.15+25)}{1.602\times10^{-19}}\ln\left(\frac{0.8}{2\times10^{-9}}\right)=0.5087\ \text{V}$$

Substituting the numerical values in (9.80) yields to:

$$\left(1+\frac{V_{mp}}{0.025683}\right)\cdot \exp\left(\frac{V_{mp}}{0.25683}\right)=0.4\times10^{9}$$

Solving the above equation numerically yields the voltage at maximum power point:

$$V_{mp}=0.46587\ \text{V}$$

Substituting this value in (9.81) yields to the current at maximum power point:

$$I_{mp} = 0.8 - 2 \times 10^{-9}\left[\exp\left(\frac{0.46587}{0.25683}\right) - 1\right] = 0.64906 \text{ A}$$

Hence, the maximum power of the cell and the fill factor are:

$$P_{max} = V_{mp}I_{mp} = 0.46587 \times 0.64906 = 0.30238 \text{ W}$$

$$FF = \frac{P_{max}}{V_{OC}I_{SC}} = \frac{0.30238}{0.5087 \times 0.8} = 0.74$$

Assuming that the solar cell behaves as an ideal diode, the fill factor can be expressed as a function of open-circuit voltage V_{oc}, as given by:

$$FF = \frac{v_{OC} - \ln(v_{OC} + 0.73)}{v_{OC} + 1} \tag{9.83}$$

Here, $v_{OC} = \frac{q \cdot V_{OC}}{kT}$ is the normalized voltage. Equation (9.83) is a good approximation for normalized voltage values higher than 10. However, *FF* does not change drastically with a change in V_{oc}, because large variations in V_{oc} are not common. For example, at standard illumination condition, a typical commercial solar cell a maximal *FF* is about 0.85. The conversion efficiency is calculated as the ratio between the maximal generated power and the incident power. The irradiance value P_{in} of 1,000 W/m^2 for the *AM1.5* spectrum has become a standard for measuring the conversion efficiency of solar cells, is:

$$\eta_{pv} = \frac{P_{\max}}{P_{IN}} = \frac{V_{OC}I_{SC} \cdot FF}{P_{IN}} \tag{9.84}$$

Typical conversion efficiency lies in the range of 15%–20% for commercial solar cells. An ideal photovoltaic cell can be described by a current source in parallel with diode. This simple equivalent circuit is well suited to describe the behavior of an irradiate solar cell. This simple equivalent circuit (as shown in Figure 9.7(a)) is sufficient in many applications. The current source generates the light (photocurrent), IL, which depends on the irradiance (solar radiation intensity), *E* and a coefficient, C_0 as:

$$I_L = C_0 \cdot E \tag{9.85}$$

An ideal solar cell can be modeled by a current source, representing the photogenerated current I_L, in parallel with a diode, representing the ideal *p-n* junction of a solar cell. In a real solar cell, there exist other effects, not accounted for by the ideal model. The differences between calculated and measured characteristics of the solar cells are in the range of few percent. However, the only extended solar cell equivalent circuit can describe its behavior over an extended range of operating conditions. Charge carriers in an actual solar cell are experiencing voltage drop on their way through the junction to external contacts. A series resistance, R_S expresses this voltage drop, and in addition a parallel resistance, R_P is included to describe the leakage currents at the cell edges. These two extrinsic effects are summarized as (1) current leaks proportional to the terminal voltage of a solar cell and (2) losses of semiconductor itself and of the metal contacts with the semiconductor. The first is characterized by a parallel resistance R_P accounting for current leakage through the cell, around the edge of the device, and between contacts of different polarity (Figure 9.7(b)). The second is characterized by a series resistance R_S, which causes an extra voltage drop between the junction voltage and the terminal voltage of the solar cell for the same flow of current. The series resistance of real cells is in the range of milliohms (mΩ), while the parallel resistance is usually higher than 10 Ω. The mathematical model of a solar cell is described by the following equations:

$$I = I_L - I_0\left(\exp\left(\frac{qV_D}{mkT}\right) - 1\right) - \frac{V_D}{R_P} \tag{9.86a}$$

And

$$V_C = V_D - R_S \cdot I \tag{9.86b}$$

where I_L is proportional to the sunlight intensity, m is the diode ideality factor ($m =$ 1 for an ideal diode), the diode reverse saturation current I_0 depends on temperature. At 25 °C and standard insolation testing condition, (9.86a) becomes:

$$I = I_L - I_0(\exp(38.9 \cdot V_D) - 1) - \frac{V_D}{R_P} \tag{9.87}$$

Important solar cell characteristics are the output current, power, and output voltage. Several factors affect the PV cell, causing variations from the theoretical behavior. The most important factors are the temperature and the solar radiation. Increasing the solar irradiance increases the magnitude of the light current consequently increases the short-circuit current and the one-circuit voltage, so the cell output power. Cell temperature affects linearly the thermal voltage V_T, while the saturation current, I_0 and the light current, I_L have nonlinear temperature dependence. The net result is that the open-circuit voltage is reduced when the temperature increases. However, cell performances vary in temperature not only because ambient temperatures change but also because insolation on the cells changes. Even the cell current increases with the temperature, V_{OC} is falling with temperature, leading lower power output, because the voltage dominates. An indicator of the temperature effects on the solar cell is the nominal operating cell temperature (NOCT), the cell temperature at 20 °C, solar irradiance of 800 W/m^2, and wind of 1 m/s. To account for other ambient conditions, the following expression may be used:

$$T_{\text{PVCell}} = T_{amb} + \frac{NOCT - 20°\text{C}}{800} \cdot S \tag{9.88}$$

where $T_{PV\ cell}$ is cell temperature (°C), T_{amb} is ambient temperature (°C), and S is solar insolation (W/m^2). For Si solar cell, the open-circuit voltage, V_{OC} drops by 0.37% per Celsius degree increases, while the output power drops about 0.5% per degree.

Example 9.22: Estimate cell temperature, open-circuit voltage, and maximum power output for a 150-W PV module operating at 32 °C and 1,000 W/m^2 insolation. The NOCT of this module is 48 °C and the open-circuit voltage is 43.5 V.

Solution: The cell operating temperature is:

$$T_{PV\ Cell} = 32 + \frac{48 - 20}{800} \cdot 1,000 = 67°C$$

The open-circuit voltage and power are computed as:

$$V_{OC} = 43.5 \cdot (1 - 0.0037(67 - 25)) = 37.74 \text{ V}$$

And

$$P_{OUT} = 150 \cdot (1 - 0.005 \cdot (67 - 25)) = 118.5 \text{ W}$$

9.7.1 PV cell manufacturing technologies

The most common material for the production of solar cells is silicon. Silicon is obtained from sand and is one of the most common elements in the Earth's crust, so there is no limit to the availability of raw materials. Current solar cell manufacturing technologies are: monocrystalline, polycrystalline, bar-crystalline silicon, and thin-film technology. Cells made from crystal silicon (Si), are made of a thinly sliced piece (wafer), a crystal of silicon (mono-crystalline) or a whole block of silicon crystals (multicrystalline); their efficiency ranges between 12% and 19%.

Monocrystalline Si cells have conversion efficiency for this type of cells ranges from 13% to 17%, and can generally be said to be in wide commercial use. In good light conditions it is the most efficient photovoltaic cell. This type of cell can convert solar radiation of 1.000 W/m^2 to 140 W of electricity with the cell surface of 1 m^2. The production of monocrystalline Si cells requires an absolutely pure semiconducting material. Monocrystalline rods are extracted from the molten silicon and sliced into thin chips (wafer). Such type of production enables a relatively high degree of usability. Expected lifespan of these cells is typically 25–30 years and, of course, as well as for all photovoltaic cells, the output degrades somewhat over the years. Multicrystalline Si cells are converting solar radiation of 1.000 W/m^2 to 130 W of electricity with the cell surface of 1 m^2. The production of these cells is economically more efficient compared to monocrystalline. Liquid silicon is poured into blocks, which are then cut into slabs. During the solidification of materials, crystal structures of various sizes are being created, at whose borders some defects may emerge, making the solar cell to have a somewhat lower efficiency, ranging from 10% to 14%. The expected lifespan is up to 25 years. Ribbon silicon, in its production process, has the advantage for not needing a wafer cutting (which results in loss of up to 50% of the material in the process of cutting). However, the quality and the possibility of production of this technology is not making it a leader in the near future. Their efficiency is around 11%.

In the thin-film technology, the modules are manufactured by piling extremely thin layers of photosensitive materials on a cheap substrate, such as glass, stainless steel, or plastic. The process of generating modules in thin-film technology has resulted in reduced production costs compared to crystalline silicon technology, which is somewhat more intense. Today's price advantage in the production of a thin-film is balanced with the crystalline silicon due to lower efficiency of the thin-film, which ranges from 5% to 13%. The share of thin-film technology on the market is 15% and constantly increasing, it is also expected an increase in years to come and thus reduce the adverse market ratio in relation to the photovoltaic module of crystalline silicon. Lifespan is around 15–20 years. There are four types of thin-film modules (depending on the active material) that are now in commercial use. Amorphous Si Cells, with efficiency is around 6%, a cell surface of 1 m^2 can convert 1.000 W/m^2 of solar radiation to about 50 W of electric energy. Progresses in research of this type of module have been made and it is expected a greater efficiency in the future. If a thin film of silicon is put on a glass or another substrate it is called amorphous or thin layer cell. The layer thickness is less than 1 μm, therefore the lower production costs are in line with the low cost of materials. However, the efficiency of amorphous cells is much lower compared to other cell types. It is primarily used in equipment where low power is needed or, more recently, as an element in building facades. Cadmium tellurium (CdTe) cells, with an efficiency of around 18%, a cell surface of 1 m^2 can convert solar radiation of 1.000 W/m^2 to 160 W of electricity in laboratory conditions. Cadmium teleurid is a fusion of metal cadmium and tellurium semimetal. It is suitable for use in thin photovoltaic modules due to the physical properties and low-technology manufacturing. However, it is not widely used due to cadmium toxicity and suspected carcinogenicity. Copper indium gallium selenide (CIS, CIGS) cells have the highest efficiency among the thin-film cells, which is about 20%. This cell type can convert solar radiation of 1.000 W/m^2 to 160 W of electricity in laboratory conditions. Thermo sensitive solar cells and other organ cells (DSC) are yet in the development stage, since it is still testing and it is not increasingly commercialized. Cell efficiency is around 10%. The tests are going in the direction of using the facade integrated systems, which has proven to be high-quality solutions in all light radiation and all temperature conditions. Also, a great potential of this technology is in low cost compared to silicon cells. There are other types of photovoltaic technologies that are still developing, while others are to be commercialized.

Regardless of the lifespan, the warranty period of today's most common commercial PV modules is 10 years at 90% power output, and 25 years at 80% power output. The period of energy depreciation of photovoltaic cells is the time period that must pass using a photovoltaic system to return the energy that has been invested in the construction of all parts of the system, as well as the energy required for the breakdown after the lifetime of a PV system. Of course, the energy depreciation time is different for different locations at which the system is located, thus it is a lot shorter on locations with a large amount of irradiated solar energy, up to 10 or more times shorter than its lifetime.

9.7.2 PV modules and arrays

Since an individual cell produces only about 0.5–0.7 V, and about 1 W power output there are rare applications for which just a single cell is needed to provide the required power. To increase the power ratings, the PV cells are connected in series and parallel configurations. The series connection increases the overall output voltage, while the parallel connection increase the overall output current. The interconnected PV cells are called PV module or panel, the basic building block for PV applications. A typical PV module consists of a number of pre-wired PV cells in series, all encased in tough, weather resistant packages. A typical PV module has 36 cells in series, designated as a "12-V PV module," even though it is capable of delivering higher voltages than that. Some 12-V modules have only 33 cells, which, as will be seen later may, be desirable in certain very simple battery charging systems. Large 72-cell modules are now quite common, some of which have all of the cells wired in series, in which case they are referred to as 24-V modules. Some 72-cell modules can be field-wired to act either as 24-V modules with all 72 cells in series or as 12-V modules with two parallel strings having 36 series cells in each. Multiple modules can be wired in series to increase voltage and in parallel to increase current, to provide the required power. The interconnected PV modules form a PV array, while several PV arrays forms a PV system. An important element in PV system design is deciding how many modules should be connected in series and how many in parallel to deliver whatever energy is needed. Such combinations of modules are referred to as a PV array. Figure 9.8 shows this distinction between PV cells, PV modules, and PV arrays. Several of these PV arrays form a PV system. In order to maximize the power output of a PV system tracking devices to follow the Sun throughout the day are mounted, they are tilting the PV arrays to maximize the solar cell exposure to the solar radiation, thus increasing the system power output.

A PV module I–V curve has the same set of operation points as solar cells, which are critical in order to properly install and troubleshoot PV power systems: Short-circuit current (I_{sc}), the maximum current generated by a PV module and is measured when no load (resistance) is connected (i.e., the module is shorted).

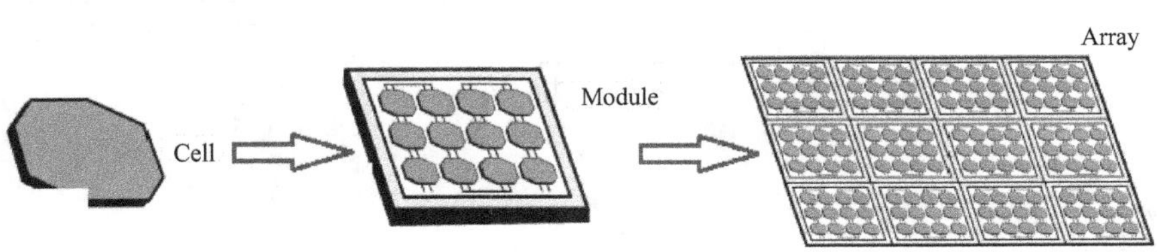

Figure 9.8 Diagrams of PV cells, modules, and arrays

Its value depends on the cell surface area and the solar radiation incident upon the surface. I_{sc} is used for all electrical ampacity design calculations. Nameplate current production is given for a PV cell or module at standard reporting condition (SRC). The SRC commonly used by the PV industry is for a solar irradiance of 1,000 W/m^2, a PV cell temperature of 25 °C, and a standardized solar spectrum referred to as an air mass 1.5 spectrum ($AM = 1.5$), which is the standard test condition (STC). However, in reality, unless one is using PV in a relatively cold climate, the modules are operating at higher temperatures (often 50 °C or more), which reduces their power performances. As module operating temperature increases, module voltage drops while current essentially holds steady. PV module operating voltage is reduced on average for crystalline modules approximately 0.5% for every degree Celsius above STC (i.e., 25°). In general, when sizing terrestrial PV systems, we expect a 15%–20% drop in module power from STC. This is important to remember when calculating daily actual energy production. One may ask why the industry does not use a more realistic operating temperature for defining STC conditions, and indeed, many module manufacturers will provide a more realistic 45 °C or other rating. Open-circuit voltage (V_{OC}) is the maximum voltage generated by the module, measured when no external circuit is connected to the PV module, similar as for the PV cells. Similar, the rated maximum power voltage (V_{mp}) corresponds to the maximum power point on the module I–V curve. Maximum power (P_{max}) is the maximum power available from a PV module, occurring at the maximum power point on the I–V curve, the product of the PV current (I_{mp}) and voltage (V_{mp}). If a PV module operates outside its maximum power value, the amount of power delivered is reduced and represents needless energy losses. Thus, this is the desired point of operation for any PV module or system. Manufacturers are providing PV module specifications, such as ones shown in Table 9.5.

The PV cell model developed before can be used to compute the values of the PV module, array, and system if the cell parameters and the environmental conditions are known. When the solar cells are connected in series and parallel, we are making the assumption that the cell parameters and the environmental conditions are the same for every cell in the module, array or system. Modules must be fabricated so the PV cells and interconnects are protected from moisture and are resistant to degradation from the ultraviolet radiation. Since the modules are

Table 9.5 Sample of PV module specifications

Parameters	PV Module (36 cells)	PV Module (72 cells)
Operating point	Model BP VLX-53	NE-Q5E2U
P_{mp}	53 Wp (peak W)	165 Wp
V_{mp}	17.2 V	34.6 V
I_{mp}	3.08 A	4.77 A
V_{oc}	21.5 V	43.1 V
I_{sc}	3.5 A	5.46 A
Standard test conditions (STCs)	1,000 W/m^2, 25 °C	1,000 W/m^2, 25 °C

usually exposed to a wide range of temperatures, they must be designed so that thermal stresses are not causing delamination. Modules must also be resistant to blowing sand, salt, hailstones, acid rain, and other unfriendly environmental conditions, and must be electrically safe over the long period. The electrical characteristics of the PV array are the same to the individual modules, with the power, current, and voltage modified according to the number of modules connected in series, parallel or series-parallel configuration. However, the module or array efficiencies are usually less than the constituting cell or modules, unless the cells or modules are perfectly similar. For modules in series, *I–V* curves are simply added along the voltage axis. For modules in parallel, the same voltage is across each PV module and the total current is the sum of the individual currents. When high power is needed, the array usually consists of a combination of series and parallel PV modules for which the total *I–V* curve is the sum of the individual module *I–V* curves. There are two ways to imagine wiring a series/parallel combination of modules: (a) the series modules are wired as strings and the strings wired in parallel, and (b) the parallel modules are wired together first and those units combined in series. The total *I–V* curve is the sum of the individual module curves, being the same in either case when everything is working. However, the wiring of strings in parallel is preferred, for the reason that if an entire string is removed from service, the array is still delivering the needed voltage to the load, though the current is diminished, which is not the case when a parallel group of modules is removed. When photovoltaics are wired in series, they all carry the same current, and at any given current their voltages add. For a PV module having n cells in series, the current is simply calculated using (9.86a) or (9.87), while the voltage, by using (9.86b), is then:

$$V_{Module} = n(V_D - R_S \cdot I) \tag{9.89}$$

Example 9.23: A PV module consist of 36 identical cells, all wired in series. Calculate the module voltage, current, and the delivered power considering 1,000 W/m^2 insolation. Each cell has short-circuit current I_{SC} of 3.5 A, its reverse saturation current is $I_0 = 6 \times 10^{-10}$ A at 25 °C, junction voltage 0.5 V, a parallel resistance R_P equal to 7.50 Ω and series resistance R_S equal to 0.005 Ω.

Solution: By using (9.86a) the current is:

$$I = 3.5 - 6 \times 10^{-10} \cdot (\exp(38.9 \times 0.5) - 1) - \frac{0.5}{7.5} = 3.265 \text{ A}$$

Then the module voltage is calculated by using (9.89), as:

$$V_{Module} = 36 \cdot (0.5 - 0.005 \cdot 3.265) = 17.4 \text{ V}$$

The power delivered by this module is:

$$P_{Module} = V_{Module} \times I = 17.4 \times 3.265 = 56.86 \text{ W}$$

9.7.3 PV system configuration and sizing

PV systems are made up of a variety of components, which include arrays, wires, fuses, controls, batteries, trackers, and inverters. What components are included into the system depend on the type of application. PV systems are modular by nature; thus, systems can be readily expanded and components easily repaired or replaced if needed. PV systems are cost effective today for many remote power applications, as well as for small stand-alone power applications in proximity to the existing electric grid. These systems should use good electrical design practices, such as the National Electrical Code (NEC) or its equivalent. In a PV system, balance of system refers to all of the system components except the PV modules, accounting for half of the system cost and most of the system maintenance. These components include fuses and disconnect switches to protect the systems, structures, enclosures, wire connectors to link different hardware components, switchgear, ground fault detectors, charge controllers, general controllers, batteries, inverters, PV system metering, and monitoring devices. The selection of good components is as important as the selection of PV modules. Low-quality components are often responsible for many avoidable PV system maintenance problems, especially in remote areas, leading usually to premature system failure. The PV industry goal is to provide PV systems with operational life spans of 25 or more years. In order to size and design a solar energy system, it is necessary to conduct assessment of the solar energy potential and energy requirements that the system needs to meet. With this information, a reasonable and informed estimate for the size of a PV system required to supply the energy needed can be made.

To compute accurately solar energy system size requires understanding the local solar resource, that vary tremendously depending on location. The solar energy resources are available almost everywhere on the globe and are more than adequate in most temperate and tropical locations to be utilized successfully. Locations where complete cloud cover occurs continuously during weeks at a time (e.g., tropical mountain rain forests) can present challenges and PV systems need to be larger to meet energy needs. Power can be generated even under overcast conditions, but it is only a fraction (~10%) of what is available during sunny, clear-sky conditions. Concentrating solar energy systems only work where direct sunlight is available. Regions within the tropics have a less variable solar resource over the course of a year as compared to higher latitude temperate regions with long summer days and short winter days. The measure of available sunlight is the peak sun hour (psh). If the sunlight intensity is measured in kW/m^2, then if the sunlight intensity is integrated from sunrise to sunset over 1 m^2 of surface, the result is expressed in kWh. This is the common unit used for insolation, the kWh/m^2, called also a *sun-hour*. The insolation value that most closely fits the project location must be used, being a good idea to be conservative (i.e., use fewer sun-hours). The solar energy system should be designed to fit the need with the seasonal solar resource. For an off-grid home, one may want, for example to design a PV system for the winter season when there is less sunlight. For practical solar energy system design and sizing, the average energy available over a day (the insolation) is used. It corresponds to the accumulated irradiance over time. Insolation is typically provided in kWh/m^2. Insolation is the key parameter for solar energy system design. The main factors affecting the amount of insolation incident upon a solar surface are orientation, mounting angle with respect to the horizontal plane, and climatic conditions. Because the insolation received by a surface depends on orientation and inclination with respect to the apparent position of the Sun, the *solar resource* of a designated site is specified as the amount of insolation measured upon the horizontal plane. Drawing on data for insolation upon the horizontal plane, insolation values can be estimated for surfaces set at specific azimuths and angles of elevation.

As discussed in detail in the previous section, the total solar radiation consists of direct (beam), diffuse, and reflected components. In regions with strong direct components of sunlight, it is advantageous to have a PV array mount to track the Sun. Such tracking systems can improve the daily performance of a PV array by more than 20% in certain regions. In cloudy regions, tracking is less advantageous. The Sun position in the sky can be uniquely described by the azimuth and the altitude angles. The azimuth is the deviation from true south. Another convenient, but redundant, angle, is the hour angle. The Sun thus appears to move along its arc 15° toward the west each hour. Another important angle that is used to predict the Sun's position is the declination, the apparent position of the Sun at solar noon with respect to the equator. Solar radiation calculations are showing that for optimal annual performance of a PV array, it to face directly south and to be tilted at an angle approximately equal to the location latitude, L. However, for best summer performances, the PV panel tilt must be at $L -15°$, while for best winter performance, the array must be tilted at an angle of $L +15°$. While calculations and estimates of the solar angles can be used to predict the Sun location in the sky at any time on any day at any location, they cannot be used to predict the degree of cloud coverage, which can significantly affect the PV system performances. Cloud cover can only be predicted on a statistical basis for any region, and thus the amount of sunlight available to a collector will also depend upon cloud cover. The most common PV system configurations are the grid connected and standalone systems, as ones shown in Figure 9.9.

9.8 Chapter summary

All the Earth's renewable energy sources are generated from solar radiation, which can be converted directly or indirectly to energy using various technologies. This radiation is perceived as white light since it spans over a wide spectrum of wavelengths, from the short-wave infrared to ultraviolet. Such radiation plays a major role in generating electricity either producing high temperature heat to power an engine mechanical energy which in turn drives an electrical generator or by directly converting it to electricity by means of the PV effect. Wind and solar energy have the potential to play an important role in future energy supply and generation mix in many areas and countries of the world. There are huge resources in solar

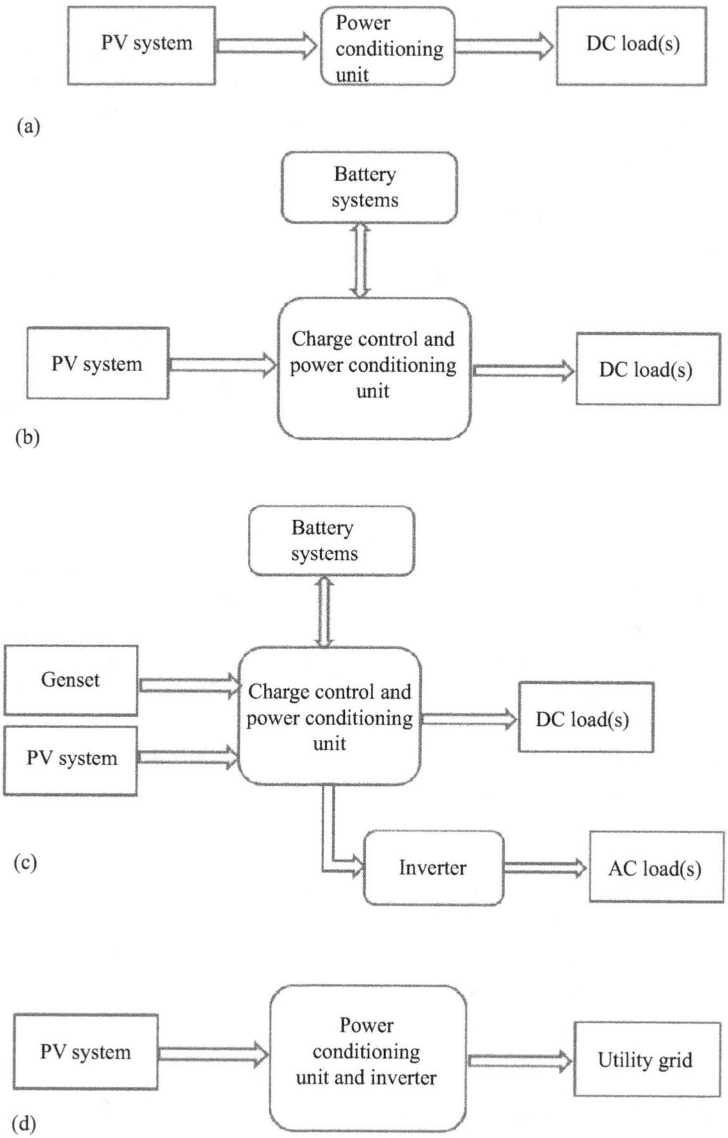

Figure 9.9 PV system configurations: (a) direct-coupled DC system, (b) DC system with battery backup, (c) hybrid DC and AC system with battery backup and conventional diesel engine-generator, and (d) grid-connected system

radiation with the average incident solar power more than 5,000 times current world power consumption and demand. The computation of solar radiation data are very important for scientists and engineers involved in creating applications implementing models for solar energy, building design, astronomy, agronomy, and agro-meteorology. Wind regime is ultimately a consequence of the Sun energy being determined by the global synoptic circulation and by the local flows and topography. The most important characteristics of wind are its variability and intermittency on a broad range of spatio-temporal scales. Wind regime knowledge and characteristics are important for assessment and analysis of wind energy potential for an area or location, exploitation of wind energy, design, management or operation of wind energy conversion systems, as well as in other engineering and technology branches. There several statistical and data analysis methods used to characterize the wind regime and to assess the wind energy potential for a specific application. PV conversion is the direct conversion of sunlight into electricity without any heat engine to interfere. PV devices are rugged and simple in design requiring very little maintenance and their biggest advantage being their construction as stand-alone systems to give outputs from microwatts to megawatts. Many modern products incorporate PV cells or modules in order to operate independently of other electrical supplies. Electricity produced from PV systems has a far smaller impact on the environment than traditional methods of electrical generation. During their operation, PV cells need no fuel, give off no atmospheric or water pollutants and require no cooling water. The use of PV systems is not constrained by material or land shortages and the Sun is a virtually endless energy source. The cost of PV systems has decreased more than 20 times since the last decades of twentieth century, and research continues on several different technologies in an effort to reduce costs to levels acceptable for wide scale use. Current PV cells are reliable and already cost effective in certain applications, such as remote power, with stand-alone PV plants built in regions not reached by the utility networks. In summary, the merits of photovoltaic technology relative to other power generation technologies include noiseless, relatively environmentally benign, proven, long life (e.g., 20–30 years for crystalline silicon modules) and low maintenance.

References and further readings

[1] P. Gipe, Wind Energy Comes of Age, John Wiley & Sons, New York, 1995.

[2] T. Burton, D. Sharpe, N. Jenkins, and E. Bossanyi, Wind Energy Handbook, Wiley, Chichester, UK, 2001.

[3] G.L. Johnson, Wind Energy Systems, Prentice Hall, NJ, 2006.

[4] O. Anaya-Lara, N. Jenkins, J. Ekanayake, P. Cartwright, and M. Hughes, Wind energy generation: modelling and control, Wiley, Chichester, UK, 2009.

[5] M.R. Patel, Wind and Solar Power Systems: Design, Analysis, and Operation (2nd ed.), CRC Press, Boca Raton, FL, 2015.

[6] R. Belu and D. Koracin, 2015. Wind energy analysis and assessment, Adv. Energy Res. Vol. 20(1), pp. 1–55.

[7] D.Y. Goswami, F. Kreith, and J.F. Kreider, Principles of Solar Engineering (2nd ed.), CRC Press, Boca Raton, FL, 2000.

[8] J.A. Duffie, and W.A. Beckman, Solar Engineering of Thermal Processes (3rd ed.) Wiley, Hoboken, NJ, 2006.
[9] J. Andrews and N. Jelley, Energy Science, Principles, Technology and Impacts, Oxford University Press, Oxford, UK, 2007.
[10] E.L. McFarland, J.L. Hunt, and J.L. Campbell, Energy, Physics and the Environment (3rd ed.), Cengage Learning, Australia, 2007.
[11] M.A. El-Sharkawi, Electric Energy—An Introduction, CRC Press, Boca Raton, FL, 2009.
[12] H.J. Wagner, J. Mathur, Introduction to Hydro Energy Systems—Basics, Technology and Operation, Springer-Verlag, Berlin, Heidelberg, 2011.
[13] G. Boyle, Renewable Energy—Power for a Sustainable Future, Oxford University Press, Oxford, UK, 2012.
[14] B. Everett and G. Boyle, Energy Systems and Sustainability: Power for a Sustainable Future (2nd ed), Oxford University Press, Oxford, UK, 2012.
[15] S.A. Kalogirou, Solar Energy Engineering, Academic Press, Oxford, UK, 2013.
[16] V. Nelson, and K. Starcher, Introduction to Renewable Energy (Energy and the Environment), CRC Press, Boca Raton, FL, 2015.
[17] A. Vieira da Rosa, Fundamentals of Renewable Energy, Academic Press, Oxford, UK, 2009.
[18] E.E. Michaelides, Alternative Energy Sources, Springer-Verlag, Berlin, Heidelberg, 2012.
[19] S.M. Muyeen (editor), Wind Energy Conversion Systems: Technology and Trends, Springer, London, 2012.
[20] V. Lyatkher, Wind Power: Turbine Design, Selection, and Optimization, John Wiley & Sons, Inc., Hoboken, NJ, 2014.
[21] G.N. Tiwari, A. Tiwari, and Shyam, Handbook of Solar Energy: Theory, Analysis and Applications, Springer, New York, 2016.
[22] R. Belu, Wind Turbines: Dynamic Modeling and Control, in Encyclopedia of Energy Engineering and Technology (ed., Dr. Sohail Anwar), Vol. 4, CRC Press/Taylor and Francis, Boca Raton, FL, 2014.
[23] R. Belu, Wind Energy Conversion and Analysis, in Encyclopedia of Energy Engineering and Technology (ed., Dr. Sohail Anwar), Vol. 4, CRC Press/Taylor and Francis, Boca Raton, FL, 2014.
[24] A.R. Jha, Wind Turbine Technology, CRC Press, Boca Raton, FL, 2011.
[25] IEC 61400-Part 1-to-21 Standard Series, Wind Turbine Design, Generators, Installation, Safety, 2005.
[26] IEA Expert Group Study: Recommended Practices for Wind Turbine Testing and Evaluation, Power Performance Testing (2nd ed.), 1990
[27] D. Wood, Small Wind Turbines – Analysis, Design, and Application, Springer, London, 2011.
[28] H. Haberlin, Photovoltaics System Design and Practice, Wiley, Chichester, 2012.

Questions and problems

1. What is the wind power class for your hometown or city?
2. List the main regions of the solar spectrum.
3. What is the air density difference between sea level and a height of 1,000 m and 2,500 m?
4. Solar power potential is around 1 kW/m^2. What wind speed gives the same power potential?
5. The area of New Mexico is 121,666 mi^2. The average annual insolation (hours of equivalent full sunlight on a horizontal surface) is approximately 2,000 h. If one-third of the area of New Mexico is covered with solar panels of 10% efficiency, how much electricity can be generated per year? What percentage of the US energy needs can be satisfied?
6. Using appropriate equations and relationships, determine the date on which the Sun-Earth distance is maximum, and the date on which this distance will be at a minimum.
7. Calculate the noon Sun angle for your location and the equinoxes and solstices.
8. What is the difference between the zenith and incidence angles?
9. What is the air density difference between sea level and a height of 2,500 m?
10. A wind power plant has 15 wind turbines, each one rated for 1.0 MW. The capacity factor is 37.5%. What is the wind power plant's annual energy yield?
11. Calculate the power, in kilowatts, across the following areas for wind speeds of 5, 15, and 25 m/s. Use the turbine diameters of 5, 10, 50, and 100 m for the rotor area. Assume the standard air density of 1.225 kg/m^3.
12. A wind turbine is rated at 300 kW in a 10 m/s wind speed in air at standard atmospheric conditions. If assume that the power output is directly proportional to air density, what is the power output of the turbine in a 10 m/s wind speed at elevation 1,500 m above sea level at a temperature of 20 °C?
13. Calculate the factor for the increase in wind speed if the original wind speed was taken at a height of 10 m. New heights are at 30, 60, and 90 m. Use the power law with an exponent equal to: (a) 0.1, and (b) 0.20.
14. At a wind farm, very high winds with gusts of over 60 m/s were recorded. An average value for 15 min was 40 m/s with a standard deviation of 8 m/s. What was the TI?
15. Calculate the wind speed distribution using the Rayleigh distribution for an average wind speed of 6.3 m/s. Use 2 m/s bin widths. Calculate the wind energy density, the monthly energy density availability, the most frequent wind velocity, and the wind velocity corresponding to the maximum energy, for Rayleigh wind speed distribution of previous problem, assuming and air density 1.225 kg/m^3.
16. Compute the wind power density for a site located at an elevation of about 500 m when the air temperature is 30 °C and the wind speed is equal to 10 m/s, 12.5 m/s, and 15 m/s.
17. Calculate the wind speed distribution for a Weibull distribution for c = 7.2 m/s and $k = 1.8$. Use 1 m/s bin widths. How many hours per day the wind

speed is between 5 and 12 m/s, the cut-in speed and the rated speed of a medium-size wind turbine.

18. Calculate the wind speed distribution for a Weibull distribution for $c =$ 7.5 m/s and $k = 2.7$.
19. Determine the solar altitude and azimuth angles at 11:00 a.m. local time at Bucharest, Romania, on July 15.
20. Calculate the factor for the increase in wind speed, if the original wind speed was taken at the standard level of 10 m. New heights are at 60, 80, 100, and 120 m. Use the power law with an exponent of 0.20.
21. In many areas of Mid-west there are wide temperature differences between summer (100°F) and winter (–20°F). What is the difference in air density? What is the change in the available average wind power? Assuming the same elevation, the same average wind speed, and pressure is the same.
22. Calculate the increase in wind speed if the original wind speed was taken at a height of 10 m. New heights are at 30, 60, and 80 m. Use the power law with exponents 0.147 and 0.21. Compare with (9.10) for $z_0 = 1.15$.
23. The Weibull parameters at a given site are $c = 5.5$ m/s and $k = 1.7$. Estimate the number of hours per year that the wind speed will be between 6.0 and 9.0 m/s. Estimate the number of hours per year that the wind speed is greater than or equal to 12.5 m/s.
24. Calculate the wind energy density, the monthly energy density availability, the most frequent wind velocity, and the wind velocity corresponding to the maximum energy, for Weibull wind speed distribution of previous problem, assuming and air density 1.225 kg/m^3.
25. A large wind turbine with a cut-in velocity of 5 m/s and a cut-out velocity of 25 m/s is installed at a site where the Weibull coefficients are $k = 2.2$ and $c = 8.5$ m/s. How many hours in a 24 h period will the wind turbine generate power?
26. Repeat the above problem for a small wind turbine with cut-in and cut-off speeds, 3 m/s and 18 m/s, respectively installed at the same site.
27. For a wind turbine having a rotor diameter of 60 m, for incoming winds of 6, 9, and 12 m/s, calculate: (a) the power of incoming wind, (b) the theoretical maximum power extracted by the wind turbine, and (c) if the gearbox, generator, and electrical transmission and processing efficiencies are: 0.85, 0.96, and 0.95, respectively what is the turbine effective converted power. Assume standard air density.
28. What is the solar time in El Paso, Texas (latitude: 31.8° N, and longitude: 106.4° W), at 10 a.m., and 3 p.m., Mountain Standard Time on March 15, July 15, and October 1?
29. Find the altitude angle and azimuth angle of the Sun at the following (solar) times and places:
    (a) March 1st at 10:00 a.m. in New Orleans, latitude 30° N.
    (b) August 10th at 2:00 p.m. in London, Ontario, Canada, latitude 43° N
    (c) July 1st at 5:00 p.m. in San Francisco, latitude 38° N.
    (d) December 21st at 11 a.m. at latitude 68° N.

30. Find the solar altitude and azimuth angles at solar noon in Gainesville, Florida, on February 28 and August 8. Also find the sunrise and sunset times in Gainesville on that day. Repeat the calculations for Sydney, Australia.
31. Calculate the time of sunrise, solar altitude, zenith, solar azimuth, and profile angles for a 60° sloped surface facing 25° west of south at 10:00 a.m. and 3:00 p.m. solar time on March 22 at latitude of 43°, north and south. Also calculate the time of sunrise and sunset on the surface.
32. Calculate the number of hours with the Sun above the horizon on March 22, August 8, and December 12, and April at Cleveland, Ohio, USA.
33. Find the solar altitude and azimuth angles in San Juan, Porto Rico on (a) June 1 at 7 a.m. and (b) December 1 at 2 p.m. Also find the sunrise and sunset times on these days.
34. Consider a solar cell that is having the light current 0.025 A/cm^2, the saturation current $1.05 \cdot 10^{-11}$ A/cm^2, and the operating temperature 25 °C. For the diode factor values of 1.2, 1.4, 1.6, and 1.8, calculate the cell open-circuit voltages.
35. Current output from a solar module is proportional to what variable? What kind of current produces a PV module?
36. An ideal solar cell with reverse saturation current of 10^{-9} A is operation at 32 °C. The solar light current at this temperature is 0.8 A. Compute the open-circuit voltage for this cell.
37. Compute the maximum power and fill factor for PV cell of the previous problem, assuming an ideal diode operating at 30 °C.
38. Given an *I–V* curve (select a typical one from any manufacturer), find I_{mp}, V_{mp}, I_{sc}, and V_{oc}. Calculate the fill factor (FF). Given P_{in}, 1,000 W/m^2, calculate the conversion efficiency.
39. Estimate the cell temperature and power delivered by a 120 W PV module, assuming 0.5% per degree power loss, an NOCT of 56 °C, ambient temperature of 25 °C, insolation of 1 kW/m^2.
40. Using the 36 cell PV module specifications from Table 9.5, determine the nominal current and voltage outputs for a PV system that is wired four strings in parallel, each having five PV modules in series.

Chapter 10

Geothermal energy, small hydropower, and bioenergy

Objectives and abstract

This chapter is focusing on geothermal energy, small hydro-power systems, and a very brief description of biomass suitable for power generation or in industrial processes, building, and other large commercial and applications. Geothermal energy sources are providing thermal energy to the industrial processes, buildings and eventually used to generate electricity, having a significant potential to contribute substantially to the world energy demands. Water energy originates from sources, such as the oceans, seas, rivers, and waterfalls. From water systems, the mechanical energy can be harvested either in kinetic or potential energy from waterfalls, rivers, currents, tides, or waves that eventually is used for power generation. The thermal energy from the temperature differences between ocean's warm and cold deeper layers can also be used for electricity generation having a huge potential and availability. However, ocean thermal energy is not discussed in this chapter, being beyond the scope of this book. Hydropower, the most and the largest renewable energy source for electricity generation, is derived from the energy of moving water from higher to lower elevations or from water kinetic energy. Hydropower systems require relatively high initial investment, but have the advantage of very low operation and maintenance costs and a long lifespan. Hydropower technology is the most advanced and mature renewable energy technology and provices an important portion of the electricity generation in many countries. Small- and mini-hydropower systems mean, the systems that can be applied to the sites ranging from a tiny scheme to electrify a single home, to a few hundred kilowatts or even few megawatts for selling it to the grid. Small-scale hydropower is one of the most cost-effective and reliable energy technologies to be considered for providing clean electricity. Hydroelectric power plants use minimal resources to generate electricity, nor do they pollute the air, land, or water, as other types of power plants may. A reference to the resource estimates and analysis are also included here. Characteristics, advantages, and disadvantages of these renewable energy sources, their operation and characteristics, as well as their major applications are presented in this chapter and discussed in details.

10.1 Introduction

Sustainable development requires new and sustainable energy sources, distributed generation, energy conservation, and systems having higher efficiency. Over the last few decades, developed and developing countries increasingly focused on the energy supply expansion and diversification, using local renewable energy, while underscoring the needs for reevaluating all energy alternatives, particularly those that are large and well-distributed nationally. Such options, quite often ignored are the geothermal energy, from conventional hydrothermal and enhanced (engineered) geothermal system sources, small hydropower, biomass, biofuels, or energy recovery from municipal or industrial waste. Today the hydropower potential exploited is over 2,800 TWh/year, or about 20% of the world's electricity. However, all other renewable when combined, provide less than 2% of global electricity consumption. North America and Europe have developed most of their large-scale hydropower potential, but huge resources remain in Asia, Africa, and South America. Small hydropower (up to 10 MW) currently contributes over 40 GW of world power capacity. The global small hydropower potential is believed to be in excess of 100 GW, while in many countries hydropower is one of the largest contributors to electricity generation. Our planet, from its center to its surface, is a massive thermal energy storehouse. Earth is formed from a core of molten metal which is slowly transferring heat to its outside layer, and finally to the outer layer, the crust. Additional heat is generated by the decay of naturally occurring radioactive materials beneath the surface. The Earth also acts as a very large collector of the Sun's energy. There are estimates that quantity of this energy may be available for human use and these quantities, regardless of the estimation method is enormous. For example, it has been estimated that the total heat available within the upper 5 km of the Earth's surface is about 140×10^6 EJ. If only 1% of this could be used at the current rate of world energy consumption of about 500 EJ/year, this would provide the world with all its energy for 2,800 years. However, this energy is also replaced by heat from the various energy sources listed earlier, and even if we were able to provide all our energy requirements from geothermal energy, it would still be fully sustainable.

Water energy and hydropower, like wind and solar energy have been used for centuries, for sailing, or as source of mechanical power for grinding grain, for sawing wood, or in primitive textile shops. Like most of the renewable energy sources, all types of water energy are driven ultimately by the solar energy. Unlike most other renewable energy sources, hydropower, large or small is a major contributor to the world energy supplies. Hydropower currently represents worldwide a significant source of electrical energy and compared to fossil and nuclear fuel, hydro resources are found almost everywhere, providing close to 20% of the world's electricity supply. The hydropower golden age was the first half of the 20th century, before oil took over as the dominant force in energy supply. Europe and North America built dams and hydropower stations at a rapid rate, exploiting up to 50% of the technically available hydro energy potential, while the equipment providers sprung up to supply this booming market. Whereas the large hydropower manufacturers have since managed to maintain their business on export markets, in particular to the developing countries, the small hydropower industry has been on the decline since 1960s. Hydropower, large and small, remains by far the most important of the "renewables" for electrical power generation worldwide, being available in a broad range of projects, scales and types. Projects are usually designed to suit particular needs and specific site/location conditions. They are classified by project type, purpose, and installed capacity. Size wise categories are different worldwide due to varying development policies in different countries. The hydropower types are: run-of-river, reservoir-based, and hydropower pumped energy storage. Hydropower impacts are well known both from the environmental and social perspectives. Experience gained during past decades in combination with new sustainability guidelines, innovative planning based on stakeholder consultations, and scientific know-how is promising to achieve a high sustainability performance for future hydropower projects. The world's technically feasible hydro potential is estimated at 14,370 TWh/year, which equates to 100% of today's global electricity demand. The economically feasible proportion of this is currently considered to be 8080 TWh/year.

10.2 Geothermal energy

Geothermal energy is advantageous because it is renewable, reliable, and efficient, and as a group, geothermal power plants can generate electricity more than 95% of the time. The use of geothermal energy for heating, electricity generation or industrial processes is believed to significantly increase in the near future. However, the way in which geothermal energy resources are utilized, ultimately determine whether or not is a sustainable utilization. Geothermal resources span a wide range of heat Earth's sources, which include easily developed, currently economic hydrothermal resources, the Earth's deeper, stored thermal energy, that is present everywhere. Conventional hydrothermal resources are used effectively for electric and/or nonelectric applications; however, they are limited to their location and ultimate potential for supplying electricity. Earth's geothermal resources are theoretically more than adequate to supply world energy needs, however, only a very small fraction may or can be economically exploited. Geothermal energy is derived from: (1) steam trapped deep into the Earth, brought to the surface, used to drive steam turbine-generator units to produce electricity, and (2) water pumped and heated through deep hot rocks, to provide heat or steam for buildings or industrial processes. Geothermal energy, in the form of natural steam and hot water, has been exploited for long time for space heating and industrial processes or to generate electricity. The Earth is giving the impression that it is dependably constant, because over the human life time scale, little seems to change. However, the Earth is quite a dynamic entity, with time scales spanning for seconds for the earthquakes, a few years that volcanoes appear and grow, over millennia that landscapes slowly are evolving, and to over millions of years the continents rearrange themselves on the planet's surface. The energy source driving these processes is heat, with a constant flux from every square meter of the Earth's surface. The average heat flux for the Earth is 87 mW/m^2, or for the Earth surface of 5.1 × 108 km^2 is equivalent to about 4.5 × 10^{13} W. For comparison, it is estimated that the total annual world power demand is approximately 1.6 × 10^{13} W. Clearly, the geothermal energy has the huge potential to significantly contribute to the human energy needs. The geothermal energy, contained in the rock and fluid into the globe layers is linked with the Earth's internal structure and composition and associated physical processes. Despite the fact it is present in huge, inexhaustible quantities into the Earth's crust or deeper layers, it is unevenly distributed, seldom concentrated, and most often at depths too great to be economically or even technological possible for exploitation. There are almost 4,000 miles from the Earth' surface to its center and the deeper it is the hotter it gets. The outer layer of the Earth, the Crust, about 35 miles thick, insulates the surface from the hot interior. The inner generated heat flows toward the surface where it dissipates, the Earth temperature increases with depth, a geothermal gradient of about 30 °C/km of depth exists. During the last century, many countries started to use geothermal energy, as it becomes economically competitive with other energy sources. Moreover, the geothermal energy is in some regions, the only energy source available locally. Geothermal energy is coming from two main sources:

1. Heat that flows upward and outward across the entire Earth' surface from the very deep, mantle, and core radioactive decay of uranium, thorium, and potassium. However, usually this energy flux is too small to be commercially useful for any application.
2. The localized heat resulting from the movement of magma into the crust. In some areas, this localized heat, with higher temperatures and heat fluxes can be found between the surface and about 3,500 m (about 10,000 ft.) depth. Where these heat fluxes meet the requisite conditions, geothermal energy can be used for multiple purposes, power generation, providing heat, or hot water for buildings or industrial processes.

Example 10.1: Estimate the available power for two 2,000 km^2 areas, one having the average geothermal flux, and the second one (an active geothermal area) has the geothermal flux of 200 mW/m^2.

Solution: The available power for the average geothermal flux area and for the active area are:

$$P_{active} = 2 \times 10^9 \times 200 \times 10^{-3} = 400\ \mathrm{MW}_t$$

$$P_{average} = 2 \times 10^9 \times 87 \times 10^{-3} = 164\ \mathrm{MW}_t$$

Geothermal energy can provide heat and hot water for homes, greenhouses or industrial processes, dry vegetables, or generate electricity. Some of these applications can be pursued anywhere, while others require special circumstances, being restricted to specific areas. In order to use these energy resources in a way that is both economical and environmentally sound requires that their characteristics to be known, through the description of the Earth compositional and physical structure. Earth is compositionally inhomogeneous, consisting of an iron–nickel core, a dense rocky mantle, and a thin, low-density rocky crust. Because of its diverse and complex composition, there are large differences in its physical or mechanical properties. The Earth's radius is about 6,370 km. Extending outward from the Earth's center, significant systematic changes occur in both composition and rheological behavior (material physical or mechanical properties). The Earth structure and its interior are shown in Figure 10.1, consisting of several layers, *the crust*, a relatively thin region of low-density silicates, *the mantle*, a thick region of higher-density iron-rich silicates, and the core, a *central region* of iron mixed with various impurities, being usually depicted as concentric spheres, in ultra-simplified schematics. However, the interfaces are likely so irregular and the boundaries so fuzzy that such a representation is misleading. The crust has continental regions, made of even lower-density aluminum-rich silicates and oceanic regions, made of denser iron-rich silicates. The mantle is divided into upper mantle within which the iron-rich silicates are gradually compressed from lower-density more open mineral structures, to higher density more compact mineral structures, and the lower mantle, where the mineral structures are compacted to their densest forms of increased density. The core, extending from the center to a depth of about 2,900 km, with the temperature of about 6,000 °C consists of a molten (liquid) outer core layer, primarily of iron, which is lowering its melting temperature, and a solid inner core, consisting of almost certainly of a crystalline mixture of iron and nickel. Overlying the core is the mantle, made of partly rock and partly magma, which extends from a depth of about 2,900 km to less than 100 km. Its volume makes up the largest part of Earth's interior, and its temperature decreases upward from about 5,000 °C to less than 1,500 °C. The last layer is the Earth's crust, consisting of a thin shell, varying

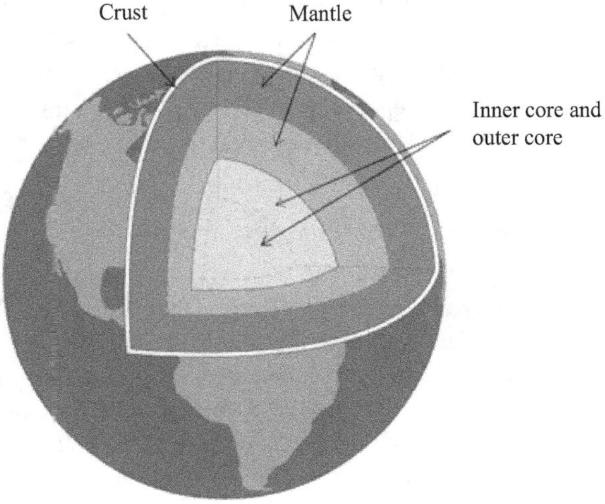

Figure 10.1 Earth internal structure and layers

from 70 to 80 km thick under the continents to less than a few kilometers thick under the ocean floor. Wells give us direct access only to the upper crust, and to depths up to 12 km. The denser, oceanic crust is made of basalt, whereas the continental crust is often referred to as being largely granite. The mantle lies closer to the Earth's surface beneath the ocean, where the crust is significantly thinner than beneath the continents, where the curst in much thicker. The crust and the uppermost mantle layer together form the lithosphere, the outer shell of the Earth that is relatively rigid and brittle. The lithosphere is divided into lithospheric plates, large blocks at continental scale or bigger. The lithosphere is about 70 km thick beneath the oceans and up to 125 km thick beneath the continents. The lithosphere seems to be in continual movement, likely as a result of the underlying mantle convection, and brittle lithospheric plates seems to move easily over the asthenosphere.

First experimental power generation was installed in Larderello, Italy, on July 4, 1904, about 40 years later than the commercial electricity uses, and the first commercial geothermal power plant (250 kW) in 1913, and the first large-power installation in 1938 (69 MW). It would be 20 years before the next large geothermal power installation was built in Wairakei, New Zealand, commissioned in 1958, that grew to 193 MW of installed capacity by 1963. In the United States, the installation of the first unit of 11 MW at Geysers, in Sonoma, California in 1960, eventually becomes later the world largest geothermal power complex, with a capacity of 1,890 MW. Plant retirements and declining steam supply have since reduced its generation capacity to an annual average of 1,020 MW from 1,421 MW of installed capacity, still the largest geothermal plant. Since those early efforts, a total of 2,564 MW of geothermal power generation capacity is currently installed in the United States, generating approximately 2,000 MW each year. In 2005, worldwide annual geothermal power was estimated at 56,875 GW h from 8,932 MW of installed capacity. Geothermal energy is also utilized in direct-heat uses for space heating, recreation and bathing, and industrial and agricultural uses. Geothermal energy in direct-use applications is estimated to have an installed capacity of 12,100 MW in thermal capacity, with annual average energy usage of 48,511 GWh, excluding the ground-coupled heat pumps (GCHPs). In the US, the first district geothermal heating was installed in Boise, Idaho, in 1892, still in operation today. GCHP units are reported to have 15,721 MW of installed capacity and 24,111 GWh, representing 56.5% worldwide direct use, respectively.

10.2.1 Geothermal energy origins and resources

Geothermal energy originates from the planet formation and from radioactive decay of materials, roughly in equal proportions. The Earth's heat flow, measured in mW/m^2 varies on its surface and with the time at any particular place. This heat flow originates from the primordial heat, generated during the Earth's formation, or due to the decay of radioactive isotopes. The average heat flow through the continental crust is 57 mW/m^2, through the oceanic crust is 99 mW/m^2, while the globe average heat flow is 82 mW/m^2, while the total global output is over 4×10^{13} W, four times than the present world energy consumption. Continental heat flow seems to originate mainly from the radiogenic decay within the upper crust, the heat generated in the most recent magmatic episode and the heat from the mantle. In the oceanic crust, the concentration of radioactive isotopes is very low, so the heat flow is largely derived from heat from the mantle. The geothermal gradient, due to the core-surface temperature differences, drives continuous thermal energy conduction in the form of heat to the surface. The crust base temperature is about 1,100 °C, the temperature gradient between the surface (~20 °C) and the crust bottom is 31.1 °C/km (the normal temperature gradient). Good geothermal sources occur where the thermal gradient is several times greater than the normal one. The rate of natural heat flow per unit area, the normal heat flux, is roughly 1.2×10^6 cal/cm^2 · s, in nonthermal Earth areas. The Earth conductive heat flow is the product of the geothermal gradient and the thermal conductivity of rocks. The geothermal gradient is measured in wells, while the conductivity of rocks is measured in laboratory, on samples taken from the well where the gradient was measured. The two heat transfer forms occurring within the Earth are conduction and convection, the former being more efficient in heat transfer. Earth thermal behavior studies imply the determination of temperature variations with depth, and how such temperature variations may have changed throughout geological time. The thermal gradient values lower as 10 °C/km are found in ancient continental crust, while very high values of about 100 °C/km are found in active volcanic areas. Once the gradient is measured, it can be used to determine the upward heat rate through a particular area. As the heat moves upward through solid impermeable rock, the principal heat transfer mechanism is conduction. The heat flow rate, proportional with the geothermal gradient and thermal conductivity of rocks is defined as the amount of heat conducted per second through unit area, for a temperature gradient 1 °C/m perpendicular to that area. If the gradient is expressed in °C/km and conductivity in W/(m °C), then the heat flow rate is in mW/m^2. Earth's thermal energy is distributed between the constituent host rock and the natural fluids, contained in hot rock fractures and pores at temperatures above ambient levels, mostly water with varying dissolved salts, being present as a liquid phase or sometimes a saturated, liquid-vapor mixture or superheated steam vapor. Notice that the amounts of hot rock and contained fluids are larger and more widely distributed in comparison to oil and natural gas contained in sedimentary rock formations.

Geothermal fluids have been used for cooking and bathing since before the beginning of recorded history; but it was not until the early twentieth century that geothermal energy was harnessed for commercial purposes. Since the Larderello experiment, other geothermal developments, such as the steam field at the Geysers, California, the hot-water systems at Wairakei, New Zealand, Cerro Prieto, Mexico, Reykjavik, Iceland, and others led to an installed world generating capacity of about 10 GWe and a direct-use, nonelectric capacity of more than 100 GWt. The source and transport mechanisms of geothermal heat are unique to this energy source. With continually technology developments, the geothermal energy can become a major factor in solving the world complex energy equation. The Earth heat flux is estimated to be equivalent to 42 million MW, far greater than the coal, oil, gas, and nuclear energy combined. It is estimated that a recovery of even a small fraction of this heat would supply the world's energy needs for centuries.

Geothermal energy is also permanently available, unlike the solar and wind energy sources, which are dependent upon factors, like weather variations and daily and seasonal fluctuations. Electricity from geothermal energy is more consistently available, once the resource is tapped. Several different types of geothermal systems can be exploited: (a) convective or hydrothermal systems, (b) enhanced geothermal systems (EGS), (c) conductive sedimentary systems, (d) hot water produced from oil and gas fields, (e) geo-pressured systems, and (f) magma bodies. Convective hydrothermal systems have seen several decades of commercial exploitation for electric generation in several countries to date but with limited distribution limited. There are two basic convective system classes, depending on the thermal energy source type: volcanic and nonvolcanic. A volcanic convective system drives its thermal energy from the convecting magma, while a non-convective system drives its thermal energy from meteoric water that has heated up by deep circulation in Earth high heat flow areas, with no associated magmatic body. The installed power capacity that exploits such systems totals about 10 GW worldwide and 3,000 MW in the US only, with a reserve base only in the US of about 20 GW. It has been suggested that a positive correlation exists between the geothermal resource potential available from volcanic convective systems and the number of active volcanoes in the country. However, an exploitable geothermal resource base may exist in the form of nonvolcanic convective systems. The heat moves from the Earth interior toward the surface where it dissipates, although this is generally not noticed, being aware of its existence as the depth increased. There are areas of the Earth's crust which are accessible by drilling, and where the gradients are well above the average. This occurs when, a few kilometers below surface, there are magma bodies undergoing cooling, in a fluid state or solidification process, releasing heat. In other areas, where there is no magmatic activity, the heat accumulation is due to particular crust geological conditions, such that the geothermal gradient reaches anomalously high values.

The extraction and utilization of large heat quantities require a suitable carrier to transfer it to accessible depths beneath the surface. The heat is transferred from depth to sub-surface regions by conduction and convection processes, with geothermal fluids acting as the carrier in the former. These fluids are rain water that has penetrated into the crust from the recharge areas, being heated by the contact with the hot rocks, and accumulated into aquifers, occasionally at high pressures and temperatures (above 300 °C). These aquifers (reservoirs) are the essential parts of many geothermal fields, usually covered with impermeable rocks, preventing the hot fluids to easily reach the surface and keeping the fluids under pressure. To obtain industrial superheated steam, steam mixed with water, or only hot water depends on the local hydro-geology and the temperature of the rocks. Wells are drilled into the reservoir to extract the hot fluids, and the usage depends on the fluid temperature and pressure. Electricity generation requires higher temperatures, while space heating and industrial processes are often run at the lower temperature range. Geothermal fields are usually systems with continuous heat and fluid circulations, where fluid enters the reservoir from the recharge zones, leaving through discharge areas, hot springs or wells. During the exploitation the fluids are recharged to the reservoir by re-injecting through wells the fluids from the utilization plants. Reinjection process may also compensate, at least part of the extracted fluid, prolonging to a certain limit the field lifetime. Geothermal energy is therefore to a large extent a renewable energy source, hot fluid production rates tend; however, to be larger than recharge rates. Enhanced geothermal systems imply a man-made reservoir created by hydro-fracturing impermeable or very "tight" rock through wells. By injecting into wells normal temperature water, in such artificially fractured reservoir and extracting heated water through other wells for industrial uses. The EGSs represent conductive systems that have been enhanced their flow and storage capacity by hydro-fracturing, and in theory can be developed anywhere by drilling deep enough to encounter attractive temperature levels. However, this technology is still experimental and posing technical challenges, such as: (a) creating a pervasively fractured large rock volume, (b) securing commercially productivity, (c) minimizing heated water cooling rate, (d) minimizing the losses, and (e) minimizing any induced micro-seismicity. Another geothermal energy resource type, considered for exploitation is the heat contained in the water produced from deep oil and gas wells, and co-produced with petroleum or from the abandoned oil or gas wells. While there are no significant challenges to exploiting this resource, the energy cost may not be always attractive due to relatively water low temperature and production rates. Other geothermal energy resource types of quite restricted distributions worldwide are the "geo-pressured" systems. These are confined sedimentary reservoirs with pressures much higher than the local hydrostatic pressure, allowing the exploitation of the kinetic energy of the produced water in addition to its thermal energy. Furthermore, because of its high pressure, such a system may contain methane gas dissolved in the water that can be used to generate electricity or for other uses. Such geo-pressured wells can provide thermal, kinetic, and gas-derived energy production. However, there are technical challenges to make such energy systems commercial viable.

Geothermal energy can be utilized as either direct heat not lonely for electricity generation. Most of the geothermal energy resources are inaccessible because of the depths and other area characteristics. However, along the plate boundaries, geothermal activity is close enough to the surface to be accessible, while the zones with the high earthquake activity are the most suitable for geothermal power generation. Geothermal resources, characterized by the thermal and compositional characteristics, are divided in four categories: *hydrothermal* (*geo-hydrothermal*) *resources*, *geo-pressurized, molted rocks* (*magma*), and *enhanced* (*hot, dry rock*) *geothermal systems*. Hydrothermal resources are the most limited type among the four classes. However, they are the easiest to harvest. In these resources, water is heated and/or evaporated by direct contact with hot porous rock or permeable rock, and bounded with low permeability rock. Water flows through the porous rocks, heated (perhaps evaporated) and discharged to the surface. Hydrothermal systems producing steam only are called vapor-dominated, and if they are producing hot water and steam mixture are called liquid-dominated. Geo-pressurized energy resources include sediment-filled and hot water confined under pressure reservoirs. The fluid temperature is range is 150–180 °C, and the pressures up to 600 bars.

In many of these systems the fluid contains methane called *geothermal brine*, a highly corrosive mixture. Magma or molten rock systems, under active volcanoes at accessible depths, have temperatures in excess of 650 °C. Hot dry rock (HDR) has the temperature in the excess of 200 °C, and as the name implies, contains small liquid amount. The method for harvesting this resource is through EGS, by directing water under the rock and rejecting the heated water back for various uses.

Further the geothermal systems are classified as: convective, liquid- and vapor-dominated hydrothermal reservoirs, lower temperature aquifers, and conductive, hot rock, and magma over a wide range of temperatures. Lower temperature aquifers contain deeply circulating fluids in porous media or fracture zones, with no localized heat source, being further subdivided into systems at hydrostatic pressure and systems at pressure higher than geo-pressured systems. Resource utilization technologies are grouped under types for electrical power generation and for direct heat uses. GHPs are a subset of the direct use, and EGS, where the fluid pathways are engineered by rock fracturing, are a subset under both types. A geothermal system requires heat, permeability, and water, the EGS techniques make up for reservoir deficiencies in any of these areas. EGS technologies enhance existing rock fracture networks, introduce water or another working fluid, or otherwise build on a geothermal reservoir that would be difficult or impossible to derive energy from by using conventional technologies. Currently, the most widely exploited geothermal systems for power generation are hydrothermal systems of continental subtype. In areas of magmatic intrusions, temperatures above 1,000 °C often occur at less than 10 km depth. Magma typically involves mineralized fluids and gases, mixed with deeply circulating ground-water. Typically, a hydrothermal convective system is established whereby local surface heat-flow is significantly enhanced. Such shallow systems can last hundreds of thousands of years, and the gradually cooling magmatic heat sources can be replenished periodically with fresh intrusions from a deeper magma. Finally, geothermal fields with temperatures as low as 10 °C are also used for direct heat pumps. Subsurface temperatures increase with depth according to the local geothermal gradient, and if hot rocks within drillable depth can be stimulated to improve permeability, using hydraulic fracturing, chemical, or thermal stimulation methods, they form a potential EGS that can be used for power generation and/or direct use applications. EGS resources may occur in all geothermal environments but are likely to be economic in geological settings where the heat flow is high enough to permit exploitation at depths of less than 5 km.

Example 10.2: Calculate the geothermal power potential of a site that covers 30 km^2 with a thermal crust of 2 km, where the temperature gradient is 240 °C. At this depth, the specific heat of the rocks is 2.5 MJ/m$^3 \cdot$ °C, and the mean surface temperature is 10 °C. Assuming that only 2% of the available thermal energy could be used for electricity generation, how much it takes to produce 5×10^4 MWh?

Solution: First the slab volume is calculated as:

$$V = 2\text{ km} \times 30\text{ km}^2 = 60\text{ km}^3 \text{ or } 60 \times 10^9\text{ m}^3$$

Then the slab stored heat is:

$$Q = V \cdot C \cdot \Delta T = 60 \times 10^9 \times 2.5 \times 10^6 (240 - 10) = 34.5 \times 10^{18}\ \text{J}$$

10.2.2 Surface geothermal technology and reservoir characteristics

Once a reservoir is found and characterized, the surface technology, power plant, and related infrastructure is designed and the equipment selected to optimize the resource use and sustainability. The goal is to construct an energy efficient, low cost, minimal environmental impact power plant. Geothermal fluid, a hot, mineral-rich liquid or vapor, is the carrier that brings geothermal energy up through wells from deeper layers to the surface. This hot water and/or steam extracted from an underground reservoir, isolated during production, flowing up through wells is converted into electricity at a geothermal power plant or is used in direct use systems for heating, cooling, or providing hot water to buildings and industrial facilities. Once used, the water and the condensed steam are re-injected into the geothermal reservoir to be reheated. It is separated from groundwater by encased pipes, making the facility virtually water pollution-free. Such resources, using the existing appropriate hot water and steam accumulations of appropriate are the *hydrothermal* resources. While other geothermal resources exist, all US geothermal power is using hydrothermal resources. The use of natural steam for electricity generation is not the only possible geothermal energy application. Hot waters, which are present in large continental areas, can be exploited, for space heating and industrial processes. The geothermal energy distribution for nonelectric applications is: (a) 42% for bathing and swimming pool heating, (b) 23% for space heating, (c) 12% for geothermal heat pumps (GHP), (d) 9% for greenhouse heating, (e) 5% for industrial applications, and (f) 9% for fish farm pond, agricultural drying, snow melting, air conditioning, or other uses.

Direct-use of geothermal energy is one of the oldest, most versatile and common forms of utilizing geothermal energy. Unlike geothermal power generation, in which heat energy is converted to electricity, direct-use applications use heat energy directly to accomplish a broad range of uses. The temperature range of these applications is from about 10 °C to about 150 °C. Given the ubiquity of this temperature range in the shallow subsurface, these types of applications of geothermal energy have the potential to be installed almost anywhere that has sufficient fluid available. Approximately 5.4×10^{27} J of thermal energy is available worldwide, of which nearly a quarter is available at depths less than 10 km. For direct use, the heat must be significantly above ambient surface temperatures and transferrable efficiently. Such conditions are satisfied in areas where hot springs emerge at the surface or in locations where high thermal gradients allow shallow drilling to access heated waters. Such sites are quite restricted in their distribution, being concentrated in volcanic activity area or where continent rifting has occurred. For such reasons, a relatively small fraction of this large amount of continental heat contained can be economically employed for geothermal direct-use applications. The fraction of the heat, readily available is not well known because thorough assessment efforts to quantitatively map the distribution of such resources have thus far been limited. As drilling technology improves and fluid circulation to support heat harvesting at depth improves, the continental thermal resource that can be accessed will significantly expand. As of 2010, approximately 122 TWh/year of thermal energy was used for direct-use purposes, worldwide, which was derived from an installed capacity of 50,583 MW. For comparison, global consumption of electricity in 2006 was 16,378 TWh/year. The growth in installed capacity of direct-use applications reflects a rapid growth in international development of this type of system. In 1985, 11 countries reported using more than 100 MW of direct-use geothermal energy, while by 2010, the number increased to 78. The global distribution of such systems reflects the diversity of applications for which they have been engineered.

A geothermal system that can be developed for beneficial uses requires heat, permeability, and water. When hot water or steam is trapped in cracks and pores under impermeable rock layers, a geothermal reservoir is formed. The exploration of a geothermal reservoir for potential developments includes exploratory drilling and testing for satisfactory conditions to produce useable energy, particularly the resource temperature and flows, water being a critical system component. The hot water, which comes from the geothermal system, is re-injected then into the reservoir to maintain reservoir pressure and to prevent reservoir depletion. However, rainwater and snowmelt usually continue to feed underground thermal aquifers, naturally replenishing geothermal reservoirs. Re-injection keeps the mineral-rich, saline water found in geothermal systems separate from ground water and fresh water sources to avoid cross-contamination. Injection wells are encased by thick borehole pipe and are surrounded by cement. Once the water is returned to the geothermal reservoir, it is reheated by the Earth's hot rocks and can be used over and over again to produce electricity or to provide heat. The key ingredients for geothermal energy production can be summarized by the following equation:

$$P_{conv} = C_P \cdot \left(T_{Rsvr} - T_{Rjec}\right) \cdot F_{rate} \cdot \eta - P_{Loss} = C_P \cdot \Delta T \cdot F_{rate} \cdot \eta - P_{Loss} \tag{10.1}$$

where C_P is the specific heat of the working fluid, F_{rate} is the flow rate from the production well (in Kg/s), ΔT is the sensible heat that can be extracted from the fluid produced by the production hole ($T_{reservoir} - T_{rejection}$), η is the efficiency with which the heat energy can be used, and P_{Loss} represents the fluid transfer and conversion losses. The goals of geothermal system developments are to optimize these parameters to increase electrical or heat output relative to the investment capital costs of the development. Based on the current experience in power generation from convective hydrothermal resources, the minimum amount of net energy produced by a well is about 4 MW. For most of the geothermal systems the working fluid is water with varying salinity or other dissolved materials. The specific heat is almost constant for all types of geothermal resources. The ΔT is often in the order of 50 °C to 150 °C, and the efficiency of current power cycles is about

10%. Based on these numbers and ignoring parasitic losses, a well needs to flow at a minimum of 70 kg/s to be viable. This rate is orders of magnitude higher than average flows in the US oil industry, and at the upper end of production rates for water wells, particularly at the depths needed to access high temperatures. The flow problem is not as significant in convective hydrothermal resources as these typically produce steam rather than water. Although the specific heat and density of steam is lower than water, high flow rates are achieved because of steam's low viscosity and density, allowing wells to produce without pumping. Exploration targets are the two components of (10.1), ΔT and F_{rate}. High temperature differences (ΔT) are targeted by looking for areas with very high thermal gradients. Flows are targeted by looking for areas with high natural permeability or with characteristics that are suitable for EGS techniques. The aims are to reduce the risks and therefore costs of discovering a geothermal resource. However, the exploration for any natural resource is inherently a high risk one. The risks can be reduced by the selection of suitable exploration targets. Targets need to be developed at all scales to enable the appropriate basin selection, selection of tenements within these basins, the development of prospects within tenements and the locating exploration wells within the prospects.

10.2.3 Direct use of geothermal energy

Direct use of geothermal energy includes the hydrothermal resources of low to moderate temperature that are utilized to provide direct heating in residential, commercial, and industrial sector, which include among others: space, water, greenhouse, and aquaculture heating, food dehydration, laundries, and textile processes. These applications are commonly used in Iceland, the United States, Japan, and France. Unlike geothermal power generation, direct-use applications use heat directly to accomplish a broad range of purposes. The temperature range of these applications is from 10 °C to 150 °C. Given the ubiquity of this temperature range in the shallow subsurface, these types of applications of geothermal energy have the potential to be installed almost anywhere that has sufficient fluid available. Geothermal resources are also used for agricultural applications or to warm greenhouses. The heat from the geothermal water is used for industrial purposes, such as drying fish, fruits, vegetables or timber products, washing wool, dying cloth, in paper and food industries. Geothermal heated water can be piped under sidewalks and roads to keep them from icing, during cold weather. Thermal waters are also used to help extract gold and silver from ore and even for refrigeration and ice-making. Geothermal, ground-source heat pumps have the largest energy use and installed capacity worldwide, accounting for 70.95% of the installed capacity and 55.30% of the annual energy use. The installed capacity is 49,898 MWt and the annual energy use is 325,028 TJ/year, with a capacity factor of 0.21 (in the heating mode). Most of the installations occur in North America, Europe, and China. The energy use reported for the heat pumps was deduced from the installed capacity (if it was not reported), based on an average coefficient of performance (COP) of 3.5, which allows for one unit of energy input (usually electricity) to 2.5 units of energy output, for a geothermal component of 71% of the rated capacity. The cooling load was not considered as geothermal as in this case; however, it has a significant role in the substitution of fossil fuels and pollutant emission reductions.

Geothermal heat is used directly, without involving a power plant or a heat pump, for a variety of applications, such as space heating and cooling, food preparation, hot spring bathing, balneology, agriculture, aquaculture, provide heat and hot water to greenhouses, and industrial processes. Uses for heating and bathing are traced back to Roman times. From about 5.4×10^{27} J of the available thermal energy in the Earth continents about a quarter is estimated to be available at depths less than 10 km. In order to be directly used, the source temperature must be significantly above ambient temperatures and to be easily and efficiently transferred to the designed premises. Such conditions are satisfied in areas where hot springs emerge at the surface, or in locations where high thermal gradients allow shallow drilling to access hot waters. Such sites have relatively restricted distributions to the regions with significant volcanic activity or near continental rifting. For these reasons, only a small fraction of the large heat amount contained within the continents is economically useable for geothermal direct applications. The geothermal energy resources that are readily available are not well known since their assessments to quantitatively map the distribution of such resources is quite limited. The main geothermal direct uses are: (1) swimming, bathing, and balneology, (2) space heating and cooling, (3) district heating, (4) agricultural, aquaculture, and industrial applications, and (5) geothermal (ground-source) heat pumps (GHP). The annual growth for the geothermal direct-use is about 8.3%, with the largest annual increase in the GHP technology. The overall global geothermal direct use is estimate close to 75 GWt, while the contribution of shallow reservoirs for small commercial, industrial or domestic applications is difficult to estimate. The capacity factors for such applications are in the range of 15%–75%. Direct use of geothermal energy is about 12.5 GWt, in the US, 32.0 GWt in China and 20 GWt in UE. The oldest geothermal direct use is bathing and therapy by using the hot spring or surfacing heating hot water. Another traditional geothermal direct use is the space heating. The hot water, 60 °C or higher from geothermal reservoirs is pumped into the building heating system, through the heat exchangers to provide heat to the building. The water is then re-injected into the geothermal reservoir for reheating. For geothermal direct use, a well is drilled into the geothermal reservoir and pumps are used to bring the hot water to the surface for specific use.

Thermal equilibrium is achieved when coexisting systems or system components are reaching the same temperature. Therefore, heat is spontaneously transferred from a hot to a colder body. This fundamental principle upon which all direct use applications rely, being also the mechanism that results in unwanted heat losses as heat is transferred from one place to another. The ability to manage heat transfer by minimizing unwanted losses and maximizing useful heat is required to set and operate an efficient direct use application. The heat transfer processes are conduction, convection, radiation, and evaporation. Heat transfer by conduction occurs at microscopic level through the exchange of vibrational energy by the atoms and molecules, while at the macroscopic level this process is manifest as changes in temperature when two bodies at different temperatures, are placed in contact with each other at a certain time t_1. If T_1 and T_2 represent the initial temperatures of the bodies 1 and 2, respectively, and T_3 is the equilibrium temperature they eventually achieve at some time. Note that T_3 is not half way between T_1 and T_2, being set by the heat capacity of each body material. Conductive heat transfer is described by the relationship:

$$\frac{dQ_{cnd}}{dt} = k \cdot A \cdot \frac{dT}{dx} \tag{10.2}$$

where dQ_{cnd}/dt (J/s or W) is the rate at which heat transfer occurs by conduction over the area A, k is the thermal conductivity (W/m · K), and dT/dx is the temperature gradient over the distance x (m). Equation (10.1) is known as Fourier's law of heat conduction. As (10.2) indicates, the rate of heat transfer can be increased by increasing the area over which heat transfer will occur or decreasing the distance. The thinner the plate, the greater is the temperature gradient across the plate and hence the greater the heat loss rate. Table 10.1 lists the thermal conductivities of some the most common materials, used in direct use applications. Notice, that there is a difference of two orders of magnitude or higher (four orders of magnitude) in the rate of heat loss for many common materials. Although large temperature changes are not associated with most of the direct use applications, it is worth to mention that thermal conductivity is a temperature-dependent material characteristic. Accurate heat transfer rate computations must take into account thermal conductivity temperature dependence, being critical to know the thermal conductivities and spatial geometry of the materials used for engineering direct use applications, or which are encountered when constructing a facility. Incorrect data on these parameters can result in seriously under-sizing thermal insulation,

Table 10.1 Thermal conductivities of selected materials used in direct use of geothermal energy

Material	Thermal conductivity (W/m · K)
Aluminum	202
Copper	385
Iron	73
Carbon steel	43
Marble	2.90
Magnesite	4.15
Quartz	6.50
Glass	0.78
Concrete	1.40
Sandstone	1.83
Air	0.0240
Water	0.5560
Water vapor	0.0206
Ammonia	0.0540

inadequately sizing piping, and underestimating the rate of heat losses, all of which can seriously compromise the efficient operation of a direct use system.

Example 10.3: A spherical tank reservoir of 2 m diameter, having the wall thickness of 10 cm is filled with liquefied natural gas (LNG) at −150 °C. The tank is insulated with a 4-cm thickness of insulation ($k = 0.015$ W/m °C). The ambient air temperature is 20 °C. How long does it takes for the temperature of the LNG to decrease to −120 °C. Neglect the thermal resistance of the steel tank and assume only conduction thermal heat transfer to the tank insulation. The density and the specific heat of LNG are 420 kg/m^3 and 3.48 kJ/kg °C, respectively.

Solution: The LNG reservoir heat loss is:

$$\Delta Q = mC\Delta T = \rho\frac{\pi D^3}{6}C\left(T_{final} - T_{initial}\right)$$

$$= 420 \times \frac{3.14 \times 2^3}{6} \times 3480(-120 - (-150)) = 183.577 \times 10^6 \text{ J}$$

The time is the estimated from (10.1), for a spherical reservoir as:

$$\Delta t = \frac{\Delta Q}{k \cdot A \cdot \dfrac{\Delta T}{\Delta x}} = \frac{183.577 \times 10^6}{0.015 \times 3.14 \times 2^2 \dfrac{20 - (-150)}{0.04}} = 229270.6 \text{ s or } 63.686 \text{ days}$$

Heat transfer by convection is a complex process that involves the movement of mass that contains a quantity of heat. Convection is the heat transfer due to the bulk motion of a fluid. Convective heat transfer also occurs at interfaces between materials, as when air is in contact with warm water or is forced to flow at high velocity through a heat exchanger. In such cases, buoyancy effects, the flow character, the boundary layers, the effects of momentum and viscosity, and the effects of the surface properties and the geometry of the flow pathway are influencing the heat transfer. For example, we assume that a cool air at temperature T_2 flows over a warm water body at temperature T_1. Viscous and frictional forces act to slow the air flow near the surface of the water body, forming a boundary layer, a region where a velocity gradient develops between the interface and the main air mass that has velocity v. At the interface the velocity approaches zero. The boundary layer characteristics are dependent upon the fluid properties, velocity, temperature, and pressure. Heat is transferred by diffusive processes from the water surface to the fluid at the near-zero velocity boundary layer base, causing its temperature to approach that of the water, T_1, resulting in a temperature gradient, in addition to the velocity gradient, between the main mass of moving fluid and the water–air interface. This thermal gradient becomes the driving force behind thermal diffusion that contributes to heat transfer through this boundary layer. Advective transport of heated molecules provides an additional mechanism for heat to move through the boundary layer and into the main air mass, resulting in an increase in the temperature of the air. The rate at which convective heat transfer occurs follows Newton's law of cooling (Equation (1.33)), which is expressed here as:

$$\frac{dQ_{cnv}}{dt} = h \cdot A \cdot dT \tag{10.3}$$

where dQ_{cnv}/dt is the rate at which heat transfer occurs by convection, h is the convection heat transfer coefficient (J/s-m^2 K), A is the exposed surface area (m^2), and dT is the temperature difference between the warm boundary and the overlying cooler air mass, $(T_1 - T_2)$. Values for h are strongly dependent on the properties of the materials involved, the pressure and temperature conditions, the flow velocity and whether flow is laminar or turbulent, surface properties of the interface, the geometry of the flow path, and the orientation of the surface with respect to the gravitational field. As a result, h is highly variable and specific to a given situation. Determining values for h requires geometry-specific experiments, or access to functional relationships that have been developed for situations that are closely analogous to those of a given application. For example, convective heat loss from a small pond over which air is flowing at a low velocity can be reasonably accurate represented by:

$$\frac{dQ_{cnv}}{dt} = 9.045 \cdot v \cdot A \cdot dT \tag{10.4}$$

Here v is the velocity of the air and the effective units of the coefficient, 9.045 are kJs/m^3 · h °C.

In the ideal case, radiation heat transfer is represented by considering a so-called ideal black body. A black body radiator emits radiation that is strictly and absolutely dependent only on temperature, so the wavelength of the emitted radiation is strictly inversely proportional to the temperature. At room temperature, for example, an ideal black body would emit primarily infrared radiation while at very high temperatures the radiation would be primarily ultraviolet. However, real materials emit radiation in more complex ways that depend upon both the surface properties of an object and the physical characteristics of the material of which the object is composed In addition, when considering radiative heat transfer from one object to another, the geometry of the heat source, as seen by the object receiving the radiation, must also need to be taken into account. For most considerations involving radiative heat transfer in direct geothermal energy applications, interfaces are commonly flat plates or enclosed bodies in a fluid, the geometrical factor of minimal importance, and the following relationship is used:

$$\frac{dQ_{rad}}{dt} = \varepsilon \cdot \sigma \cdot A \cdot \left(T_2^4 - T_1^4\right) \tag{10.5}$$

where ε is the emissivity of the radiating body (ε equals 1.0 for a perfect black body), at temperature T_1, σ is the Stefan–Boltzmann constant, equals 5.669 10^{-8} W/m^{2}K^4, T_1 and T_2 are the respective temperatures of the two involved objects, and A is its effective surface area.

Heat transfer trough evaporation can be an efficient energy transport mechanism. The factors that influence evaporation rate are temperature and pressure of the vapor, overlying the evaporating fluid, the exposed area, the fluid temperature, the equilibrium vapor pressure, and the wind velocity. Each of these properties are relatively simple to formulate individually, however, the evaporation process is affected by factors similar to those that influence convective heat transfer becoming quite complex. A complicating factor is that the boundary layer behavior with respect to the partial pressure varies both vertically away from the interface and along the interface due to turbulent flow mixing, so the ambient vapor pressure is not represented rigorously. Moreover, the evaporation rate is affected by the temperature gradient above the interface, which is affected by the boundary layer properties, which in turn, influence the equilibrium vapor pressure. Since the gradient is the driving force for diffusional processes, the rate at which diffusion transports water vapor from the surface is affected by the local temperature conditions as well as the fluid velocity. These complications have led to an empirical approach for establishing evaporation rates in which various functional forms are fit to data sets that span a specific range of conditions. One such common heat rate relationship due to evaporation used in direct geothermal energy use is:

$$\frac{dQ_{evp}}{dt} = \frac{a \times (P_{water} - P_{air})^b \times H_w}{2.778 \times e^{-7}} \tag{10.6}$$

where P_{air} and P_{water} are the water vapor saturation pressures (kPa) at the water and the air temperatures, respectively, H_w is the, a and b, two empirical constants, determined by the velocity of the fluid moving over the interface (m/s), that often are calculated as:

$$a = 74.0 + 97.97 \times v + 24.91 \times v^2$$
$$b = 1.22 - 0.19 \times v + 0.038 \times v^2$$

Example 10.4: Estimate the convection, conduction, radiation, and evaporation heat transfer rates for a rectangular water pool, constructed of concrete walls 15 cm in the ground, having the dimensions, 15 m by 40 m and 2 m depth, assuming the external air temperature 20 °C, the wind speed is 5 m/s and the pool has the temperature of 37 °C.

Solution: The heat transfer rates due to the conduction, convection, radiation, and evaporation are computed by applying (10.2), (10.4), (10.5), and (10.6). The total area for conduction heat transfer is calculated considering the lateral walls and the bottom:

$$A_1 = 2 \times (15 + 40) \times 2 + 15 \times 40 = 880 \text{ m}^2$$

$$\frac{dQ_{cnd}}{dt} = 1.4 \frac{W}{m \cdot K} \times 880 \text{ m}^2 \times \left(\frac{17}{0.2 \text{ m}} \right) = 104,720 \text{ W(J/s)}$$

Convection and radiation heat losses are taking place only on the pool upper (open) surface.

$$\frac{dQ_{cnv}}{dt} = 9.045 \cdot 5(\mathrm{m/s}) \cdot (15 \times 40)\mathrm{m}^2 \cdot (37 - 20)(^\circ\mathrm{C}) = 461{,}295\ \mathrm{J/s(W)}$$

$$\frac{dQ_{rad}}{dt} = 0.99 \times 5.669 \times 10^{-8} \frac{W}{m^2\,K^4} \times 600\ \mathrm{m}^2 \left(310^4\ \mathrm{K}^4 - 293^4\ \mathrm{K}^4\right)$$

$$= 62{,}807.1\ \mathrm{W(J/s)}$$

And the rate of evaporation heat loss, assuming the partial pressure to 3.7 kPa and 1.23 kPa is then:

$$a = 74.0 + 97.97 \times 5 + 24.91 \times 5^2 = 1{,}186.6$$

$$b = 1.22 - 0.19 \times 5 + 0.038 \times 5^2 = 1.22$$

$$\frac{dQ_{evp}}{dt} = \frac{1{,}186.6 \times (3.7 - 1.23)^{1.22} \times 1}{2.778 \times e^{-7}} = 14{,}411{,}639\ \mathrm{W(J/s)}$$

10.2.3.1 Assessing feasibility of direct use applications

The amount of heat, Q_{Load}, required to operate the function of the designed installation depends upon the specific process and the operation size and the external conditions. Assuming that load, Q_{Load} is constant over time, then the geothermal resource must be of sufficient temperature and flow rate to satisfy:

$$\frac{dQ_{Geotherm}}{dt} > \frac{dQ_{Load}}{dt} + \frac{dQ_{Losses}}{dt} \tag{10.7}$$

From the discussion of heat transfer mechanisms the total heat losses, Q_{Losses} that are needed to be accounted for in any application are the sum of all relevant heat loss mechanisms, conduction (Q_{cnd}), convection (Q_{cnv}), radiation (Q_{rdt}), and evaporation (Q_{evp}). and is expressed as:

$$\frac{dQ_{Losses}}{dt} = \frac{dQ_{cnd}}{dt} + \frac{dQ_{cnv}}{dt} + \frac{dQ_{rad}}{dt} + \frac{dQ_{evp}}{dt} \tag{10.8}$$

This equation is used to estimate the heat losses that are assumed for specific application operating conditions. The heat, Q_{Load}, required to perform the function of the designed installation (load) depends upon the specific processes and the operation size. For most applications seasonal variability influences the total heat losses through changes in the air temperature and other weather parameters, wind, humidity, or solar radiation. For this reason, the design load concept was developed, in which the most severe condition set that a facility can likely to experience are used to maximize the heat losses. Hence, when evaluating the feasibility of a potential direct use geothermal project, it is important to establish whether the resource is sufficient to meet the maximum demanding conditions that are likely to occur. Strategies for conducting such an analysis are varied. In some instances, where an abundant resource is available, it is suitable to size the facility in such a way that the resource meets all energy demands. In other instances, for economic reasons, it may turn out to be sufficient to design the facility such that the geothermal resource is meeting the demand of some percentage of the probable events, while the remainder is addressed with other energy sources. However, direct heating in all its forms is more efficient than electricity generation, placing less demanding temperature requirements on the resources. Heat may come from cogeneration with a geothermal electrical plant or from smaller wells or heat exchangers buried in shallow ground. As a result, geothermal heating is economical over a greater geographical range than geothermal electricity generation. Where natural hot springs are available, the hot water is piped directly into radiators. If the ground is hot but dry, tubes or downhole heat exchangers are collecting the heat. But even in areas where the ground is colder than room temperature, heat can still be extracted with a GHP, a more cost-effectively and cleanly than it can be produced by conventional systems. These devices draw on much shallower and colder resources than traditional geothermal techniques, being frequently combining other functions, including air conditioning, seasonal energy storage, solar energy collection, and electric heating. GHPs can be used for space heating essentially anywhere in the world.

Example 10.5: For the pool of Example 10.3 estimate the geothermal inflow rate from a 50 °C reservoir to keep the pool temperature at 37 °C.

Solution: The total heat losses, due to the conduction, convection, radiation, and evaporation are given by (10.8):

$$\frac{dQ_{Losses}}{dt} = 104,720 + 461,295 + 62,807 + 14,411,639 = 1,638,929\ \text{W(J/s)}$$

The geothermal inflow rate is computed as:

$$q_{\text{inflow}} = \frac{dQ_{Losses}/dt}{C_P(T_{Geothermal} - T_{Water})} = \frac{1,638,929}{4183.3 \times (45 - 37)} = 489.7\ \text{kg/s}$$

Notice the evaporation heat transfer losses are the most important mechanism for heat losses for the open pools or ponds.

10.2.3.2 District heating

Approximately 5.4×10^{27} J of geothermal energy is available, of which nearly a quarter is at depths less than 10 km, and in many areas have temperatures higher than the ambient ones, making such resources useable for direct and district heating applications. The basic district heating requirements are a source of warm geothermal fluid, a pipe network to distribute the fluid, a control system, and a reinjection system. These conditions have traditionally been satisfied in areas with hot springs or in locations where high thermal gradients allow shallow drilling to access hot waters. For these reasons, only a relatively small fraction of the large heat amount contained within the crust can be economically employed district heat applications. However, for example the space heating accounted for about 60,000 TJh/year of the total 273,372 TJh/year of energy consumed through direct use applications, the third largest worldwide usage of the geothermal fluids for direct use. The majority of these heating systems involve district heating systems in which multiple users are linked into a network that distributing the heat. District heating distributes the hydrothermal water through piping system to the buildings. Such system design requires matching the distribution network size to the available resources. The resource attributes are the sustainable flow rate (usually between 30 kg/s and 200 kg/s, depending on the district heating system size and the resource temperature. For example, district heating system in Boise, Idaho, the first modern district heating, were 271 communities with geothermal resources are using such systems. There are three typical components of a district heating system: a production facility, a mechanical system, and a disposal system. A production system is the well(s) needed to bring the hydrothermal water/heat energy from the geothermal reservoir. A mechanical system is a system that delivers the hydrothermal water/heat energy to the process. A disposal system is a medium that receives the cooled geothermal fluid. It can be a pond, river, or an injection fluid system. The geothermal power (Q_{Geo}) that must be provided from a resource:

$$Q_{Geo} > m \times C_P \times (T_{Geo} - T_{Rtn}) \tag{10.9}$$

Here m is the mass flow rate (kg/s), C_P is the constant pressure heat capacity of the fluid (J/kg-K), T_{Geo} and T_{Rtn} are the temperatures (K) of the water from the geothermal source, and temperature of the return water after it has been through the network, respectively.

District heat basic requirements and components are a source of warm geothermal fluid, a network of pipe to distribute the heated fluid, a control system, and a disposal or reinjection system. System design requires matching the size of the distribution network to the available geothermal resource. The geothermal water from the production well(s) passes through heat exchangers, where it transfer its heat/enthalpy to a cleaner secondary fluid, often fresh water, distributed via insulated pipes to the district, where it transfer the energy to rise the temperature to the houses and commercial facilities (Figure 10.2). Fluid valves control the amount of geothermal water at heat exchangers and balance the heating demands with the supply. Because the geothermal water does not exit the heat exchangers, dissolved gases or solids are re-injected into reservoir, so do not posing almost any environmental problems. The resource characteristics are a sustainable flow rate, depending on the size of the district heating system and the resource temperature. The geothermal power, P_G, provided from a resource is estimated by:

$$P_G = q_m C_P (T_{GW} - T_{RW}) \tag{10.10}$$

where q_m is the mass flow rate (kg/s), C_P is the constant pressure heat capacity of the fluid (J/kg-K), and T_{GW} and T_{RW} are the temperature (K) of the water from the geothermal source and temperature of the return water after it has been through the network, respectively. The heat demand, Q_{Load}, which is imposed on the system, is

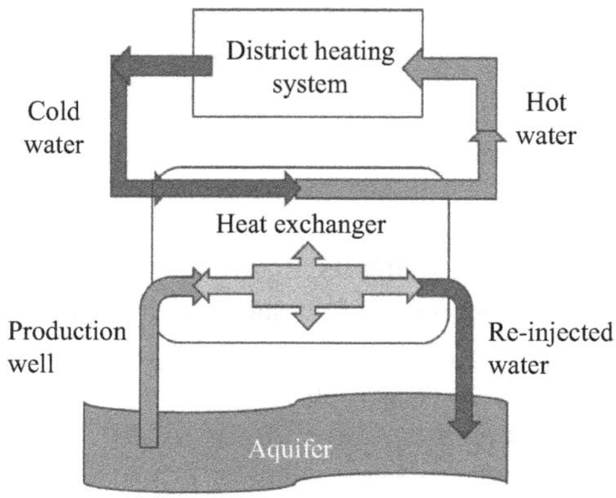

Figure 10.2 Block diagram of a district heating system

a complex function of time. During the day, the load can vary by up to a factor of 3, and is affected by the seasonal climatic variability. From above equation, it is clear that the only variable that can be controlled, affecting P_{GW} is the return water temperature, T_{RW}, since the other variables are set by the properties of the natural system. Maximizing the temperature drop across the network thus becomes a means to increase the power output of the system. However, how this is addressed depends upon the operating mode that can be employed for the system.

Example 10.6: Assuming that a geothermal well is providing a flow rate of 300 kg/s with the difference of temperature of water between the geothermal source and return of 75 °C, what is the geothermal power of this application? If the total losses are 30% of the geothermal power, what is the maximum load power?

Solution: The geothermal power of the well is given by (10.10), as:

$$P_G = 300 \times 4{,}180 \times 80 = 100.32 \text{ MW}$$

The maximum usable power is:

$$Q_{Load(Max)} = (1 - 0.30) \times 100{,}320{,}000 = 70.224 \text{ MW}$$

10.2.4 Geothermal heat pumps

Heat pump technology is considered to be one of the most sophisticated and beneficial engineering accomplishments of the twenty century. Heat pumps are devices, using a compressor-pump system to extract heat from a low-temperature reservoir and reject it a higher temperatures. They are, in essence Carnot cycle applications, operating at highest efficiency levels, by employing transport heat that already exist, without generation, for both heating and cooling, while using only a small fraction of energy amount that they move. These devices can be used in low-temperature geothermal schemes to maximize heat extraction from fluids, while their specific function in any scheme depends on the used fluid temperature. A GHP is a heat pump that uses the earth thermal capacity as an energy source to provide heat to a system or as an energy sink to remove heat from a system, cooling the system. There are three major GHP types: ground-coupled GHPs, ground water GHPs, and hybrid GHPs. The ground-coupled GHPs are of two types: vertical closed-loop and horizontal closed-loop, based on the heat pump piping system shape. In the case of a moderate fluid temperature range (50 °C–70 °C), the extracted heat depends primarily on the heat exchanger, and how the heat pump is connected to extract additional heat from the geothermal fluid. For low temperature range (less than 50 °C, usually 40 °C or lower), heat extraction is almost impossible and the heat pump is connected to ensure all heat transfer. GHPs take advantage of the Earth's relatively constant temperature at depths of about a few meters to about 100 m (or 10 ft. to about 300 ft.). GHPs can be used almost everywhere in the world, as they do not have the requirements of fractured rock and water as are needed for a conventional geothermal reservoir. GHPs circulate water or other liquids through pipes buried in a continuous loop, either horizontally or vertically, under a landscaped area, parking lot, or any number of areas around the building, being one of the most efficient heating and cooling systems available. To supply heat, the system pulls heat from the Earth through the loop and distributes it through a conventional duct system. For cooling, the process is reversed; the system extracts heat from the building and moves it back into the earth loop. It can also direct the heat to a hot water tank, providing another advantage, free hot water. GHPs reduce electricity use 30%–60% compared with traditional heating and cooling systems, because the electricity which powers them is used only to collect, concentrate, and deliver heat, not to produce it.

Direct geothermal systems for heating and cooling typically comprise a primary circuit that exchanges heat with the ground, a heat pump that exchanges and enhances heat transfer between the primary circuit and the secondary circuit, and a secondary circuit that circulates heat within the building. Space heating systems usually require higher temperatures than that of the ground. At first glance, the use of the cooler ground to heat a building may appear a contravention of the second law of thermodynamics (heat flows from hot to cold). Heat pumps overcome this apparent restriction by enhancing the ground-sourced energy with electrical or mechanical work. Refrigerators are every day-life common example of a heat pump. Figure 10.3 gives a schematic of a GHP diagram and the basic operating principle. Heat transfer occurs in fluids when they change temperature and/or phase, while the heat transfer associated with phase change is much greater than that corresponding to only temperature change, and heat pumps make use of the properties of refrigerants (which can change phase at suitable operating temperatures and pressures) to achieve efficient heat transfer.

The heat pump principle and operation consist of a cooled, liquid refrigerant that is pumped into the heat exchanger (evaporator), where it absorbs thermal energy from the ambience as a result of the temperature differential, then through the compressor to a condenser where is released the heat, as is shown in Figure 10.4. During this process, the refrigerant then changes state, becoming gas, and then the gaseous refrigerant is recompressed in the compressor, resulting in a temperature increase. A second heat

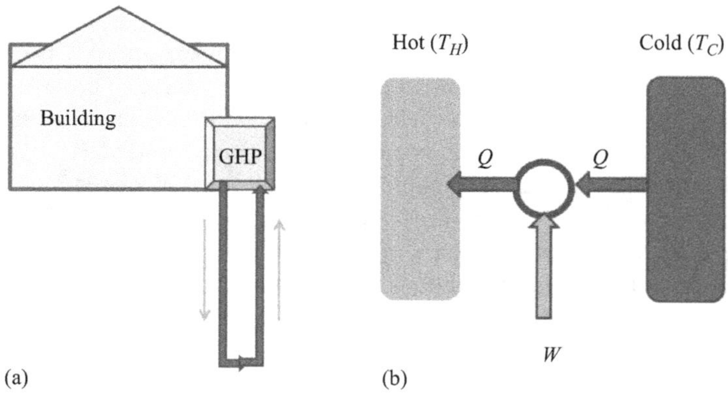

Figure 10.3 (a) Geothermal heat pump, and (b) operating principle

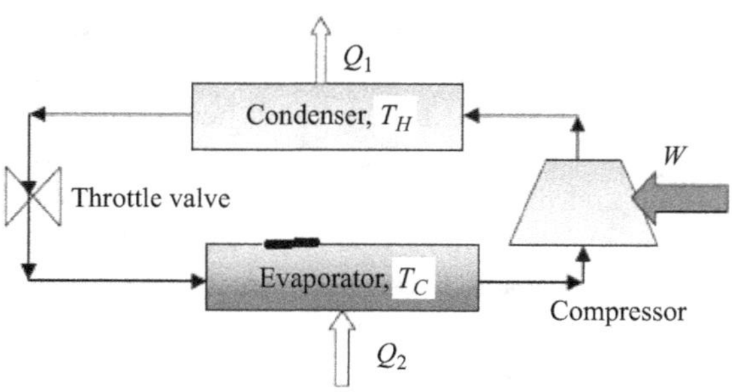

Figure 10.4 Heat pump schematic and operating diagram

exchanger (condenser) transports this thermal energy to the heating system and the refrigerant reverts to a liquid, and the refrigerant pressure is reduced again in the expansion valve. Basically, the liquid refrigerant absorbs heat from a heat source and evaporates, the refrigerant, cooler than the heat source, has boiling point below the heat source temperature, the refrigerant gas then passes through compressor, which increases its pressure and temperature and the hot and high pressure refrigerant gas from the compressor is hotter than the heat sink (so that heat flows from the refrigerant to the heat sink). At this higher pressure, the refrigerant gas condenses at a higher temperature than at which it boiled. Thus when the refrigerant gas reaching the condenser condenses and releases heat. The hot, high pressure liquid refrigerant then passes through an expansion valve which returns the pressure and temperature of the liquid to its original conditions prior to the cycle. For GHPs in heating mode, refrigerant evaporation occurs where the heat pump joins the primary circuit and condensation occurs where the heat pump joins the secondary circuit. Refrigerant evaporation cools the circulating fluid in the primary circuit is then re-heated by the ground. This process is reversed in cooling mode as refrigerant condensation heats the circulating fluid in the primary circuit, which is re-cooled by the ground. It is important to note that GHPs require energy input (the compressor and to the pumps that circulate fluid) to move heat around the system. However, the energy input required is typically small compared to the heat output: GHPs typically produce around 3.5–5.5 kW of thermal energy for every 1 kW of electricity used. Heat pump efficiency also increases as the temperature difference between the heat sink and heat source decreases. GHPs are thus more efficient than air source heat pumps the seasonally averaged ground temperature is closer to the desired ambient building temperature than the air is. In order to transport heat from a heat source to a heat sink, heat pumps need external energy. Theoretically, the total heat delivered by the heat pump is equal to the heat extracted from the heat source, plus the amount of drive energy supplied to it. It is simply a heat engine running in reverse mode, as shown in Figure 10.4. It accepts heat Q from the sink at T_C (at lower temperature), rejecting heat Q into the source which is at a higher temperature, T_H in doing so it consumes work W, from an external source.

From thermodynamic point of view, a heat pump thermal cycle is a revere thermal engine cycle, so the input work provides the heat transfer to rise the temperature. The basic relationships governing the heat pumps are the laws of thermodynamics and are independent of the working fluid, cycle type, and the form of heat transfer. The heating efficiency of a heat pump is given by is coefficient of performance (COP) defined as the ratio of heat provided per energy input:

$$COP = \frac{\text{Heat output}}{\text{Input work}} = \frac{Q_H}{W} < \frac{T_H}{T_H - T_C} \tag{10.11}$$

However, it is usually to define overall efficiencies for actual (real) heat engines and heat pumps, according to the Carnot cycle, as:

$$W = \eta \cdot Q_H$$

with $\eta < \eta_C$, for heat engines, and $\eta > \eta_C$ for heat pumps, where η_C is the Carnot efficiency, operating between the same temperatures. Previous equation is usually rewritten for heat pumps as:

$$W = \frac{Q_H}{COP_{HP}} = \frac{Q_C}{COP_{RF}} \tag{10.12}$$

Here COP_{HP} is the so-called "coefficient of performance" based on heat output Q_H from a heat pump or vapor recompression system, and COP_{RF} is the coefficient for a refrigeration system based on heat absorbed Q_C from the process. The coefficients of performance are defined for real (actual) systems, with temperature expressed in Kelvin degrees (the absolute temperature) as:

$$COP_{HP} = \frac{\text{Desired output}}{\text{Required input}} = \frac{\text{Heating effect}}{\text{Input work}} = \frac{Q_H}{W} = \eta_{mech}\frac{T_H}{T_H - T_C} \tag{10.13a}$$

And

$$COP_{RF} = \frac{\text{Desired output}}{\text{Required input}} = \frac{\text{Cooling effect}}{\text{Input work}} = \frac{Q_C}{W} = \frac{Q_H - W}{W} = COP_{HP} - 1 \tag{10.13b}$$

Hence, the heat pump as a low temperature lifting device ($T_C \rightarrow T_H$) gives higher COP_{HP}, which is inversely proportional to the temperature stretch and a large amount of upgraded heat per unit power. However, lower values of T_C reduce COP_{RF}, so refrigeration systems need more power per unit for upgraded heat as the absolute temperature falls. With the above understanding, one can start analyzing the GHP problems. Theoretically, heat pumping can be achieved by many more thermodynamic cycles and processes. These include Stirling cycle, single-phase cycles (e.g., with air, CO_2 or noble gases), solid-vapor sorption systems, hybrid systems (combining the vapor compression and absorption cycle), and electromagnetic and acoustic processes. Some of these systems are close to entering the market or have reached technical maturity, and can become significant in the future. Almost all heat pumps fall on two categories, i.e., either based on a vapor compression, or on an absorption cycle. Heat pumps are used for two purposes, for refrigeration/cooling below-ambient temperature, or as a heat recovery system. However, the equipment used in both cases is very similar.

Example 10.7: Compare the heating efficiencies (maximum COP) of the same heat pump installed in New Orleans, Louisiana and in Cleveland, Ohio.

Solution: In New Orleans, the climate is milder with higher average temperatures, we can assume that T_H (summer) is about 20 °C (or 70 °F) and that T_C (winter season) is about 5 °C (or about 40 °F). In Cleveland, assume that T_H is the same but that T_C (the outside temperature) is much lower, on average is about −10 °C (or about 14 °F). Since the heat pump is used as a heater, after the conversion to absolute temperatures, the maximum COP at each of the two locations is calculated as:

$$\text{Cleveland: } COP_{\max} = \frac{T_H}{T_H - T_C} = \frac{293}{293 - 263} = 9.77$$

$$\text{New Orleans: } COP_{\max} = \frac{T_H}{T_H - T_C} = \frac{293}{293 - 278} = 19.53$$

The most important benefits of GHPs is that they are using about one third or even over half less electricity than the conventional heating and cooling systems, having a good potential for energy consumption reductions in any area. The cooling efficiency is defined as the ratio of the heat removed to the input energy, or the energy efficiency ratio (EER). Good GHP units must have a COP of 3 or greater and an EER of 13 or greater.

10.2.5 Electricity from geothermal energy sources

Geothermal power plants use the natural hot water and/or steam to turn turbine-generator units for electricity generation. Unlike fossil fuel-based power plants, no fuel is burned in these plants. Geothermal power plants byproduct is water vapors, with no smoky emissions. Geothermal power plants are for the base load power as well as the peak load demand units. Geothermal electricity has become competitive with conventional electricity generation in many world regions. Main types of geothermal power plants are listed and discussed here. *Dry steam power plants* are the simplest and most economical technology, and therefore are widespread. It is suitable for sites where the geothermal steam is not mixed with water. Geothermal wells are drilled down to the aquifer, and the superheated and pressurized steam (180 °C to 350 °C) is brought to the surface and passed through a steam turbine to generate electricity. In simple power plants, the low pressure steam output from the turbine is vented to the atmosphere. However, usually after passing the turbine-generator unit, the steam is condensate, resulting almost pure water, which is re-injected into the aquifer or used for other purposes. This can improve the overall plant efficiency, while avoiding the environmental problems associated with the direct steam release into the atmosphere. Italy and United States have the largest dry steam geothermal resources. This type of resources is also found in Indonesia, Japan, and Mexico. In *single flash steam technology*, hydrothermal resource is in a liquid form. The fluid is sprayed into a flash tank, held at a much lower pressure than the fluid, causing it to vaporize (flash) rapidly to steam. The steam is then passed through a turbine coupled to a generator. To prevent the geothermal fluid flashing inside the well, the well is kept under high pressure. Flash steam plant generators range from 10 to 55 MW, with a standard power size of 20 MW is used in several countries. *Binary cycle power* plants are used where the geothermal resource is insufficiently hot to produce steam, or where the resource contains too many chemical impurities to allow flashing. In addition, the fluid remaining in the tank of flash steam plants can be utilized in binary cycle plants (e.g., Kawerau in New Zealand). In the binary cycle process, the geothermal fluid is passed through a heat exchanger. The secondary fluid (e.g., isobutene or pentane) which has a lower boiling point than water is vaporized and expanded through a turbine to generate electricity. The working fluid is condensed and recycled for another cycle. The geothermal fluid is then re-injected into the ground in a closed-cycle system. Binary cycle power plants can achieve higher efficiencies than flash steam plants and allow the utilization of lower temperature resources. In addition, corrosion problems are also avoided.

The world geothermal electrical capacity installed in the year 2000 was about 8 GWe with the generation in that year of 49.3 billion kWh, while in 2015 was about 12 GWe, generating about 80 TWh/year. In the industrialized countries, where the installed electrical capacity reaches very high levels, in the range of tens or even hundreds of thousands of MWe, the geothermal energy is unlikely to account for more than 2%, at most, of the total in the near future. On the other side, in the developing countries, with quite limited electrical consumption but good geothermal prospects, geothermal electricity generation could make quite a significant contribution to the total, with estimates of about or over 15% in countries like Philippines, El Salvador, Nicaragua, or Costa Rica. The efficiency of the generation of electricity from geothermal steam ranges from 10% to 20%, about three times lower than the efficiency of nuclear or fossil-fuelled plants. Geothermal power plants have the lower efficiencies due to the low steam temperature, 250 °C or lower. Furthermore, geothermal steam has a different chemical composition than the pure water vapor, containing usually noncondensible gases, that reduces the overall system efficiency. The simplest and cheapest of the geothermal cycles used to generate electricity is the direct-intake noncondensing cycle. Steam from the geothermal well is passed through a turbine and exhausted to the atmosphere, with no condensers at the outlet of the turbine. Such cycles consume about 20 kg of steam per kWh. Noncondensing systems can be used if the content of noncondensible gases in the steam is very high, greater than 50% in weight, and are used in preference to the condensing cycles for gas contents exceeding 15%, because the high energy required extracting these gases. In power plants where electricity is produced from dry or superheated steam, vapor-dominated reservoirs, steam is piped directly from the wells to the turbine. This is a well-developed, commercially available technology, with typical turbine-size units in the 20–120 MWe capacity range. Recently, a new trend of installing modular standard generating units of 20 MWe has been adopted in Italy. Vapor-dominated systems are less common in the world, steam from these fields has the highest enthalpy (energy content), generally close to 670 kcal/kg (2800 kJ/kg). At present these systems have been found only in Indonesia, Italy, Japan, and the USA. These fields produce about half of the geothermal electrical energy of the world. Water-dominated fields are much more common. Flash steam plants are used to produce energy from these fields that are not hot enough to flash a large proportion of the water to steam in surface equipment, either at one or two pressure stages.

If the geothermal well produces hot water instead of steam, electricity can still be generated, provided the water temperature is above 85 °C, by means of binary cycle plants. These geothermal plants operate with a secondary, low boiling-point working fluid (Freon, Isobutane, Ammonia, etc.) in an organic Rankine cycle. The working fluid is vaporized by the geothermal heat in the vaporizer, and then passes through the organic vapor turbine, coupled to the generator. The exhaust vapor is then condensed in a condenser and is recycled to the vaporizer by a fluid cycle pump. The efficiency of these cycles is quite low, up to 6%. Typical unit sizes are from 1 to 3 MWe. However, the binary power plant technology has emerged as the most cost-effective and reliable way to convert large amounts of low temperature geothermal resources into electricity, such large low-temperature reservoirs at accessible depths existing, in almost any world areas. The power rating of geothermal turbine-generator units tends to be smaller than in conventional thermal power stations, with common power levels of 55, 30, 15, and 5 MWe or even smaller. One of the main advantages of geothermal power plants is that they can be built economically in relatively much smaller units than, for example, hydropower stations. In developing countries with a small electricity market, geothermal power plants with units from 15 to 30 MWe can be more easily adjusted to the annual increase in electricity demand than larger hydropower or fossil fuel power plants. The reliability of geothermal power plants is very good, the annual load factor and availability factor are commonly about 90%, and geothermal fields are not affected, by annual or monthly fluctuations in rainfall or weather, since the essentially meteoric water has a long residence time in geothermal reservoirs. There are basically three types of geothermal power plants, which are determined primarily by the nature of the geothermal resource at the site. The first ones are direct steam geothermal plant, used where the geothermal resource produces steam directly from the well. These are the earliest types of plants developed in Italy and in the US. However, such steam resources are one of less common of all the geothermal resources, existing in only a few places. Obviously steam plants are improper to the low-temperature resources. Flash steam plants are employed in cases where the geothermal resource produces high-temperature hot water or a combination of steam and hot water. The fluid from the well is delivered to a flash tank where a part of the water flashes to steam and is directed to the turbine. The remaining water is directed to the disposal. Depending on the resource temperatures it may be possible to use two stages of flash tanks, in which the water separated at the first stage tank is directed to a second stage flash tank where more (but lower pressure) steam is separated. Remaining water from the second stage tank is then directed to disposal. The *double flash* plant delivers steam at two different pressures to the turbine. Again, this type of plant cannot be applied to low-temperature resources. The third type of geothermal power plant is the binary geothermal power plant. The name derives from the fact that a second fluid in a closed cycle is used to operate the turbine rather than geothermal steam. Geothermal fluid passes through a heat exchanger (boiler or vaporizer), in some plants, two heat exchangers in series ae used, the first a preheater and the second a vaporizer where the heat in the geothermal fluid is transferred to the working fluid causing it to boil. Past working fluids in low temperature binary plants were CFC (Freon type) refrigerants, while the new ones use hydrocarbons (isobutane, pentane, etc.), as refrigerants with the specific fluid chosen to match the geothermal resource temperature. The working fluid vapor passes to the turbine where its energy is converted to mechanical energy, delivered to the generator. The vapor exits the turbine to the condenser where it is converted back to a liquid. In most plants, cooling water is circulated between the condenser and a cooling tower to reject this heat to the atmosphere. An alternative is to use so called "dry coolers" or air cooled condensers which reject heat directly to the air without the need for cooling water. This design essentially eliminates any consumptive use of water by the plant for cooling. Dry cooling is operating at higher temperatures (in the summer season) than cooling towers does result in lower plant efficiency. Liquid working fluid from the condenser is pumped back to the higher pressure pre-heater/vaporizer by the feed pump to repeat the cycle. The binary cycle is the type of plant which would be used for low temperature geothermal applications, while the processes in binary geothermal plant are very similar to ones in steam turbine based power generation. Currently, the binary equipment is available in modules of 200–1,000 kW.

The process of generating electricity from a low temperature geothermal heat source (or from steam in a conventional power plant) involves a process engineers refer to as a Rankine Cycle. A conventional power plant includes a boiler, turbine, generator, condenser, feed water pump, cooling tower, and cooling water pump. Steam is generated in the boiler by burning a fuel (coal, oil, gas, or uranium). The steam is passed to the turbine where, in expanding against the turbine blades, the heat energy in the steam is converted to mechanical energy causing rotation of the turbine. This mechanical motion is transferred, through a shaft to the generator where it is converted to electrical energy. After passing through the turbine the steam is converted back to liquid water in the condenser of the power plant. Through the process of condensation, heat not used by the turbine is released to the cooling water. The cooling water, delivered to the cooling tower where, and the "waste heat" from the cycle is often rejected to the atmosphere. However, the modern power plants are using advanced heat recovering technologies and combined heat and power generation in order to increase the overall system efficiency. Steam condensate is delivered to the boiler by the feed pump to repeat the process. In summary, a power plant is simply a cycle that facilitates the conversion of energy from one form to another. In this case the chemical energy in the fuel is converted to heat (at the boiler), and then to mechanical energy (in the turbine) and finally to electrical energy (in the generator). EGS systems are referring to the creation of the artificial conditions at a site or a location where a reservoir has the potential to produce geothermal energy. A geothermal system requires heat, permeability, and water, so EGS techniques make up for reservoir deficiencies in any of these areas. These systems involve injecting water into the source and circulating it through the dry rocks. Because of the low thermal conductivity of the rocks large surface areas are necessary for such systems.

10.3 Small hydropower

The main components of a hydroelectric system may be classified into two groups: (1) *the hydraulic system* components that include the turbine, the associated conduits-like penstocks, tunnel and surge tank-and its control system, and (2) *the electric system* components formed by the electrical generator, usually synchronous type and the control system. Hydraulic power can be captured wherever a flow of water falls from a higher level to a lower level. Hydraulic turbines are defined as prime movers that transform the kinetic energy of the falling water into mechanical energy of rotation and whose primary function is to drive a electric generator. The vertical fall of the water, the *head* is essential for hydropower generation; fast-flowing water on its own does not contain sufficient energy for useful power production except on a very large scale, such as offshore marine currents. Hydropower is essentially a controlled method of water descent usefully utilized to generate power. Hydroelectric plants utilize the energy of water falling through a head that may vary from a few meters to up to 1,500 or even 2,000 m. To manage this wide range of heads, many different kinds of turbines are employed, which differ in their working components.

Two quantities are required, for hydropower: a water flow rate Q, and a head H. However, it is better to have more head than more flow, since this keeps the equipment relatively smaller. *The gross head*, H_g, is the maximum available vertical fall in the water, from upstream to downstream level. The actual head seen by a turbine is less than the gross head due to losses incurred when transferring the water into and away from the turbine. This reduced head is known as the *net head*. In a hydropower plant,

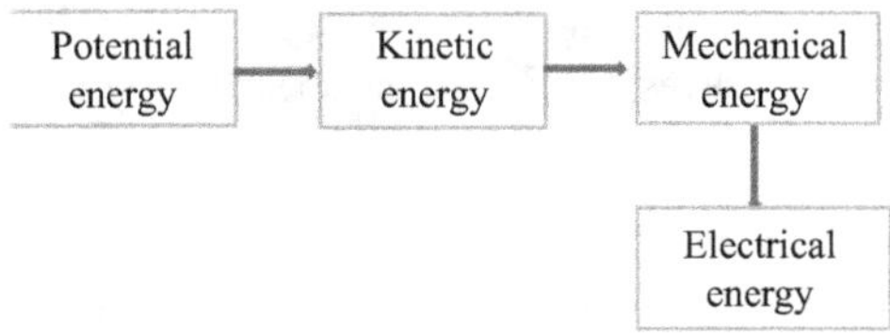

Figure 10.5 Hydropower energy conversion process

the water potential energy is first converted to equivalent amount of kinetic energy. Thus, the water height is used to calculate its potential energy and this energy is converted to water speed at the turbine intake, and is calculated by balancing the water potential and kinetic energy. Turbines are devices designed to extract energy from a flowing fluid. The geometry of turbines is such that the fluid exerts a torque on the rotor in the direction of its rotation. The generated shaft power is then available to derive generators or other devices. Based on the conservation of energy, and laws of dynamics, the hydropower energy conversion is summarized in Figure 10.5. However, a thorough study of each conversion level is needed in order to explain how hydropower is converted into electricity, to estimate the power output, conversion efficiency, and how it can be used as sustainable energy source. The performance of hydraulic turbines is strongly influenced by the hydraulic system, topography characteristics of water conduit that feeds the turbine, including the water inertia effect, water compressibility and penstock characteristics. The energy in water can be potential energy from a height difference, used to generate electricity, which is what most people think of in terms of hydro-energy from water stored in dams. However, there is also kinetic energy due to water flow in rivers and ocean currents. The potential energy of a mass of water stored in a reservoir, having a volume V, having a head H, is expressed as:

$$P.E. = W = \rho g V H \tag{10.14}$$

Example 10.8: Find the potential energy for 5,000 m^3 of water at a height of 100 m.

Solution: By using (10.14), the potential energy is:

$$P.E. = \rho g V H = 10^3 \cdot 9.806 \cdot 5000 \cdot 50 = 4.903 \times 10^9 \text{ J}$$

The specific energy of a hydropower plant is the quantity of potential and kinetic energy which one unit mass of water delivers when passing through the plant from an upper to a lower reservoir. The expression of the specific energy is Nm/kg or J/kg and is designated as [m^2/s^2]. In a hydropower plant, the difference between the level of the upper reservoir z_{res} and the level of the tail water z_{tw} at the penstock exit to the turbine is defined as the gross head, $H_g = z_{res} - z_{tw}$. This is one of the most hydropower fundamental parameters, the difference between the maximum and minimum water heights, between the upstream side of the turbine and downstream of the turbine at the draft tube outlet. Potential energy of the water is direct proportional to the reservoir head level. A high head level would mean that the potential energy for that system is higher. The corresponding gross specific available energy is then:

$$E_g = gH_g$$

These considerations lead to the definition of the net head across the turbine as described in (10.14), according to the IEC41 standard were the draft tube outlet loss is regarded to be a power plant loss and not a turbine loss. Note that the flow rate $Q = A \cdot v$ (m^3/s), A is the cross section of draft tube outlet and v is the flow velocity. To express the energy for the hydropower system, we are using the Bernoulli equation, expressed as:

$$E = gH = \left(gh_0 + \frac{v_0^2}{2} + gz_0\right) - \left(gh_{tw} + \frac{v_{tw}^2}{2} + gz_{tw}\right) \tag{10.15}$$

Here "0" refers the penstock intake (reservoir level) while "*tw*" to the turbine input level. When a water discharge (flow rate) Q [m^3/s] passes through the hydropower plant, the delivered power is determined as in the following. The effective head is the difference between the energy head at the turbine entrance and the energy head at the exit of the draft tube. When the volume of waters moves from the maximum level at dV_1 to the minimum level of dV_2 for a height of H, the flow rate or discharge is the difference between them, a mechanical work is produced and is given by the equation:

$$\begin{aligned} W_g &= \rho g V H_g \\ V &= dV_1 - dV_2 \end{aligned} \tag{10.16}$$

By using the equation for work, theoretical power output of the hydropower system can be calculated:

$$P_g = \frac{dW_g}{dt} = \rho g Q H_g \tag{10.17}$$

In (10.13)–(10.15), g is acceleration due to gravity (9.806 m/s), ρ is the water density (1,000 kg/m^3), H_g is the gross head (m), while Q is the volumetric flow rate through the turbine. Power is measured in units of Watts. However, the net head, H, defined in (10.15) is used to estimate the theoretical energy of which the majority will be utilized and transformed to mechanical energy by the turbine. The rest of the energy will be converted to losses, i.e., mainly increased heat energy in the water and a very small negligible part as a heat flux to its surroundings. The available power for the turbines in a hydropower plant project depends on the available flow Q m^3/s and is define by:

$$P = \rho E Q \tag{10.18}$$

The available energy stored in a reservoir is determined by multiplying the mass of water (i.e., the stored volume of water multiplied by the density) with the net head. The equation for stored energy of a water volume V and a net head H, i.e., $E = g \cdot H$, yields:

$$W = \rho \cdot V \cdot E \tag{10.19}$$

Example 10.9: For a reservoir with a volume of $V = 500 \cdot 10^6$ m^3 and a head of $H = 400$ m, calculate the available storage of energy which can be utilized by the turbine when ignoring head losses in tunnels and the losses in turbines and generators:

Solution: From (10.19) the stored available energy is:

$$W = \rho \cdot V \cdot E = 10^3 \times 500 \cdot 10^6 \times 9.806 \times 400/\left(3{,}600 \times 10^3\right)$$

$$= 10.9 \cdot 10^8 \text{ kWh} = 1.09 \text{ TWh}$$

The actual power output is the theoretical power output time the efficiency, η, expressed as:

$$P = \eta \rho g Q H \tag{10.20}$$

Example 10.10: Estimate the power output from a reservoir with net head of 80 m and the volume flow rate of 30 m/s and efficiency 90%.

Solution: From (10.19), the power output is:

$$P = \eta \rho g Q H = 0.90 \times 9.806 \times 1{,}000 \times 30 \times 80 = 21.181 \text{ MW}$$

The hydraulic efficiency of a turbine excludes friction losses on the outside of the runner, water leakage loss, not passing through the runner blades and mechanical friction losses. The hydraulic efficiency of a well-designed turbine is 98%–99% and can be developed as shown in the following description.

Example 10.11: Estimate the flow rate required for a 85% hydropower plant efficiency with a net head of 60 m to satisfy the electrical needs of a load of 15 kW.

Solution: From (10.17), the flow rate can be found as:

$$Q = \frac{P}{\eta \rho g H} = \frac{15{,}000}{0.85 \times 1{,}000 \times 9.80 \times 60} = 0.030 \text{ m}^3/\text{s}$$

As the water hits the impulse vanes of hydro turbine, a dynamic force will exist in order for the vanes or buckets to start rotating. The rotation of the vanes converts the potential energy to kinetic energy. The force on the moving vane or bucket by a jet of water is derived as the equation of force, and the torque that is exerted on the vanes by the jet of water, the product of the force, can be computed. Once the value for torque has been obtained, it is now possible to calculate the theoretical power that is exerted by the vanes. Power is simply the product of torque and angular velocity of the runner in rad/s.

$$P = T \cdot \omega \tag{10.21}$$

If the water weight is low, it will reduce the force on the bucket, torque and power. If the buckets are placed too far apart, water would flow through and only very little energy will be extracted from the system. In order for the runner to perform efficiently, the water should leave the runner in an axial direction. However, this will not be possible as to do so would mean obtaining completely axial flow at all the gate openings. Hence, the absolute velocity of the water is considered as the water exits from the runner to be equal to the water discharged divided by the area of the draft tube. Hydropower turbines are used to extract energy from a fluid and by this decrease the total energy of the fluid. The total energy in the fluid is measured by a total head that composes of various forms of energy as follows:

$$H_{tot} = H_{pr} + H_{st} + H_{vl} + H_{fr} \tag{10.22}$$

Terms in (10.22) are defined below.

$H_{pr} = \dfrac{p_2 - p_1}{g\rho}$ – Pressure head

$H_{st} = H_{res} - H_{tw}$ – Static head

$H_{vl} = \dfrac{v_2^2 - v_1^2}{2g}$ – Velocity head

H_{fr} – Friction head

The friction head reflects the losses in a system and is commonly expressed in meters, while the static head is basically the earlier defined gross head. A turbine system denotes a system, in which a turbine is used to extract energy from a fluid. The system consists of pipes (or ducts) on the pressure and the discharge side of the turbine as well as eventually valves, reservoirs, or other devices. A turbine decreases the total head in a system. This implies that there is high-energy fluid available at the inlet of a turbine and that the fluid leaves the turbine with reduced energy content. Depending on the application, the primary contribution of the high-energy source might be different:

1. Hydro turbine driven by high-velocity fluid which results from a great difference in elevation: the primary high energy source is static head that is transformed into velocity head by flow acceleration.
2. Hydro turbine that is driven by the flow in a river: the primary high energy source is velocity head.

Hydropower generation involves the water storage, conversion of the potential energy of the fluid into kinetic energy, using hydraulic turbines, and conversion of the mechanical energy to electrical energy in electric generators. Hydroelectric units have been installed in capacities ranging from a few kilowatts to nearly 1 GW. Hydroelectric power plants are of three major types (a) run-of-the-river in which small amounts of water storage is used to generate electricity, with a very little control of the flow through the plant, (b) hydropower plants with a large storage, consisting of an artificial basin (created by a dam on a river course) that allows storing water and thus controlling the flow through the plant on a daily or seasonal basis, having several hydro turbine-generator units, (c) pumped hydropower storage designed to operate during off-peak hours, the water is pumped (by means of reversible pump-turbines or dedicated pumps) from a lower reservoir to an upper reservoir, so that the energy is thus stored for later production during peak hours. Net head is lower than gross head due to energy losses in the penstock:

$$H = H_g - Losses = H_g - H_{Losses} \tag{10.23}$$

Penstock efficiency is the ratio of net and gross head, and head losses, H_{Losses} is given by:

$$\eta_p = \frac{H}{H_g} = 1 - \frac{H_{Losses}}{H_g} \tag{10.24}$$

We are defining the efficiency of a pump or turbine to impart energy to or extract energy from water, as the ability of hydraulic structure or element to conduct water with minimum energy losses. The hydraulic efficiency is defined as the ratio of power delivered to the runner, gH to the power supplied at the inlet, W, expressed as:

$$\eta_y = \frac{W}{gH} \tag{10.25}$$

The volumetric efficiency is the ratio of the net flow rate (discharge) to the input flow rate, as:

$$\eta_v = \frac{Q_{net}}{Q_{in}} \tag{10.26}$$

The input power in the turbine-generator unit is then given by:

$$P_{in} = \eta_y \eta_v \rho g Q_{net} H \tag{10.27}$$

Overall hydropower turbine efficiency is computed by multiplying the efficiency of turbines, efficiency of generators, and is on average for most of the turbine-generator units around 60%–70%. Net or output power hydropower turbine output, the difference in gross power and mechanical and auxiliary losses of the turbine-generator unit, and is expressed as:

$$P_{out(el)} = P_{in} - P_{mech} - P_{aux} \tag{10.28}$$

The overall efficiency of the hydropower generation system is:

$$\eta_{gen} = \frac{P_{out(el}}{P_{in}} = 1 - \frac{P_{mech} + P_{aux}}{P_{in}}$$

Net power output of the hydropower generation unit is:

$$P_{out(el)} = \eta_y \eta_v \eta_{gen} \rho g Q H_g = \eta_t \rho g Q H_g \qquad (10.29)$$

Here, η_t is the total turbine-generator unit efficiency. The overall plant efficiency is the product of the overall generation unit efficiency and the penstock the efficiency, and is expressed as:

$$\eta = \eta_p \eta_t = \frac{P_{out(el)}}{\rho g Q H_g} \qquad (10.30)$$

Hydraulic turbines extract mechanical energy from water which has a high head. The power of a rotary hydraulic machines, such as a hydropower depends upon the working fluid density, ρ, speed of rotation, N, the characteristic diameter, D, the change in the head, ΔH, the flow rate or discharge, Q, and acceleration due to the gravity, g. The functional relationship of power is expressed as:

$$P = f(\rho, N, D, \Delta H, Q, g) \qquad (10.31)$$

There are two main types, reaction and impulse, the difference being in the manner of head conversion. In reaction turbines, the water fills the blade passages and the head change or pressure drop occurs within the impeller. They can be of radial, axial, or mixed flow configurations. Impulse turbines convert first the high head through a nozzle into a high velocity jet that strikes the blades at one position as they pass by. Reaction turbines are smaller because water fills all the blades at one time. For hydraulic impulse turbines, the pressure drop across the rotor is zero, the all pressure drop across the turbine stages occurs in the nozzle row. Pelton wheel is the classical example of impulse turbines. A high-speed jet of water strikes the Pelton wheel buckets and is deflected, while the water enters and leaves the control volume surrounding the wheel as free jet. Ideally, the fluid enters and leaves the control volume with no radial component of velocity. For impulse turbines the total fluid head is converted into a large velocity head at the exit of the supply nozzle. The pressure drop across the bucket (blade) and the fluid relative speed change across the bucket are negligible, while the space surrounding the rotor is not completely filled with fluid. The individual jets of fluid striking the buckets are generating the torque, eventually transferred to the generator. For reaction turbines, there is both a pressure drop and a fluid relative speed change across the rotor. Guide vanes act as nozzle to accelerate and turn the flow in the appropriate direction as the fluid enters the rotor, and the pressure drop occurs across the guide vanes and across the rotor, being best suited for higher flow rate and lower head such as are often encountered in hydroelectric stations associated with a dammed river. The working fluid completely fills the passageways through which it flows. The fluid angular momentum, pressure, and velocity decrease as it flows through the turbine rotor, and the turbine rotor extracts energy from the fluid.

In order to estimate the water power potential fluid mechanics principles and laws are applied. For any stretch of a watercourse or river, characterized by a difference in level (head) of H meters, conveying a flow rate or discharge of Q (m^3/s), the theoretical (potential) power, P_{st}, can be expressed by:

$$P_{st} = \rho QH = 1,000\, QH\left(\frac{kg \cdot m}{s}\right) = 13.405\, QH(HP) = 9.786\, QH(kW) \quad (10.32)$$

If the rate of flow changes along a stretch, the mean value of the discharges or flow rates pertaining to the two terminal sections of the stretch or current is to be substituted in (10.32). The theoretical power resources of any river or river system are given by the total of the values computed for the all individual stretches or flow currents, and is given by the expression:

$$P_{theor} = 9.786 \sum QH(kW) \quad (10.33)$$

Potential water power resources can be characterized by different values according to the discharge taken as basis of computation. The conventional discharges or flow rates, used to characterize hydropower potential of a river are listed here. Minimum potential power, or theoretical capacity of 100%, is the term for the value computed from the minimum flow observed, $P_{teor100}$. Small potential power represents the theoretical capacity of 95% can be derived from the discharge of flow rate of 95% duration as indicated by the average flow duration curve, P_{teor95}. Median or average potential power is the theoretical capacity of 50% can be computed from the discharge or flow rate of 50% duration as represented by the average flow duration curve, P_{teor50}. Mean potential power represents the value of theoretical mean capacity can be ascertained by taking into account the average of mean flow. The average of mean flow is understood as the arithmetic mean of annual mean discharges for a period of 10–30 years. The annual mean discharge is the value that equalizes the area of the annual flow duration curve.

An important aspect of the hydropower assessment and analysis consists of good measurements on the hydropower site and surrounding areas, secondary rivers, etc. For power extraction purpose from a stream or river, it is important to measure the water flow rate and head. Head can be measured as vertical distance (feet or meters) or as pressure (e.g., lb./sq. ft., N/m^2). Regardless of the size of the stream, higher head will produce greater pressure and therefore higher output at the turbine. An altimeter can be useful in estimating head for preliminary site evaluation but should not be used for the final measurement. Low-cost barometric altimeters can reflect errors of 150 ft. (46 m) or more, GPS altimeters are often less accurate. Topographic maps can be used to give an estimate of the vertical drop of a stream. But two methods of head measurement are accurate for design: direct height measurement and water pressure. The second major step in evaluating a site's hydropower potential is measuring the flow of the stream. Stream levels change through the seasons, so it is important to measure flow at various times of the year. Three popular methods are used for measuring flow in small hydropower applications: container, float, and weir. Determining the turbine performances requires measurement of hydraulic power (what is available to the turbine) consisting of flow inlet water level,

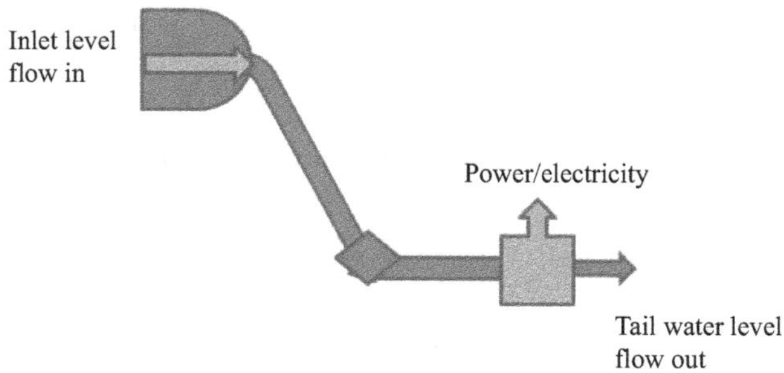

Figure 10.6 Flow-penstock's configuration

tail-water lever, turbine-generator output, which is the one sent to the grid. Absolute flow measurement methods include, area-velocity (current meters) method, ultrasonic (transit times of ultrasonic pulses) method, and dye-dilution (change in concentration of an injected tracer) method. Dye dilution method requires long length of conduit (penstock) for mixing, or injection manifold, careful handling and preparation of dye and equipment calibration (Figure 10.6). Once the procedure is underway, testing is relatively quick, about 15 min per operating point, and has no impact to plant operation (no dewatering, etc., to implement). Ultrasonic method is suitable for long and short penstocks, requiring site-specific installation. It is relatively fast measurement, has impact to plant operation (dewatering, etc., to install), and once installed, future testing is easy to perform. When assessing a micro-hydro site, we are interested in quantifying the available head and the flow-rate, since both are necessary in determining power. Of the two attributes head is usually considered more desirable because it results in smaller diameter pipes and fittings, reducing overall system costs. However, when working with a resource as site specific as hydropower, one must accept whatever is available.

10.3.1 Small and mini hydropower

Hydroelectric power comes from water at work, water in motion. It can be seen as a form of solar energy, Sun powering the hydrologic cycle that gives water to the Earth. In this cycle, atmospheric water reaches the Earth's surface as precipitation. Some of this water evaporates, but much of it either percolates into the soil or becomes surface runoff. Water from rain and melting snow eventually reaches ponds, lakes, rivers, or oceans where evaporation is constantly occurring. Water vapor passes into the atmosphere by evaporation then circulates, condenses into clouds, and some returns to earth as precipitation, completing the water cycle. Nature ensures that water is a renewable resource. Hydropower is the most economical way to generate electricity today, no other energy source, renewable or not, is comparable to it. Producing electricity from hydropower is inexpensive because, once a dam or the equivalent has been built and the equipment installed, the energy source, the flowing water is free. Hydroelectric plants also produce power at a minimal cost due to their sturdy structures and simple equipment. From a thermal perspective, some of the gravitational energy associated with the decrease in height is not converted to hydropower. That energy is converted into heat, which increases the temperature of the water and the surroundings. The maximum temperature rise is approximately one Celsius degree per 400 m of height decrease. Existing hydro plants have been described in three categories: small, micro, and pico. Small hydro represents hydroelectric power on a scale serving a small community or an industrial facility. The definition of a small hydro project varies, with a generating capacity of up to 10 MW generally accepted as the upper limit of what can be termed small hydro. Micro-hydro is a term used for hydroelectric power installations that typically produce up to 100 kW of power. These installations can provide power to a small community or may be connected to electric power networks. Pico-hydro is a term used for hydro-electric power generation of under 5 kW. It is useful in small, remote communities that require only a small amount of electricity. The key advantages of small hydro are:

1. High efficiency (70%–90%), by far the best of all energy technologies;
2. High capacity factor (typically >50%);
3. High level of predictability, varying with annual rainfall patterns
4. Slow rate of change; the output power varies only gradually from day to day (not from minute to minute);
5. A good correlation with demand, i.e., output is maximum in winter;
6. It is a long-lasting and robust technology; systems can readily be engineered to last for 50 years or more; and
7. It is also environmentally benign.

Small hydro is in most cases "run-of-river"; in other words any dam or barrage is quite small, usually just a weir, and little or no water is stored. Therefore run-of-river installations do not have the same kinds of adverse effect on the local environment as large-scale hydropower systems. Small-scale, mini, micro or pico hydroelectric systems can be constructed in many options, such as on dam-toe, canal drops, and return canals of thermal power stations and also in the flowing small river as well as small revolute which are flowing usually nearby villages. Area required for the construction work is small as canal already exists. It requires very small gestation period and such power stations can be ready for generation within 3 years, in contrast to the large hydro schemes. Small hydro power plants are usually up to capacity of 25 MW whereas mini-hydro plants are above 100 kW but below 1 MW either stand-alone scheme or more often feeding into the grid. Micro hydroelectric plants (MHPs), ranging from a few hundred Watts for battery charging or food processing application up to 100 kW usually provided power for small community or rural industry in remote area away from the grid [34–50]. The most important steps in establishing a small scale hydropower are consisting of:

1. **Site selection**: MHP are to be situated in hilly areas where there are natural falls, on the canal drops or at the dam-toe, long range studies are not required for such site selection. Wind energy conversion system should be located

preferably in the areas where the wind are strong and persistent, where daily wind flow is variable but monthly and annual average speed should be remarkably constant from year to year.

Similarly energy harvesting by other renewable energy systems small hydropower is very much customized by site specific characteristics.

2. **Grid connection issues**: In MHP input power is almost constant. Quantum of power fed to the grid remains constant in a season. Operation of grid connected small or mini hydro-electric station is rather smooth comparing with other renewable energy conversion systems, without facing many problems like production of harmonics, abnormalities in voltage, and frequency etc.
3. **Operation, maintenance, and control issues**: Operation of small, mini, or micro hydropower systems is smooth, maintenance is free, and easy to operate whereas for wind or wave power there are problem of noise pollution, teething troubles, power fluctuations, poor performance due to operation and maintenance problems. However, the development of appropriate instrumentation for signal conditioner, computer interfacing mechanism, and software for different aspects of system operation is a major challenge. A small scale hydropower facility generates power through the kinetic energy of moving water as it passes through a turbine. Most small scale hydropower facilities are "run-of-river," means that the natural flow of the river is maintained, and that a dammed reservoir is not created in order to generate power. Without a permanent dam to block river flow, nor a large reservoir to flood arable land and disrupt river temperature and composition levels, many of the negative riverine effects of traditional large scale hydropower are avoided with a small scale hydropower plant. However, some of the small, mini or micro hydroelectric power system may consist of a small reservoir, a section of a stream river, or an irrigation canal, used to provide water to turbine, governor, generator and power electronic interface and eventually energy storage. The water is passed from reservoir to turbine through pen-stock. When water strikes at the blades of the turbine, it converts hydraulic energy into mechanical energy. Head is described as the vertical distance, or as a function of the characteristics of the channel or pipe. Most SHP sites are categorized as a low or high head. Low head refers to a change in elevation of less than 10 ft. (3 m). A vertical drop of less than 2 ft. (0.6 m) will probably make a small-scale hydroelectric system unfeasible. The net head (H_0) of an SHP can be created in quite number of ways, being the most known the following two types: building a dam across a stream in order to increase the water level just above the plant; or diverting part of the stream, with a minimum of head loss, to just above the plant. Figure 10.7 shows the main components of a hydropower scheme. The basic hydropower principle is based on the conversion of a large part of the gross head, H_G (m), (i.e., net head H_0 (m)) into mechanical and electrical energy:

$$H_0 = H_G - \Delta H_{AB} \tag{10.34}$$

Here head losses along the total conversion system are expressed by ΔH_{AB} (m).

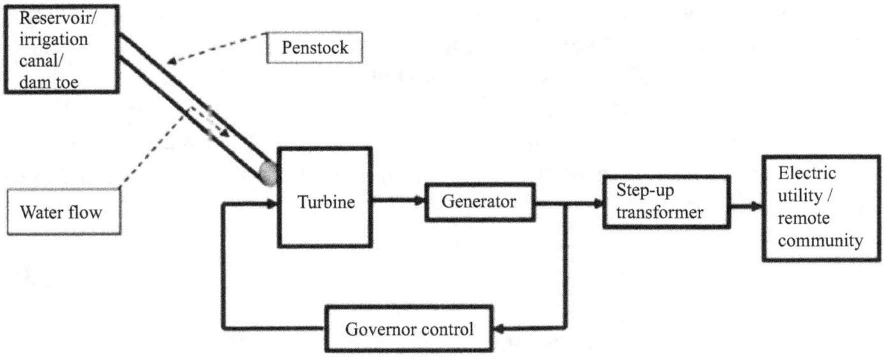

Figure 10.7 Schematic diagram of a mini or small hydroelectric power plant

---

Example 10.12: A small hydropower facility has a head of 100 m. Determine (a) the water speed entering the turbine, (b) if the volume flow rate is 25 m^3/s determine the penstock diameter, and (c) if the head loss due to the friction is 10% of the static head, determine the actual velocity approaching the turbine and the required penstock diameter.

Solution:

(a) The water velocity entering the hydropower turbine is given by:

$$v = \sqrt{2gH} = \sqrt{2 \cdot 9.80 \cdot 100} = 44.3 \text{ m/s}$$

(b) Applying continuity equation for the given flow rate, the penstock diameter is then:

$$D = \sqrt{\frac{4\pi v}{Q}} = \sqrt{\frac{4 \times 3.14 \times 44.3}{25}} = 4.72 \text{ m}$$

(c) The actual head for 10% friction losses, using (10.9) is:

$$H_{effective} = H - 0.1H = 0.9H = 90 \text{ m}$$

$$v_{eff} = \sqrt{2gH_{effective}} = \sqrt{2 \cdot 9.80 \cdot 90} = 42 \text{ m/s}$$

And the required penstock diameter is:

$$D_{req} = \sqrt{\frac{4\pi v_{eff}}{Q}} = \sqrt{\frac{4 \times 3.14 \times 42}{25}} = 4.60 \text{ m}$$

The head in (10.21), (10.22) or (10.33) is the gross head, which does not account for the losses due to turbulence and friction in the piping. The effective head, which is the head at the turbine inlet in the form of hydraulic pressure, is the gross head minus the head losses. Head losses are a function of the pipe length, diameter, surface texture, flow-rate, and the number and type of fittings between the intake and the turbine. Typically pipe losses are separated into two parts: the losses due to the pipe itself and the losses due to the fittings. For a straight pipe, friction is proportional to the velocity of the water and to the ratio of the pipe's length with respect to its diameter. This relationship is expressed mathematically by the Darcy equation, given by:

$$H_{Lmajor} = f \frac{L \cdot v^2}{2g \cdot D} \tag{10.35}$$

where H_{Lmajor} represents the head loss due to friction, L is the length of the pipe or penstock, D the pipe or penstock diameter, v the average flow velocity, g acceleration due to gravity, and f the friction factor. The friction factor is specific to the material and construction of the pipe and the flow characteristics, laminar or turbulent. The Darcy equation can be used to calculate energy loss due to friction in long, straight sections of round pipe. Pipe losses due to turbulence, such as from fittings and valves, are called minor losses, and are calculated using the following relationship:

$$H_{L\,minor} = \frac{k_L \cdot v^2}{2g} \tag{10.36}$$

where H_{Lminor} represents the head loss due to turbulence, and k_L is the loss coefficient. The effective head is therefore the gross head minus the head losses, expressed as:

$$H_{eff} = H_G - H_{Lmajor} - H_{Lminor} \tag{10.37}$$

One should always attempt to minimize the length of the pipe as well as the number of elbows, valves, and other fittings in the flow path, as their combined losses can be significant. Any reduction in the effective head will reduce the power output proportionately. It is not uncommon for a micro-hydro system to have an effective head that is as much as 30% less than the gross head. However, the pipe losses are not the only losses that must be considered. The efficiencies of the turbine, generator and electronics also rob power from the system. Small water turbines rarely achieve efficiencies greater than 80%. Combined with losses in the generator and power electronics, the overall efficiency is likely to be closer to 50%. To account for these equipment losses, the power equation is modified by the total efficiency coefficient, so:

$$P_{Out} = \eta_{tot} \cdot \rho \cdot g \cdot Q \cdot H_{eff} \tag{10.38}$$

This equation provides a reasonable estimate of the power output of a hydroelectric system regardless of its size or construction. The relationship in this form will be used throughout this analysis.

An SHP design involves a multi-disciplinary engineering team work, hydrologic, hydraulic, structures, electric, mechanical, geologic, and environmental experts. SHP layout schemes are usually characterized through different intakes and diversion structures depending on the type of the conveyance system. The SHPs include some essential components are penstock, power house, tailrace, generating plant, and allied equipment. Figure 10.7 is showing a small hydro system. The two small-scale hydropower systems with capacities below 100 kW are referred to as micro-hydropower (MHP) systems, and sites with capacity between 101 kW and 1 MW are referred to as SHP systems. MHP systems, which use cross flow turbines and Pelton wheels, can provide both direct mechanical and electrical energy. In a small hydropower system, the function of governor is to control generator speed so that its electric frequency remains constant, while the gate position of turbine is controlled through servomotor, which adjusts water flow to produce power according to the load connected. Hydropower turbines convert water pressure into mechanical shaft power, which can be used to drive an electricity generator, or other machinery. The general formula for a hydro system's energy generated over a certain interval of time (Δt) are given by the following relationship:

$$E_{hydr} = P \cdot \Delta t = \eta \cdot \rho \cdot g \cdot Q \cdot H \cdot \Delta t \tag{10.39}$$

where P_{hydr} is the mechanical power produced at the turbine shaft (W), h is the hydraulic efficiency of the turbine, ρ is the density of water (kg/m^3), g is the acceleration due to gravity (m/s^2), Q is the volume flow rate passing through the turbine (m^3/s), and H is the effective pressure head of water across the turbine (m). The potential energy in water is converted into mechanical energy in the turbine as a result of the water pressure which applies a force on the face of the runner blades and then decreases as it passes through the reaction turbine. The relation between the electrical, mechanical, and hydraulic powers can be obtained by using the hydraulic turbine, power-train, and generator efficiencies, η_h, η_{mech}, and η_t, expressed in the following relationship:

$$P_{electric} = \eta_{mech} P_{mech} = \eta_{mech} \eta_h P_{hydro} = \eta_t P_{hydro} \tag{10.40}$$

The best turbines can have hydraulic efficiencies in the range 80% to over 90% (higher than most other prime movers), although this will reduce with size. Micro-hydro systems tend to be in the range 60%–80% efficient. Water is taken from the river by diverting it through an intake at a weir. The weir is a man-made barrier across the river which maintains a continuous flow through the intake. Before descending to the turbine, the water passes through a settling tank or fore-bay in which the water is slowed down sufficiently for suspended particles to settle out. The fore-bay is usually protected by a rack of metal bars which filters out water-borne debris which might damage the turbine such as stones, timber, or man-made litter. The mini-hydro or micro-power power plants mainly consist of a small reservoir or irrigation canal, governor, turbine and generator. The water is passed from reservoir to turbine through penstock. When water strikes at the blades of turbine it converts hydraulic energy into mechanical energy. Currently mini hydro schemes employ conventional equipment which have resulted them an un-economical option. In order to make mini hydro schemes a cost effective technology different new designs have been proposed in almost every component of mini hydro power plant. The new designs include penstock, hydraulic turbines, generators and governor controller. In a run-of-river or diversion-type small hydropower systems, the power generating capacity of a water mass, flowing through the river at a velocity, v is computed as the kinetic energy over time:

$$P_w = \frac{1}{2}\left(\frac{m}{t}\right)v^2 = \frac{1}{2}\rho Q v^2 \tag{10.41}$$

For the flow rate through an opening with the area, A ($Q = A \cdot v$), the power density (power per unit of area) is expressed as:

$$\frac{P_W}{A} = \frac{1}{2}\rho v^3 \tag{10.42}$$

This equation is similar to the one of wind power density per unit rotor area but the water density is substantially much higher than the air density. The converted electric power depends on the turbine efficiency, turbine blades power coefficient and generator efficiency:

$$P_{el} = P_W\left(C_P \eta_{tr} \eta_{gen}\right) \tag{10.43}$$

Here, C_P is the turbine blades power coefficient, η_{tr}, η_{gen} are turbine and generator efficiencies.

Example 10.13: A run-of-river system installed on a small river, 25 m wide and 4.5 m deep, with water flowing at 3.20 m/s, diverts 25% of the flow. If overall efficiency is 75%, what is the power output?

Solution: The river volumetric flow rate is:

$$Q = A \cdot v = 25 \times 4.5 \times 3.25 = 360\ \text{m}^3/\text{s}$$

The flow diverted to run-of-river system is the:

$$Q_{r-r} = 0.25 \times 360 = 90\ \text{m}^3/\text{s}$$

From (10.9) and (10.0) the power output is:

$$P_{electric} = \eta_t P_{hydro} = 0.85 \cdot 0.5 \cdot 90 \cdot 1000 \cdot (3.2)^2 = 391.68\ \text{kW}$$

10.3.2 Small hydroelectric power technology

The role of the hydropower plants is to capture the energy in flowing water and convert it to usable energy. The hydropower potential depends on the availability of suitable water flow and then, where the resource exists, these plants can provide cheap, clean, and reliable electricity. Moreover, small scale hydropower plants, when designed taking into account of surroundings without interfere significantly with river flows, have minimal negative environmental impacts; also because they don't need a reservoir, being in large part run of the river, respect to the large hydropower systems. A hydroelectric turbine converts the energy from falling water into rotating shaft power. The selection of the best turbine for any particular hydro site depends upon the site characteristics, the head and flow available. Selection also depends on the desired running speed of the generator or other device loading the turbine. Other considerations, such as whether the turbine will be expected to produce power under reduced flow conditions, also play an important role in the selection. All turbines have a power-speed characteristic, and an efficiency-speed characteristic. They tend to run most efficiently at a particular speed, head and flow. Hydroelectric power plants are of three major types:

1. Impoundment system is a large hydroelectric power system, using a dam to store river water in a reservoir that is used to generate electricity.
2. Diversion facility channel is a portion of a river through a canal or penstock, which may not require the use of a dam.
3. Run-of-river system uses water with in the natural flow range and it requires little or no impoundment.

Hydroelectric turbines can be classified as high-head, medium-head, or low-head machines. However, this is relative to the size of machine: what is low head for a large turbine can be high head for a small turbine; for example, a Pelton turbine might be used at 50 m head with a 10 kW system but would need a minimum head of 150 m to be considered for a 1 MW system. The main reason that different types of turbine are used at different heads is that electricity generation requires a shaft speed as close as possible to 1,500 rpm to minimize the speed change between the turbine and the generator. Turbines used in hydroelectric systems have runners of different shapes and sizes. There are two main categories of hydroelectric turbines in use: **impulse and reaction turbines**. Figure 10.8 is showing the schematic diagrams of the impulse and reaction hydropower turbines. The selection of any type of hydropower turbine for a project is based on the head and the flow or volume of water at the site. However, other deciding factors include how deep the turbine must be set, efficiency, and cost.

The speed of any given type of turbine tends to decline with the square-root of the head, so low-head sites need turbines that are faster under a given operating condition. The reaction turbine rotor is fully immersed in water and is enclosed in a pressure casing. The runner blades are profiled so that pressure differences across

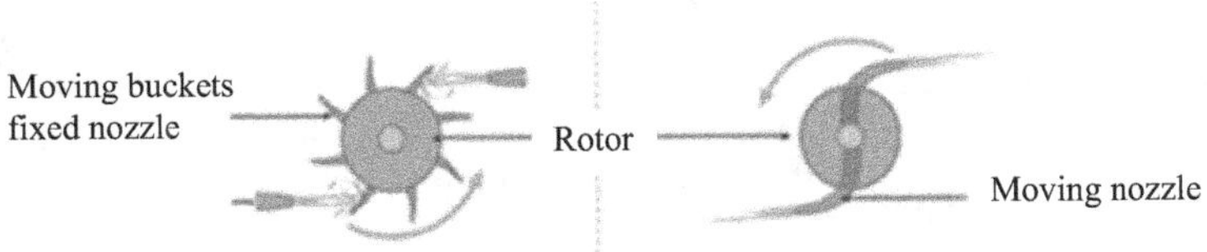

Figure 10.8 Schematic diagrams and the operation principles of the impulse turbine (left panel) and the reaction turbine (right panel)

them impose lift forces, akin to those on aircraft wings, which cause the runner to rotate. In contrast an impulse turbine runner operates in air, driven by a jet (or jets) of water, and the water remains at atmospheric pressure before and after making contact with the runner blades. Turbines used in hydroelectric power systems have usually runners of different shapes and sizes. The selection of any type of hydro-power turbine for a specific application is based on the head and the flow or volume of water at that site. However, other deciding factors include how deep the turbine must be set, efficiency and cost. The choice of turbine for hydro plant depends upon the available water potential and local conditions. Turbine comes in wide variety of designs and sizes. The hydro-electric turbines are also classified according to the available energy at the turbine inlet, flow direction through vanes, head at the turbine inlet and specific turbine speed. According to specific turbine speed there are three major classes: (1) low specific speed and high head turbine (Pelton); (2) medium specific speed and medium head turbine (Francis); and (3) high specific speed and low head turbine (Kaplan and Propeller). Depending upon specific site conditions, turbines are designed as Dam Base, Canal Fall, Run-of- River, and Hilly Region.

In order to improve the overall system efficiency the pipe length, the number of elbows, valves, and other fittings in the flow path, as their combined losses can be significant need to be minimized. Any reduction in the effective head reduces the power output proportionately. It is not uncommon for a micro-hydro system to have an effective head that is as much as 30% less than the gross head. However, the pipe losses are not the only losses that must be considered. The efficiencies of the turbine, generator, and electronics must be considered. Small water turbines rarely achieve efficiencies greater than 80%. Combined with losses in the generator and power electronics, the overall efficiency is usually closer to 50%. To account for these equipment losses, the power equation is modified by the total efficiency coefficient, so:

$$P_{Out} = \eta_{tot} \cdot \rho \cdot g \cdot Q \cdot H_{eff} \tag{10.44}$$

This equation provides a reasonable estimate of the power output of a hydro-electric system regardless of its size or construction. The relationship in this form is used throughout the small hydroelectric system analysis. For a reservoir type small hydroelectric power the analysis of the potential electricity are taking into account the input power (the potential energy), the losses into the penstock, turbine blades power coefficient, and the turbine and generator losses. Equation (10.21) can be re-written in this case to estimate the output electric power, as:

$$P_{Out} = gQ_{mass}H_{eff} \times \left(C_P \eta_{pen} \eta_{tr} \eta_{gen} \right) \tag{10.45}$$

10.3.2.1 Impulse turbines

Impulse turbine uses the water kinetic energy to drive the runner and discharges to atmospheric pressure, being moved by the water jet at atmospheric pressure before and after making contact with the runner blades. Water that falls into tail after striking the buckets has little energy remaining, so the turbine has light casing serving only for the purpose of protecting the surroundings. Impulse turbines are usually used in systems with high head and low flow. There are three common types of impulse turbines: the Pelton, the Cross-flow and Turgo hydropower turbine. Pelton turbine consists of a wheel with a series of split buckets (vanes) set around its rim, and a high velocity water jet is directed tangentially at the wheel, hits each bucket and is split in half, so that each half is deflected back almost through 180°. Nearly all the water energy goes into propelling the bucket and the deflected water falls into a discharge channel. The jets are through nozzles, each with an axis in the runner plane and a needle (or spear) valve to control the flow. To stop the turbine, in case the turbine approaches the runaway speed due to load rejection, the jet is deflected by a plate, no longer impinging on the buckets, so the valve is closed slowly keeping the over-pressure surge to an acceptable minimum. The water kinetic energy leaving the runner is lost and so the buckets are designed to keep exit velocities to a minimum. This turbine does not require draft tubes since the runners are positioned above the maximum tail water to permit operation at atmospheric pressure. Pelton turbines are usually applied in systems with large water heads. Unlike the Francis turbine, Pelton, and cross flow turbines can operate at high efficiencies even when running below their design flow.

A Turgo turbine is similar to the Pelton but the jet is designed to strike the plane of the runner at an angle (typically 20°) so that the water enters the runner on one side and exits on the other. Therefore, the flow rate is not limited by the discharged fluid interfering with the incoming jet (as for Pelton turbines), so Turgo turbines can have smaller diameter runner than a Pelton for an equivalent power. Turgo turbines have also different shape of the buckets, with the water jet entering the runner through one side and exits through the other side. A Turgo turbine has a higher running speed which makes a direct coupling of turbine and generator more likely, thus increasing the overall efficiency and decreasing maintenance costs. Turgo turbines operate effectively in systems with large water heads. Turgo turbine based system is capable of handling varying seasonal flows and can operate efficiently in variety of different heads. These small turbines are ideal for connecting to ranchers existing gravity fed irrigation systems, streams and creeks, for remote home-sites to reduce electrical consumption through net metering.

The Cross-flow turbine has a drum-like rotor with a solid disk at each end and gutter-shaped "slats" joining the two disks. A jet of water enters the top of the rotor through the curved blades, emerging on the far side of the rotor by passing through the blades a 2nd time. The shape of the blades is such that on each passage through the periphery of the rotor the water transfers some of its momentum, before falling away with little residual energy. Cross-flow turbine has a drum-like rotor and uses an elongated, rectangular-section nozzle, which is directed against curved vanes on a cylindrically shaped runner. Cross-flow turbines are less efficient than the modern-day turbines, but can accommodate larger water flows and lower heads. A jet of water enters the turbine, thus gets directed through the guide-vanes at a transition piece upstream on the runner which is built from two or more parallel disks connected near their rims by a series of curved blades. The flow is directed to a limited portion of the runner by the guide vane at the entrance to the turbine. The turbine allows water to flow twice through the blades. In the first stage water flows from the outside of the blades to the inside; in the second stage the water passes from the inside back out. The flow leaves the first stage attempts to cross the open center of the turbine but as the flow enters the second stage, a compromise direction is achieved which causes significant shock losses.

10.3.2.2 Reaction turbines

This type of hydropower turbine generates electricity from the mutual action of pressure and by moving water. Reaction turbines exploit the incoming water flow to generate hydrodynamic lift forces to propel the runner blades. They are having a runner that always functions within a completely water-filled casing. The reaction turbine operates when the rotor is fully submerged in water and is enclosed in a pressure casing. The runner blades are profiled so that pressure differences across them impose lift forces, akin to those on aircraft wings, causing the runner to rotate. Reaction turbines are generally appropriate for sites with lower head and higher flow rates compared with the impulse turbines. The runner blades are profiled so that pressure differences across them impose lift forces, akin to those on aircraft wings or wind turbine blades, which cause the runner to rotate. Reaction turbines are generally appropriate for sites with lower head and higher flows compared with the impulse turbines. Typical reaction turbine types are Propeller, Francis, and Kinetic ones. Reaction turbines have a diffuser known as a 'draft tube' below the runner through which the water discharges. The draft tube slows the discharged water and reduces the static pressure below the runner and thereby increases the effective head. The two main types of reaction turbines with a few variants are the propeller (with Kaplan variant) and Francis turbines. Propeller-type turbines are similar in principle to the propeller of a ship but operating in reversed mode. Various configurations of propeller turbine exist; a key feature is that for good efficiency the water needs to be given some swirl before entering the turbine runner. With good design, the swirl is absorbed by the runner and the water that emerges flows straight into the draft tube with little residual angular momentum. Methods for adding inlet swirl include the use of a set of fixed guide vanes mounted upstream of the runner with water spiraling into the runner through them. Another method is to form "snail shell" housing for the runner in which the water enters tangentially and is forced to spiral in to the runner. When guide vanes are used, these are often adjustable so as to vary the flow admitted to the runner. In some cases, the blades of the runner can also be adjusted, in which case the turbine is called a Kaplan. The mechanics for adjusting turbine blades and guide vanes can be costly and tend to be more affordable for large systems but can greatly improve efficiency over a wide range of flows.

A propeller hydroelectric turbine generally has an axial flow runner with three to six blades depending on the designed water head. For higher efficiency, the water needs to be given some swirl before entering the turbine runner. Propeller turbines are suitable for systems with low water heads. There are several different types of propeller turbines: bulb turbine, Kaplan, Straflo, and tube turbine. The

Kaplan turbine has adjustable blade pitch and it can achieve high efficiency under varying power output conditions. The methods used for adding inlet swirl include fixed guide vanes mounted up stream of the runner and snail-like shell housing for the runner, in which the water enters tangentially and is forced to spiral into the runner. In the case of the Kaplan turbine, the blades of the runner are adjusted. Adjustment of the turbine blades and guide vanes can greatly improve efficiency over a wide range of flows; however, it is costly and so can only be economical in larger systems. The unregulated propeller turbines are commonly used in micro-hydro systems where both the flow and head remain practically constant.

The Francis turbine is essentially a modified form of propeller turbine in which water flows radially inwards into the runner and is turned to emerge axially. The runner is most commonly mounted in a spiral casing with internal adjustable guide vanes. Reaction turbines require more sophisticated fabrication than impulse turbines because they involve the use of more intricately profiled blades and profiled casings, making them less attractive for use in micro-hydro in developing countries. Nevertheless, because low head sites are far more common and often more closer to where the power is needed, work is being undertaken to develop propeller machines which are simpler to construct. Francis type is the most common type of hydropower turbine in use. This turbine generally has radial or mixed radial/axial flow runner which is most commonly mounted in a spiral casing with internal adjustable guide vanes. Water flows radially inwards into the runner and emerges axially, causing it to spin. In addition to the runner, the other major components include the wicket gates and draft tube. The runners with smaller diameter are made of aluminum bronze casting, while the larger runners are fabricated from curved stainless steel plates that are welded to cast steel hub. Francis turbines are applied in hydroelectric systems with medium head size and their efficiency can be above 90% but tend to have higher cost.

10.3.2.3 Pump as hydroelectric turbine

Alternative options to the conventional hydroelectric turbines are the use of pumps, in reverse mode of operation as prime mover. The initial cost of small, mini, or micro-hydropower plants largely depends upon the cost of equipment. One way to reduce this cost of equipment is to use centrifugal pump as a turbine (PAT). PAT can be considered as a cost-effective alternative and viable option for small-, mini-, micro- or pico-hydropower generation especially in rural hilly and complex terrain areas or in agricultural land areas. PAT being mass produced all over the world is a standard product available in variety of sizes of different head and flow rates. These are cheaper and easily available in the market. Their repair and maintenance can be easily carried out by local technicians and in local workshops. The PAT cost is about 50% less than the cost of conventional hydroelectric turbine of similar characteristics. PATs are available in wide variety of power from about 1.7–160 kW range. PAT is basically a pump which can operate in a turbine mode if the direction of flow is reversed, with higher efficiency in turbine mode operation. With suitable conditions, PAT can cover the range of multi-jet turbines, cross-flow turbines, and small Francis hydro turbines. Standard pumps can be easily used for electrical power generation operated in the reverse mode. Axial, radial, and mixed centrifugal pumps operating in reverse mode can operate as hydroelectric turbines. Despite of having many advantages over conventional turbine, PAT has one major drawbacks, prediction of turbine characteristics of the centrifugal pump is very difficult. PAT selection for a specific mini hydro site is a major problem, while the turbine characteristics of pumps cannot be generalized. However, despite intensive research on adapting methods for determining the optimum behavior of pumps from analytical analysis, simulation, experimental work, computational studies optimum results still have not been found. One of the reason is that manufacturers of pumps do not provide their characteristics curves and it is very essential to know about the characteristics of pumps for successful operation of pumps as a hydroelectric turbine.

10.3.3 Generators and control

The power generation industry almost exclusively uses large synchronous generators (SGs), as they have the advantage of a variable reactive power production, i.e., voltage control. There are two most common used types of generators: synchronous and asynchronous electric machines. However, DC (direct current) electric generators are sometimes used, especially for pico-hydroelectric applications. An electric generator is a device that converts mechanical energy to electrical energy. A generator forces electric current to flow through an external circuit. The source of mechanical energy may be a reciprocating or turbine steam engine, water falling through a turbine or waterwheel, an internal combustion engine, a wind turbine, a hand crank, compressed air, or any other source of mechanical energy. Generators provide nearly all of the power for electric power grids. An electric generator uses the rotor shaft speed and torque to convert mechanical energy to electrical energy with the use of electromagnetic fields. The main concern with generators in hydroelectric turbines is that they must produce electricity compatible with that in the electrical grid, if the system is grid connected for the given site. In America, the grid is at 60 Hz, while in most parts of Europe the grid runs at 50 Hz. The electricity must have the same characteristics, be of sufficient quality, and be connected in such a way as to not interrupt the existing current flow. The function of an electrical generator is providing a means for energy conversion between the mechanical torque from the hydro rotor turbine, as the prime mover, and the local load or the electric grid. Different types of generators are being used with hydroelectric system, induction (asynchronous), synchronous or occasionally DC electric generators.

SGs are standard in electrical power generation and most commonly used in most power plants. They are used especially in large power plant applications either there is significant research for their applications in lower power range. However, more often in renewable energy induction generators are employed. Asynchronous generators are more commonly known as induction generators. A great deal of research has been carried out, over the last two decades regarding the application of the induction generator as an alternative option for electricity generation over the synchronous generator in small-, mini-, or micro-hydropower schemes for making these schemes cost effective. The induction generators have lower cost per unit of KWh as compared to SGs. Besides, induction generators are more robust and have easy starts and control mechanisms, self-protection against faults. It has the ability for generating power at changing speeds and can be used in an off-grid or in connected mode with synchronous generator for load sharing. It can also be operated as generator when its stator winding is connected with capacitor and rotor is driven by prime mover. In that case the magnetizing lagging reactive power is provided by the capacitor. This capacitor establishes the air-gap flux. This configuration enables the induction machine to work as a self-excited induction generator. Induction generators excited with capacitor are emerging as a suitable candidate for renewable energy power generation operating in stand-alone mode. Despite all these advantages, induction generators have encountered problem in maintaining the frequency and voltage within its range.

The induction generator is a standard three-phase induction motor, wired to operate as a generator. Capacitors (C) are used for excitation, by connecting unequal excitation capacitance across the windings of the motor, converting a three-phase motor into a single-phase generator. This is cost-effective and has been popular for smaller off-grid systems below 10–15 kW. They have the advantage of being rugged, cheaper than SGs, robust and widely available and can withstand over-speed and overload. For MHP induction generators, with capacitive VAR controllers, are used for both standalone and grid connected mode. SGs are also used preferably for standalone mode. Although most induction generators in operation are employed in wind power, such machines have also been used in medium-size hydro and thermal plants. However, most distributed generation systems employ SGs, which can be used in thermal, hydro, or wind power plants. In the electromagnetic transient simulations, the SGs were represented by an eight-order model, which was reduced to a sixth-order model in the transient stability simulations. Usually, SGs connected to distribution networks are operated as constant active power sources, so that they do not take part in the system frequency control.

Induction generators need very little auxiliary equipment and can be run in parallel with generator without hunting at any frequency. For IG speed variation of prime mover is less important. A self-protective feature of this is that if there is terminal short circuit, excitation fails and so does the generator output. Its disadvantage is that it draws considerable amount of lagging KVAR from the supply for excitation, the efficiency is comparatively poor and can operate only at leading power factor. An Induction generator controller for micro-hydropower systems using the induction generator has also been developed. ELG is not suitable for an induction generator. This is the generator of choice if micro-hydropower system is to supplying to the grid. Induction generators are increasingly used these days because of their relative advantageous features over conventional SGs. These features are brushless and rugged construction, low cost, maintenance and operational simplicity, self-protection against faults, good dynamic response, and capability to generate power at varying speed. The later feature facilitates the induction generator operation in stand-alone/isolated mode to supply far flung and remote areas where extension of grid is not economically viable; in conjunction with the synchronous generator to fulfill the increased local power requirement, and in grid-connected mode to supplement the real power demand of the grid by integrating power from resources located at different sites. The reactive power requirements are the disadvantage of induction generators. This reactive power can be supplied by a variety of methods, from simple capacitors to complex power conversion systems.

SGs are the most commonly used machine for generation of electrical power for commercial purpose is the synchronous generator or alternator. Typical alternators use a rotating field winding excited with direct current, and a stationary (stator) winding that produces alternating current. Since the rotor field only requires a tiny fraction of the power generated by the machine, the brushes for the field contact can be relatively small. In the case of a brushless exciter, no brushes are used at all and the rotor shaft carries rectifiers to excite the main field winding. SGs driven at low speeds by prime-movers like water turbines will have salient pole construction having large number of projected poles. SGs in standalone mode: When SG is independently supplying load, increasing its excitation will increase the no load voltage. This will raise the whole load characteristics and increase the value of terminal voltage. Permanent magnet synchronous generator is a solution that is appreciated in small wind turbines but it cannot be extended be extended to large-scale power because it involves the use of big and heavy permanent magnets. However, the SGs, used in the large hydropower plants are not well suited for renewable and nonconventional energy generation. New technologies and configurations are under research of in testing for commercial application. One of them is represented by the six-phase synchronous generator, used initially for uninterruptible power supply. In the last decade, the application of six-phase generators for renewable energy started to increase. However, in the past none has adapted six-phase synchronous generator in small, mini or micro hydropower plants to show its economic viability. Several studies were conducted to find additional advantages possessed by six-phase synchronous generator over conventional three-phase generator. Usually, the three-phase synchronous generator was formed the six-phase by adapting a phase belt splitting. It should be noted that electrical displacement is of great interest because it allows recombination of three-phase power in step-up transformer bank. Experiments were carried out on constant voltage as well as on constant frequency or speed operation. Six-phase has an advantage that it can supply two three-phase loads separately. Another advantage of using this type of generator is that by employing a transformer having six-phase to three-phase windings, output of two three-phase winding can be supplied to a single three phase loads. The use of transformer has also the advantage that in case of failure in any of three-phase winding set, the system will not lead to shut down but load continued to be supplied through remaining healthy winding thus improving system reliability and robustness, while reducing the cost.

DC generators or dynamos are electrical machines that produce direct current with the use of a commutator. Dynamos were the first electrical generators capable of delivering power for industry, and the foundation upon which many other later electric-power conversion devices were based, including the electric motor, the alternating-current alternator, and the rotary converter. Today, the simpler alternator dominates large scale power generation, for efficiency, reliability, and cost reasons. A dynamo has the disadvantages of a mechanical commutator. Also, converting alternating to direct current using power rectification devices (vacuum tube or more recently solid state) is effective and usually economic.

Small scale hydropower plants due to their complex nature and nonlinear behavior of hydraulic turbine encountered very often frequency and voltage variation in the system. Different control techniques are proposed and needed for maintaining voltage and frequency within the prescribed range. Voltage can be controlled by varying the excitation, while frequency can be controlled by making generation equal to load through governor control scheme. In the past, mechanical hydraulic governors were used, today being replaced by electrohydraulic, PI/PID governors and control schemes. Fundamental mathematical models of all these governors, control schemes and turbines suitable for hydroelectric power plants are intensively discussed in the literature and are beyond the scope of this book. Different governing control techniques for frequency stabilization have been discussed and proposed over the years, covering the classical PID controller to the intelligent control techniques. PID controllers have better response in dealing governor control because it provides three functions to control the system. Control for water flow is also important for a number of applications as in hydropower, tank water level, etc. Turbine gate opening and water flow plays an important role for generation and in maintaining the system frequency. At present, water is controlled by conventional control techniques. However, they can be controlled by using robustly approaches with fuzzy logic control schemes.

10.4 Bioenergy, biofuel, biomass, and waste energy

Bioenergy is a general term that covers energy derived from a wide variety of materials of plant or animal origin. Strictly speaking, this term includes also fossil fuels but, generally, the term is used to mean renewable energy sources, such as wood and wood residues, agricultural crops and residues, animal fats, and animal and human wastes, all of which can yield useful fuels either directly or after some form of conversion or processing. There are technologies for bioenergy using liquid and gaseous fuel, as well as traditional applications of direct combustion. The conversion process can be physical, such as drying, size, reduction or densification, thermal, such as in carbonization or chemical, such as in biogas production. The end result of the conversion process may be a solid, liquid, or gaseous fuel and this flexibility of choice in the physical form of the fuel is one of the advantages of bioenergy over other renewable energy sources. The basis for all these applications is organic matter, in most cases plants and trees. There is a trend toward purposefully planted biomass energy crops, although biomass can also be collected as a by-product and residue from agricultural, forestry, industry, and household waste. Bioenergy can be used for a great variety of energy needs, from heating and transport fuel to power generation. There are numerous commercially available technologies for the conversion processes and for utilization of the end-products. Although the different types of bioenergy have features in common, they exhibit considerable variation in physical and chemical characteristics, which influence their use as fuels. There is such a wide range of bioenergy systems that this module does not aim to cover and describe each one.

Biomass provided about 10.2% (50.3 EJ/year) of the annual global primary energy supply in 2008, from a wide variety of biomass sources feeding numerous sectors of society. The biomass feed-stocks used for energy is more than 80% are derived from wood (trees, branches, residues) and shrubs. The remaining bioenergy feed-stocks came from the agricultural sector (energy crops, residues, and by-products) and from various commercial and post-consumer waste and by-product streams (biomass product recycling and processing or the organic biogenic fraction of municipal solid waste). Until the nineteenth century, biomass was the predominant fuel for providing heat and light all over the world. In industrialized countries, it was then displaced by coal and later by petroleum but in developing countries, it remains the most important fuel. Some strengths and weaknesses of bioenergy, in general, are summarized here. Conversion technologies are available in a wide range of types and forms, fuel production, and conversion technology are indigenous, being available in almost all developing or developed countries, generating more jobs than other than other renewable energy systems of a comparable size, conversion can be to gaseous, liquid or solid fuel, while environmental impacts are usually low (overall no increase in carbon dioxide) compared with conventional energy sources. Among bioenergy weaknesses are the production can create land use competition range of power levels at different levels of technological complexity, often large areas of land are required (usually low energy density), production can have high fertilizer and water requirements, may require complex management system to ensure constant supply of resource, which is often bulky adding complexity to handling, transport and storage, resource production may be variable depending on local climatic/weather effects, i.e., drought and are likely to be uneven resource production and distribution throughout the year or region.

Transformation of waste materials into energy various uses can generally be accomplished through biological, thermal, and chemical processes. Low-efficiency traditional biomass, such as wood, straws, dung and other manures are used for cooking, lighting, and space heating, generally by the poorer populations in developing countries. This biomass is mostly combusted, creating serious negative impacts on health and living conditions. High energy efficiency biomass conversion is found typically in the industry sector (with a total consumption of ~7.7 EJ/year) associated with the pulp and paper industry, forest products, food, and chemicals. Examples are fiber products (e.g., paper), energy, wood products, and charcoal for steel manufacture. Industrial heating is primarily steam generation for industrial processes, often in conjunction with power generation. Although biomass encompasses several kinds of different organic matter, fibrous plant materials that are characterized as solid materials, being carbonaceous fuels of high volatile content and heating value of about 18 MJ/kg. Either direct combustion or thermal gasification can be used to transform the chemical energy of such materials into electric power. Direct combustion releases heat that can be used in Stirling engines or Rankine steam power cycles. Thermal gasification yields flammable gases suitable for firing in internal combustion engines, gas turbines, or fuel cells and eventually to generate electricity. Global bioenergy use has steadily grown worldwide in absolute terms in the last 40 years, with large differences among countries. Bioenergy production interacts with food, fodder, and fiber production as well as with conventional forest products in complex ways. Bioenergy demand constitutes a benefit to conventional plant production in agriculture and forestry by offering new markets for biomass flows that earlier were considered to be waste products; it can also provide opportunities for cultivating new types of crops and integrating bioenergy production with food and forestry production to improve overall resource management. However, biomass for energy production can intensify competition for land, water, and other production factors, and can result in overexploitation and degradation of resources. The anaerobic digestion process, carried out in the absence of oxygen, involves the use of microorganisms for the conversion of biodegradable biomass material into energy, in the form of methane gas and a stable humus material. Anaerobic digestion can occur under control conditions in specially designed vessels (reactors), semi-control conditions, such as in a landfill or under uncontrolled conditions as it does in the environment.

10.5 Chapter summary

Geothermal energy has tremendous potential to provide many areas and regions of the world with reliable, base-load, dispatchable, and clean renewable energy for centuries to come. Geothermal growth and development of electricity generation has increased significantly over the past 30 years, with higher rates in the early part of this period, and lower rates in the last 10 years due to lower overall economic increasing in many countries and the low price of competing fuels. Geothermal heat can be used for different applications, not only electricity generation, based primarily on its temperature range. The other key aspect is the type of geothermal resource. Direct geothermal energy use has remained fairly steady over the 30-year period at 10% growth annually. The majority of the increase has been due to GHPs. Through direct use and ground source heat pump applications, a significant amount of generation capacity can be offset. For example, ground source heat pumps are a well-established technology with a proven track record internationally and could be applied in many regions of US or many other countries. As the exploitation and use of EGS and HSA resources is established domestic and internationally, it is quite likely that the costs is reduced as more efforts are spent on research and improving the component and system technologies. The equivalent is also true for direct use geothermal applications. Lower temperature resources can be harnessed and can reduce the electricity demand in applications that require heat and hot water. These include space heating and cooling, industrial process heat, hot water, and desalination. Hydro-electricity is the term referring to electricity generated by hydropower. The production of electrical power "arises" through the use of the gravitational force of falling or flowing water. It is the most widely used form of renewable energy. Once a hydroelectric complex is constructed, the project produces no direct waste and has a considerably lower output level of the greenhouse gas carbon dioxide (CO_2) than any of the fossil fuel powered energy plants. The major advantage of hydroelectricity is the elimination of the cost of fuel. The cost of operating a hydroelectric plant is not affected by increases in the cost of fossil fuels such as oil, natural gas, or coal, and no imports are required. These plants also have long lives, with some plants still in service after more than 50 years.

Further readings

[1] J. Andrews and N. Jelley, Energy Science, Principles, Technology and Impacts, Oxford University Press, Oxford, UK, 2007.

[2] F.P. Incropera and D.P. DeWitt, Introduction to Heat Transfer, Wiley, New York, 2002.

[3] E.L. McFarland, J.L. Hunt, and J.L. Campbell, Energy, Physics and the Environment (3rd ed.), Cengage Learning, Australia, 2007.

[4] J. Andrews and N. Jelley, Energy Science, Principles, Technology and Impacts, Oxford University Press, Oxford, UK, 2007.

[5] F. Kreith, and D.Y. Goswami, Handbook of Energy Efficiency and Renewable Energy, CRC Press, Boca Raton, FL, USA, 2007.

[6] R. DiPippo, Geothermal Power Plants: Principles, Applications and Case Studies (2nd ed.) Elsevier Advanced Technology, Oxford, England, 2008.

[7] M.A. El-Sharkawi, Electric Energy—An Introduction, CRC Press, Boca Raton, FL, 2009.

[8] A. Vieira da Rosa, Fundamentals of Renewable Energy, Academic Press, 2009.

[9] B.K. Hodge, Alternative Energy Systems and Applications, Wiley, Hoboken, NJ, 2010.

[10] I.W. Johnston, G.A. Narsilio, and S. Colls, 2011. Emerging geothermal energy technologies, KSCE J. Civ. Eng. Vol. 15, pp. 643–653.

[11] E.E. Michaelides, Alternative Energy Sources, Springer-Verlag, Berlin, Heidelberg, 2012.

[12] H.J. Wagner, J. Mathur, 2010. Introduction to hydro energy systems—basics, technology, KSCE J. Civil Eng., Vol. 15(4), pp. 643–653.

[13] G. Boyle, Renewable Energy—Power for a Sustainable Future, Oxford University Press, Oxford, UK, 2012.

[14] B. Everett and G. Boyle, Energy Systems and Sustainability: Power for a Sustainable Future (2nd ed.), Oxford University Press, Oxford, UK, 2012.

[15] R.A. Dunlap, Sustainable Energy, Cengage Learning, Stamford, CT, 2015.

[16] V. Nelson and K. Starcher, Introduction to Renewable Energy (Energy and the Environment), CRC Press, Boca Raton, FL, 2015.

[17] W.E. Glassley, Geothermal Energy: Renewable Energy and the Environment, (2nd ed.), CRC Press, Boca Raton, FL, 2015.

[18] D.R. Boden, Geologic Fundamentals of Geothermal Energy, CRC Press, Boca Raton, FL, 2017.

[19] R. Belu, Hydroelectric Power Systems, Small, in Encyclopedia of Energy Engineering and Technology (ed., Dr. Sohail Anwar), Vol. 2, CRC Press/Taylor and Francis, Boca Raton, FL, 2014.

Questions and problems

1. How much geothermal energy is used internationally?
2. How much geothermal energy is used in the US? How much is the potential of using geothermal resources in the US?
3. What geological areas are most likely to have higher heat flows?
4. On the map of your country or of a US state, select the most suitable site for a geothermal energy development; describe the tectonic setting and likely types of rocks and geologic structures that would be found. Justify your conclusions.
5. What regions in your country are best suited for geothermal power facilities? Why? What regions are the least suited?
6. What is the global average heat flow at the surface of the Earth? What is its range? What controls heat flow?
7. How does a conventional geothermal power plant work? How does geothermal energy benefit local economies?
8. What factors influence the cost of a geothermal power plant?
9. List the main advantages and disadvantages of the direct-use of geothermal energy.
10. What nonconventional technologies are used or will be used in the next decade for geothermal energy generation?
11. What are the factors determining the geothermal resource sustainability?
12. List the most promising technologies that will expand geothermal energy uses in the near future.
13. List the most important potential environmental benefits of using geothermal energy.
14. How does geothermal energy benefit the US economy?
15. Identify locations in the US that are suitable for geothermal energy.
16. If we are assuming that the US geothermal energy sources generated some 15 billion kWh of electricity. How many conventional 1,000-MW power plants are needed to produce this electricity, if they operate 9,000 h/year?
17. It is accepted that geothermal fluids must about or exceed 200 °C in order to be practical for electricity generation. Assuming a surface temperature of 25 °C, how deep must one drill to hit this temperature at a location having a normal temperature gradient?
18. The average geothermal heat flux is about 0.050208 J/m^{2}s in the continental regions. Assuming that three-fourths of this is attributed to the crust, by using the average thermal conductivity for rock, calculate the temperature at 5000 m.
19. If a geothermal well of a good thermal aquifer is producing a flow rate of 250 kg/s of saturated water at 225 °C, what is the maximum power that this well may produce, assuming the surface temperature of 25 °C and the overall conversion efficiency of 36%?
20. An average house in a certain mid-latitude town needs 2.4 kW in winter months. A geothermal district heating is proposed for this town, consisting of 1,500 houses. How much heat power is needed? If the geothermal water is entering in the district heating system at 75 °C and is re-injected at 45 °C, how much flow rate in kg/s is required for this system?

21. If a GHP has a COP of 3.50 estimates the electrical energy need for heating (mid-November to mid-April) for a 200 m^2 house in Detroit, Michigan.
22. Repeat the problem above for cooling season (June–August).
23. A drilling in a certain location intercepts fluid at 250 °C at 1 km depth with a very good flow rate. What types of geothermal applications are suitable for this site? Justify your answer.
24. Estimate the available geothermal power form a well having the following characteristics, 250 m^3/h flow rate, the well bottom, and out (top) temperatures are 170 °C and 120 °C, respectively.
25. Give classification of the hydro-electric power plants. Write advantages, disadvantages, and application of hydro-electric power plant.
26. A Pelton turbine works with a head of 100 m and a water flow equal to10 m^3/s. What is the power output if the efficiency of the plant is 67%?
27. A village needs 20 m^3/day for water, collected from a waterfall with a head of 18 m. Estimate the hydraulic power.
28. Discuss the factors for site selection for a hydroelectric plant.
29. The Three Gorges Dam that spans the Yangtze River in China is the world's largest power station in terms of installed capacity. (a) In a normal year, the river flow rate is equal to 30,000 m^3/s. The dam has a head of 80 m. Calculate the power output, if the overall efficiency of the power plant is 90%. (b) In a drought year, the river flow rate can decrease to 50% of its normal value, and the head to 75 m. Calculate the dam power output if the hydro-power plant efficiency remains the same.
30. A water turbine converts power from a river to the rotation of a shaft which is used to run an electric generator. If the efficiency of the turbine is $\eta_t = 80\%$ and that of the generator is $\eta_g = 90\%$, what is the overall efficiency η_{tg} of the turbine-generator combination? If 100 kW of electrical power is produced, what is the rate of energy loss?
31. Water falls from a head 85 m at a flow rate of 30 m^3/s, and the overall conversion to electricity is 70%. How many typical single-family homes can this hydropower system provide with electricity?
32. An artificial reservoir is 20 km long, 1.8 km wide and has an average depth of 80 m. 1.5 % of the reservoir volume drops 100 m and passes through a hydroelectric conversion system with the overall efficiency of 0.82%. What is the electrical power output?
33. List the classification of hydraulic turbines.
34. What the main parts of Pelton and Turgo hydropower turbines.
35. Define the following terms: (a) gross head; (b) net head; (c) hydraulic efficiency; (d) volumetric efficiency; (e) mechanical efficiency; and (f) overall efficiency.
36. Explain the principles and operation of impulse turbine, as well as the reaction turbine.
37. A remote location in Montana needs 20 m^3/day, if the dynamic head is 18 m, what is the hydro-dynamic power?

38. A reservoir-type small hydropower system has a penstock of 1.8 m diameter, the water speed at the penstock exist is 12 m/s. Assuming the power coefficient of 36%, turbine efficiency 85%, and the generator-electronics efficiency 93%, compute the generator output electric power. If the penstock efficiency is 93% what is the water head?
39. Consider a turbine system as the one depicted below. The turbine is connected to an upper reservoir on the pressure side and discharges to a lower reservoir. The following is given, static pressure $p = 1\ 80$ kPa, volume flow rate at position 2 is 10 m^3/s, height $H = 400$ m, and pipe diameter at turbine inlet $d_2 = 0.6$ m, pipe diameter $d_3 = d_2$, static pressure p_3 is 105 kPa. The gravitational constant is g 9.81 m/s^2. Working medium is water with a density of 1,000 kg/m^3. Friction is neglected. Determine the following: flow speed at turbine inlet, total head at turbine inlet, difference in total head over turbine, and maximum possible power produced.

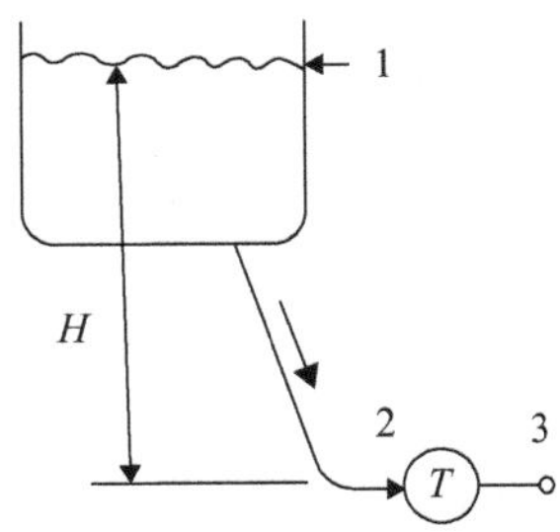

40. A site has a head of 100 m with a variable flow rate, as shown in the table below. Calculate the output power if a) the efficiency is (a) 100% and (b) 50%. Plot the results.

Month	**Jan**	**Feb**	**Mar**	**Apr**	**May**	**Jun**	**Jul**	**Aug**	**Sep**	**Oct**	**Nov**	**Dec**
Flow rate (m^3/s)	10	12	100	80	35	28	25	17	22	60	45	14

Chapter 11
Energy storage systems

Objectives and abstract

Over the last few decades several innovative ideas have been explored in the energy storage areas, ranging in size, capacity, design complexity, and targeted applications. Some of them are designed for large scale power system applications, others for small- or medium-scale renewable energy or hybrid power systems, while the others are designed to perform short-term energy storage ride through for critical infrastructure (communication systems, hospitals, military facilities, etc.). Energy storage has become an enabling technology for renewable energy applications, grid integration and enhancing power quality and stability in the power transmission and distribution, having a great potential to improve power grid quality and stability and to provide an alternative to fossil fuel-based energy generation. The major constraints for renewable energy penetration are the availability, intermittency, and variability, which can be addressed through energy storage. The energy storage choice depends on specific usage requirements, often incorporating several energy storage systems in order to increase system reliability, capacity, and supply security. In the electric power system, the renewable energy promise lies in its potential to increase grid efficiency, reliability, or in optimizing power flows and supporting variable power supplies. The parameters used in comparisons of various energy storage technologies include efficiency, energy capacity and density, run time, costs, system s response time, lifetime in years and cycles, self-discharge, and maturity of each energy storage technology. The most common energy storage technologies include compressed air, pumped hydro, batteries, fuel cells, flywheels, and super-capacitors. The last four are suitable for the medium scale applications. The chapter discussed energy storage technologies and gives an up to date comparative summary of their performances. *After completing this chapter, the readers are able to understand the role, importance, configurations and topologies of energy storage systems, operation principles, characteristics, performances, and operation of major energy storage systems used in power systems, buildings, and industrial facilities.* Another benefit is that readers are able to understand the critical role and necessity of energy storage systems in power and renewable energy systems, the differences between large-, medium- and small-scale energy storage systems, and how a system is selected on specific applications based on system characteristics and performances. Major energy storage technologies discussed in this chapter are compressed air energy storage, pumped hydropower storage systems, batteries, flywheels, hydrogen energy storage, fuel cells, supercapacitors, and superconducting energy storage systems. Thermal energy storage systems are covered in detail in the next chapter. This chapter provides comprehensive reviews of the energy storage technologies and gives an up to date comparative summary of their performances, characteristics, and applications.

11.1 Introduction and energy storage importance

Electrical power systems are in a period of considerable changes, driven by several interconnected factors, ageing power networks, increased use of distributed generation (DG) and renewable energy, pollutant reductions, regulatory incentives, and new power and energy system technologies. In this climate, energy storage use has emerged as an area of considerable interest due to extended applications in industrial and building energy systems and as a critical player into the future smart grids. The transition period end is signaled by the successful establishment of the technology and practices that must go together to create what is termed the Smart Grid (SG). There also are considerable interests in the optimization, energy conservation and in improving the efficiency of industrial or building energy systems, in which energy storage applications are critical. Energy storage systems (ESS), designed to provide support for long-term applications and dynamic performance enhancement, can provide better balancing between the electricity demand and the supply, allowing the increased asset utilization, facilitating the renewable energy penetration, and improving the overall grid flexibility, stability, reliability, and efficiency. Almost all renewable energy sources are characterized by the generation variability, intermittency and discontinuity, the generation is not controlled by the system operator, making their integration into power systems difficult. Over the past few decades, many new and innovative ideas have been explored in the energy storage areas, ranging in size, capacity and design complexity. Some of them are designed for large scale power system applications, others for small- or medium-scale renewable energy systems, while the others are designed for short-term energy storage ride through capabilities for critical infrastructure (military, hospitals, or communication facilities). Energy storage has become an enabling technology for renewable energy applications and for enhancing power quality and power system stability, having a great potential to improve power grids and to provide alternatives to conventional energy generation. One of the major constraints for increasing renewable energy penetration is their availability and intermittency, which can be addressed through energy storage. The energy storage system choice depends on specific requirements, usually several ESS are used in order to increase capacity and improve supply security. The aim energy storage presentation is to provide the basis for development of new energy storage possibilities in buildings, industrial, and commercial facilities and in power distribution. Among the parameters used in comparisons of various energy storage technologies are efficiency, energy capacity and energy density, run time, costs, the response time, lifetime in years and cycles, self-discharge rate, and technology maturity. The most common energy storage technologies for medium- and low-scale applications are batteries, fuel cells, flywheels, capacitors, and superconductive ESS.

Energy storage technology has been in existence for a long time and has been utilized in many forms and applications from a flashlight to spacecraft systems. Today energy storage is used to make the electric power systems more reliable as well as making the broader renewable energy use a reality. ESSs become critical in enabling renewable energy applications, providing the means to make the non-dispatchable resources into dispatchable ones. In order to match power demand, in the renewable energy intermittent and non-dispatchable context and the fairly predictable electrical demand, the ESS are critical, allowing de-coupling of generation from consumption, reducing the needs for constant monitoring and energy demand predictions. Energy storage also provides economic benefits by allowing a plant energy generation reduction to meet average demands rather than peak power demands. The current status shows that several drivers are emerging and are spurring growth in the energy storage demands, such as the renewable energy growth, increasingly strained grid infrastructure as new lines lag well behind demands, or the microgrid emergence as part of smart grid architecture, higher energy supply reliability and security, improved and optimized building and industrial process energy use and efficiency. However, issues regarding the optimal integration (operational, technical, and market) of energy storage into the electric grid are still not fully developed, tested, and standardized. The ESSs integration and further energy conversion unit development, including renewable energy must be based on the existing electric infrastructure, requiring optimal integration of ESS.

Energy storage becomes a critical factor that can solve the problems described earlier. A renewable energy system with its corresponding energy storage system can behave as a conventional power plant, at least for time intervals in the order of half an hour up to a day, depending on the storage capacity. Electricity generated from renewable sources can rarely provide immediate response to demand as these sources do not deliver a regular supply easily adjustable to consumption needs. Thus, the growth of this decentralized production means greater network load stability problems and requires energy storage, as a potential solution. However, lead batteries cannot withstand high cycling rates, nor can they store large amounts of energy in a small volume. The energy storage is also a crucial element in the energy management from renewable sources, allowing energy to be released into the grid during peak hours when it is more valuable. In the conventional integration and operation planning process of bulk power plants, normally a top-down strategy, coming from an energy consumption point of view down to a stepwise detailed description is used. In this strategy, the planning horizon is subdivided in long-, medium-, and short-term planning task. The discretion of the time scale in each planning step is a compromise between accuracy and the number of technical and economical boundaries. This planning strategy is driven by economical consideration in a unidirectional electrical power supply chain. In these planning strategies, the detailed control functionality can be only figured out when the system model in the planning approach is detailed enough. This means the detailed accuracy of the model inside of each planning stage (long, medium, short-term, quasi-stationary, dynamic) defines the optimal layout of those conventional supply systems. In the DG case, the technical boundaries are initially more important for the planning process to get an economical optimal supply configuration under stable operation conditions. When this is not taken into conflict potential between the energy supplier and the operator of electricity distribution networks, especially the ones based on renewable energy sources. Therefore, the need arises for planning and integration strategy which already includes clear specifications and definitions of the system inherent in forefront. This means, the process of system modeling and structuring for optimal integration of ESS and distributed energy resources (DERs) must already include ESS and DER control functions, limits, and boundaries. These requirements lead to bottom-up strategy starting from DG up to the centralized conventional electrical supply system. With bottom-up strategy, clear definition of supply and control functionality between ESS, DER, and the electrical grid can be derived.

There are several well-established electricity storage technologies, as well as a large number in the process of development offering significant application potential. Economically viable energy storage requires conversion of electricity or energy storage into other energy form, which can be converted back to electricity when needed. All storage methods need to be feasible, efficient, and environmentally safe. ESS can be separated in four major classes: *mechanical, electrical, thermal,* and *chemical* based systems. Each class contains several technologies with specific characteristics and applications. Mechanical storage includes pumped hydropower storage, compressed air energy storage, and flywheels. Electrochemical storage includes batteries, hydrogen based energy storage, and fuel cells. Here also is included the thermochemical energy storage, such as solar-hydrogen energy storage or solar-metal energy storage techniques, which are not discussed here. Electromagnetic energy storage includes supercapacitors and superconducting magnetic energy storage. Thermal energy storage includes two broad categories: low-temperature and high-temperature energy storage. Figure 11.1 summarizes the most common energy storage technologies. Despite the opportunity offered by the energy storage to increase the stability and reliability of the intermittent energy sources, there were only very few installed ESS (batteries, pumped hydropower storage, compressed air energy storage, and thermal storage) with a capacity exceeding 100 MW. The opportunities for significant improvement in development of the energy storage are strong, with the most appropriate applications to power quality management, load shifting, and energy management.

The ESSs integration and further developments must be based on the existing electric supply system infrastructure, requiring optimal integration of ESS. Renewable energy systems with optimum energy storage can behave as conventional power plants, at least for short-time intervals of order of half-hour to a day, depending on the storage system capacity. Electricity generated from renewable sources can rarely provide immediate response to demands as these sources do not have capabilities to easily adjust to the consumption needs, being in this regard

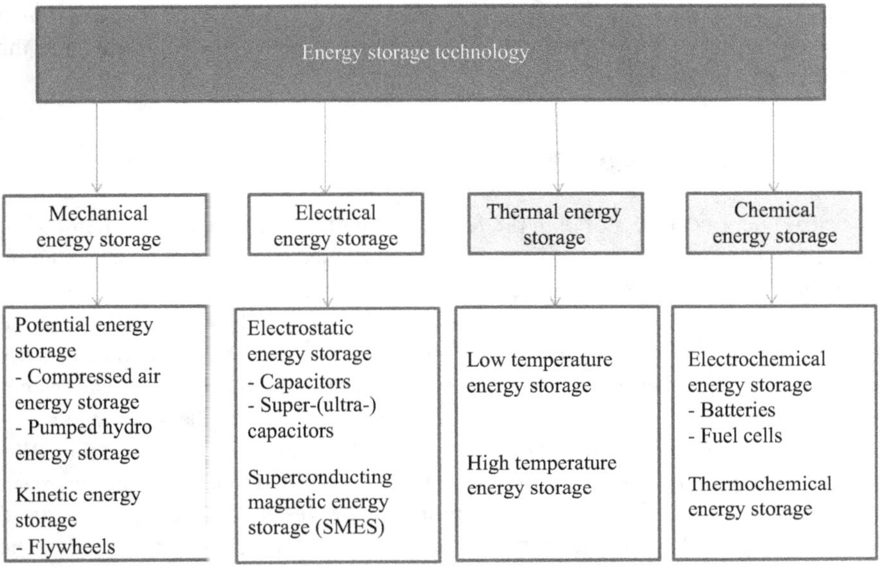

Figure 11.1 Classification of the major energy storage technologies

low-inertia systems. Growth of decentralized power production may result in greater network load stability problems, requiring energy storage, as one of the potential solutions. Energy storage is crucial in the energy management from renewable energy sources, allowing energy to be released into the grid during peak hours when it is more valuable. In the conventional integration and operation planning process of bulk power plants, normally a top-down strategy, coming from an energy consumption point of view down to a stepwise detailed description is used. In this strategy, the planning horizon is subdivided in long-, medium-, and short-term planning tasks. The time scale in each planning step is a compromise between accuracy and the number of technical and economical boundaries. Planning strategy is driven by economic considerations in a unidirectional electrical power supply chain. In these planning strategies, the detailed control functionality can be only figured out when the system model in the planning approach is detailed enough. This means the detailed accuracy of the model inside of each planning stage (long-, medium-, short-term, quasi-stationary, or dynamic) defines the optimal layout of those conventional supply systems. In the DG case, the technical boundaries are very important for the planning process to get an economical optimal supply configuration under stable operation conditions. When this is not taken into account a conflict potential between the energy supply and the operation of electricity networks, especially the ones based on renewable energy sources may exist. Therefore, the needs arise for planning and integration strategy, which includes clear system specifications and definitions, meaning a system structure for optimal integration of energy storage and DERs, that include ESS and DER control functions, limits, and boundaries. These requirements lead to bottom-up strategy starting from DG up to the centralized conventional electrical supply system, which a clear definition of supply and control functionality between ESS, DER, and the electrical grid is derived.

11.2 Energy storage functions and applications

Electrical energy can be stored in the forms of kinetic, potential, electrochemical or electromagnetic energy, and transferred back into electrical energy when required or needed. The conversion of electrical energy to different forms and back to electrical energy is done by a specific conversion process. The electricity generated during off-peak periods can be stored and used to meet the loads during peak periods when the energy is more expensive, improving the power system economics and operation. Compared to conventional generators, the storage systems have faster ramping rates to respond to the load fluctuations. Therefore, the EESs are a perfect spinning reserve, providing a fast load following and reduces the need for conventional and more expensive spinning reserves. EESs were initially used only for load leveling applications, while now are seen as tools to improve the power quality, stability, to ensure a reliable and secure power supply, and to black start the power systems. Breakthroughs are reducing dramatically the EES costs and are driving significant changes into the power system design, structure, and operation. Peak load problems could be reduced, stability improved, power quality issues reduced, or even eliminated. Storage can be applied at the power plant level, in support of the transmission system, at various power distribution system points, and on particular equipment on the customer side. Figure 11.2 shows how the new electricity value chain is changing supported by the integration of ESS. EESs in combination with advanced power electronics can have a great technical role and

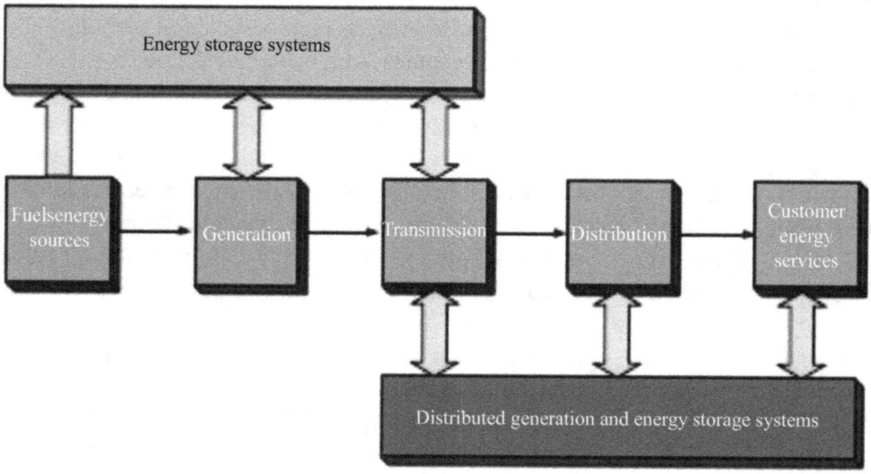

Figure 11.2 New chain of electricity, with energy storage as the sixth dimension

can lead to many benefits. EESs are also suitable for particular applications, primarily due to their potential energy storage capacities. Therefore, to provide a fair comparison between the various energy storage technologies, they are grouped, based on the size of power and energy storage capacity. There are four categories: devices with large power (>50 MW) and energy storage (>100 MWh) capacities, medium power (1–50 MW) and storage capacities (5–100 MWh), medium power or medium storage capacities but not both, and small-scale ESS, with power less than 1 MW and energy capacity less than 5 MWh. Storage systems, such as pumped-hydroelectric energy storage (PHES) have been in use since 1929 for daily load leveling. Today with the smart grid advent, energy storage is a realistic option for: (1) electricity market restructuring; (2) integrating renewable energy resources; (3) improving power quality; (4) shifting toward DG; and (5) helping network operate under more stringent environmental requirements.

ESSs are needed by the electricity industry, because unlike any other successful commodities, the conventional electricity industries have little or almost no energy storage facility. The electricity transmission and distribution systems are operated through a simple one-way transportation from large power plants to the consumers, so the electricity must always be used precisely when produced. However, the electricity demand varies considerably emergently, daily, weekly, and seasonally, while the maximum demand may only last for very short periods, leading to inefficient, overdesigned, and expensive peak power plants. ESS allows energy production to be de-coupled from the supply, self-generated, or purchased. Having large-scale electricity storage capacities available over any time, system planners need to build only sufficient generating capacity to meet average electricity demands rather than peak demands. This is particularly important to large utility generation systems, e.g., nuclear power plants, which must operate near full capacity for economic reasons. Therefore, ESS can provide substantial benefits including load following, peaking power and standby reserve, higher overall efficiencies of thermal power plants, reducing the harmful pollutant emissions. Furthermore, ESS is regarded as an imperative technology for DER systems. The traditional electricity value chain has been considered to consist of five links: fuel/energy sources, generation, transmission, distribution, and customer-side energy service as shown in Figure 11.2. Supplying power when and where is needed, ESS is on the brink of becoming the *sixth link* by integrating the existing segments and creating a more responsive market. Storage technologies are various and covering a full spectrum from larger scale, generation and transmission, to those related to power distribution and even *beyond the meter*, into the end-user sites.

ESS performances include cycle efficiency, cost per unit capacity, energy density, power capacity, lifecycle, environmental effect including end-of-life disposal cost. An ideal ESS system is one exhibiting the best possible performances so that it will have the minimum amortized (dollar or environmental) cost during its whole lifetime. Unfortunately, no single ESS type can simultaneously fulfill all the desired characteristics of an ideal ESS system, and thus minimize the amortized lifetime cost of ESS. Capital cost, one of the most important criteria in the ESS design, can be represented in the form of cost per unit of delivered energy ($/kWh) or per unit of output power ($/kW). Capital cost is an especially important concern when constructing hybrid energy storage (HES) systems. Such systems often consist of several ESS elements with relatively low unit cost (e.g., lead–acid batteries) and ESS elements with relatively high unit cost (e.g., supercapacitors). The overall HES system cost is minimized by allocating the appropriate mix of low-cost versus high-cost ESS elements while meeting other constraints, such as cycle efficiency, total storage capacity, or peak output power rate. The ESS cycle efficiency is defined by the "roundtrip" efficiency, i.e., the energy efficiency for charging and then discharging. The cycle efficiency is the product of charging efficiency and discharging efficiency, where charging efficiency is the ratio of electrical energy stored in an ESS element to the total energy supplied to that element during the entire charging process, and discharging efficiency is the ratio of energy derived from an ESS element during the discharging process to the total energy stored in it. Charging-discharging efficiency is significantly affected by the charging/discharging profiles and the ambient conditions. The ESS state of health (SOH) is a measure of its age, reflecting the ESS general condition and its ability to store and deliver energy compared to its initial state. During the ESS lifetime, its capacity or "health" tends to deteriorate due to irreversible physical and chemical changes which taking place. Term "replacement" implies the discharge, meaning the use, until the ESS is no longer usable (its end-of-life). To indicate the rate at which SOH is deteriorating, the lifecycle may be defined as the number of ESS cycles performed before its capacity drops to a specific capacity threshold, being one of the key parameters and gives an indication of the expected ESS working lifetime. The lifecycle is closely related to the replacement period and full ESS cost. The self-discharge rate is a measure of how quickly a storage element loses its energy when it simply sits on the shelf, being determined by the inner structure and chemistry, ambient temperature and humidity, and significantly affect the sustainable energy storage period of the given storage element.

Deferred from the conventional power system which has large, centralized units, DERs are installed at the distribution level, close to the consumers, and generate lower power typically in the range of a few kW to a few MW. The electric grid is undergoing the change to be a mixture of centralized and distributed subsystems with higher and higher DER penetration. However, more drastic load fluctuations and emergent voltage drops are anticipated, due to smaller capacity and higher line fault probability than in conventional power system. ESS is a key solution to compensate the power flexibility and provide a more secure power supply, being also critically important to the integration of intermittent renewable energy. The renewable energy penetration can displace significant amounts of energy produced by large conventional power plants. A suitable ESS could provide an important (even crucial) approach to dealing with the inherent RES intermittency and unpredictability as the energy surplus is stored during the periods when generation exceeds the demand and then used to cover periods when the load demands are greater than the generation. Future development of the RES technologies is believed to drive the energy storage cost down. Nonetheless, the widespread deployments are facing the fundamental difficulty of intermittent supplies, requiring demand flexibility, backup power sources, and enough energy storage for significant time. For example, the EES applications to enhance wind energy generation are (i) transmission curtailment, mitigating the power delivery constraint due to insufficient transmission capacity; (ii) time-shifting, firming, and shaping of wind generated energy by storing it during the off-peak interval (supplemented by power from the grid when wind generation is inadequate) and discharging during the peak interval; (iii) forecast hedge mitigating the errors in wind energy bids prior to required delivery, reducing the price volatility and mitigating consumer risk exposure to this volatility; (iv) grid frequency support through the energy storage during sudden, large decreases in wind generation over a short discharge interval; and (v) fluctuation suppression through stabilizing the wind farm generation frequency by suppressing fluctuations (absorbing and discharging energy during short duration variations in output). The key storage energy applications should be equally relevant to all intermittent renewables.

1. ***Grid voltage support,*** the additional power provided to the electrical distribution grid to maintain voltages within the acceptable range.
2. ***Grid frequency support,*** real power is provided to the grid to reduce any sudden, large load-generation imbalance to keep the frequency within the permissible tolerance for up 30-min periods.
3. ***Grid angular (transient) stability,*** the reduction of the power oscillations (due to rapid events) by real power injection and absorption.
4. ***Load leveling (peak shaving)***, consisting of rescheduling certain loads to lower power demands, and/or the energy generation during off-peak periods for storage and use during peak demand periods.
5. ***Spinning reserve,*** defined as the amount of generation capacity that is used to produce active power over a given period which has not yet been committed to the production of energy during this period.
6. ***Power quality improvement,*** which is basically related to the changes in magnitude and shape of voltage and current, energy storage can help to mitigate the power quality problems.
7. ***Power reliability,*** defined as the percentage/ratio of interruption in delivery of electric power (may include exceeding the threshold and not only complete loss of power) versus total uptime. Distributed energy storage systems (DESSs) can help provide reliable electric service to consumers.
8. ***Ride through support,*** the electric unit staying connected during system disturbance (voltage sag); ESSs have the potential of providing energy and support to ride-through.
9. ***Unbalanced load compensation,*** which is done by injecting and absorbing power individually at each phase to supply unbalanced loads.

The advanced electric energy storage technologies, when utilized properly, would have an environmental, economic, and energy diversity advantages to the system. These include:

1. *Matching electricity supply to load demand:* energy is stored during periods when production exceeds consumption (at lower cost possible) and the stored energy is utilized at periods when consumption exceeds production (at higher cost level), so the electricity production need not be scaled up and down to meet the demand variations, instead the production is maintained at a more constant and economic level. This has the advantage that fuel-based power plants (i.e., coal, oil, gas) are operated more efficiently and easily at constant production levels, while maintaining a continuous power to the customer without fluctuations.
2. *Reducing the risks of power blackouts:* energy storage technologies have the ability to provide power to smooth out short-term fluctuations caused by interruptions and sudden load changes. If applied properly, real long-term energy storage can also provide power to the grid during longer blackouts.
3. *Enabling renewable energy generation:* solar and wind energy systems are intermittent sources, expected to produce about 20% of the future electricity, generating during off-peak, when the energy has a low financial value, and the energy storage can smooth out their variability, allowing the electricity the dispatch at a later time, during peak periods, make them cost effective, and more reliable options.
4. *Power Quality:* it may cause poor operations or failures of end-user equipment. Distribution network, sensitive loads, and critical operations suffer from outages and service interruptions, leading to financial losses to utility and consumers. Energy storage, when properly engineered and implemented, can provide electricity to the customer without any fluctuations, overcoming the power quality problems, such as swells/sags or spikes. A summary of energy storage applications and requirements are given in Table 11.1.

Table 11.1 Energy storage applications in power systems

Applications	**Matching supply and demand**	**Providing backup power**	**Enabling renewable technology**	**Power quality**
Discharged power	1–100 MW	1–200 MW	20 KW–100 MW	1 KW–20 MW
Response time	< 10 min.	< 10 ms (Quick) < 10 min. (Conventional)	< 1 s	
Energy capacity	1–1,000 MWh	1–1,000 MWh	10 KWh–200 MWh	50–500 kWh
Efficiency needed	High	Medium	High	Low
Life time needed	High	High	High	Low

11.2.1 Summary of benefits from energy storage

The electricity supply chain is deregulated with clear divisions between generation, transmission system operators, distribution network operators, and supply companies. Energy storage applications to power distribution can benefit the customer, supply company, and generation operator (conventional and DG) in several ways. Major areas where ESS can be applied can be summarized as follows:

1. Voltage control, supporting a heavily loaded feeder, providing power factor correction, reducing the need to constrain DG, minimizing on-load tap changer operations, mitigating flicker, sags, and swells.
2. Power flow management, by redirecting power flows, delay network reinforcement, reduce reverse power flows, and minimize losses.
3. Restoration, by assisting voltage control and power flow management in a post-fault reconfigured network.
4. Energy market, through the arbitrage, balancing market, reduced DG variability, increased DG yield from nonfirm connections, replacing the spinning reserve.
5. Commercial/regulatory, by assisting the compliance with energy security standard, reducing customer minutes lost, while reducing generator curtailment.
6. Network management, by assisting islanded networks, support black starts, switching ESS between alternative feeders at a normally open point.

It is evident that developing a compelling EES installation at power distribution level in today's electricity market with present technology costs is difficult, if value is accrued from only a single benefit. The importance of understanding the interactions between various objectives and quantifying the individual benefits brought is a critical issue in ESS potential evaluation. EES systems can contribute significantly to meeting the needs for more efficient and environmentally benign energy use in buildings, industry, transportation, and utilities. Overall the ESS uses often results in significant benefits as: reduced energy costs and consumption, improved air and water quality, increased operation flexibility, and reduced initial and maintenance costs, reduced size, more efficient and effective utilization of the equipment, fuel conservation, by facilitating more efficient energy use and/or fuel substitution, and reduced pollutant emissions. EESs have significant potential to increase the effectiveness of energy-conversion equipment use and for facilitating large-scale fuel substitutions. EESs are complex and cannot be evaluated properly without a detailed understanding of energy supplies and end-use considerations.

11.3 Energy storage system types

The energy storage technologies can be classified in four categories depending on the type of energy stored, mechanical, electrical, thermal, and chemical energy storage technologies each offering different opportunities but also consisting of own disadvantages. In the following, each method of electricity storage is assessed and the characteristics of each technology, including overall storage capacity, energy density (energy stored per kilogram), power density (energy transfer time rate per kilogram), and round trip efficiency of energy conversion are compared. Energy storage can be defined as the conversion of electrical energy from a power network into a form in which it can be stored until converted back to electrical energy. The ESS are currently characterized by (a) disagreement on the role and design of ESS, (b) common energy storage uses, (c) new available technologies are still under demonstration and illustration, (d) no recognized planning tools/models to aid understanding of storage devices (e) system integration including power electronics must be improved, and (f) it seems small-scale storage will have great importance in the future. However, several ESS are available nowadays with different characteristics, capabilities, and applications. In the subsection of the next chapter, a quite comprehensive presentation of various energy storage technologies is presented.

11.3.1 Pumped hydroelectric energy storage (PHES)

PHES is the most mature and largest energy storage technique available. It consists of two large reservoirs located at different elevations and a number of pump/turbine units (as shown in Figure 11.3). During off-peak electrical demand, water is pumped from the lower reservoir to the higher reservoir, and stored until it is needed. Once required, the upper reservoir water is released through penstocks and the turbines, connected to generators producing electricity. Therefore, during generation, a PHES operates similarly to a conventional hydroelectric system. The efficiency of modern PHES facilities is in the range of 70%–85%. However, variable speed machines are now being used to improve it. The efficiency is limited by the pump/turbine unit efficiency used. Until recently, PHES units have always

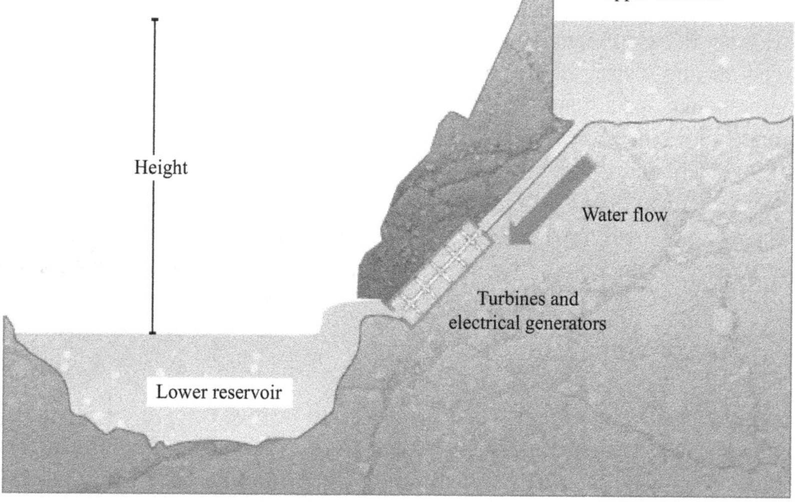

Figure 11.3 Typical pumped hydroelectric energy storage

used fresh water as the storage medium. A typical PHES facility has 300 m of hydraulic head (the vertical distance between the upper and lower reservoir). The power capacity (kW) is a function of the flow rate and the hydraulic head, while the energy stored (kWh) is a function of the reservoir volume and hydraulic head. To calculate the mass power output of a PHES facility, the following relationship can be used:

$$P_C = \rho g Q H \eta \tag{11.1}$$

where P_C *is the* power capacity, in W, ρ is the mass density of water in kg/m^3, g is the acceleration due to gravity in m/s^2, Q *is the* discharge through the turbines in m^3/s, H is the effective head in m, and η is the efficiency. And to evaluate the storage capacity of the PHES, the following must be used:

$$S_C = \frac{\rho g H V \eta}{3.6 \times 10^9} \tag{11.2}$$

where S_C is the storage capacity in megawatt-hours (MWh), V is the volume of water that is drained and filled each day in m^3. It is evident that the power and storage capacities are both dependent on the head and the volume of the reservoirs. However, facilities should be designed with the greatest hydraulic head possible rather than largest upper reservoir possible, being cheaper to construct a facility with a large hydraulic head and small reservoirs, than to construct a facility of equal capacity with a small hydraulic head and large reservoirs because (1) less materials are removed for the required reservoirs, (2) smaller piping is necessary, hence, smaller boreholes during drilling, and (3) the turbine is physically smaller.

Example 11.1: A pumped storage facility has a head of 450 m, and an efficiency of 93%. What is the flow rate needed to generate 100 MW for 3.5 h/day? Assume the water density is 10^3 kg/m^3 and the acceleration due to gravity 9.80 m/s. What is the required working volume per day?

Solution: From (11.1), the flow rate is:

$$Q = \frac{P_C}{\eta \rho g H} = \frac{10^8}{0.93 \times 9.806 \times 10^3 \times 450} = 24.383 \text{ m}^3/\text{s}$$

Working volume can be estimated as:

$$V = Q \cdot \Delta t = 24.383 \times 3.5 \times 3600 = 307{,}225.8 \text{ or } \simeq 307{,}226 \cdot 10^3 \text{ m}^3$$

Today, there is over 90 GW in more than 240 PHES facilities in the world, roughly 3% of the world's global generating capacity. Each individual facility can store from 30 to 4,000 MW of electrical energy. Pumped storage has been commercially implemented for load balancing for over 80 years. The pumped energy storage is classified as real long-term response energy storage and typically used for applications needing to supply power for periods between hours and days (power outages). In the US only, there are 38 pumped energy storage facilities, providing a total power capacity of about 19 GW. Pumped hydroelectric energy storage usually comprises the following parts: an upper reservoir, waterways, a pump, a turbine, a motor, a generator, and a lower reservoir, as shown in Figure 11.3. As in any system, including hydraulic ones, in a hydroelectric energy storage system there are losses during operation, such as frictional losses, turbulence, and viscous drag, and the turbine itself is not 100% efficient. However, the efficiency of large-scale water-driven turbines can be quite high, even over 95%, the efficiency of the dual-cycle reversible hydropower storage system typically is about 80%. There also are other losses, such as water evaporation from the reservoirs, and leakage around the turbine. The water retains some kinetic energy even when it enters the tailrace. For the final conversion of hydropower to electricity, the turbine-generator losses need to be taken into account. Therefore, the overall PHES efficiency is as the ratio of the energy supplied, while generating, E_{Gen}, and the energy consumed while pumping, E_{Pump}, depending on the pumping and generation efficiencies:

$$\eta_{PHES} = \frac{E_{Gen}}{E_{Pump}} = \eta_{Gen} \times \eta_{Pump} \tag{11.3}$$

The energy used for pumping a volume V of water up to height h with a pumping efficiency, η_{Pump}, and the energy supplied to the grid or load while generating with generating efficiency η_{Gen} are given by:

$$\begin{aligned} E_{Pump} &= \frac{\rho g h V}{\eta_{Pump}} \\ E_{Gen} &= \rho g h V \cdot \eta_{Gen} \end{aligned} \tag{11.4}$$

The volumetric energy density for a pumped hydroelectric energy storage system will therefore depend on height h and is given by:

$$W_{PHES} = \frac{E_{Pump}}{V} = \frac{\rho g h}{\eta_{Pump}} \tag{11.5}$$

At first pumped hydro storage stations usually copied conventional hydroelectric design in having the power transformation (extraction) system located outdoors close to the lower reservoir. The increase in power capacity ratings and pumping heads, combined with the higher rotational speeds of turbines, has required the hydraulic unit to be set at considerable depths below the minimum tail-water level in order to avoid cavitation. To meet these requirements, a massive concrete construction is needed to withstand the external water pressure and resist hydrostatic uplift, so the system has become increasingly expensive.

Example 11.2: A hydropower pumping storage station has the following characteristics, the effective storage capacity of the upper reservoir 375 × 10^3 m^3, the generation efficiency 0.91, the average head 200 m, the pumping efficiency 0.87. Find the overall efficiency and the volumetric energy density?

Solution: From the overall hydropower efficiency, given by (11.3) is:

$$\eta_{PHES} = \eta_{Gen} \times \eta_{Pump} = 0.91 \times 0.87 = 0.79 \text{ or } 79\%$$

And the volumetric energy density, assuming the water density 1,000 kg/m^3, given by (11.5) is:

$$W_{PHES} = \frac{\rho g h}{\eta_{Pump}} = \frac{10^3 \times 9.81 \times 200}{0.87} = 1.2552 \times 10^6 \frac{J}{m^3}$$

A variant of pumped hydro-power storage are the underground PHES (UPHES) facilities, having the same operating principle as PHES system: two reservoirs with a large head between them. The only major difference between the two designs is the reservoir locations. In conventional PHES, suitable geological formations must be identified to build the facility, while in UPHES facilities have the upper reservoir at ground level and the lower reservoir deep below the earth's surface. The depth depends on the amount of hydraulic head required for a specific application. UPHES has the same disadvantages as PHES (large-scale required, high capital costs, etc.), with one major exception. As stated previously, the most significant problem with PHES is their geological dependence. As the lower reservoir is obtained by drilling into the ground and the upper reservoir is at ground level, so no stringent geological dependences. The major disadvantage for UPHES is its commercial youth. To date there is a very few, if any, UPHES facilities in operation, being difficult to analyze and to trust the performance of this technology. However, the UPHES has a very bright future if cost-effective excavation techniques can be identified for its construction. Its relatively large-scale storage capacities, combined with location independence, provide storage with unique characteristics. The cost, design, power and storage capacities, and environmental impacts need to be investigated, in order to prove that UPHES systems are a viable option.

11.3.2 Compressed air energy storage (CAES)

CAES is an established energy storage technology, being in the grid operation since late 1970s. The energy stored mechanically by compressing the air, being released to the grid when the air is expanded. Compressed air energy storage is achieved at high pressures (70 atm) and near the ambient temperatures, which means less volume and smaller reservoirs. Large caverns made of high-quality rock deep in the ground, ancient salt mines, or underground natural gas storage

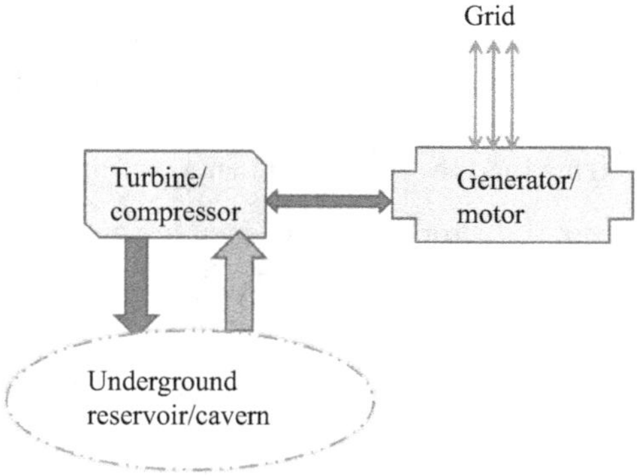

Figure 11.4 Diagram of typical compressed air energy storage

caves are the best CAES options, as they benefit from geostatic pressure, which facilitates the containment of the air mass (Figure 11.4). Studies have shown that the air could be compressed and stored in underground, high-pressure piping (20–100 bars). A CAES facility may consist of a power train motor that drives a compressor, compressing the air into the cavern, high pressure and low pressure turbines, and a generator. In a gas turbine (GT), 66% of the energy is used is required to compress the air. Therefore, the CAES is pre-compressing the air using off-peak electrical power, taken from the grid to drive a motor (rather than using GTs) and stores it in large storage reservoirs. CAES may use the energy peaks generated by renewable energy plants to run compressors that are compressing the air into underground or surface reservoirs. The compressed air is used into turbine-generator units to generate electricity, during peak demand. The energy storage capacity depends on the compressed air volume and storage pressure. When the grid is producing electricity during peak hours, the compressed air, stored by using cheaper off-peak electricity, from the storage facility is used instead of using more expensive natural gas. However, when the air is released from the reservoir it is mixed with small amounts of gas before entering the turbine, to avoid air temperature and pressure issues. If the pressure using air alone is high enough to achieve a significant power output, the air temperature would be far too low for the materials and connections to tolerate. The required gas amount is very small so a GT working with CAES can produce three times more electricity than operating alone, using the same natural gas amount. The reservoir can be man-made, the expensive choice, or by using suitable natural geological formations, such as: salt-caverns, hard-rock caverns, depleted gas fields or an aquifer, selected to suit specific requirements. In a slat-cavern fresh water is pumped into the cavern and left until the salt dissolves and saturates the fresh water, then transferred at surface to remove salt, and the cycle is repeated until the required cavern volume is created. This process is expensive and can take up to two years to complete. Hard-rock caverns are even more expensive, about 60% higher than salt-caverns. Finally, the aquifers cannot store the air at higher pressures, having relatively lower energy capacities. CAES efficiency is difficult to estimate, especially when is using both electrical energy and natural gas, with estimated efficiencies based on the compression and expansion cycles in the range of 68%–75%. CAES systems with typical capacities between 50 to 300 MW are used for large and medium scale applications. Their life-times are far longer than existing gas turbines and the charge/discharge ratio is dependent on the compressor and reservoir size and pressure. With assumption of ideal gas and isothermal process, the energy stored by compressing m amount of gas at constant temperature from initial pressure, P_i to final pressure, P_f is given by:

$$E_{fi} = -\int_{P_i}^{P_f} VdP = mR_g \ln\left(\frac{P_i}{P_f}\right) = P_f V_f \ln\left(\frac{P_i}{P_f}\right) = P_f V_f \ln\left(\frac{V_i}{V_f}\right) \tag{11.6}$$

Here, air is assumed an ideal gas which specific heat is constant, R_g is the ideal gas constant (8.31447 J/K/mol), V_i and V_f are the initial and final volumes of the compressed air. The compression power, P_C depends on the air flow rate, Q (the volume per unit time) and the compression ratio (P_f/P_i), expressed as:

$$P_C = \frac{\gamma}{\gamma - 1} P_i \cdot Q\left[\left(\frac{P_f}{P_i}\right)^{\frac{\gamma}{\gamma-1}} - 1\right] \tag{11.7}$$

where γ is the ratio of air specific heat coefficients ($\gamma \approx 1.4$), P_i, P_f are the initial and final pressure, the atmospheric and compressed state, respectively. A 290-MW CAES (300,000 m^3 and 48,000 Pa), Hundorf, Germany was built in 1978, and a 110-MW CAES (540,000 m^3 and 53,000 Pa) was built in 1991 at McIntosh, Alabama. Other CAES projects are implemented or in process to be built in Canada, U.S. and E.U., using different technologies, geological structures and approaches at various power capacities.

Example 11.3: A CAES has a volume of 450,000 m^3, and the compressed air pressure range is from 75 bars to atmospheric pressure. Assuming isothermal process and an efficiency of 30%, estimate the energy and power for a 3-h discharge period.

Solution: From (11.6), assuming the atmospheric pressure 1 bar, the energy is:

$$E = 45 \times 10^4 \times 75 \times 10^5 \ln\left(\frac{75}{1}\right) = 1457.2 \times 10^4 \text{ MJ}$$

The average power output is then:

$$P = \frac{\eta \cdot E}{\Delta t} = \frac{0.3 \times 1457.2 \times 10^{10}}{2 \times 3,600} = 404.76 \text{ MW}$$

Compression of a fluid, air in CAES generates heat, while the after the decompression the air is colder. If the heat generated during compression can be stored and used again during decompression, the overall system efficiency improves considerable. CAES is achieved through *adiabatic*, *diabatic*, and *isothermal* processes. An adiabatic storage has basically no heat exchange during the compressions–expansion cycle, heat being stored in fluids, such as oil or molten salt solutions, while in diabatic storage the heat is dissipated with intercoolers. In an isothermal process, the operating temperature is maintained constant (or rather quasi-constant) through heat exchange with the environment, being practical only for low power levels and some heat losses are unavoidable, compression process not being truly isothermal. For an isothermal processes, the maximum energy that can be stored and released is given by:

$$E_{fi} = nRT \ln\left(\frac{V_i}{V_f}\right) \tag{11.8}$$

Here T is the absolute temperature (K) and n is the number of moles of the air in the reservoir.

Example 11.4: Determine the maximum stored energy, if a mass of 2,900 kg of air is compressed isothermally at 300 K from 100 to 1,500 kPa, assuming a heat loss of 33 MJ and ideal gas.

Solution: With the ideal gas assumption, molar mass for air 29 kg/Kmol and R 8.314 kJ/kmol·K, the number of moles of air, is:

$$n = \frac{2,900}{29} = 100 \text{ kmol}$$

The mechanical energy stored, by using (11.8) is then:

$$E_{fi} = -nRT \ln\left(\frac{V_i}{V_f}\right) = -100 \times 8.314 \times 300 \cdot \ln\left(\frac{100}{1,500}\right) = 675,441.9 \text{ kJ}$$

Net stored energy is then:

$$E_{net} = E_{fl} - \text{Heat losses} = 675,441.9 - 33,000 = 642,441.9 \text{ kJ}$$

CAES systems, the only very large-scale energy storage other than PHES have fast reaction time with plants usually able to go from 0% to 100% in less than 10 min, 10%–100% in approximately 4 min and from 50% to 100% in less than 15 s. As a result, it is ideal for acting as a large sink for bulk energy supply, being also able to undertake frequent start-ups and shut-downs. CAES do not suffer from excessive heat when operating on partial load, as traditional gas turbines. These flexibilities mean that CAES can be used for ancillary services, such as load following, frequency regulation, and voltage control. As a result, CAES has become a serious contender in the wind energy industry. A number of possibilities, being considered, such as integrating a CAES facility with several wind farms within the same area, so the excess off-peak power from the wind farms is stored into CAES facilities. CAES advantages are high energy and power capacity, long lifetime, while the major disadvantages are low efficiencies, some adverse environmental impacts, and difficulty of the siting. CAES coupled with natural gas storage is based on the idea of coupling underground natural gas storage with electricity storage. The pressure difference between high-pressure gas storage in reservoirs deep underground and the gas injected into the conduits leads to the energy consumption for compression that can be released in the form of electricity during decompression. The natural gas and air liquefaction requires a large amount of energy that can be uses through two storage reservoirs for liquefied natural gas and liquid air, regenerative heat exchangers, compressors, and gas turbines for energy storage. Such systems are still under development.

11.3.3 Electrochemical energy storage

Chemical energy storage is classified into electrochemical and thermochemical energy storage. The electrochemical energy storage refers to conventional batteries, such as lead–acid (LA), nickel–metal hydride, and lithium-ion (Li-ion), and flow batteries (zinc/bromine (Zn/Br) and vanadium redox) and metal air batteries. Electrochemical energy storage is also achieved through fuel cells (FCs), most commonly hydrogen fuel cells but also includes direct-methanol, molten carbonate, and solid oxide fuel cells. Batteries and fuel cells are common energy storage devices used in power systems and in several other applications. In batteries and fuel cells, electrical energy is generated through chemical energy conversion, via redox reactions at the anode and cathode. Reactions at the anode take place at lower electrode potentials than at the cathode, so terms negative and positive electrode are used. The difference between batteries and fuel cells is the locations of energy storage and conversion. Batteries are closed systems, with the anode and cathode being the charge-transfer medium, taking an active role in the redox reaction as *active masses*, the energy storage and conversion occurring in the same compartment. Fuel cells are open systems where the anode and cathode are just charge-transfer media and the *active masses* undergoing the redox reaction are delivered from outside the cell, either from the environment, e.g., oxygen from air, or from a tank, e.g., hydrogen or hydrocarbons. Energy storage (in the tank) and energy conversion (in the fuel cell) are thus separated. However, to date, the only portable and rechargeable electrochemical ESSs (with the exception of the capacitors) are the batteries.

Battery or fuel cell basic configuration consists of two electrodes and an electrolyte, placed together in a container and connected to an external device (source or load). Usually a battery consists of two or more electric cells joined together, in series, parallel or series–parallel configurations in order to achieve the desire voltage, current and power. The cells, consisting of positive and negative electrodes joined by an electrolyte are converting chemical energy into electrical energy, through a chemical reaction between the electrodes and the electrolyte, generating DC electricity. In the case of secondary (rechargeable) batteries, the chemical reaction is reversed by reversing the current and the battery returned to a charged state. Battery types are of two forms, *disposable* or *primary batteries* and *rechargeable* or *secondary batteries*. A primary battery is a cell or group of cells, intended to be used until exhausted and then discarded, and are assembled in the charged state, while the discharge is the process during operation. A secondary battery is a cell or group of cells for the generation of electrical energy in which each cell, after being discharged, may be restored to its original charged condition by an electric current flowing in the opposite direction to the current when the cell was discharged. Other terms for this battery type are rechargeable battery or accumulator. Such batteries are assembled in the discharged state, being first charged before the use through secondary processes. Mature secondary battery chemistries are: (a) lead–acid (LA), (b) nickel–cadmium (Ni–Cd), (c) nickel–metal hydride (Ni–MH), (d) lithium-ion (Li-ion), (e) lithium-polymer/lithium metal (Li-polymer), (f) sodium–sulfur (Na–S), (g) sodium–nickel chloride, and (h) lithium-iron phosphate. The flow batteries are discussed later. With the increased demand growing from electric vehicles and portable consumer products, significant funds are spent by companies on the research of new battery technologies, such as zinc-based chemistries and silicon as a material for improving battery properties and performances. Higher energy density and life cycle, environmental friendliness, and safer operation are among the general design research targets for secondary batteries. Primary batteries are a reasonably mature technology, in terms of chemistry but still there is research to increase the energy density, reduce self-discharge rate, increase the battery life, or to improve the usable temperature range. To complement these developments, many semiconductor manufacturers continue to introduce new integrated circuits for battery power management. Other chemical energy storage options, such as the thermochemical energy storage, although promising, are still at an infant development stages and are not included in this entry.

Chemical energy storage is usually achieved through accumulators and batteries, characterized by a double function of storage and release of the electricity by alternating the charge–discharge phases. They transform chemical energy through electrochemical reactions into electrical energy and vice versa, without almost any harmful emissions or noise, while requiring little maintenance. There is a wide range of battery technologies in use, and their main assets are their energy densities (up to 2,000 Wh/kg for lithium-based types) and technological maturity. Chemical energy storage devices and electrochemical capacitors (ECs) are among the leading EES technologies today. Both technologies are based on electrochemistry, the fundamental difference being that batteries store energy in chemical reactants capable of generating charges, whereas electrochemical capacitors store energy directly as electric charges. Although the electrochemical capacitor is a promising technology for energy storage, especially considering its high power capabilities, the energy density is too low for large scale energy storage. The most common rechargeable battery technologies are lead–acid, sodium–sulfur, vanadium redox, and lithium-ion types. During discharge, electrochemical reactions at the two electrodes generate an electron flow through an external circuit. Vanadium redox batteries, having good prospects because they can be scale up to much larger storage capacities, are showing great potential for longer lifetimes and lower per-cycle costs than conventional batteries requiring refurbishment of electrodes. Li-ion batteries are also displaying high potential for large-scale energy storage. A battery consists of one or more electrochemical cells, connected in series, in parallel or series-parallel configuration to provide the desired voltage, current, and power. The anode (the electronegative electrode) from which the electrons are generated to do the external work. The cathode is the electropositive electrode to which positive ions migrate inside the cell and the electrons migrate through the battery external electrical circuit. The electrolyte allows the ions and electrons flow, from one electrode to another, is commonly a liquid solution containing a dissolved salt, and must be stable in the presence of both electrodes. The current collectors allow the transport of electrons to and from the electrodes, typically are made of metals and must not react with the electrode or electrolyte materials. The cell voltage is determined by the chemical reaction energy occurring inside the cell. The anode and cathode are, in practice, complex composites, containing, besides the active material, polymeric binders to hold together the powder structure and conductive diluents, such as carbon black to give the whole structure electric conductivity so that electrons can be transported to the active material. In addition, these components are combined to ensure sufficient porosity to allow the liquid electrolyte to penetrate the powder structure and permit the ions to reach the reacting sites. During the charging process, the electrochemical reactions are reversed via the application of an external voltage across the electrodes.

The battery technologies range from the mature, long-established lead–acid type through to various more recent and emerging systems and technologies. Newest technologies are attracting an increased interest for possible use in power systems, having achieved market acceptance and uptake in consumer electronics in the so-called 3Cs sector (cameras, cellphones, and computers). Batteries have the potential to span a broad range of energy storage applications, in part due to their portability, ease of use, large power storage capacity (100 W up to 20 MW), and easy to connected in series–parallel combinations to increase their power capacity for specific applications. Major battery advantages include standalone operation, no need to be connected to an electrical system, easily to expand and reconfigure, while the disadvantages are cost, limited life-cycle, and maintenance. These systems could be located in any place, buildings or industrial facilities, near the demand point, can be rapidly installed, less environmental impacts of other ESS technologies. Grid connected BESS uses an inverter to convert the battery DC voltage into AC grid-compatible voltage. These units present fast dynamics with response times near 20 ms and efficiencies ranging from 60% to 80%. The battery temperature change during charge and discharge cycles must be controlled because it affects its life expectancy. Depending on how the battery and cycle are, the BESS can require multiple charges and discharges per day. The battery cycle is normal while the discharge depth is small but if the discharge depth is high, the battery cycle duration could be degraded. The expected useful life of a Ni–Cd battery is

20,000 cycles, if the discharge depth is limited to 15%. Example of large-scale BESSs installed today are 10 MW (40 MWh) Chion system, California and 20 MW (5 MWh) Puerto Rico. Their main inconvenient; however, is their relatively low durability for large-amplitude cycling. They are often used in emergency back-up, renewable-energy system storage, etc. The minimum discharge period of the electrochemical accumulators rarely reaches less than 15 min. However, for some applications, power up to 100 W/kg, even a few kW/kg, can be reached within a few seconds or minutes. As opposed to capacitors, their voltage remains stable as a function of charge level. Nevertheless, between a high-power recharging operation at near-maximum charge level and nearing full discharge, the voltage can easily vary by a ratio of two.

11.3.4 Battery operation principles and battery types

Batteries convert the chemical energy contained in its active materials into electrical energy through an electrochemical oxidation–reduction reversible reaction. Battery fundamental principles and operation, regardless the battery type can be explained by using the so-called galvanic element or electrochemical cell. A galvanic cell consists of three main components: the *anode*, the *cathode*, and the *electrolyte*. In this case of a galvanic cell, the electrons needed for conduction are produced by a chemical reaction. From thermodynamics, we know that the work done by the electrochemical cell comes at a cost, which in turn implies that the chemical reactions taking place within the cell must lead to a decrease in free energy. In fact for a reversible process at constant temperature and pressure, the maximum work done by the system, W, is equal the free energy (Gibbs free energy) change $-\Delta G$. The work performed when transporting an electric charge e (in C) through a potential difference E (in V) is simply the product of eE. Here of interests is to express this work in a per mole basis. The total charge carried by one mole of positively charged ions of valence +1 is 96,487 C and this number, denoted by F, the Faraday's constant. Thus, the work produced by the electrochemical cell is:

$$W = -\Delta G = \xi F \cdot E \Rightarrow E_{\max} = \frac{-\Delta G^0}{\xi F} \tag{11.9}$$

where ξ is the valence of the ions produced in the chemical reaction. The electric potential difference across the cell electrodes, E, is the electromotive force, or EMF, of the galvanic cell. It is clear from (11.9) that in order to have higher work and potential, we need to find reactions with the highest driving force, the free energy change $-\Delta G$.

Example 11.5: Estimate the maximum output voltage that a Zn–Cu electrochemical cell can generate.

Solution: The Zn–Cu reaction taking place in this electrochemical cell is:

$$Zn + Cu^{2+} \rightarrow Cu + Zn^{2+}$$

By using (11.9) and from the data tables of the book appendixes (for Zn–Cu cell ξ is equal to 2, and ΔG^0 is equal to 216,160 kJ/kmole), the maximum voltage is:

$$E_{\max} = \frac{-\Delta G^0}{\xi F} = \frac{216,160\ \mathrm{kJ/kmole}}{2 \times 96,500.0} = 1.12\ \mathrm{V}$$

To understand the role of each of the electrochemical cell components and its operation, it is best to refer to specific examples. If, for example, a galvanic element, made of a zinc (Zn) electrode, a copper (Cu) electrode, and an electrolyte (e.g., $CuSO_4$), the two electrodes are electrically connected (Figure 11.5), the Zn^{2+} ions flow from Zn electrode, through electrolyte, while the electrons migrate from Zn, and eventually combine with Cu^{2+} ions, residing into the electrolyte and from copper atoms, increasing the Cu electrode volume. The direction of electrons is determined by the potential difference, between metal and electrolyte, $\Delta\varphi_{\text{metal–electrolyte}}$, the electron transport is in such way that the metal (Zn, here) separate the electrons and ions because it has a lower metal–electrolyte potential difference than that of the Cu electrode against electrolyte. The metal (and electrode) with higher potential is serving as positive electrode (and cell terminal) of a galvanic element. The metals are usually arranged in voltaic sequences, in a way that all metals (e.g., Fe, Cd, Ni, Pb, or Cu) in the right side of a certain metal (e.g., Zn) to for a positive pole in a combination of electrodes, chosen based of the metal–electrolyte potential difference values of Table 11.2. For example, Zn is a

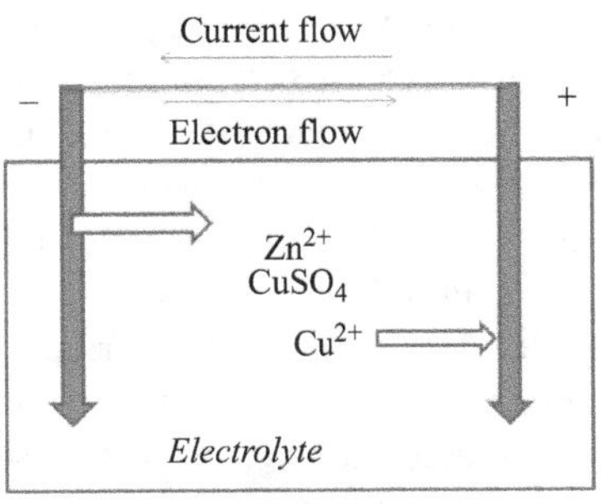

Figure 11.5 Galvanic element diagram (consisting of Zn and Cu electrodes, and $CuSO_4$ electrolyte). Current flows are also shown here

Table 11.2 Metal voltaic series

Metal electrode	Li	K	Na	Mg	Zn	Fe	Cd
$\Delta\varphi_{\text{metal–electrolyte}}$ (V)	−3.02	−2.92	−2.71	−2.35	−0.762	−0.44	−0.402
Metal electrode	Ni	Pb	H_2	Cu	Ag	Hg	Au
$\Delta\varphi_{\text{metal–electrolyte}}$ (V)	−0.25	−0.126	0.0	+0.345	+0.80	+0.86	+1.50

negative pole with respect of Fe, Cd, Ag, or Au electrodes, while Li electrode form a negative pole with respect to K, Na, Mg, Zn, or Fe electrodes. The potential difference (external voltage) between the cell (galvanic element) terminals is the voltage difference, for this galvanic element, existing between Zn electrode and electrolyte ($CuSO_4$) and between Cu electrode and electrolyte. Notice that is not possible to directly measure the potential difference between the metal electrode and electrolyte, only the metal–metal potential difference can be measured.

Example 11.6: Determine the potential differences between Li–Au, Li–Ag, and Pb–H_2, using the values in Table 11.2.

Solution: The required potential differences are:

$$\Delta\phi_{Li-Au} = -3.02 - 1.50 = -4.52 \text{ V}$$

$$\Delta\phi_{Li-Au} = -3.02 - 0.86 = -3.88 \text{ V}$$

and

$$\Delta\phi_{Ni-Cu} = -0.25 - 0.345 = -0.595 \text{ V}$$

Lead–acid batteries (Pb–acid/LA) are the most common electrical energy storage device used at the present, especially in transportation, renewable energy, and standalone hybrid power systems. Its success is due to its maturity (research has been ongoing for about 150 years), low cost, long lifespan, fast response, and low self-discharge rate. They are used for both short-term and long-term applications. A lead–acid battery works somewhat differently than series galvanic elements, employing the positive and negative charge polarizations. Lead and lead-dioxide are good electrical conductors, the electrolyte contains aqueous ions (H^+ and SO_4^{-2}), and the conduction electrolyte mechanism is via migration of ions through diffusion or drift. Two lead plates of a cell are immersed in a dilute sulfuric acid solution, each covered with a $PbSO_4$ layer, basically consisting of spongy lead anode and lead acid cathode, with lead as the current collector. The lead–acid batteries for electric vehicles are using a gel rather than a liquid electrolyte, in order to withstand deep cycle that is required here, for that reason being more expensive than regular ones. The sulfuric acid combines with the lead and lead oxide to produce lead sulfate and water, and the electrical energy is released during the process. The overall lead–acid battery reaction is:

$$Pb + PbO_2 + 2H_2SO_4 = 2PbSO_4 + 2H_2O \tag{11.10}$$

This reaction proceeds from left to right, during battery discharge, and in opposite direction during the battery charging process. During the discharging process, the electrons flow from the cathode to anode (current is flowing in opposite direction), until the chemical process is reversed and $PbSO_4$ forms on the anode and cathode, while the anode is covered with PbO_2 which gets reduced as the discharging process progresses. The voltage across of the two lead plates is $e_0 = 2.02$ V. If six sets of lead plates are connected in series, then the total terminal voltage is $E^0 = 6 \cdot e_0 = 12.12$ V. The efficiency of a lead–acid battery is from 80% to 90% of the charged energy. During the discharging the electrolyte gradually is loses the sulfuric acid and becomes more dilute, while during charging process, the electrolyte revert to lead and lead dioxide, and also recovering the its sulfuric acid and concentration increases.

There are two types of lead–acid battery: (a) flooded lead-acid (FLA) and (b) valve-regulated lead–acid (VRLA). From application point of view, the lead–acid batteries split into stationary, traction, and car batteries. Stationary batteries ensure uninterruptible electric energy supply, in case of the power system failure, undergoing usually only few cycles, so have about 20 years life-time. Traction batteries, used for power supply of electric vehicles, electro-mobiles, or industrial tracks, etc. are working in deep cycle charging–discharging regime with the life-span of about 5 years (about 1,000 charge–discharge cycles). Automotive (car) batteries, used to crank car engine, being able to supply short and intense discharge current, and also to support car electrical devices, when the engine is not running. FLA batteries, described earlier are made up of two electrodes (lead plates), immersed in a mixture of water (65%), and sulfuric acid (35%). VRLA batteries have the same operating principle, with the difference that they are sealed with pressure-regulating valves, to prevent the air entering the cells and hydrogen venting. VRLA batteries are operating with the help of an internal oxygen cycle, emitted in the later charging stages and during overcharging on the positive electrode, traveling through a gas separator to the cathode where is reduced to water. This oxygen cycle is making the cathode potential less negative, so the rate of hydrogen evolution decreases. Part of the generated electricity is used by the internal recombination oxygen cycle and is converted to heat. VRLA batteries have lower maintenance costs, less weight and space but these advantages are coupled with higher initial costs and shorter lifetime. During discharge, lead sulfate is the product on both electrodes. Sulfate crystals become larger and difficult to break up during recharging, if the battery is over-discharged or kept discharged for a prolonged time period Hydrogen is produced during charging, leading to water losses if overcharged. Their popularity is result of their wide availability, robustness, and reasonably low cost. The disadvantages of the lead–acid batteries are their weight, low specific energy and specific power, short cycle life (100–1,000 cycles), high maintenance requirements, hazards associated with lead and sulfuric acid during production and disposal, and capacity drop at low temperatures. FLA batteries have two primary applications: starting and ignition, short bursts of strong power, e.g., car engine batteries, and deep cycle, low steady power over long-time applications. VRLA batteries are very popular for backup power, standby power supplies in telecommunication centers, and for uninterruptible power supply. Among the major reasons for failure of the lead–acid batteries include positive plate expansion, positive mass fractioning, water loss, acid stratification, incomplete charging causing active mass sulfating, positive grid corrosion, and negative active mass sulfating.

Nickel–cadmium (Ni–Cd) and nickel–metal hydride (Ni–MH) batteries use nickel oxyhydroxide for the cathode and metallic cadmium as the anode with a potassium hydroxide as an electrolyte. Other possible systems are Ni–Fe, Ni–Zn, and Ni–H_2. These types of batteries were very popular between 1970 and 1990 but have been largely superseded by Ni–MH due to their inferior cycle life, memory effect, energy density and toxicity of the cadmium in Ni–Cd, compared to Ni–MH batteries. Ni–MH also have the advantage of improved high rate capability (due to the endothermic nature of the discharge reaction), and high tolerance to over-discharge. Ni–MH use nickel oxyhydroxide for the cathode, with a potassium hydroxide electrolyte and a hydrogen-absorbing alloy, alloys of lanthanum and rare earths serving as a solid source of reduced hydrogen which can be oxidized to form protons. Ni–Cd and Ni–MH batteries are mature technologies, rugged with higher energy density, low maintenance requirements and higher cycle life than lead–acid batteries but more expensive than lead–acid batteries, with limitations on the long-term cost reduction potential due to material costs. In addition, Ni–Cd batteries contain toxic components (cadmium) making them environmentally challenging. With a higher energy density, longer cycle life and low maintenance requirements, Ni–Cd batteries have a potential edge over lead–acid batteries. The drawbacks of this technology include the toxic-heaviness of cadmium and higher self-discharge rates than lead–acid batteries, while Ni–Cd batteries may cost up to ten times more than a lead–acid battery.

Ni–Cd batteries are produced in a wide commercial variety from sealed free-maintenance cells (capacities of 10 mAh to 20 Ah) to vented standby power units (capacities of 1,000 Ah or higher), longer cycle life, overcharge capacity, high charge–discharge rates, almost constant discharge voltage, and they can operate at low temperature. Depending on the construction, Ni–Cd batteries have energy densities in the range 40–60 Wh/kg, and cycle life above several hundred for sealed cells to several thousand for vented cells. There are two main construction types of Ni–Cd batteries. First construction type are using pocket plate electrodes, in vented cells, where the active material is found in pockets of finely perforated nickel plated sheet steel, positive, and negative plates are separated by plastic pins or ladders and plate edge insulators. Second type are using sintered, bonded or fiber plate electrodes, for both vented and sealed cells. Active material is distributed within the pores. The electrolyte is an aqueous solution of KOH, with 20%–28% weight concentration and density of 1.18–1.27 g/cm^3 at 25 °C, with a 1%–2% of LiOH usually added to minimize the coagulation of the NiOOH electrode during the charge–discharge cycling. An aqueous NaOH electrolyte may be used in cells operating at higher temperatures. The overall Ni–Cd cell discharge reaction is:

$$2NiOOH + Cd + 2H_2O \rightarrow 2Ni(OH)_2 + Cd(OH)_2 \quad (11.11)$$

The Ni–Cd cell voltage is $E^0 = +1.30$ V. The Ni–Cd batteries are designed as positive limited oxygen cycle use, amount of water decreases during discharge. The oxygen evolved at anode during charging diffuses to the cathode and reacts with cadmium, forming $Cd(OH)_2$, and carbon dioxide from air and react with KOH in electrolyte to form K_2CO_3, and $CdCO_3$ at cathode, increasing internal resistance and reducing capacity of the Ni–Cd batteries. Sealed Ni–MH cell has as active negative material hydrogen absorbed in a metal alloy that increases its energy density and making it more environmentally friendly compared with Ni–Cd cells. However, these cells have higher self-discharge rates and are less tolerant to overcharge than Ni–Cd cells. The hydrogen absorption alloys are made up of two metals (first metal absorbs hydrogen exothermically and the second one endothermically), and also the second metal is serving as catalyst for the absorption of the hydrogen into the alloy lattice. A hydrolytic polypropylene separator is used in Ni–MH cells. The electrolyte, used in Ni–MH cells is potassium hydroxide, and the cell voltages range from 1.32 V to 1.35 V. Their energy density is 80 Wh/kg (25% higher than that of Ni–Cd cells), life cycle over 1000 cycles, and a high self-discharge rate of 4%–5% per day. Ni–MH batteries are used in electric vehicles, consumer electronics, phones, medical instruments, and other high-rate and longer life-cycle applications. The overall Ni–MH cell reaction during discharge cycle is:

$$NiOOH + MH \rightarrow Ni(OH)_2 + M \tag{11.12}$$

Ni–Cd and Ni–MH cells suffer, both from memory effects, a temporary capacity reduction but reversible caused by the shallow charge–discharge cycling. After a shallow cycling, there is a voltage step during discharge, means that the cell remembers the depth of the shallow cycling. The voltage reduction size depends on the number of preceding shallow cycles and the discharge current. However, the cell capacity is not affected, if the cell is fully discharged and then charged, then a deep discharge is showing a normal discharge curve.

Lithium-ion (Li-ion) batteries are largely cobalt or phosphate based. In both embodiments lithium ions flow between the anode and cathode to generated current flow. Li-ion batteries have a high energy to weight ratio (density), no memory effect, light weight, high reduction potential, low internal resistance, and low self-discharge rates. Prices may be high and increasing penetration may push prices higher as limited lithium resources are depleted. Lithium-ion battery technology has progressed from developing mental and special-purpose status to a global mass-market product in less than 20 years. The technology is especially attractive for low-power and portable applications because Li-ion batteries offer high-power densities, typically 150–200 Wh/kg and generally acceptable cycle life, about 500 cycles, very low self-discharge rate (less than 10% per month) and high voltage, about 3.6 V. The operation of the lithium ion cells involves the reversible transfer of lithium ions, between electrodes, during charge and discharge. During charging, lithium ions move out (de-intercalate) from the lithium metal oxide cathode and intercalate into the graphite-based anode, with the reverse happening during the discharge reaction. The non-aqueous ionic conducting electrolyte takes no part in the reaction except for conducting the lithium ions during the charge and discharge cycles. Notwithstanding their significant advantages, lithium ion systems must be maintained within well-defined operating limits to avoid permanent cell damage or failure. The technology also possesses no natural ability to equalize the charge amount in its component cells. This, and the closely defined operational envelope of lithium-ion batteries, essentially dictates the use of relatively sophisticated management systems. Most of the commercial Li-ion cells have anodes of cobalt oxide, or manganese oxide. Negative electrode is made of carbon, in graphite form (light weight a low price) or an amorphous material with a high surface-area. Electrolyte is made of an organic liquid (ether) and a dissolved salt, and the positive and negative active mass is applied to both sides of thin metal foils, aluminum on the positive side and copper on the negative. A micro-porous polymer is used as separator between positive and negative electrode. The positive electrode reaction in a Li-ion cell is:

$$\mathrm{LiCoO_2} \rightarrow \mathrm{Li_{1-\mathit{x}}CoO_2} + x\mathrm{Li^+} + xe^- \tag{11.13a}$$

And the negative electrode reaction is:

$$xLi^+ + xe^- + C_6 \rightarrow Li_xC_6 \tag{11.13b}$$

Here the index, x moves on the negative electrode from 0 to 1, and on the positive electrode from 0 to 0.45. Li-ion batteries are manufactured in coin format, cylindrical, and prismatic shapes. On the other hand, the application of the technology to larger-scale systems is relatively limited to date, although various developments are in hand in relation to the automotive, power utility, submersible, and marine sectors. Lithium-polymer (Li-pol) batteries are employing the polymer property (containing a hetero-atom, i.e., oxygen or sulfur) to dissolve lithium salts in high concentrations, leading at higher temperatures to a good electric conductivity (0.1 S/m at 100 °C), allowing to such polymers to be used as electrolyte for lithium batteries. Polymer electrolyte, not being flammable is safer than the liquid electrolyte.

11.3.5 Battery fundamentals, parameters, and electric circuit models

Battery or cell capacity (*CAP*(*t*)) means an integral of current, *i*(*t*) over a defined time period as:

$$CAP(t) = \int_0^t i(t)dt \tag{11.14}$$

The above relationship applies to either battery charge or discharge, meaning the capacity added or capacity removed from a battery or cell, respectively. The capacity of a battery or cell is measured in milliampere-hours (mAh) or ampere-hours (Ah). This basic definition is simple and straight; however, several different forms of capacity relationship are used in the battery industry. The distinctions between them reflect differences in the conditions under which the battery capacity is measured. Standard capacity measures the total capacity that a relatively new but stabilized production cell or battery can store and discharge under a defined standard set of application condition, assuming that the cell or battery is fully formed, that it is charged at standard temperature at the specification rate, and that it is discharged at the same standard temperature at a specified standard discharge rate to a standard end-of-discharge voltage (EODV). The standard EODV is subject to variation depending on discharge rate. When the application conditions differ from standard ones, the cell or battery capacity changes, so the term actual capacity includes all nonstandard conditions that alter the amount of capacity the fully charged new cell or battery is capable of delivering when fully discharged to a standard EODV. Examples of such situations might include subjecting the cell or battery to a cold discharge or a high-rate discharge. That portion of actual capacity that is delivered by the fully charged new cell or battery to some nonstandard EODV is called available capacity. Thus, if the standard EODV is 1.6 V/cell, the available capacity to an EODV of 1.8 V/cell would be less than the actual capacity. Rated capacity is defined as the minimum expected capacity when a new, fully formed, cell is measured under standard conditions. This is the basis for C rate (defined later) and depends on the standard conditions used which may vary depending on the manufacturers and the battery types. If a battery is stored for a period of time following a full charge, some of its charge will dissipate. The capacity which remains that can be discharged is called retained capacity. In most of the practical engineering applications, the battery capacity, C, for a constant discharge rate of I (in A) as:

$$C = I \times t \tag{11.15}$$

It is clear from the earlier equation that the capacity of a battery is reduced, if the current is drawn more quickly. Drawing 1 A for 10 h does not take the same charge from a battery as running it at 10 A for 1 h. Notice that, the relationship between battery capacity and discharge current is not linear, and less energy is recovered at faster discharge rates. This phenomenon is particularly important for electric vehicles, as in this application the currents are generally higher, with the result that the capacity might be less than is expected. It is important to be able to predict the effect of current on capacity, both when designing electric vehicles, or when are designing and making instruments to measure the charge left in a battery, the so-called battery fuel gauges. The best way to do this is by using the Peukert model of battery behavior. Although this model is not very accurate at low currents, as for higher currents where it models battery behavior well enough. The starting point of this model is that there is a capacity, called the Peukert capacity, which is constant, and is given by the equation:

$$C_{Pkt} = I^k \cdot t \tag{11.16}$$

where C_{Pkt} is the amp-hour capacity at a 1 A discharge rate, I is the discharge current in Amperes, t is the discharge time, in hours, and k is the Peukert coefficient, typically with values from 1.1 to 1.3.

Example 11.7: A lead–acid battery has a nominal capacity of 50 Ah at a rate of 5 h, and a Peukert coefficient of 1.2. Estimate the battery Peukert capacity.

Solution: This battery of a capacity of 50 Ah if discharged at a current of:

$$I = \frac{50\text{ Ah}}{5\text{ h}} = 10\text{ A}$$

If the Peukert coefficient is 1.2, then the Peukert capacity is:

$$C_{Pkt} = 10^{1.2} \cdot 5 = 79.3\text{ Ah}$$

The capacity of a battery, sometimes referred to as C_{load} or simply C, is an inaccurate measure of how much charge a battery can deliver to a load. It is an imprecise value because it depends on temperature, age of the cells, state of the charge, and on the rate of discharge. It has been observed that two identical, fully charged batteries, under the same circumstances, will deliver different charges to a load depending on the current drawn by the load. In other words C is not constant and the value of C is for a fully charged battery is not an adequate description of the characteristic of the battery unless it is accompanied by an additional information, *rated time of discharge* with the assumption that the discharge occurs under a constant current regime. Usually, lead–acid batteries are selected as energy storage for the building DC microgrids, because of relatively low cost and mature technology. The ESS are usually operated by current closed-loop control, while the storage power is controlled by supervision unit which calculates the corresponding power reference. The storage state of charge (SoC) must be respected to its upper (maximum) and lower (minimum) SoC limits, SoC_{max} and SoC_{min} respectively, to protect the battery from over-charging and over-discharging, as given in (11.17a), below. SoC is then calculated with (11.17b), where SoC_0 is the initial SoC at t_0 (initial time), CREF is the storage nominal capacity (Ah) and V_S is the storage voltage.

$$SoC_{\min} \le SoC(t) \le SoC_{\max} \tag{11.17a}$$

And

$$SoC(t) = SoC_0 + \frac{1}{3,600 \times V_S \times CREF} \int_{t_0}^{t} (P_{SC} - P_{SD}(t))dt \tag{11.17b}$$

Accurate battery models are required for the simulation, analysis, and design of energy consumption of electric vehicles, portable devices, or renewable energy and power system applications. The major challenges in modeling a battery are the nonlinear characteristics of the equivalent circuit parameters, which depend on the battery SoC, and are requiring complete and complex experimental and/or numerical procedures. The battery itself has internal parameters, which need to be taken care of for modeling purposes, such as internal voltage and resistance. All electric cells in a battery have nominal voltages which gives the approximate voltage when the cell is delivering electrical power. The cells can be connected in series to give the overall voltage required by a specific application.

Example 11.8: A battery bank consists of several cells, connected in series. Assuming that the cell internal resistance is 0.012 Ω and the cell electrochemical voltage is 1.25 V, if the battery bank needs to delivery 12.5 A at 120 V to a load determine the number of cells.

Solution: With cells in series, the number of cell can be estimated as:

$$N(cells) = \frac{V_L}{V_{OC}(cell)} = \frac{120}{1.25 - 12.5 \times 0.012} = 109.09$$

We are choosing 110 cells, round off to upper integer, meaning a higher terminal voltage than 120 V. However, the terminal voltage is decreasing, over the battery bank lifetime.

There are three basic battery models, most used in engineering applications: the ideal, linear, and Thevenin models. The battery ideal model, a very simple one is made up only by a voltage source (Figure 11.6(a)), and ignores the internal parameters and, hence. The battery linear model (Figure 11.6(b)), a widely used battery model in applications, and consists of an ideal battery with open-circuit voltage, V_0, and an equivalent series resistance, R_S, while V_{Out} represents the battery terminal (output) voltage. This terminal voltage is obtained from the open-circuit tests as well as from load tests conducted on a fully charged battery. Although this model is

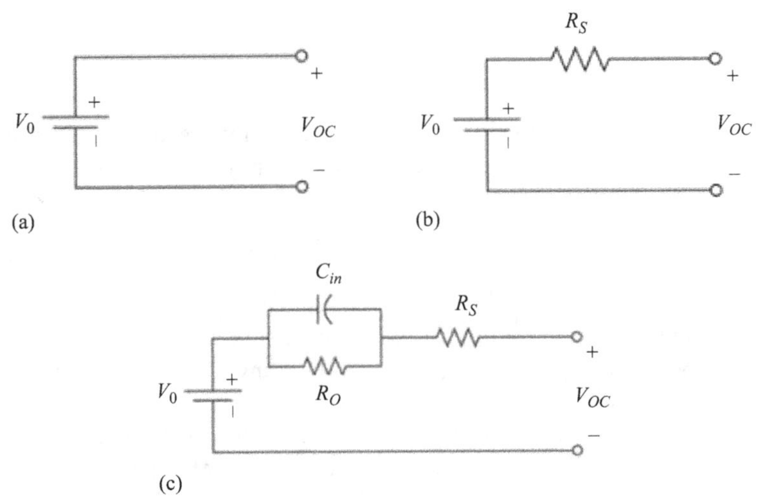

Figure 11.6 Battery electric diagrams, steady-state conditions, (a) battery ideal model, and (b) battery linear model, (c) Thevenin battery model

quite widely used, it still does not consider the varying characteristics of the internal impedance of the battery with the varying SoC and electrolyte concentration. The Thevenin model (Figure 11.6(c)), a more complete, accurate and complex model, consists in addition to the open-circuit voltage and internal resistance, of an internal capacitance, C_{in}, and the overvoltage resistance, R_O. The capacitor C_{in} accounts for the capacitance of the parallel plates and resistor R_O accounts the nonlinear resistance offered by the plate to the electrolyte. All the elements in this model are assumed to be constants. However, in reality, they depend on the battery conditions. Thus, this model is also not the most accurate, but is by far the most widely used. In this view, a new approach to evaluate batteries is introduced. Any battery, regardless the type in the first approximation (for steady-state operation) works as constant voltage's source with an internal (source/battery) resistance, by considering battery linear model as shown in Figure 11.6(b). These parameters, the internal voltage (V_{OC}) and resistance (R_S) are dependent of the discharged energy (Ah) as:

$$V_{OC} = V_0 - k_1 \times DoD \tag{11.18a}$$

And

$$R_S = R_0 + k_2 \times DoD \tag{11.18b}$$

Here, V_{OC}, known also as the open-circuit or electrochemical voltage decreases linear with depth of discharge (DoD), while the internal resistance, R_S increases linear with DoD. V_0 and R_0 are the values of the electrochemical (internal) voltage and resistance, respectively, when the battery is fully charged, DoD is 0, and when full discharged DoD is 1.0. The constant, k_1 and k_2, are determined form the battery test data, through curve fitting or other numerical procedures. DoD is defined from the battery SoC, as:

$$DoD = \frac{\text{Ah drained form battery}}{\text{Battery rated Ah capacity}} = 1 - SoC \tag{11.19}$$

While, SoC (as defined in (11.17)) is computed, for practical application by a much simpler relationship, that can be estimated from the battery monitoring (test data) as:

$$SoC = 1 - DoD = \frac{\text{Ah remaining in the battery}}{\text{Rated Ah battery capacity}} \tag{11.20}$$

Notice, the battery terminal voltage is lower and the internal resistance is higher in a partially battery discharge state (i.e., any time when $DoD > 0$). All electric battery cells have nominal voltages, which gives the approximate voltage when the battery cell is delivering electrical power. Cells can be connected in series to give the required voltage. The terminal voltage, V_L of a partially discharged battery, with notation of Figure 11.7 is expressed as:

$$V_L = V_0 - I \cdot R_S = V_0 - k_1 \times DoD - I \cdot R_S \tag{11.21}$$

The load delivered power is I^2R_L, the battery internal loss is I^2R_S, dissipated as heat inside the battery. In consequence, as the battery discharge internal resistance, R_L increases, and more heat are generated.

Example 11.9: A 12-V lead–acid car battery has a measured voltage of 11.2 V when delivers 40 A to a load. What are the load and the internal battery resistances? Determine its instantaneous power and the rate of sulfuric acid consumption.

Solution: For the battery equivalent circuit of Figure 11.7, the voltage across the load is:

$$V_L = V_0 - I \cdot R_S$$

And the internal battery and load resistances are:

$$R_S = \frac{12.0 - 11.2}{45} = 0.02\ \Omega$$
$$R_L = \frac{V_0}{I} - R_S = \frac{12}{40} - 0.02 = 0.3 - 0.02 = 0.28\ \Omega$$

The power delivered by the battery is:

$$P = V_L \times I = 11.2 \times 40 = 448\ \text{W}$$

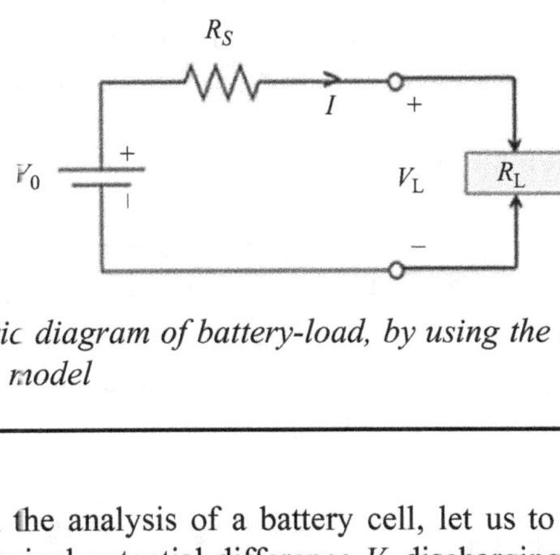

Figure 11.7 Electric diagram of battery-load, by using the steady-state battery linear model

---

In order to perform the analysis of a battery cell, let us to consider a reversible battery cell with terminal potential difference V, discharging a current I through a variable load resistance. The electron flow through the load produces the instantaneous electrical power, $I{\cdot}V$. At low current levels, the cell voltage is close to the cell EMF, which is the open-circuit voltage and the external work done and power output are small. To estimate the sulfuric acid (reactant) consumption rate, we suppose that the battery consumes N_{Rct} moles of reactant per second, and the released electrons in reaction flow through electrodes and external circuit is proportional to the reaction rate, jN_{Rct}, the rate of flow of electrons from the cell (in mole per second), while j ($j = 2$ for lead–acid batteries) is the number of moles of electrons released per mole of reactant. Suppose that chemical reaction in the battery consumes reactant at an electrode at a rate of N_C moles per second. Electrons released in the reaction flow through the electrode to the external load at a rate proportional to the rate of reaction, $j_{rct}{\cdot}N_C$, where *jrct* is the number of moles of electrons released per mole of reactants. For example, for a lead–acid battery, two moles of electrons are freed to flow through the external load for each mole of lead reacted, *jrct* is equal to 2. Thus, $j_{rct}.N_C$ is the cell electron flow rate, in moles of electrons per second. There are 6.023×10^{23} electrons per gram-mole of electrons, and each has a charge of 1.602×10^{-19} C, and their product, F is the Faraday constant and is equal to: 96,488 C/g-mol. The electric current from a cell may then be related to the rate of reaction in the cell as:

$$I = jN_{Rct}F(\mathrm{A}) \tag{11.22}$$

F is the Faraday constant equal to the Avogadro number times the electron charge:

$$F = 6.023 \times 10^{23} \cdot 1.609 \times 10^{-19} = 96{,}488\mathrm{C/g-mole}$$

The instantaneous power delivered by the cell is then:

$$P = jN_{Rct} \cdot F \cdot V \tag{11.23}$$

The battery cell electrode and electrolyte materials and cell design determine the maximum cell voltage. Equations (11.22) and (11.23) are showing that the nature and rate of chemical reaction that are controlling the cell current and maximum power output of a cell. Moreover, it is clear that the store of consumable battery reactants sets a limit on battery capacity.

Example 11.10: For the battery of previous example, 40 A current estimate the acid consumption rate.

Solution: For the battery in this example, I equal to 40 A, the sulfuric acid consumption rate, calculated form (11.22) is:

$$N_{Rct} = \frac{I}{jF} = \frac{40}{2 \times 96488} = 2.0728 \times 10^{-4}\ \mathrm{g-mole/s}$$

Because the sulfuric acid is consumed at the same rate at both anode and cathode, the above value is multiplied by 2, and it can be expressed in kg/h, by using the sulfuric acid molecular weight

$$N_{Rct} = 2 \times 2.0728 \times 10^{-4} \cdot 10^{-3} \cdot 98 \times 3{,}600 = 0.1463\ \mathrm{kg/h}$$

11.3.5.1 Summary of battery parameters

1. ***Cell and battery voltage***: All battery cells have a nominal voltage which gives the approximate voltage when delivering power. Cells are connected in series to give the overall battery voltage required. Figure 11.6 is showing one of the equivalent battery electric circuits.
2. ***Battery charge capacity*** (***Ahr***): The most critical parameter is the electric charge that a battery can supply, and is expressed in Ampere-hour (Ahr). For example, if the capacity of a battery is 100 Ahr, the battery can supply 1 A for 100 h. The storage capacity, a measure of the total electric charge of the battery, and is an indication of the capability of a battery to deliver a particular current value for a given duration. Capacity is also sometimes indicated as energy storage capacity, in watt-hours (Wh).
3. ***Cold-cranking-amperage*** (***CCA***), for a battery composed of nominal 2 V cells is the highest current (A) that the battery can deliver for 30 s at a temperature of 0°F and still maintain a voltage of 1.2 V per cell. The CCA is commonly used in automotive applications, where the higher engine starting resistance, in cold winter conditions is compounded by reduced battery performance. At 0°F, the cranking resistance of a car engine may be increased more than a factor of two over its starting power requirement at 80°F, while the battery output at the lower temperature is reduced to 40% of its normal output. The battery output reduction is due to the decrease of chemical reaction rates with temperature decreases.
4. ***The storage capacity,*** a measure of the total electric charge of the battery, is usually quoted in ampere-hours rather than in coulombs (1 Ahr = 3,600 C), being an indication of the capability of a battery to deliver a particular current value for a given duration. Thus a battery that can discharge at a rate of 5 A for 20 h has a capacity of 100 Ahr. Capacity is also sometimes indicated as energy storage capacity, in watt-hours. The energy stored in a battery depends on the *battery voltage*, and the *charge* stored. The SI unit for energy is Joule (J), however this is an inconveniently small unit, and in practical application Watt-hour (Wh) is used instead. The energy is expressed in Wh as:

$$\text{Energy (Wh)} = \text{V} \times \text{A} \cdot \text{hr} \tag{11.24}$$

5. ***Specific energy*** is the amount of electrical energy stored in a battery per unit battery mass (kg), and in practical applications is expressed in Wh/kg.
6. ***Energy density*** is the amount of electrical energy per unit of battery volume, expressed in practical applications and engineering in Wh/m^3.
7. ***Charge (Ahr) efficiency*** of actual batteries is less than 100%, and depends on the battery type, temperature, rate of charge, and varies with the battery SOC. An ideal battery will return the entire stored charge to a load, so its charge efficiency is 100%.

8. ***Energy efficiency*** is another important parameter, defined as the ratio of the electric energy supplied by the battery to the electric energy required to return the battery at its state before discharge. In other words, is the ratio of the energy delivered by a fully charged battery to the recharging energy required to restore it to its original SoC.
9. ***Self-discharge rate*** refers to the fact that most of the battery types are discharging, when left unused, the self-discharge, an important battery characteristic, meaning that some batteries must be recharged after longer periods. The self-discharge rate varies with battery type, temperature, and storage conditions.
10. ***Battery temperature, heating, and cooling***, either most of the battery types are running at ambient temperatures, there are battery types that need heating at start and then cooling when in use. For some battery types, performances vary with temperature. The temperature effects, cooling, and heating are important parameters that designers need to take in consideration.
11. Most of the rechargeable batteries have limited ***number of deep cycles*** of 20% of the battery charge, in the range of hundreds or thousands cycles, which is also determine the ***battery life***. The number of deep cycles depends on the battery type, design details, and the ways and conditions that a battery is used. This is a very important parameter in battery specifications, reflecting the battery lifetime, and its cost.

11.3.6 Flow batteries and special battery types

A flow battery is a type of rechargeable secondary battery in which energy is stored chemically in liquid electrolytes. Secondary batteries are using the electrodes as an interface for collecting or depositing electrons and as a storage site for the products or reactants associated with the battery's reactions. Consequently, their energy and power densities are set by the size and shape of the electrodes. Flow batteries, on the other hand, store and release electrical energy by means of reversible electrochemical reactions in two liquid electrolytes. Their electrolytes contain dissolved electro-active species that flow through a power cell that converts chemical energy to electricity. Simply stated, flow batteries are fuel cells that can be recharged. From a practical view, storage of the reactants is very important. The storage of gaseous fuels in fuel cells requires large, high pressure tanks or cryogenic storage that is prone to thermal self-discharge. There are some advantages of the flow battery over a conventional secondary battery, such as: the system capacity is scalable by simply increasing the amount of solution, leading to cheaper installation costs as the systems get larger, the battery can be fully discharged with no ill effects and has little loss of electrolyte over time.

Flow battery flow cells (or redox flow cells), convert electrical energy into chemical potential energy through a reversible electrochemical reaction between two electrolyte solutions. In contrast to conventional batteries, they are storing the energy in the electrolyte solutions, and the power and energy ratings of redox flow cells are independent variables. Their power rating is determined by the active area

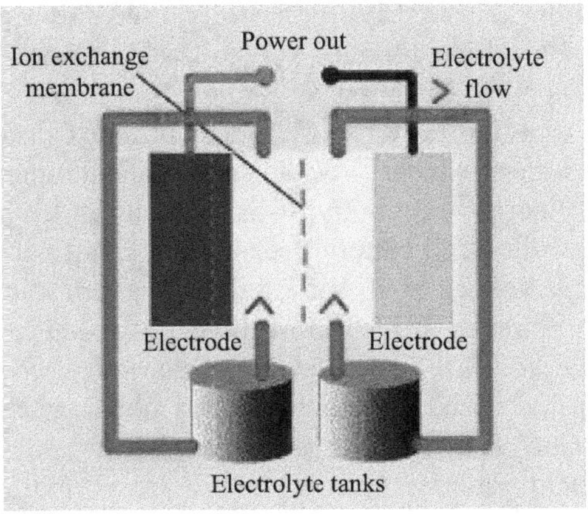

Figure 11.8 Flow battery schematic diagram

of the cell stack assembly and their storage capacity by the electrolyte quantity. Figure 11.8 depicts a flow cell energy storage system, having two compartments, one for each electrolyte, physically separated by an ion exchange membrane, basically two flow loops. Electrolytes flow into and out of the cell through separate manifolds and undergo chemical reaction inside the cell, with ion (proton) exchange through the cell membrane and electron exchange through the external electric circuit. The electrolytes' chemical energy is turned into electrical energy and vice versa. All flow batteries work in the same way but are varying in the chemistry of electrolytes. Because the electrolytes are stored separately and in large containers (with a low surface area to volume ratio), flow batteries show promise to have some of the lowest self-discharge rates of any energy storage technology available. The development activities have been centered on four principal electrochemistry combinations for flow batteries: vanadium–vanadium, zinc–bromine, polysulfide–bromide and zinc–cerium, although others are under development, too. Production of flow cell-based ESS, proceeds at a slow pace, via the activities of a relatively small number of developers and suppliers. There are two costs associated with flow batteries the power and the energy costs, as they are independent of each other. Their major advantages are: high power capacity, long life, while the main disadvantages are: low energy density, and efficiency. Flow batteries can be used for any energy storage applications, including load leveling, peak shaving, and renewable energy integration. Under normal conditions, flow batteries can charge or discharge very fast, making them useful for frequency response and voltage control. However, poor energy densities and specific energies remand these battery types to utility-scale power shaping and smoothing, although they might be adaptable for distributed-generation use. Nevertheless, the Tennessee Valley Authority released a finding of no significant impact for a proposed 430 GJ facility and deemed it safe. There are three types of flow batteries that are closing in on commercialization: *vanadium redox, polysulfide bromide*, and *zinc bromide*.

However, installations to date have principally used the vanadium redox and zinc–bromine. The polysulfide–bromide system was developed for grid-connected, utility-scale storage applications at power ratings from 5 MW upwards.

Zinc–bromine (Zn–Br) batteries are a type of redox flow battery, operating by using a pump system which circulates reactants through the battery. One manufacturer, ZBB Energy, builds Zn–Br batteries in 50 kWh modules made of three parallel connected 60-cell battery stacks. The modules are rated to discharge at 150 A at an average voltage of 96 V for 4 h. The battery stack design allows for individual stacks to be replaced instead of the entire module. Zinc–bromine battery life time is rated at 2,500 cycles. The electrochemical charging and discharging reaction is characterized as reversible and nondestructive, meaning it is capable of 100% *DoD*. The Zn–Br battery uses less toxic electrolytes when compared to lead–acid batteries making them a more environmentally friendly choice. Zinc–bromine flow batteries lack technological maturity, so there are only a few examples of real installations. The overall chemical reaction during discharge:

$$Zn + Br_2 \rightarrow ZnBr_2, \text{and with the theoertical voltage} : E_0 = 1.85\ \text{V} \quad (11.25)$$

During discharge product of reaction, the soluble zinc bromide is stored, along with the rest of the electrolyte, in the two loops and external tanks. During charge, bromine is liberated on the positive electrode and zinc is deposited on the negative electrode. Bromine is then mixed with an organic agent to form a dense, oily liquid poly-bromide complex system. It is produced as droplets and these are separated from the aqueous electrolyte on the bottom of the tank in positive electrode loop. During discharge, bromine in positive electrode loop is again returned to the cell electrolyte in the form of a dispersion of the poly-bromide oil. Vanadium redox batteries have the advantage of being able to store electrolyte solutions in non-pressurized vessels at room temperature. Vanadium based redox flow batteries hold great promise for storing electric energy on a large scale (theoretically, they have infinite capacity).

Vanadium redox battery (VRB) **flow battery** was pioneered at the University of New South Wales, Australia, and has shown potentials for long cycle life and energy efficiencies of over 80% in large installations. A VRB is another type of a flow battery in which electrolytes in two loops are separated by a proton exchange membrane (PEM). The VRB uses compounds of the element vanadium in both electrolyte tanks. The electrolyte is prepared by dissolving of vanadium pentoxide (V_2O_5) in sulfuric acid (H_2SO_4). The electrolyte in the positive electrolyte loop contains $(VO_2)^+$—(V^{5+}) and $(VO)^{2+}$—(V^{4+}) ions, the electrolyte in the negative electrolyte loop, V^{3+} and V^{2+} ions. Chemical reactions proceed on the carbon electrodes, while the reaction chemistry at the positive electrode is:

$$\text{VO}_2^+ + 2\text{H}^+ + e^- \leftrightarrow \text{VO}^{2+} + \text{H}_2\text{O}, \ \text{with}\ E_0 = +1.00\ \text{V} \quad (11.26a)$$

And at the negative electrode, the reaction is:

$$V^{2+} \leftrightarrow V^{3+} + e^-, \text{ with } E_0 = -0.26 \text{ V} \tag{11.26b}$$

Under the actual VRB cell conditions, an open circuit voltage of 1.4 V is observed at 50% SoC, while a fully charged cell produces over 1.6 V at open-circuit, and a fully discharged cell of 1.0 V. The extremely large capacities possible for VRB batteries make them well suited in large grid applications, where they could to average out the production of highly variable wind and/or solar power sources. Their extremely rapid response times make them suitable for uninterruptible power system applications, where they can be used to replace lead–acid batteries. Disadvantages of vanadium redox batteries are low energy density of about 25 Wh/kg of electrolyte, low charge efficiency (need to use pumps) and a high price.

Polysulfide–bromide (PSB) battery was developed in Canada for utility-scale storage applications at power ratings from 5 MW upwards. A PSB battery utilizes two salt solution electrolytes, sodium bromide (NaBr), and sodium polysulfide (Na_2S_x). PSB electrolytes are separated in the battery cell by a polymer membrane that only passes positive sodium ions. The chemical reaction at the positive electrode is:

$$NaBr_3 + 2Na^+ + 2e^- \leftrightarrow 3NaBr \tag{11.27a}$$

And at the negative electrode the reaction is:

$$2Na_2S_2 \leftrightarrow Na_2S_4 + 2Na^+ + 2e^- \tag{11.27b}$$

This technology is expected to attain energy efficiencies of approximately 75%. Although the salt solutions themselves are only mildly toxic, a catastrophic failure by one of the tanks could release highly toxic bromine gas. Nevertheless, the Tennessee Valley Authority released a finding of no significant impact for a proposed 430-GJ facility and deemed it safe. PSB batteries have a very fast response time; it can react within 20 ms if electrolyte is retained charged in the stacks (of cells).

Sodium–Sulfur (Na–S) battery was originally developed by the Ford Motor Company in the 1960s. It contains sulfur at the positive electrode and sodium at the negative electrode. These electrodes are separated by a solid beta alumina ceramic. Through an electrochemical reaction, electrical energy is stored and released on demand. Na–S batteries have an operating temperature between 300 and 360 °C. The primary manufacturer of Na–S batteries, NGK Insulators, LTD, builds the batteries in 50 kW modules which are combined to make MW power battery systems. Na–S batteries exhibit higher power and higher energy density, higher columbic efficiency, good temperature stability, long cycle life, and low material costs. Their energy density is approximately three times that of traditional lead–acid batteries, with a high DC conversion efficiency of approximately 85%, making them an ideal candidate for the use in future DC distribution networks. Na–S batteries can be used for a wide variety of applications including peak shaving, renewable energy grid integration, power quality management, and emergency power units. They have the ability to discharge above their rated power, which makes them ideal to operate in both a peak shaving and power quality management environment.

11.3.7 Fuel cells and hydrogen energy

Fuel cells are an interesting alternative for power generation technologies because of they have higher efficiencies and quite very low environmental effects. A fuel cell is an electrochemical conversion device that has a continuous supply of fuel, such as hydrogen, natural gas, or methanol and an oxidant, such as oxygen, air, or hydrogen peroxide. It can have auxiliary parts to feed the device with reactants as well as a battery to supply energy for start-up. In conventional power generation systems, fuel is combusted to generate heat and then heat is converted to mechanical energy before it can be used to produce electrical energy. The maximum efficiency that a thermal engine can be achieved, when it operates at the Carnot cycle, which is related to the ratio of the heat source and sink absolute temperatures. Fuel cell operation is based on electrochemical reactions and not fuel combustion, by avoiding the chemical energy conversion into mechanical energy, through thermal phase enables fuel cells to achieve higher efficiency than that of conventional power generation technologies. A fuel cell can be considered as a "cross-over" of a battery and a thermal engine, resembling an engine because theoretically it can operate as long as it is fed with fuel. However, its operation is based on electrochemical reactions, resembling batteries providing significant advantages for fuel cells. On the other hand, batteries are devices that when their chemical energy is depleted, they must be replaced or recharged, whereas fuel cells can generate electricity as long are fueled. However, fuel cells resemble rechargeable batteries, while their theoretical open-circuit voltage is given by (11.9). The open-circuit voltage of single fuel cell is about 1.2 V.

11.3.7.1 Hydrogen storage and economy

Hydrogen is advocated for quite a long time as and environmentally friendly and a powerful energy storage medium and fuel. Among the most important advantages of using the hydrogen as energy storage medium are: the lightest element, very stable compound, on volumetric basis can store several times more energy than compressed air, reacting easily with oxygen to generate energy, forming water, harmless to the environment, can be easily used in fuel cells, and has a long industrial application history. However, the hydrogen has a few disadvantages as energy storage medium, is flammable and explosive, requiring special containers and transportation, highly diffusive, due to its specific energy content, and being the lightest element high-pressure and large containers must be used for significant mass storage. Hydrogen (H_2) can be produced with electrolysis, consisting of an electric current applied to water separates it into components O_2 and H_2. The oxygen has no inherent energy value, but the higher heat value (HHV) of the resulting hydrogen can contain up to 90% of the applied electric energy.

This hydrogen can then be stored and later combusted to provide heat or work, or to power fuel cells. Compression to a storage pressure of 350 bar, the value usually assumed for automotive technologies, consumes up to 12% of the hydrogen's HHV if performed adiabatically, although the loss approaches a lower limit of 5% in a quasi-isothermal compression. Alternatively, the hydrogen can be stored in liquid form, a process that costs about 40% of HHV, using current technology, and that at best would consume about 25%. Liquid storage is not possible for automotive applications, because mandatory boil-off from the storage container cannot be safely released in closed spaces. Hydrogen can also be bonded into metal hydrides using an absorption process. The storage energy penalty may be lower for this process, which requires pressurization to only 30 bars. However, the density of the metal hydride can be between up to100 times the density of the stored hydrogen. Carbon nanotubes have also received attention as a potential hydrogen storage medium. A major issue of the use of hydrogen as energy medium is the hydrogen embrittlement (grooving), consisting of hydrogen diffusion through metal matrices that can lead to small cracks compromising the hydrogen storage container quality. However, despites of its disadvantages, its higher storage capacity, abundance, and relative easy way to be produced through electrolysis from water make the hydrogen a strong candidate for energy storage medium. For these reasons, some scientists are suggesting that the widespread of the hydrogen use may transform our economy into the *hydrogen economy*.

In the context of the hydrogen economy, hydrogen will be an energy carrier, rather than a primary energy source. It may be generated through electrolysis or other industrial processes, using the energy harnessed by wind or solar power conversion systems, or chemical methods and used as fuel with almost no harmful environmental impacts. However, the establishment of the hydrogen economy is requiring that the current hydrogen storage and transportation are solved and suitable materials for storage are readily available. A hydrogen economy may lead to widespread of the use of renewable energy based power generation, avoiding the use of expensive and environmentally harmful fossil fuels. Weather a hydrogen economy is evolving in the near or far future is strongly dependent on the technological advances in hydrogen-based storage and transportation. Proponents of the hydrogen economy are making the cause based on the fact that hydrogen is the cleanest end-user energy source, especially in transportation applications and one of the most abundant elements in the nature. Almost every country can become energy independent in this scenario. Critics of this full transition to the hydrogen economy are arguing that the transition cost is prohibitive, and a transition in intermediate steps may be more economically viable, with the transition focusing on more locally than regionally, entire country or globally. A future and extensive hydrogen infrastructure to be established in the same ways as the energy distribution networks were established in the early twenty-first century. The options exist for the storage of hydrogen: compressed gas, chemical compounds, liquid hydrogen, and metallic hydrides. Since the cost of liquefaction is considerable, the most attractive concept for the bulk storage of hydrogen produced from substantial non-oil-based primary energy sources is compressed gaseous hydrogen in underground caverns, where it can be stored in a similar way to natural gas. Moreover, the high diffusivity of gaseous hydrogen has likely little effect on leakage since most rock structures are sealed in their capillary pores by water.

11.3.7.2 Fuel cell principles and operation

Fuel cells are electrochemical devices that are producing electricity form paired oxidation or reduction reactions, being in some way batteries with flows/supplies of reactants in and products out. Fuel cells are hardly a new idea, being invented in about 1840, but they are really making their mark as a power source for electric vehicles, space applications, and consumer electronics, only in the last part of twentieth century and their time is about to come. A battery has all of its chemicals stored inside, and it converts those chemicals into electricity too. This means that a battery eventually "goes dead" and you either throw it away or recharge it. With a fuel cell, chemicals constantly flow into the cell so it never goes dead—as long as there is a flow of chemicals into the cell, the electricity flows out of the cell. Fuel cells, similar to batteries exhibit higher efficiency at part load than at full load and with less variation over the entire operating range, having good load-following characteristics. Fuel cells are modular in construction with consistent efficiency regardless of size. Reformers, however, perform less efficiently at part load so that overall system efficiency suffers when used in conjunction with fuel cells. Fuel cells, like batteries, are devices that react chemically and instantly to changes in load. However, fuel cell systems are comprised of predominantly mechanical devices each of which have their own response time to changes in load demand. Nonetheless, fuel cell systems that operate on pure hydrogen tend to have excellent overall response. Fuel cell systems that operate on reformate using an on-board reformer, which can be sluggish, particularly if steam reforming techniques are used. The fuel cells are distinguished form the secondary rechargeable batteries by their external fuel storage and extended life-time. Fuel cells have the advantages of high efficiency, low emissions, quite operations, goo reliability, and fewer moving parts (only pumps and fans to circulate coolant and reactant gases) over other energy generation systems. Their generation efficiency is very high in fuel cells, higher power density, and lower vibration characteristics. A fuel cell is a DC voltage source, operating at about 1 V level. However, this might be set to change over the next 20 or 30 years. The basic principle of the fuel cell is that it uses hydrogen fuel to produce electricity in a battery-like device, as discussed later. The fuel cell basic chemical reaction is:

$$2H_2+O_2 \rightarrow 2H_2O + \text{Energy} \tag{11.28}$$

The reaction products are thus water, and energy, the sole reaction product of a hydrogen–oxygen fuel cell is water, an ideal product from a pollution standpoint. In addition to electrons, heat is also a reaction product. This heat must be continuously removed, as it is generated, in order to keep the cell reaction isothermal. A fuel cell-based vehicle can be described as zero-emission, running off a fairly normal chemical fuel (hydrogen), a very reasonable energy can be stored, and its range is quite satisfactory, thus offering the only real prospect of a silent zero-emission vehicle with a range and performance broadly comparable with IC engine vehicles. It is not surprising then that there have, for many years, been those who have seen fuel cells as a technology that shows great promise, and could even make serious inroads into the internal combustion engine domination. There are quite a few problems and challenges for fuel cells to overcome before they become a commercial reality as a vehicle power source or other power applications. The main problems and issues are listed here. Fuel cells are currently far more expensive than IC engines, and even hybrid IC/electric systems, so the cost reduction is a priority. Water management is another important and difficult issue with automotive fuel cells. The thermal management of fuel cells is actually rather more difficult than for IC engines. Hydrogen is the preferred fuel for fuel cells, but hydrogen is very difficult to store and transport. There is also the vital question of "where does the hydrogen come from" these issues are so difficult and important, with so many rival solutions. However, there is great hope that these problems can be overcome, and fuel cells can be the basis of less environmentally damaging transport. We have seen that the basic principle of the fuel cell is the release of energy following a chemical reaction between hydrogen and oxygen. The key difference between this and simply burning the gas is that the energy is released as an electric current, rather that heat. There are for main parts of a fuel cell: *the anode, the cathode, the catalyst, and the proton exchange membrane (PEM)*. The catalyst is a special material that facilitates the reaction of oxygen and hydrogen, made usually of platinum powder very thinly coated onto carbon paper or cloth. The catalyst is rough and porous so that the maximum surface area of the platinum can be exposed to the hydrogen or oxygen. The platinum-coated side of the catalyst faces the PEM. The electrolyte is the proton exchange membrane, being a specially treated material that only conducts positively charged ions, and the membrane blocks the electrons. In order to understand the fuel cell operation, the separate reactions taking place at each electrode must be considered. These important details vary for different types of fuel cell, but if we start with a cell based on an acid electrolyte, we shall consider the simplest and the most common type. At the anode of an acid electrolyte fuel cell the hydrogen gas ionizes, releasing electrons and creating H+ ions:

$$2H_2 \rightarrow 4H^+ + 4e^- \tag{11.29}$$

During this reaction energy is released. At the cathode, oxygen reacts with electrons taken from the electrode, and H^+ ions from the electrolyte, to form water.

$$O_2 + 4H^+ + 4e^- \rightarrow 2H_2O \tag{11.30}$$

In order, that these reactions proceed continuously, electrons produced at the anode must pass through an electrical circuit to the cathode. Also H^+ ions must pass through the electrolyte, so an acid, having free H^+ ions serves this purpose very well. Certain polymers can also contain mobile H^+ ions. These reactions may seem simpler, but they are not in normal circumstances. Also, the fact that hydrogen has to be used as a fuel is a disadvantage. To solve these and other problems many fuel cell types have been researched. The different types are usually distinguished by the used electrolyte, though there are other important differences as well. In fuel cells, similar to batteries, the electrode reactions are surface phenomena, occurring at a liquid–solid or gas–solid interface and therefore proceed at a rate proportional to the exposed solid areas. For this reason, porous electrode materials are used, often porous carbon impregnated or coated with a catalyst to speed the reactions. Thus, because of microscopic pores, the effective area of the electrodes is very large. Any phenomenon that prevents the gas from entering the pores or deactivates the catalyst must be avoided in order that a fuel cell to function effectively over longer periods. Because of the reaction rate–area relation, fuel cell current and power output increase with increased cell area. The surface power density (W/m^2) is an important parameter in comparing fuel cell designs, and the fuel cell power output can be scaled up by increasing its surface area. The electrolyte acts as an ion transport medium between electrodes. The passage rate of the positive charge through the electrolyte must match the rate of electron arrival at the opposite electrode to satisfy the physical requirement of electrical neutrality of the discharge fluids. Impediments to the ion transport rate through the electrolyte can limit current flow and the power output. Thus care must be taken in design to minimize the length of ion travel path and other factors that are retarding the ion transport. The theoretical maximum energy of an isothermal fuel cell (or other isothermal reversible control volume) is the difference in free energy (Gibbs) functions of the cell reactants and the products. The drop in free energy is mainly associated with (11.30), expressing free energy of the chemical potential, e.g., of the H^+ ions dissolved in the electrolyte, which related to the maximum (open-circuit) voltage of a fuel cell through (11.9). The maximum voltage of a fuel cell is about 1.23 V, as estimated in the example below. The maximum work output of which a fuel cell is able to perform is given by the decrease in its free energy, ΔG. The fuel cell conversion efficiency, η_{fc}, is defined as the electrical energy output per unit mass (or mole) of fuel to the corresponding heating value of the fuel consumed, the total energy drawn from the fuel, given by the maximum energy available from the fuel in an adiabatic steady-flow process, which the difference in the inlet and exit enthalpies, ΔH. The thermal fuel cell efficiency is expressed as:

$$\eta_{fc}(thermal) = \frac{\Delta G}{\Delta H} \tag{11.31}$$

Example 11.11: Determine the open-circuit voltage, maximum work, and the thermal efficiency for a direct hydrogen–oxygen fuel cell at standard reference conditions (see Table C3 Appendix C). Consider two cases when the product is liquid water and vapor.

Solution: The maximum work of the fuel cell is the difference of the free energy of the products and the free energy of the reactants. The maximum work for the liquid–water product and the water–vapor product are, respectively:

$$W_{\max}(liquid) = G_r - G_p(l) = 0 - (-237,141) = 237,141 \text{ kJ/kg} \cdot \text{mole}$$
$$W_{\max}(vapor) = G_r - G_p(v) = 0 - (-228,582) = 228,582 \text{ kJ/kg} \cdot \text{mole}$$

The theoretical ideal one-circuit voltages, in the two cases are computed with (11.9), as:

$$V_{\max}(liquid) = \frac{W_{\max}(l)}{nF} = \frac{237,141 \times 10^3}{2 \times 96,400 \times 10^3} = 1.2299 = 1.23 \text{ V}$$

$$V_{\max}(vapor) = \frac{W_{\max}(v)}{nF} = \frac{228,582 \times 10^3}{2 \times 96,400 \times 10^3} = 1.1855 = 1.19 \text{ V}$$

The change in the enthalpies for water–liquid product and water–vapor product (Table C3 of Appendix C) are: 285,830 kJ/kg·mole and 241,826 kJ/kg·mole, respectively. The thermal maximum efficiencies in the two cases, computed with (11.31), are:

$$\eta_{fc}(liquid) = \frac{\Delta G}{\Delta H} = \frac{237,141}{285,830} = 0.83 \text{ or } 83\%$$

$$\eta_{fc}(vapor) = \frac{\Delta G}{\Delta H} = \frac{228,582}{241,826} = 0.95 \text{ or } 95\%$$

However, the actual efficiencies are lower than in this example, while the actual open-circuit voltage (potentials) of 0.7–0.9 V are typical for most of the fuel cell regardless the type, and losses or inefficiencies in fuel cells, often called polarizations, are reflected in a cell voltage drop when the fuel cell is under load. Three major types of polarization are: (a) the *ohmic polarization* is due to the internal resistance to the motion of electrons through electrodes and of ions through the electrolyte; (b) the *concentration polarization* due to the mass transport effects relating to diffusion of gases through porous electrodes and to the solution and dissolution of reactants and products; and (c) the *activation polarization*, which is related to the activation energy barriers for the various steps in the oxidation–reduction reactions at the electrodes. The net effect of these and other polarizations is a decline in terminal voltage with increasing current drawn by the load. This voltage drop is reflected in a peak in the fuel cell power–current characteristic. The high thermal efficiencies of the fuel cells are based on the operation of the fuel cell as an isothermal, steady-flow process. Although heat must be rejected, the transformation of chemical energy directly into electron flow does not rely on heat rejection in a cyclic process to a sink at low temperature as in a heat engine, which is why the Carnot efficiency limit does not apply. However, the inefficiency associated with the various polarizations and the incomplete fuel utilization tends to reduce overall cell conversion efficiencies to well below the thermal efficiencies predicted in the example above. Fuel cells, like other energy conversion systems do not function in exact conformance with simplistic models or without inefficiencies. Defining the voltage efficiency, η_V, as the ratio of the terminal voltage to the theoretical *EMF*, V_T/EMF, and the current efficiency, η_I, as the ratio of the cell electrical current to the theoretical charge flow associated with the fuel consumption, the overall fuel cell conversion efficiency is related to the thermal efficiency by this relationship:

$$\eta_{fc} = \frac{V_T}{EMF} \times \frac{I}{jN_{Rct}F} \times \eta_{th} = \eta_V \cdot \eta_I \cdot \eta_{th} \tag{11.32}$$

Example 11.12: The actual DC output voltage of hydrogen–oxygen fuel cell is 0.80 V. Assuming its current efficiency 1.0, what is the fuel cell actual efficiency?

Solution: From previous example, the fuel cell thermal efficiency is 0.83, and the voltage efficiency is equal to 0.80/1.23 = 0.65, then the actual fuel cell efficiency as per (11.32) is:

$$\eta_{fc} = \eta_V \cdot \eta_I \cdot \eta_{th} = 0.65 \times 1.0 \times 0.83 = 0.54 \text{ or } 54\%$$

11.3.7.3 Fuel cell types and applications

An important element of fuel cell design is that, similar to the large battery systems, are built from a large number of identical units or cells. Each has an open-circuit voltage on the order of 1 V, depending on the oxidation–reduction reactions taking place in the cell. The fuel cells are usually built in sandwich-style assemblies called *stacks*, while the fuel and oxidant cross-flow through a portion of the stack. A fuel cell stack can be configured with many groups of cells in series and parallel connections to further tailor the voltage, current, and power produced. The number of individual cells contained within one stack is typically greater than 50 and varies significantly with stack design. The basic components that comprise a fuel cell stack include the electrodes and electrolyte with additional components required for electrical connections and/or insulation and the flow of fuel and oxidant through the stack. These key components include current collectors and separator plates. The current collectors conduct electrons from the anode to the separator plate. The separator plates provide the electrical series connections between cells and physically separate the oxidant flow of one cell from the fuel flow of an adjacent cell. The channels in the current collectors serve as the distribution pathways for the fuel and oxidant. Often, the two current collectors and the separator plate are combined into a single unit called a bipolar plate. Electrically conducting bipolar separator plates serve as direct current transmission paths between successive stack cells. This modular type of construction allows research and development of individual cells and engineering of fuel cell systems to proceed in parallel. The preferred fuel for most fuel cell types is the hydrogen. Hydrogen is not readily available, however, but the infrastructure for the reliable extraction, transport or distribution, refining, and/or purification of hydrocarbon fuels is well established. Thus, fuel cell systems that have been developed for practical applications to date have been designed to operate on hydrocarbon fuels. In addition to the fuel cell system requirement of a fuel processor for operation on hydrocarbon fuels, a power conditioning, and for grid-connection of to supply an AC load an inverter is also needed. There are five major types of fuel cells being known or used in the stationary and mobile applications. All these fuel cell types have the same basic design as mentioned above, but with different chemicals used as the electrolyte. These fuel cells are:

1. Alkaline fuel cell (AFC);
2. Phosphoric acid fuel cell (PAFC);
3. Molten carbonate fuel cell (MCFC);
4. Solid oxide fuel cell (SOFC); and
5. Proton exchange membrane fuel cell (PEMFC).

All these fuel cells, regardless of the types require fairly pure hydrogen fuel to run. However, large amount of hydrogen gas is difficult to transport and store. Therefore, a reformer is normally equipped inside these fuel cells to generate hydrogen gas from liquid fuels, such as gasoline or methanol. Among these five types of fuel cell, PEMFC has the highest potential for widespread use. PEMFC is getting cheaper to manufacture and easier to handle. It operates at relatively low temperature when compared with other types of fuel cell. AFC systems have the highest efficiency and are therefore being used to generate electricity in spacecraft systems for more than 30 years. However, it requires very pure hydrogen and oxygen to operate and thus the running cost is very expensive. As a result, AFCs are unlikely to be used extensively for general purposes, such as in vehicles and in our homes. In contrast, the MCFC and the SOFC are specially designed to be used in power stations to generate electricity in large scale power systems. Nevertheless, there are still a lot of technical and safety problems associated with the use of these fuel cells (MCFC and SOFC) in the long term applications. Apart from these fuel cell types, a new type of fuel cell, the direct methanol fuel cell (DMFC), being under vigorous on-going research is coming on the market. This type of fuel cell has the same operating mechanism as PEMFC but instead of using pure hydrogen, it is able to use methanol directly as the basic fuel. A reformer is therefore not essential in this fuel cell system to reform complex hydrocarbons into pure hydrogen. Several companies around the world are presently working on DMFC to power electronic equipment. The DMFC appears to be the most promising alternative electric source to replace the battery used in portable electronics, such as mobile phones and laptop computers. Although the development of fuel cell is fast, fuel cell has not yet reached its potential level of commercial success due to high material costs (Pt electrode) and market barriers. To tackle the electrode problem, platinum or platinum-based nanoparticles are coated on the surface of carbon black. This porous electrode used in fuel cells can give higher current densities than geometric plate electrode. More importantly, it can reduce the quantity of platinum used.

11.3.8 Flywheel energy storage (FES)

A flywheel energy storage system converts electrical energy supplied from DC or three-phase AC power source into kinetic energy of a spinning mass or converts kinetic energy of a spinning mass into electrical energy. Basically, a flywheel is a disk with a certain amount of mass that can spin, holding kinetic energy. Modern high-tech flywheels are built with the disk attached to a rotor in upright position to prevent gravity influence. They are charged by a simple electric motor that simultaneously acts as a generator in the process of discharging. When dealing with efficiency; however, it gets more complicated, as stated by the rules of physics; they will eventually have to deal with friction during operation. Therefore, the challenge to increase that efficiency is to minimize friction. This is mainly accomplished by two measures: the first one is to let the disk spin in a vacuum, so there will be no air friction; and the second one is to bear the spinning rotor on permanent and electromagnetic bearings, so it basically floats. The spinning speed for a modern single flywheel reaches up to 16.000 RPM and offers a capacity up to 25 kWh, which can be absorbed and injected almost instantly. These devices are comprised of a massive or composite flywheel coupled with a motor-generator and special brackets (often magnetic), set inside a housing at very low pressure to reduce self-discharge losses. They have a great cycling capacity (a few 10,000 to a few 100,000 cycles) determined by fatigue design. For electrical power system applications high-capacity energy flywheels are needed. Friction losses of a 200-tons flywheel are estimated at about 200 kW. Using this hypothesis and an efficiency of about 85%, the overall efficiency would drop to 78% after 5 h, and 45% after one day. Long-term energy storage with flywheels is therefore not foreseeable. An FES device is made up of a shaft, holding a rotor, rotating on two magnetic bearings to reduce friction. These are all contained within a vacuum to reduce aerodynamic drag losses. Flywheels store energy by accelerating the rotor/flywheel to a very high speed and maintaining the energy in the system as kinetic energy, and release the energy by reversing the charging process so that the motor is used as a generator. As the flywheel discharges, the rotor slows down until eventually coming to a complete stop. The rotor dictates the amount of energy that the flywheel is capable of storing. Due to their simplicity, flywheel ESS have been widely used in commercial small power units (about 3 kWh) in the range of from 1 kW—3 h to 100 kW—3 s. Energy is stored as kinetic energy using a rotor:

$$E = \frac{1}{2}J\omega^2 \tag{11.33}$$

where J is the momentum of inertia and ω is the angular velocity. The moment of inertia is given by the volume integral taken over the product of mass density ρ, and squared distance r^2 of mass elements with respect to the axis of rotation:

$$J = \int \rho r^2 dV$$

However, in the case of regular geometries, the momentum of inertia is given by:

$$J = kmr^2 \tag{11.34}$$

Here, m is the flywheel total mass, and r the outer radius of the disk. Equation (11.33) then became:

$$E = \frac{1}{2}kmr^2\omega^2 = \frac{1}{2}k(\rho\Delta V)r^2\omega^2 \tag{11.35}$$

Here, ΔV is the increment of the volume. The inertial constant depends on the shape of the rotating object. For thin rings, $k = 1$, while for a solid uniform disk, $k = 0.5$. Flywheel rotor is usually a hollow cylinder, and has magnetic bearings to minimize the friction. The rotor is located in a vacuum pipe to decrease the friction even more. The rotor is integrated into a motor/generator machine that allows the energy flow in both directions. The energy storage capacity depends on the mass and shape of the rotor and on the maximum available angular velocity.

Example 11.13: A flywheel is a uniform circular disc of diameter of 2.90 m, 1,500 kg, and is rotating at 5,000 RPM. Calculate the flywheel kinetic energy.

Solution: In (11.35), $k = 0.5$, so:

$$E = \frac{1}{2}kmr^2\omega^2 = \frac{1}{2} \times 0.5 \times 1500 \times \left(\frac{2.90}{2}\right)^2 \times \left(\frac{2\pi 5000}{60}\right)^2$$

$$= 2.16142 \times 10^8 \text{J} = 216.142 \text{ MJ}$$

There are two topologies, *slow* flywheels (with speed of rotation up to 6,000 rpm) based on steel rotors, and *fast* flywheels (below 60,000 rpm), using advanced material rotors (carbon fiber or glass fiber) that present higher energy and power densities than steel rotors. The flywheel designs are modular and systems of 10 MW are possible, the efficiency of 80%–85%, with a useful life of 20 years. The advances on the rotor technology have permitted high dynamics and a durability of tenths of thousands of cycles, making them suitable for power quality applications: frequency deviations, temporary interruptions, voltage sags, and voltage swells. FES applied to the renewable energies trend is to combine them with other technologies, like micro-CAES or thermal energy storage. Flywheels store power in direct relation to the mass of the rotor but to the square of its surface speed. The amount of energy which can be stored by a flywheel is determined by the material design stress, material density and total mass, as well as flywheel shape factor K. It is not directly dependent on size or angular speed since one of these can be chosen independently to achieve the required design stress. Material properties also govern flywheel design and therefore allowable K values. In order to take maximum advantage of the best properties of highly anisotropic materials, the flywheel shape is such that lower K values have to be accepted compared with those normally associated with flywheels made from isotropic material. However, at first sight from (11.33) to (11.35), we are tempted to maximize spatial energy density by increasing the product of the three factors ω^2, ρ, and the squared distance r^2 of the relevant mass elements with respect to the axis of rotation. This approach, however, cannot be extended to arbitrarily high spatial energy densities, because the centrifugal stresses caused by the rotation also roughly increase with the product of these three factors. To avoid fragmentation of the flywheel, a certain material dependent tensile stress level must not be exceeded. Consequently, the most efficient way to store energy in a flywheel is to make it spin faster, not by making it heavier. The energy density within a flywheel is defined as the energy per unit mass:

$$\frac{W_{KIN}}{m_{FW}} = 0.5 \cdot v_l^2 = \frac{\sigma}{\rho} \tag{11.36}$$

where W_{KIN} is the total kinetic energy in Joules (J), m_{FW} is the mass of the flywheel in kg, v_l is the linear velocity of the flywheel in m/s, σ is the specific strength of the material in Nm/kg; and ρ is the density of the material in kg/m^3. For a rotating thin ring, therefore, the maximum energy density is dependent on the specific strength of the material and not on the mass. The energy density of a flywheel is normally the first criterion for the selection of a material. Regarding specific strength, composite materials have significant advantages compared to metallic materials. Table 11.3 lists some flywheel materials and their properties. The burst behavior is a deciding factor for choosing a flywheel material. However, not all the stored energy, during charging phase can be used during discharging phase. The useful energy per mass unit (the energy released during the discharge) is expressed as:

$$\frac{E}{m} = \left(1 - s^2\right) K \frac{\sigma}{\rho} \tag{11.37}$$

where s is the ratio of minimum to maximum operating speed, usually taken to be 0.2.

Table 11.3 Typical materials used for flywheels and their properties

Material	**Density (kg/m^3)**	**Strength (MN/m^2)**	**Specific strength (MNm/kg)**
Steel	7,800	1,800	0.22
Alloy (AlMnMg)	2,700	600	0.22
Titanium	4,500	1,200	0.27
GRFP*	2,000	1,600	0.80
CFRP*	1,500	2,400	1.60

* GFRP, glass fiber reinforced polymer; CFRP, carbon fiber-reinforced polymer

Example 11.14: Compute the energy density for a steel flywheel.

Solution: Using (11.36) and the values of the density and specific strength for steel, from Table 11.3, the flywheel energy density is:

$$\frac{W_{KIN}}{m_{FW}} = \frac{0.22 \times 10^6}{7800} = 28.21 \text{ J/kg}$$

The power and energy capacities are decoupled in flywheels. In order to obtain the required power capacity, you must optimize the motor/generator and the power electronics. These systems, the *low-speed flywheels*, having relatively low rotational speeds, approximately 10,000 rpm and a heavy rotor made form steel. They can provide up to 1,650 kW, for very short times (up to 120 s). To optimize the flywheel storage capacities, the rotor speed must be increased. These high-speed flywheels, spin on a lighter rotor at much higher speeds, with prototype composite flywheels claiming to reach speeds in excess of 100,000 rpm. However, the fastest flywheels commercially available spin at about 80,000 rpm. They can provide energy up to an hour, with a maximum power of 750 kW. Over the past years, the flywheel efficiency has improved up to 80%, although some sources claim efficiencies as high as 90%. Flywheels have an extremely fast dynamic response, a long life (~20 years), requiring little maintenance, and are environmentally friendly. As the storage medium used in flywheels is mechanical, the unit can be discharged repeatedly and fully without any damage to the device. Flywheels are used for power quality enhancements, uninterruptable power supply, capturing waste energy that is very useful in electric vehicle applications and finally, to dampen frequency variation, making FES very useful to smooth the irregular electrical output from wind turbines. The stored energy in flywheels has a significant destructive potential when released uncontrolled, safety being a major disadvantage. Efforts are made to design rotors such that, in the case of a failure, many thin and long fragments, having little trans-lateral energy are released, so the rotor burst can be relatively benign. However, even with careful design, a composite rotor still can fail dangerously. The safety of a flywheel system is not related only to the rotor. The housing enclosure, and all components and materials within it, can influence the result of a burst significantly. To facilitate mechanical ESS by flywheel, low-loss and long-life bearings and suitable flywheel materials need to be developed. Some new materials are steel wire, vinyl-impregnated fiberglass, and carbon fiber. However, that major advantages of the flywheels include: high power density, nonpolluting, high efficiency, long life (over 20 years), and independent operation from extreme weather conditions. Their major disadvantages are: safety, noise, and high-speed operations leading to wear, vibration and fatigue.

11.3.9 Superconducting magnetic energy storage

Superconducting magnetic energy storage (SMES) exploits advances in materials and power electronics technologies to achieve novel energy storage based on three principles of physics: (a) superconductors carry current with no resistive losses, (b) electric currents induce magnetic fields, and (c) magnetic fields are energy forms that can be stored. These principles provide the potential for the highly efficient electricity storage in superconducting coils. Operationally, SMES is different from other storage technologies in that a continuously circulating current within the superconducting coil produces the stored energy, the only conversion process in the SMES system is from AC to DC power conversion, i.e., there are no thermodynamic losses inherent in this conversion. Basically, SMESs store energy in the magnetic field created by the flow of direct current in a superconducting coil, which has been cryogenically cooled to a temperature below its superconducting critical temperature. The idea is to store energy in the form of an electromagnetic field surrounding the coil, which operating at very low temperatures, to become superconducting, which made the system a superconductor. SMES makes use of this phenomenon and—in theory—stores energy almost without any energy loss (practically about 95% efficiency). SMES was originally proposed for large-scale, load leveling, however, because of its rapid discharge capabilities, it has been implemented on electric power systems for pulsed-power and system-stability applications.

The power and stored energy in a SMES system are determined by application and site-specific requirements. Once these values are set, a system can be designed with adequate margin to provide the required energy on demand. SMES units have been proposed over a wide range of power (1–1,000 MW_{AC}) and energy storage ratings (0.3 kWh–1,000 MWh). Independent of size, all SMES systems include three parts: superconducting coil, power conditioning system (power electronics and control), and cryogenically cooled refrigerator. Once the superconducting coil is charged, the current is not decaying and magnetic energy can be stored indefinitely. The stored energy can be released back to the network by discharging the coil. The power conditioning system uses an inverter/rectifier to transform AC power to DC or convert DC back to AC. Inverter/rectifier accounts for about 2%–3% energy losses in each direction. SMES loses the least electricity amount in the energy storage process compared to any other methods of storing energy. There are several reasons for using superconducting magnetic energy storage instead of other energy storage methods. The most important advantage of SMES is that the time delay during charge and discharge is quite short. Power is available almost instantaneously and very high power output can be provided for a brief period of time. Other energy storage methods, such as pumped hydro or compressed air have a substantial time delay associated with the energy conversion of stored mechanical energy back into electricity. Thus if demand is immediate, SMESs are a viable option. Another advantage is that the loss of power is less than other storage methods because electric currents encounter almost no resistance. Additionally the main parts in a SMES are motionless, which results in high reliability. The magnetic energy stored by a carrying current coil is given by:

$$E_{SMES} = \frac{1}{2}LI^2 \tag{11.38}$$

where E is the energy, expressed in J, L is the coil inductance measured in H, and I is the current. The total stored energy, or the level of charge, can be found from the above equation and the current in the coil. Alternatively, the SMES magnetic energy density per unit volume, for a magnetic flux density B (in T) and magnetic permeability of free space, μ_0 ($= 4\pi \times 10^{-7}$ H/m) is expressed as:

$$E_{SMES} = \frac{B^2}{2\mu_0} \approx 4 \times 10^5\, B^2\ \mathrm{J/m^3} \tag{11.39}$$

Example 11.15: Find the energy density for a SMES that has a magnetic flux density of 4.5 T.

Solution: Applying (11.39), the volume energy density is:

$$E_{SMES} \approx 4 \times 10^5 (4.5)^2 = 8.1\ \mathrm{MJ/m^3}$$

The maximum practical stored energy, however, is determined by two factors: the size and geometry of the coil, which determine the inductance. The characteristics of the conductor determine the maximum current. Superconductors carry substantial currents in high magnetic fields. For example, at 5 T, which is 100,000 times greater than the earth's field, practical superconductors can carry currents of 300,000 A/cm^2. For example, for a cylindrical coil with conductors of a rectangular cross-section, with mean radius of coil R, a and b are width and depth of the conductor, f is the called form function, determined by the coil shapes and geometries, ξ and δ are two parameters to characterize the dimensions of the stored energy is a function of coil dimensions, shape, geometry, number of turns and carrying current, given by:

$$E = \frac{1}{2} R N^2 I^2 f(\xi, \delta) \tag{11.40}$$

where I is the current, $f(\xi,\delta)$ is the form function (*J/A-m*), and N is the number of turns of coil.

Example 11.16: If the current flowing in a SMES coil having 1,000 turns, a radius of 0.75 m is 250 A, compute the energy stored in this coil. Assume a form function, $f(\xi,\delta)$ to 1.35.

Solution: For the SEMS characteristics, the stored energy is:

$$E = 0.5 \times 0.75 \times \left(10^3\right)^2 (250)^2 \times 1.35 = 31.64063 \times 10^9\ \mathrm{J}$$

The superconductor is one of the major costs of a superconducting coil, so one design goal is to store the maximum amount of energy per quantity of superconductor. Many factors contribute to achieving this goal. One fundamental aspect, however, is to select a coil design that most effectively uses the material, minimizing the cost. Since not too many SMES coils have been constructed and installed, there is still little experience with a generic design. This is true even for the small or micro-SMES units for power-quality applications, where several different coil designs have been used. A primary consideration in the design of a SMES coil is the maximum allowable current in the conductor. It depends on: conductor size, the superconducting materials used, the resulting magnetic field, and the operating temperature. The magnetic forces can be significant in large coils, a containment structure within or around the coil is needed. The superconducting SMES coil must be maintained at a temperature sufficiently low to sustain the conductor superconducting state, about 4.5 K (−269 °C, or −452 °F). This thermal operating regime is maintained by a special cryogenic refrigerator that uses helium as the refrigerant, being the only material that is not a solid at these temperatures. Just as a conventional refrigerator requires power to operate, electricity is used to power the cryogenic refrigerator. Thermodynamic analyses have shown that power required removing heat from the coil increases with decreasing temperature. Including inefficiencies within the refrigerator itself, 200–1,000 W of electric power are required for each watt that is removed from the 4.5 K environment. As a result, design of SMES and other cryogenic systems places a high priority on reducing losses within the superconducting coils and minimizing the flow of heat into the cold environment. Both the power requirements and the physical dimensions of the refrigerator depend on the amount of heat that must be removed from the superconducting coil. However, small SMES coils and modern MRI magnets are designed to have such low losses that very small refrigerators are adequate.

Charging and discharging a SMES coil is different from that of other energy storage technologies, because it carries a current at any SOC. Since the current always flows in one direction, the power conversion system (PCS) must produce a positive voltage across the coil when energy is to be stored, increasing the current to increase, while for discharge, the electronics is adjusted to make it appear as a load across the coil, producing a negative voltage causing the coil to discharge. The applied voltage times the instantaneous current determines the power. SMES manufacturers design their systems so that both the coil current and the allowable voltage include safety and performance margins. The PCS power capacity determines the rated SMES unit capacity. The control system establishes the link between grid power demands and power flow to and from the SMES coil. It receives dispatch signals from the power grid and status information from the SMES coil, and the integration of the dispatch request and charge level determines the response of the SMES unit. The control system also measures the condition of the SMES coil, the refrigerator, and other equipment, maintaining system safety and sends system status information to the operator. The power of a SMES system is established to meet the requirements of the application, e.g., power quality or power system stability. In general, the maximum power is the smaller of two quantities, the PCS power rating and the product of the peak coil current and the maximum coil withstand voltage. The physical size of an SMES system is the combined sizes of the coil, the refrigerator and the PCS, each depending on a variety of factors. The overall efficiency of an SMES plant depends on many factors. In principle, it can be as high as 95% in very large systems. For small power systems, used in power quality applications, the overall system efficiency is lower. Fortunately, in these applications, efficiency is usually not a critical factor. The SMES coil stores energy with absolutely no loss while the current is constant. There are, however, losses associated with changing current during charging and discharging, and the resulting magnetic field changes. In general, these losses referred to as eddy current and hysteresis losses are also small. Major losses are in the PCS and especially in the refrigerator system. However, power quality and system stability applications do not require high efficiency because the cost of maintenance power is much less than the potential losses to the user due to a power outage.

11.3.10 Supercapacitors

Supercapacitors are very high-capacity electrolytic devices that store energy in the form of electrostatic charge. They are composed of two electrodes with a very thin separator. Energy storage capacity increases as the surface area of the electrodes increases. Energy is stored as a DC electric field in the supercapacitor, while the systems uses power electronics to both charge and discharge the supercapacitors. Supercapacitors can have very high discharge rates and could handle fast load changes in a power system or a microgrid. In electrochemical capacitors (or supercapacitors), energy may not be delivered via redox reactions and, thus the use of the terms anode and cathode may not be appropriate but are in common usage. By orientation of electrolyte ions at the electrolyte/electrolyte interface, so-called electrical double layers (EDLs) are formed and released, which results in a parallel movement of electrons in the external wire, that is, in the energy-delivering process. Capacitors consist of two conducting plates are separated by an insulator, as shown in Figure 11.9. A DC voltage is connected across the capacitor, one plate being positive the other negative. The opposite charges on the plates attract and hence store energy. The electric charge $Q(C)$ stored in a capacitor of capacitance C (*in* F) at a voltage of V (V) is given by the equation:

$$Q = C \cdot V \tag{11.41}$$

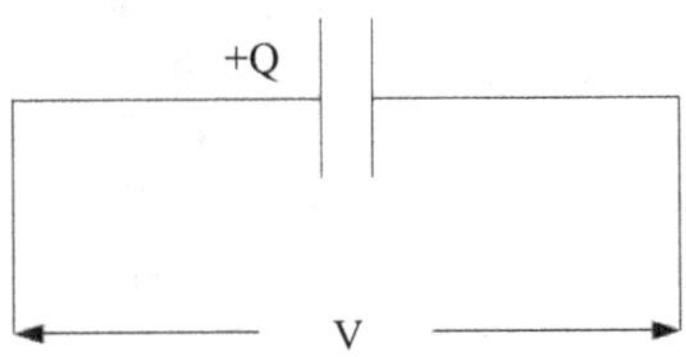

Figure 11.9 Capacitor symbol and diagram

Similar to the flywheels, the capacitors can provide large energy storage capabilities either they are usually used in small size configurations as components in electronic circuits and systems. The large energy storing capacitors with very large plate areas are the so-called supercapacitors or ultracapacitors. The energy stored in a capacitor is given by the equation:

$$E = 0.5C \cdot V^2 \tag{11.42}$$

where E(J). The capacitance C of a capacitor in Farads will be given by the equation:

$$C = \frac{\varepsilon A}{d} \tag{11.43}$$

Here ε is the electric permittivity of the material between the plates, A is the plate area and d is the separation of the plates, all expressed in standard units. The key technology of the supercapacitors is that the separation of the plates is so small, while the plate area is very large. In contrast to the conventional capacitors, the electric double layer capacitors do not have any dielectrics in general but rather utilize the phenomena typically referred to as the electric double layer. In the double layer, the effective "dielectric" thickness is exceedingly thin, and because of the porous nature of the carbon the surface area is extremely large, which translates to a very high capacitance. Generally, when two different phases come in contact with each other, positive and negative charges are set in array at the boundary. At every interface an array of charged particles and induced charges exist. This array is known as *electric double layer*. The high capacitance of an EDLC arises from the charge stored at the interface by changing electric field between anodes and cathodes. The very large capacitances arise from the formation on the electrode surface of a layer of electrolytic ions (the double layer), having huge surface areas, leading the capacitances of tens, hundreds, or even thousands of *Farads,* with the capacitor fitted into a small size container.

However, the problem with this technology is that the voltage across the capacitor can only be very low, usually lower than 3 V. Equation (11.41) severely limits the energy that can be stored for a given capacity and so the voltage. In order to store charge at a reasonable voltage, several capacitors are usually connected in series. This not only increases the cost but putting capacitors in series the total capacitance is reduced, as well as the charge equalization problem. In a string of capacitors in series, the charge on each one should be the same, as the same current flows through any series circuit. However, the problem is that always there is a certain amount of self-discharge in each one, due to the fact that the insulation between the plates of the capacitors is not perfect. This self-discharge is not equal in all the capacitors, which is an issue that needs to be corrected, otherwise there may be a relative charge build-up on some of the capacitors, and this will result in a higher voltage on those capacitors. The solution to this issue, being essential in systems of more than about six capacitors in series, is to have charge equalization circuits. These are circuits connected to each pair of capacitors that continually monitor the voltage across adjacent capacitors, and move charge from one to the other in order to make sure that the voltage across the capacitors is the same. These charge equalization circuits increase the cost and size of a capacitor energy storage system, and they also consume some energy, though very efficient designs are available. In many ways, the characteristics of supercapacitors are like those of flywheels. They have relatively high specific power and relatively low specific energy. They can be used as the energy storage for regenerative braking. Although they could be used alone on a vehicle, they would be better used in a hybrid as devices for giving out and receiving energy rapidly during braking and accelerating afterwards. Supercapacitors are inherently safer than flywheels as they avoid the problems of mechanical breakdown and gyroscopic effects. Power electronics are needed to step voltages up and down as required.

11.4 Chapter summary

The present chapter identifies the characteristics, possible applications, strengths, and weaknesses of the different energy storage concepts and technologies. ESSs are the key enabling technologies for transportation, building energy systems, conventional and alternative energy systems for industrial processes and utility applications. In particular, the extended applications of energy storage are enabling the integration and dispatch of renewable energy generation and are facilitating the emergence of smarter grids with less reliance on inefficient peak power plants. Energy storage will play a critical role in an efficient and renewable energy future; much more so than it does in today's fossil-based energy economy. Major power and ESS, with the exception of thermal energy storage, discussed in another chapter are reviewed in this chapter. It is concluded that EESs can contribute significantly to the design and optimal operation of power generation and smart grid systems, as well as the improved energy security and power quality, while reducing the overall energy costs. There is a range of options available to store intermittent energy until it is needed for electricity production. In the transportation sector, the emergence of viable onboard electric energy storage devices, such as high-power and high-energy Li-ion batteries will enable the widespread adoption of plug-in and HEVs, which will also interact with the smart grids of the future. Mature energy storage technologies can be used in several applications but in other situations, these technologies cannot fulfill the application requirements. Thus, new storage systems have appeared, passing new challenges that have to be solved by the research community. Simply stated, flow batteries are fuel cells that can be recharged. From a practical view, storage of the reactants is very important. The storage of gaseous fuels in fuel cells requires large, high pressure tanks or cryogenic storage that is prone to thermal self-discharge. Vanadium redox batteries have the advantage of being able to store electrolyte solutions in nonpressurized vessels at room temperatures. Vanadium based redox flow batteries hold great promise for storing electric energy on a large scale (theoretically, they have infinite capacity). They have many attractive features including independent sizing of power and energy capacity, longer lifetimes, high efficiency, and fast response, and relatively low cost (small initial investment and reduced operational expenditures).

References and further readings

[1] EPRI 1000419, Engineering Guide for Integration of Distributed Generation and Storage into Power Distribution Systems, Electric Power Research Institute, Palo Alto, CA, 2000.

[2] EPRI 1016097, Distribution System Losses Evaluation, Electric Power Research Institute, Palo Alto, CA, 2008.

[3] EPRI 1023518, Green Circuits: Distribution Efficiency Case Studies, Electric Power Research Institute, Palo Alto, CA, 2011.

[4] EPRI 1024101, Understanding the Grid Impacts of Plug-In Electric Vehicles (PEV): Phase 1 Study—Distribution Impact Case Studies, Electric Power Research Institute, Palo Alto, CA, 2012.

[5] D. Linden and T.B. Reddy, Handbook of Batteries (3rd ed.), McGraw-Hill, USA, 2002.

[6] C.A. Vincent and B. Scrosati, Modern *Batteries, Antony Rowe Ltd.*, Wiley, New York, 2003.

[7] F.A. Farret and M.G. Simões, *Integration of Alternative Sources of Energy*, Wiley-Interscience, New York, 2006.

[8] R.O'Hayre et al., Fuel Cell Fundamentals, Wiley, Hoboken, NJ, 2006.

[9] V. Quaschning, *Understanding Renewable Energy Systems*, Earthscan, London, UK, 2006.

[10] D.Y. Goswami, and F. Kreith, *Energy Conversion* (1st ed.), CRC Press, Boca Raton, FL, 2007.

[11] H. Chen, T.N. Cong, W. Yang, C. Tan, Y. Li, and Y. Ding, 2009. Progress in electrical energy storage system: a critical review, *Prog. Nat. Sci.* Vol. 19, pp. 291–312.

[12] B. K. Hodge, *Alternative Energy Systems and Applications*, Wiley, Hoboken, NJ, 2010.

[13] B. Everett and G. Boyle, Energy Systems and Sustainability: Power for a Sustainable *Future* (2nd ed), Oxford University Press, Oxford, UK, 2012.

[14] R.A. Dunlap, *Sustainable Energy*, Cengage Learning, Stamford, CT, 2015.

[15] V. Nelson and K. Starcher, Introduction to Renewable Energy (Energy and the Environment), CRC Press, Boca Raton, FL, 2015.

[16] R. Belu, Renewable Energy: Energy Storage Systems, in Encyclopedia of Energy Engineering and Technology (Online) (Ed: Dr. Sohail Anwar, R. Belu et al.), CRC Press/Taylor and Francis, 2014.

Questions and problems

1. What are the benefits of the ESS?
2. What are the major problems for using: (a) pumped water energy storage; (b) compressed air energy storage; and (c) the flywheels?

3. List some of the potential applications of energy storage.
4. Which of the following is not a method to store energy: (a) battery, (b) motor, (c) compressed air, (d) flywheel, and (e) all of the above store energy?
5. List the benefits of electricity energy storage for power grid operation.
6. Classify and explain each compressed-air storage systems.
7. List the essential criteria for comparing ESS.
8. List a few of the potential applications of energy storage options.
9. What gases did we use to feed the fuel cell to produce electricity? Can you name any by-products that may have been produced by fuel cells?
10. Explain the differences between shallow- and deep-cycle batteries, list main characteristics and applications for each type.
11. List the major advantages and disadvantages of fuel cells.
12. What gases did we use to feed the fuel cell to produce electricity? Can you name any by-products that may have been produced by fuel cells?
13. The lead acid battery (a secondary type cell) can be recharged and reused for a few years after which it cannot be recharged and reused. Can you think of reasons why it cannot be used any further?
14. How much energy can be delivered form a pumped storage facility of 8 million m^3, a head of 500-m and the overall efficiency of 85%?
15. What energy storage options are available for solar energy applications?
16. List the key measures of merit for any energy storage system.
17. How much kinetic energy does a flywheel (steel disk of diameter 6 in., thickness 1 in), speed of rotation 30,000 RPM have?
18. A water-pumped energy storage facility has a level difference (head) of 500 m and a working volume of 081 km^3. Estimated how much power is generated, if it is required to operate 1.5 h/day. The overall facility efficiency is 83% and the water density is 1,000 kg/m^3.
19. Estimate the volumetric energy density and the overall efficiency of a PHES, having the difference level between the lower and upper reservoirs 250 m, and the efficiencies of the pumped and generating phases, 0.85 and 0.90.
20. Estimate the energy of the CAES facilities at Hundorf, Germany and McIntosh, Tennessee, US.
21. A CAES has a volume of 500,000 m^3, and the compressed air pressure range is from 80 bars to 1 bar. Assuming isothermal process and an efficiency of 33% estimate the energy and power for a 3-h discharge period.
22. If the power generated by CAES of 300,000 m^3, in 2-h discharge period from 66 bars to the atmospheric pressure is 360 MW, what is the system efficiency?
23. An old salt mine, having a 15,000 m^3 storage capacity has been selected for pressurized air storage at 33 bar, if the temperature during the filling/charging phase is 175 °C, assuming an isothermal process and an efficiency of 35%, estimate the energy and average power for a 3-h discharge period.
24. Assuming a flow rate of 36 m^3/s, what is the power capacity of the CAES in Example 8.2?
25. An underground cavern of volume 35,000 m^3 is used to store compressed air energy isothermally at 300 K. Determine the maximum stored energy by

air compression from 100 to 1,750 kPa, assuming heat loss of 63,000 kJ. Hint: 1 mole of air occupies 22.4 L.

26. Determine the maximum available stored energy, if a 1,450 kg of air is compressed from 100 to 1,600 kPa at 27 °C, assuming isothermal condition and a heat loss of 27.5 MJ.
27. A flywheel has a weight of 20 kg, 8 m diameter, an angular velocity of 1,200 rad/s, a density of 3,200 kg/m^3, and a volume change of 0.75 m^3. Evaluate for k equal to 0.5 and 1.05, how much energy is stored in this flywheel.
28. Very high-speed flywheels are made of composite materials. If a flywheel has the following characteristics (a) a ring with radius of 2.5 m, mass of 100 kg, and speed of 25,000 RPM; and (b) a sold uniform disk, with same mass, radius, and running at the same speed of rotation. How much energy is stored in each system?
29. If the speed of rotation in the previous problem decreases to 2,500 RPM during discharge phase, what is the useful energy density for this device?
30. Estimate the energy stored in a CAES with capacity of 250,000 m^3, if the reservoir air presses is 60-bar, and the overall system efficiency is 33%. Estimate also the total generated electricity, if the air is discharged over a period of 90 min.
31. A flywheel is constructed in a toroidal shape, resembling a bicycle wheel, has a mass of 300 kg. Assuming all mass is concentrated at 1.8 m, what is its RPM to provide 800 kW for 1 min? In your opinion, is this system physically possible?
32. Compute the kinetic energy density of flywheels made of alloy, GRFP, and CRFP.
33. What is the significance of DOD, and how this parameter is affecting the battery lifecycle?
34. A battery has an internal resistance of 0.02 Ω per cell needs to deliver a current of 25 A at 150 V to a load. If the cell electrochemical voltage is 1.8 V, determine the number of cells in series.
35. A battery stack, used as an electrical energy storage system is designed for 20-MW peak power supply for duration of four hours. This energy storage system uses 600 Ah batteries operating at 420 V DC. Estimate the stack minimum number of batteries and the current in each during peak operation.
36. For a lead–acid battery having a nominal Peukert capacity of 60 Ah, assuming 1.2 Peukert coefficient plot the capacity for different discharge rates and for different currents.
37. The voltage of a 12.5-V lead–acid battery is measured as 10.5 V when delivers a current of 50 A to an external (load) resistance. What are the load and the battery internal resistances? What is the battery voltage and current through a load of 7.5 Ω?
38. An automobile storage battery with an open-circuit voltage of 12.8 V is rated at 280 Ah. The internal resistance of the battery is 0.25 Ω. Estimate the maximum duration of current flow and its value through an external resistance of 1.85 Ω.

39. Under standard (normal) operation conditions a fuel cell, hydrogen–oxygen type generates a 3.5 A current and a 0.85 V. What are the fuel cell internal resistance and its energy efficiency?
40. A 2.45 MW hydrogen–oxygen fuel cell stack, designed to produce electrical energy for a small electric network (microgrid) has an efficiency of 74%. Calculate the flow rates (kmol/s) of hydrogen and oxygen, and the fuel cell stack losses.
41. A hydrogen–oxygen has a liquid waver as product when it operates at 0.825 V. What is the electrical energy output in kJ/kg·mole of hydrogen, and the cell efficiency?
42. A hydrogen–oxygen fuel cell stack produces 60 kW of DC power at an efficiency of 63%, with water–vapor as product. What is the hydrogen mass flow rate, in g/s, and the cell voltage?

Chapter 12

Distributed generation, microgrids, thermal energy storage, and micro-combine heat and power generation

Objectives and abstract

Energy sustainability is the cornerstone to the health and competitiveness of the industries in our global economy. It is more than being environmentally responsible, means the ability to utilize and optimize multiple sources of secure and affordable energy for the enterprises, and then continuously improve the utilization through systems analysis, energy diversification, conservation, and intelligent use of these resources. Distributed energy resources (DER) and dispersed generation systems are becoming more important in the future electricity generation. A description of distributed energy resource and types, characteristics, performances, is the subject of this chapter. Brief presentations of the power system interfaces, power electronics, and control of distributed generation systems are also included. The chapter presents an overview of the key issues concerning the integration of distributed and dispersed generation systems, the role of thermal energy storage (TES) systems and the main applications. A synopsis of the main challenges and issues that must be overcome in the process of DG and DER applications and integration are presented. Particular emphasis is placed on the need to move away from the fit and forget approach of connecting DG to electric power systems to a policy of integrating DG into power system planning and operation through active management of distribution networks and application of other novel concepts. Several distributed energy systems, together with energy storage capabilities, expected to have a significant impact on the energy market are presented and discussed. Microgrid is a new approach of power generation and delivery system that considers DG, DER, and loads, often controllable loads is set as a small controllable subsystem of a power distribution network. The microgrid subsystem has characteristics, such as the ability to operate in parallel or in isolation from the electrical grid, having the capabilities and functionalities to improve service and power quality, reliability, and operational optimality. Microgrids may also be described as a self-contained subset of indigenous generation, distribution system assets, protection and control capabilities, and end user loads that may be operated in either a utility connected mode or in an isolated from the utility mode. In addition to providing reliable electric power supply, microgrids are also capable of providing a wide array of ancillary services, such as voltage support, frequency regulation, harmonic cancellation, power factor correction, spinning, and non-spinning reserves. A microgrid may be intrinsically distributive in nature including several DGs—both renewable and conventional sourced energy storage elements, protection systems, end user loads, and other elements. In order to achieve a coordinated performance of a microgrid (or several microgrids) within the scope of a distribution company, it is required to perform distributed or cooperative control.

This harvested energy through such applications can be released onto the grid, when needed, to eliminate the need for high-cost peak generators or can be used local for heat and hot water or other industrial process applications. Micro-combined heat and power (CHP) systems powering up to about 10 kWe are considered as a future key technology for the building or facility energy supplies from the viewpoints of heating system users, manufacturers, and energy suppliers. CHP plants can be based on conventional diesel, gas or biomass engines, gas turbines, Stirling engines, or fuel cells. Energy storage systems are an important component of the renewable energy technology applications. Among the storage technologies, the TES, a technology that stocks thermal energy by heating or cooling a storage medium and use the stored energy at a later time for heating, cooling and power generation. TES systems are used particularly in buildings and in industrial processes, while the main advantages of using TES in an energy system, building or industrial process include an increase in overall efficiency and better reliability, leading to the reductions in investment and running costs, and less environmental pollution of the environment. Energy storage inclusion into distributed generation systems provides the user dispatchability of DER, while improving the overall system performances and capabilities. All of the DER and DG technologies require specific power electronics and control schemes to convert the generated power into useful power that can be directly interconnected with the grid or that can be used for specific applications. This chapter presents convenient resources to understand the current state-of-the art power electronic interfaces for DER and DG applications. In this chapter, a description of TES systems and micro-CHP generation systems is presented with references to heating, ventilation, and air conditioning systems. A discussion on the major components of such systems, load analysis and methods for improving the energy efficiency of existing systems are also included in this chapter. After completing this chapter, the readers are able to understand the importance and role of the thermal energy systems and storage, energy conservation and efficiency in building electrical and mechanical systems, and in industrial energy systems and equipment. A special attention is given to the understanding and learning about micro-CHP generation systems, components and configurations of such systems, their operation, functions, and capabilities.

12.1 Introduction, distributed, and dispersed generation

Electricity and energy demands are growing; together there with a significant increase in the distributed generation (DG) penetration as one of the ways to fulfill these increased energy demands. Distributed energy (DE) systems already have and are expected to have even more significant impacts on the energy market in the near future. DE systems include, but are not limited to, photovoltaics (PV), wind energy conversion systems, microturbines, geothermal energy systems, fuel cells, cogeneration, Sterling engines, and internal combustion engines. In addition, several energy storage systems, such as batteries, fuel cell stacks, flywheels, and thermal energy storage (TES) are also under consideration to harness excess electricity produced by the most efficient generators during the off-peak and low demand periods and to use the stored energy during the peak periods or when needed. The harvested energy can be fed into grid, when is needed, to reduce the need for high-cost peak-load generators. Inclusion of the energy storage units into the DG system actually provides the dispatchability of the DERs which are often renewable energy sources, such as PV, wind energy conversion systems, and solar-thermal energy systems, having no dispatchability by their own. Modern power systems generate and supplies electricity through a complex process, consisting of electricity generation in large power plants, usually located close to the primary energy source (coal mines or water reservoir), usually far away from the large consumer centers, then delivering the generated energy to the customers using a large passive but complex power distribution infrastructure, which involves high-voltage (HV), medium-voltage (MV), and low-voltage (LV) networks. The power distribution networks are designed to operate mostly radially, in which the power flows only in one direction: from upper voltage levels down-to customers situated along the distribution feeders. In this process, there are three stages to be passed through before the electricity is reaching the final users, i.e., *generation*, *transmission*, and *distribution*. Nowadays, the technological advancements, environmental policies, and the expansion of the finance and electrical markets, are promoting significant changes into the electricity industry. New technologies allow the electricity to be generated in smaller-sized plants and in distributed generation units, located in the MV and LV grid sections, closer to the customers. Moreover, the increasing use of renewable energy sources in order to reduce the environmental impacts of power generation leads to the development and application of new electrical energy supply schemes. In this new conception, the generation is not exclusive to the grid generation end but is shifted MV and LV grid sections. Hence some of the energy-demand is supplied by the centralized generation and another part is produced by distributed generation. The electricity is going to be produced closer to the customers. Large-scale integration of distributed generators is at present the trend followed in power systems to cover the supply of some loads. These generators are of considerable smaller size than the traditional generation units.

DE systems are not new in a technological respect, however, are receiving increased attention because their ability to be used in CHP applications, to provide peak power, demand reduction, back-up power, improved power quality, and ancillary services to the power grid. A wide range of power generation technologies are currently in use or development, including small gas turbines and microturbines, small steam turbines, fuel cells, small-scale hydroelectric power units, photovoltaics, solar-thermal energy, wind turbines, energy storage systems etc., as

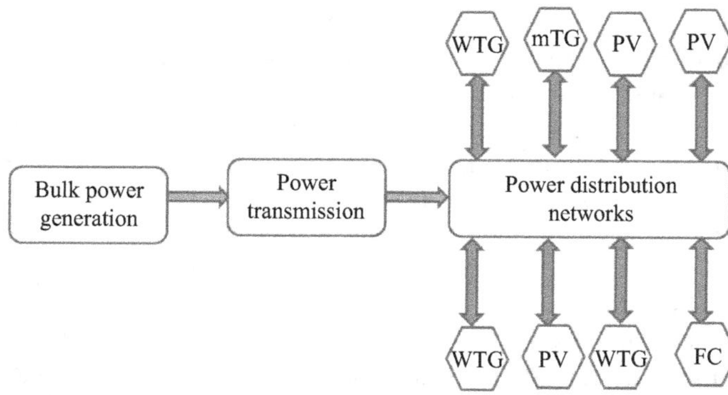

Figure 12.1 Distributed energy resources on the power distribution network

shown in Figure 12.1. The DER benefits in relation to the transmission and distribution (T&D) include the reduction in system losses, enhanced service reliability and quality, improved voltage regulation, or relieved T&D congestion. However, the DER integration into an existing power distribution system has several impacts on the system, the power system protection being one of the major issues, and the DG integration of DG into grids has associated several technical, economical, and regulatory issues. DG may cause the system to lose its radial power flow, besides the increased fault level of the system caused by the interconnection of the DG. Short circuit of a power distribution system changes when its state changes or when generators in the distribution system are disconnected. This may result in elongation of fault clearing time and hence disconnection of equipment in the distribution system or unnecessary operation of protective devices. Therefore, new protection schemes for both DG and utility distribution networks have been developed but the issue has not been properly addressed, yet. On the other sides, the power system distribution are well designed which could handle the addition of generation if there is proper grounding, transformers, and protection are provided. But there are limits to the addition of distributed generations, if it goes beyond such limits, it is critical to modify and change the existing power distributed system equipment and protection, which could result to facilitate the DER and DG integration. This addition of the equipment could involve protection relays, switchgears, change of the voltage regulation system, revised grounding, and transfer trips.

Adding CHP capabilities to DER systems or facilities can increase the overall system efficiencies to as much as 80%, as reported in the literature, a dramatic improvement over producing electricity and heat separately, which generally has efficiencies of about 45% or lower. The improved efficiency by adding CHP results of meeting the thermal loads with the waste heat produced by the electricity generation. Customers with substantial thermal loads and high electricity prices have higher gains from investing in a CHP systems. In microgrids, for example, the inclusion of CHP can provide additional economic benefits to make DER economically attractive to customers who otherwise may have no enough incentives to join the microgrid. The process of converting the form of energy input to the final output form generally comprises a number of intermediate transformations or conversions. There are, for example, several conversion stages in any energy conversion process, each stage in the process has its conversion efficiency and the overall system efficiency is found by multiplying the individual efficiencies. Clearly, the more stages there are in a conversion process the lower is the overall system efficiency. This means not only a loss of useful energy but a higher financial cost as well, so any energy conversion process must be designed with fewer possible stages. Another way to get round to the efficiency problem, is to harness the waste heat from the process and to use it elsewhere. For instance, the exhaust gases from a boiler in an agro-processing facility, containing a substantial amount of heat, can be used, via a heat exchanger, to dry the product, to generate electricity, or for other purposes. By reducing both wasted energy and the input energy, increases the overall system energy efficiency. This idea is the basis of co-generation or CHP, where the waste heat from electricity generation is used as process heat in a factory. Rather than allowing the waste heat to run off into the atmosphere, the heat is used, via a heat exchanger, to provide hot water for an industrial process, to dry products, or to generate electricity. The overall efficiency of a co-generation system can be 80% or higher. Using the energy of the original source for two or more applications is known as cascading, where the energy grade is closely matched to the available energy, so the efficiency of an energy conversion process depends not only on the equipment used but also on the input energy form. Some forms can be converted more efficiently than others, related to their actual potential to do work (or exergy). The higher the exergy content of an amount of energy, the easier is to do a certain task. After any conversion process, the exergy or ability to do useful work is less than it was before the conversion, either due to energy losses or energy quality degradation.

The chapter is structured as follows next sections and subsections are focused on a review of the thermal engineering basics, TES systems, the microgrid concepts, architecture and structure, micro-combined heat and power (CHP) generation and cogeneration concepts, structure, and applications. Other chapter sections are dedicated to the energy conservation and energy efficiency, summary, and relevant concept discussions, critical references, and end of chapter questions and problems. A chapter section is also dedicated to the microgrid challenges and issues, such as control, protection and islanding.

12.1.1 Thermal engineering basics

Thermal energy of a gas results from the kinetic energy of the microscopic movement of molecules. Temperature is a characteristic of a body thermal energy, due to the internal motion of molecules. Two systems in thermal contact are in thermal equilibrium if they have the same temperature. In general, if the phase change is not involved, the temperature of any material increases as it absorbs heat. The heat required to rise the system temperature, with *specific heat,* C by an amount ΔT is expressed as:

$$\Delta Q = m \cdot C \cdot \Delta T \tag{12.1}$$

Example 12.1: The specific heat of water is 4,180 J/(kg · °C). Calculate the energy required to increase the temperature of 1 kg of water with 25 °C.

Solution: Using (1.11) to solve for the heat gives:

$$Q = (1\,\text{kg}) \times [4{,}180\,\text{J}/(\text{kg}\,{}^\circ\text{C})] \times 25\,{}^\circ\text{C} = 1.045 \times 10^5\,\text{J}$$

However, heat and temperature are different. Heat is the energy and temperature is the potential for heat transfer from a hot to a cold place. Materials with large specific heat require a large amount of heat per unit of mass to rise their temperature by a given amount. These materials can store large amounts of thermal energy per unit of mass for a small increase in temperature, having applications in energy storage or solar thermal energy. In the heat transfer, work can also be done. Understanding of the energy concept arises from the laws of thermodynamics:

1. *Energy is conserved, cannot be created or destroyed, only transformed from on form to another*.
2. *Thermal energy, heat, cannot be transformed totally into mechanical work. Systems tend toward disorder, and in energy transformations, disorder increases*, entropy is a measure of disorder. This means that some forms of energy are more useful than other forms. In other words, the *heat naturally flow form hot place to a cold place.*

Mathematically, the process of energy transfer is described by the *first law of thermodynamics*:

$$\Delta E = W + Q \tag{12.2}$$

where E is the internal energy, W represents the work done by the system and Q represents the heat flow. By convention Q and W are positive, if heat flows into the system and work is done by the system. When the thermal efficiency of the system is high, the heat term in (12.2) is ignored:

$$\Delta E = W \tag{12.3}$$

This simplified equation represents one of the fundamental physics principles, for example, the one used to define the *joule* (J) unit. Errors often occur when working with energy, power, heat, or work. Units and quantities are mixed up frequently. Wrong usage of these quantities or units can change the statements and cause misunderstandings. A simple law understanding is considering the behavior of system formed from a hot metal and a cold metal, brought into thermal contact. The system attains thermal equilibrium by transferring heat from hot metal to cold one until the two pieces are at the same temperature. Other expression of this law is: *the entropy of the universe always increases*. Entropy is a measure of disorder. Thus, attaining thermal equilibrium increases the overall universe entropy.

Thermal energy can do work only if heat flows from the hot side to cold side of the system. An analogy is the conversion of potential energy into kinetic energy in systems. For example, if the water in a reservoir of a hydro-electric station remains in the reservoir, no electricity is generated. When the water is running down through the station generators, electricity is generated. In this process, the gravitational potential energy of the stored water is converted into kinetic energy, and subsequently into electrical energy. Similarly, the thermal energy contained in hot materials is converted into other energy forms, the principle of heat engine operation. If the heat is transferred from hot reservoir to cold reservoir, part of the thermal energy is converted into mechanical work, through the so-called *heat engine*. Examples of heat engine include steam turbines, internal combustion engines, jet engines, etc. Engine operation consists of removing heat from hot reservoir at temperature, T_H, while some of the heat is transferred to cold reservoir, at temperature T_C and part is used to develop mechanical work. If the removed heat from the hot reservoir is Q_H, and the deposited heat into cold reservoir is Q_C, then the relationship including the developed work is:

$$Q_H = Q_C + W \tag{12.4}$$

The terms *energy efficiency* and *energy conservation* have often been used interchangeably in policy discussions but they do have very different meanings. The meaning of *the energy conservation* is to reduce the energy consumption through lower quality of the energy services, e.g., lower heating levels, car speed limits, appliance consumption limits and capacity, often set by standards. It is strongly influenced by the regulations, consumer behavior, and lifestyle changes. Energy efficiency is simply the ratio of energy services out to the energy input, meaning getting the most out of every unit of input energy. It is mainly a technical process caused by stock turnover where old equipment is replaced by newer more efficient ones. Measuring energy efficiency, particularly on a macro-scale, is very difficult, there are methodological problems, while being very hard to measure between countries or sectors. *Efficiency* means different things to different professions engaged in achieving it. To the engineers, *efficiency* means a physical output/input ratio, while to the economists, *efficiency* means a monetary output/input ratio, and often confusingly, *efficiency* may refer to the economic optimality of a market transaction or process. However, the physical energy efficiency is defined as:

$$\eta = \frac{P_{OUT}}{P_{IN}} = \frac{P_{IN} - Losses}{P_{IN}} \tag{12.5}$$

Example 12.2: An electric motor consumes 100 W of electricity to obtain 87 W of mechanical power. Determine its efficiency.

Solution: Because power is the rate of energy utilization, efficiency can also be expressed as a power ratio. The time units cancel out, and we have:

$$\eta = \frac{\text{Power output}}{\text{Power input}} = \frac{87\ \text{W}}{100\ \text{W}} = 0.87 \text{ or } 87\%$$

Technical systems perform the energy conversions with various efficiencies. The ratio of Q_C and Q_H is equal to the ratio of the reservoir temperatures. For a heat engine, it can be written as:

$$\eta = 1 - \frac{Q_C}{Q_H} = 1 - \frac{T_C}{T_H} \tag{12.6}$$

The efficiency form involving cold and hot reservoir temperatures is more convenient because the temperature is much more easily measured quantity than heat. The efficiency as stated by (12.6) is known as *ideal Carnot efficiency*, after the name of French engineer Sadi Carnot, being the maximum efficiency attainable by a heat engine. Real heat engines typically operate at efficiencies lower than the Carnot efficiency. The ideal Carnot efficiency is valid for ideal processes, taking place in either direction equally, existing in the real world only for limited cases. Time reversible ideal processes require that net entropy of the S remain constant. An alternative definition of the second law of thermodynamics is that in any real process $dS > 0$. Working on the same principles, as any heat engine are the heat pumps, which are using the mechanical work to transfer heat from a cold reservoir to a hot one. Similar, conservation of energy requires that:

$$W + Q_C = Q_H \tag{12.7}$$

Example 12.3: What is the thermal efficiency of the most efficient heat engine that can run between a cold reservoir at 20 °C and a hot reservoir at 200 °C?

Solution: Carnot engine has the ideal or maximum efficiency, so:

$$\eta = 1 - \frac{T_C}{T_H} = 1 - \frac{273.15 + 20}{273.15 + 200} = 0.38 \text{ or } 38\%$$

Note also that a heat pump used to heat homes is the same device as a refrigerator or air conditioner in that heat is transferred from a lower temperature to a higher temperature at the expense of work input. Heat pumps have applications for the heat transfer from cold areas/reservoir to hot ones. They have applications in CHP generation, in recovering the wasted heat, or moving heat from the outside to the inside in a cold day. However, heat pumps maybe economically attractive for heating purposes, careful considerations of cost, local climate and other factors are necessary to assess their viability. Heat pump performances are expressed, through the *coefficient of performance* (COP), the ratio of heat deposited in the hot reservoir to the work done, is given by the following relationship:

$$COP = \frac{Q_H}{W} \tag{12.8}$$

COP is a quantity greater than 1 (or, in percent, greater than 100%). Using the relationship between heat and temperature, COP can be expressed as:

$$COP = \frac{Q_H}{Q_H - Q_C} = \frac{T_H}{T_H - T_C} \tag{12.9}$$

In the case of refrigerator, the COP expression is:

$$COP = \frac{Q_C}{Q_H - Q_C} = \frac{T_C}{T_H - T_C} \tag{12.10}$$

The only difference with the heat pump is that in a refrigerator, Q_C is the desired heat transfer (from an object, to make it colder than ambient temperature) and Q_H is waste (heat transfer to the surroundings at higher than ambient temperature, which is why your cats love to sleep behind your refrigerator), whereas with a heat pump Q_H is the desired heat transfer (to your living room) and Q_C is waste (heat transfer from the cold outside environment). Because of this, the definition of COP for a heat pump is different from that of a refrigerator, as expressed in (12.9). However, for an actual (real) heat pump, COP is less:

$$COP < \frac{T_H}{T_H - T_C} \tag{12.11}$$

The heat pump advantage compared to a simple electrical heating is that several units of heat transfer (Q_H) can be obtained for one unit of work transfer (W), whereas with simple electrical heating, only one unit of heat transfer is obtained per unit of work transfer. Of course, as the difference between T_H and T_C increases, COP decreases and thus the advantage of the heat pump decreases. Moreover, heat pumps, being functionally equivalent to air conditioners or refrigerators, are far more complicated than simple electrical heaters. For this reason, heat pumps are not in widespread use, even in locations where air conditioners are installed and you could use the same device run forward and backward for both heating and cooling. However, we have to keep in mind that heat pumps need electrical power whereas heating can also be done with natural gas which costs about 25% as much for the same energy.

There six key quantities, useful in the description of a thermal system, such as thermal power plant: temperature T, pressure p, specific volume (the inverse of the density, i.e., the volume per unit of mass), specific internal energy u, specific enthalpy h, and specific entropy s. However, only two thermodynamic quantities are strictly needed to completely specify the thermal state of a system. We already introduced the concept of internal energy (first law of thermodynamics). Here the specific quantities are referred the unit of mass. Specific enthalpy is defined as:

$$h = u + pv \tag{12.12}$$

Enthalpy is very useful in describing the heat transfer at constant pressure (e.g., in boilers, or condensers), where the change in the enthalpy is equal to the heat input, or in adiabatic ($Q = 0$) compression or expansion (e.g., compressors and turbines), where the network on the shaft is equal to the change in enthalpy. The concept of entropy arises from the second and is a measure of the degree of disorder of a system. From the thermodynamic point of view there are two types of processes: reversible and irreversible processes. In the first one, the system and the surroundings can recover their initial states, by changing the system slowly enough that it remains in a quasi-static thermal equilibrium throughout the process. In irreversible process, the system and the surroundings are changed in such way that they are not able to return to their original states. Mathematically, the entropy change is expressed as:

$$\Delta s = \frac{\Delta Q_{rev}}{T} \tag{12.13}$$

Here, ΔQ_{rev} is the heat supplied to the system reversible at absolute temperature, T. Notice that there is no change in entropy in a reversible adiabatic process, where $\Delta Q_{rev} \approx 0$, while in an irreversible process there is a net increase in entropy.

Basically there are three types of heat transfer: *conduction*, *convection*, and *radiation*. Conduction is the thermal transfer process due to the molecule random motions. The average molecule energies are proportional to the temperature. The heat flow rate, in the steady-state along the length of bar, d of cross-sectional area, A with one end at higher temperature, T_1 and the other at lower temperature, T_2 is given by the Fourier law of heat conduction:

$$Q = kA\frac{T_1 - T_2}{d} \tag{12.14}$$

where k (W/mK) is the thermal conduction of the bar material.

Example 12.4: A steel bar of 2 m and a cross-sectional area of 10 cm^2 have a temperature of 1,200 °C at one end and 200 °C at the other end. Calculate the heat thermal flow along the bar in the steady-state, ignoring the heat losses from the bar surface, if the steel thermal conductivity is 50 W/mK.

Solution: From (12.32), the heat flow along the bar is calculated as:

$$Q = 50 \times 10 \times 10^{-4}\frac{1{,}200 - 200}{2} = 25 \text{ W}$$

Convection represents the heat transfer through the fluid bulk motion, the actual movement, or circulation of a substance. It takes place in fluids, such as water or air where the materials are able to flow. Much of the heat transport that occurs in the atmosphere and ocean is carried by convection. However, the atmospheric circulation consists of vertical as well as horizontal components, so both vertical and horizontal heat transfer occurs. Considering a fluid of density ρ, temperature T, and moving with velocity v, then heat flow rate per unit of area is the product of the mass flow per unit area per second, ρv and the thermal energy per unit mass, cT:

$$\frac{Q}{A} = \rho v c T \tag{12.15}$$

When a cold fluid is forced to flow over a hot surface the heat transfer rate in this forced convection from the surface to the fluid is higher than in the case of a stationary fluid. The temperature gradient at the surface is very large, and the fluid layer above the surface is rapidly heated by thermal convection, and the heat transfer rate per unit the area is often expressed as:

$$\frac{Q}{A} = Nu\frac{k(T_S - T_\infty)}{L} \tag{12.16}$$

Here, T_S and T_∞ are the surface and the fluid temperatures, L is the characteristic length, and Nu is the dimensionless Nusselt parameter. The choice of L depends on the fluid-surface geometry, while the Nusselt parameter is function two other nondimensional parameters, the Prandtl and Reynolds numbers, depending the fluid mechanical and thermal properties, obtained from empirical correlations.

Radiative heat transfer represents the energy transfer through the electromagnetic radiation, which includes the heat transfer into the vacuum. The power per unit area transferred from a surface at temperature T is given by the Stefan-Boltzmann law:

$$P_{rad} = \varepsilon\sigma T^4 \tag{12.17}$$

Here, ε is the surface emissivity (a dimensionless number ranging from 0 to 1, depending on the surface nature), $\sigma \approx 5.67 \times 10^{-8}$ W/m^2 K^4 is the Stefan-Boltzmann constant. Opaque surfaces absorb electromagnetic radiation from the environment. The surface absorptivity and emissivity are the same. The absorption rate per unit of area, for the environment temperature T_0 is:

$$P_a = \varepsilon\sigma T_0^4$$

The net emission rate per unit of area per unit of time is then given by:

$$P_{rad} = P_e - P_a = \varepsilon\sigma\left(T^4 - T_a^4\right) \tag{12.18}$$

A blackbody is a surface that is absorbing all incident electromagnetic radiation. The outer Sun surface determines the electromagnetic radiation flux incident to the upper Earth atmosphere. Another important relationship is the one describing mathematically the relationship between the temperature (T) of the radiating body and its wavelength of maximum emission (λ_{max}), the Wien's displacement law:

$$\lambda_{max} = \frac{C}{T} \tag{12.19}$$

Here, C, the Wien's constant is equal to 2,898 μm · K.

Example 12.5: By using Wien's law, estimate the maximum wavelengths of the Sun and the Earth, assuming that the Sun temperature is 6,000 K, the one of the Earth is 300 K.

Solution: Applying Wien's law for the Sun and the Earth temperatures, we found:

$$\lambda_{max}(Sun) = \frac{2{,}898}{6{,}000} = 0.483\ \mu\text{m}$$

And

$$\lambda_{\max}(Earth) = \frac{2{,}898}{300} = 9.660\ \mu\text{m}$$

Radiation is often identified by the effects it produces when it interacts with an object. We are dividing radiant energy into categories based on our ability to perceive them; however, all wavelengths of radiation behave in a similar manner. An important difference among the various wavelengths is that the shorter wavelengths are more energetic. Sun emits all radiation forms, in varying quantities with over 95% of solar radiation emitted in wavelengths between 0.1 and 2.5 μm, with much of the energy concentrated in the visible and near-visible parts of the electromagnetic spectrum. The visible light spectrum band, wavelengths between 0.4 and 0.7 μm, represents over 43% of the total energy emitted, while the rest lies in the infrared, about 49%, and ultraviolent, about 7%. In order to have a better understanding, how the Sun radiant energy interacts with Earth's atmosphere and land-sea surface, it is helpful to have a general understanding of the basic radiation laws. However, this is beyond the scope of this book and the interested readers are directed to elsewhere in the literature. The radiation is the dominant mode energy transfer mode in the power plant furnaces. The atmosphere effects on the electromagnetic radiation transmission are very important in determining the Earth surface temperature.

Example 12.6: The temperature on the inside wall of a house is 80 °F and the outside temperature is 10 °F. The house wall is 13.5 ft. high and 360 ft. wide. Assuming for this house, a bare wall above is exposed to ambient air, what is the rate of heat transfer by radiation? The wall emissivity, ε is 0.5.

Solution: The area of the house wall, expressed in square meters is:

$$A = 13.5 \times 360 \times \left(\frac{m}{3.281\ \text{ft.}}\right)^2 = 83.605\ \text{m}^2$$

The wall total heat transfer, by using (12.18) is then:

$$\dot{Q} = \sigma\varepsilon \cdot A \cdot \left(T_{Indoor}^4 - T_{Outdoor}^4\right)$$

$$= 5.67 \times 10^{-8}\frac{\text{W}}{\text{m}^2\text{K}^4} \times 0.5 \times 83.605 \times \left((80+460)^4 - (10+460)^4\right)\left(\frac{1\ \text{K}}{1.8\ \text{R}}\right)^4$$

$$= 8{,}181\ \text{W}$$

Considering the heat transfer mechanisms, it is worth to notice that since the conduction and convection increase linearly with temperature, *radiation increases with temperature to the fourth power, thus at higher temperatures, heat transfer by radiation always exceeds that due to conduction and convection.*

12.2 Energy conservation and efficiency in building and industrial energy systems

Energy efficiency, energy conservation, and renewable energy technologies are the pillars of the sustainable energy policy. There are several reasons to pursue energy conservation and efficiency and the renewable energy use, such as increasing and diversifying our energy supply, making the existing energy supply to last longer or to have cleaner and less affected environment. Greater energy efficiency and higher level of the energy conservation are critical issues in sustainable development. For example, increased energy efficiency and energy conservation are providing to any country significant economic, environmental and energy security benefits. Higher energy efficiency is translating directly into lower costs of the energy, products, and services and higher life standard. Energy efficiency and conservation methods can also prolong the equipment life. The two concepts are closely related but they have different connotations. Efficiency involves using less energy to achieve the same results, while the conservation is putting the stress on using less energy even there may be a compromise on the results. An energy balance is a set of relationships accounting for all the energy which is produced and consumed, and matches inputs and outputs, in a system over a given time period. The system can be anything from a whole country to an area to a process in a factory. An energy balance is usually made with reference to a year, though it can also be made for consecutive years to show variations over time. Energy balances provide overviews, and are basic energy planning tools for analyzing the current and projected energy situation. The overviews aid sustainable resource management, indicating options for energy saving, or for policies of energy pricing and redistribution, etc. An energy conversion device or equipment, as one represented in Figure 12.2 has the efficiency defined as the ratio of the energy output to the energy input:

$$\text{Equipment (device) efficiency} = \frac{\text{Useful energy output}}{\text{Energy input}} \tag{12.20}$$

The meaning of the word *useful* depends strongly on the device or equipment purpose. Energy efficiency of a device, equipment, or process is a quantitatively dimensional values between 0 and 1, the larger the higher the efficiency it is. The concept of efficiency embodies the laws of thermodynamics. On the other hand, even almost of the energy on our planet comes directly from the sun, we are not yet able how to harness on large scale solar energy directly and efficiently. Instead, we have in a large measure to rely on the chemical energy of fossil fuels for most of

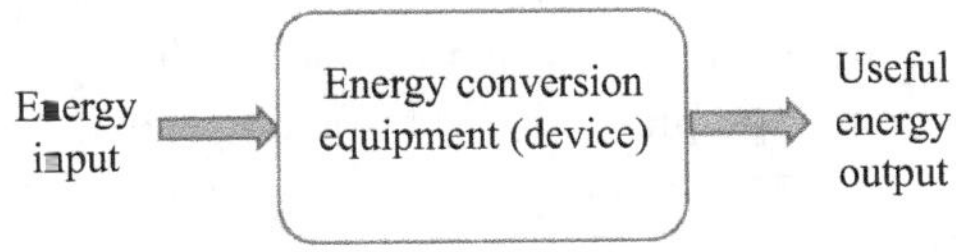

Figure 12.2 Schematic diagram of energy conversion equipment (device)

our energy needs. The problem with chemical energy is that it is a potential energy form and must be converted to other energy forms before it can be used. The only way to exploit the stored solar energy is to release it by burning the fossil fuels, through the combustion processes and eventually convert it in mechanical, electrical, or other energy forms. The conversion processes of the stored energy involve two or more conversion devices, each with its own efficiency, so it is useful to introduce the *system efficiency*. A system, consisting of two or more devices or subsystems, means a well-defined space in which two or usually more than two energy conversions take place. The efficiency of a system is equal to the product of efficiencies of the individual devices or subsystems, and the concept is exemplified in the following example.

Example 12.7: Calculate the overall efficiency of a power plant, if the efficiencies of the boiler, steam turbine, generator, and step-up transformer are 90%, 42%, 96%, and 98%, respectively.

Solution: The overall (system) efficiency of the power plant is the product of the conversion subsystem efficiencies.

$$\eta_{\text{Power plant}} = \eta_{\text{Boiler}} \times \eta_{\text{Turbine}} \times \eta_{\text{Generator}} \times \eta_{\text{Transformer}} = 0.90 \times 0.42 \times 0.96 \times 0.98$$
$$= 0.3556 \text{ or } 35.56\%$$

Note that the system efficiency is lower than any one of the efficiencies of the individual components of the system. In the case of this electric power plant, only 35% of the chemical energy input is converted to electricity, while the rest is lost to the environment, mostly as heat. However, if part of the wasted heat is recovered, via CHP methods the overall power plant (system) efficiency can be significantly improved.

The building sectors are accounting for about 40% of the total energy uses in the US, EU, Canada, and most of developed countries. However, at the same time, the buildings sectors have a documented cost-effective savings potential of up to 80%, which can be effected over the next 40 years. In order to ensure energy conservations and savings, while in the same time to use renewable energy in an optimal way, building integrated energy technologies are required. Energy demands in the commercial, industrial, residential and utility sectors vary on a daily, weekly, and seasonal basis. These demands are matched by various energy conversion systems, operating synergistically. Peak hours are the most difficult and expensive to supply, being met by conventional gas turbines or diesel generators, which are reliant on costly oil or natural gas. Energy storage provides alternative ways to supply peak energy demands or to improve the operation of DG, solar, wind facilities. Energy storage plays also significant roles in energy conservation. In processes where there is a large recoverable wasted energy, the storage can result in savings of premium fuels. TES is one of the key technologies for energy conservation, being of great importance in energy sustainability, and a well-suited technology for hot water, heating, and cooling thermal applications that can be easy embedded into distributed generation, contributing significantly to the overall system efficiency improvement. TES deals with the stored energy by cooling, heating, melting, solidifying, or vaporizing a material, while the thermal energy becomes available when the process is reversed. Large TES systems are employed for applications, ranging from solar hot water storage to building air conditioning systems. However, advanced TES technologies have only recently been developed to a point that they have significant impacts on modern industries. TES appears to be an important solution to correcting the mismatch between the energy supply and demand. TES can contribute significantly to meet the needs for more efficient, environmentally benign energy use. TES is a key component of many thermal systems, and a good TES should allow little thermal losses, leading to energy savings, while permitting the highest reasonable extraction efficiency of the stored thermal energy.

TES is a key component of many successful building and industrial energy systems, minimizing thermal energy losses, enhancing energy savings, and permitting the highest appropriate extraction efficiency of the stored energy. The design and selection criteria for TES systems are examined. Further, energy-saving techniques and applications are discussed and highlighted with illustrative examples. TES is considered by many to be an *advanced energy technology*, and there has been increasing interest in using this essential technology for thermal applications, such as hot water, space heating, cooling, air-conditioning, and so on. TES systems have enormous potential for permitting more effective use of thermal energy equipment and for facilitating large-scale energy substitutions. The resulting benefits of such actions are especially significant from an economic perspective. In general, a coordinated set of actions has to be taken in several energy system sectors for the maximum TES potential benefits to be realized. TES appears to be the best means of correcting the mismatch that often occurs between the supply and demand of thermal energy. The first step of a TES project is to determine the energy load profile of the building. Parameters influencing the building demand and load profile are the building use, internal loads, and the climatic conditions. Following steps are to determine the type and amount of storage appropriate for the particular application, the effect of storage on system performance, reliability and costs, and the storage systems or designs available. It is also useful to characterize the TES types in relation to the storage duration. Short-term storage is used to address peak loads lasting from few hours to a day in order to reduce the system size, taking advantage of the energy-tariff daily structures. Long-term storage is used when waste heat or seasonal energy loads can be transferred with a delay of a few weeks to several months. Related to the energy storage amount required, it is important to avoid both undersized and oversized systems. Undersizing the systems result in poor performance levels, while oversizing results in higher initial costs and energy waste if more energy is stored than is required. The effect of TES on the overall energy system performance should be evaluated in details. The economic justification for storage systems requires that the annualized capital and operating costs for TES be less than those required for primary generating equipment supplying the same service loads and periods. TES systems are an important element of many energy-saving programs in a variety of sectors, residential, commercial, industrial, and utility, as well as in the transportation sector. TES can be employed to reduce energy consumption or to transfer an energy load from one period to another. The consumption reduction can be achieved by storing excess thermal energy that would normally be released as waste, such as heat produced by equipment, appliances, lighting, and even by occupants. Energy-load transfer is achieved by storing energy at a given time for later use, and can be applied for either heating or cooling capacity. The consumption of purchased energy can be reduced by storing waste or surplus thermal energy available at certain times for use at other times. The demand of purchased electrical energy can be reduced by storing electrically produced thermal energy during off-peak periods to meet the thermal loads that occur during high-demand periods. There has been an increasing interest in the reduction of peak demand or transfer of energy loads from high- to low-consumption periods. The use of TES can defer the need to purchase additional equipment for heating, cooling, or air conditioning applications and reduce equipment sizing in new facilities. The relevant equipment is operated when thermal loads are low to charge the TES, and energy is withdrawn from storage to help meet the thermal loads that exceed equipment capacity.

Energy storage density, in terms of the energy per unit of volume or mass, is an important factor for optimizing solar ratio (how much solar radiation is useful for the heating/cooling purposes), efficiency of appliances (solar thermal collectors and absorption chillers), and energy consumption for space heating or cooling room consumption. Therefore, the possibility of using phase-change materials (PCMs) in solar system applications is important to be investigated. PCMs might be able to increase the energy density of small-sized water storage tanks, reducing storage volume for a given solar fraction or increasing the solar fraction for a given available volume. It is possible to consider thermal storage on the hot and/or cold side of the building, allowing the hot water storage from the collectors and the auxiliary heater to be supplied to the generator of the absorption chiller (in cooling mode) or directly to the users (in heating mode). The latter allows the storage of cold water produced by the absorption chiller to be supplied to the cooling terminals inside the building. It is usual to identify three situations as "hot", "warm", and "cold" storage based on the different temperature ranges. Typically, a hot tank may work at 80 °C–90 °C, a warm tank at 40 °C–50 °C, and a cold tank at 7 °C–15 °C. While heat storage on the hot side of solar plants is always present because of heating and/or domestic hot water (DHW) production, cold storage is justified in larger plants. Cold storages are used not only to gain economic advantages from lower electricity costs (in the case of electric compression chillers) depending on the time of day but also to lower the cooling power installed and to allow more continuous operation of the chiller. The use of thermal storage, initially, could not provide effective backup but helped the system to thermally stabilize. Consequently, thermal storage found use in solar-assisted thermal systems. Since then, studying TES technologies as well as the usability and effects of both sensible and latent heat storage in numerous applications increased, leading to a number of reviews. These reviews focused only on one side (cold or hot) or component of the system or one of its integral mechanisms.

12.3 Thermal energy storage systems

TES technologies store thermal energy for later use as required, rather than at the time of production. They are therefore important counterparts to intermittent renewable energy generation and also providing a way to use the waste process heat and reducing the energy demand of buildings, facilities, and industrial processes. The use of the TES in buildings, industrial facilities, and processes in combination with distributed generation, space heating, hot water, and space cooling is an important way to conserve and to efficiently use energy. A variety of TES techniques have been developed, including building thermal mass utilization, phase change materials (PCM), underground TES, heat pumps, and energy storage tanks. TES systems can be either centralized or distributed systems. Centralized TES systems are used in district heating or cooling systems, large industrial plants, large CHP units or renewable energy plants. Solar thermal systems are often applied in residential and commercial buildings to capture solar energy for water and space heating or cooling. In such cases, TES systems can reduce the energy demand during peak times. The economic performance depends significantly on the specific application and operational needs, and the number and frequency of storage cycles. TES systems for cooling or heating are used where there is a time mismatch between the demand and the economically most favorable energy supply. TES can provide short-term storage for peak shaving as well as long-term storage for the introduction of renewable and natural energy sources. Sustainable buildings need to take advantage of renewable and waste energy to approach ultra-low energy buildings. TES systems are methods that enable the collection and preservation of excess heat for later utilization. Practical situations where TES systems are often installed are solar energy systems, geothermal systems, DG units and other energy conversion systems where heat availability and peak use periods do not coincide. The three basic types of TES systems are *sensible heat storage, latent heat storage, and thermochemical heat storage*. Energy storage by causing a material temperature to change is the sensible heat storage, for which the system performance depends on the storage material specific heat, if the volume is important, on the density. Sensible heat storage systems usually use rocks, ground, oil, or water as the storage medium. Latent heat storage systems store energy in PCMs, with the thermal energy stored when the material changes phase, e.g., from a solid to a liquid. Latent heat storage systems use the fusion heat, needed or released when a storage medium changes phase by melting, solidifying, liquefaction, or vaporization. Thermochemical energy storage is based on chemical reactions in inorganic substances. The TES choice depends on the required storage time period, e.g., day-to-day or seasonal, and outer operating conditions. The specific heat of solidification or vaporization and the temperature at which the phase change occurs are design criteria. Both sensible and latent heat types may occur in the same storage material. Notice that the oldest TES form involves harvesting ice from lakes and rivers, storing it in well-insulated warehouses for use throughout the year for almost all tasks that mechanical refrigeration satisfies today, preserving food, cooling drinks, and air-conditioning.

The need of TES is often linked to the cases, where there is a mismatch between thermal energy supply and energy demand, when intermittent energy sources are utilized and for compensation of the solar fluctuations in solar heating systems. Possible technical solutions to overcome the thermal storage needs may be the following: building production over-capacity, using a mix of different supply options, adding back-up/auxiliary energy systems, only summer-time utilization of solar energy, and short/long-term TES. In traditional energy systems, the need for thermal storage is often short-term one and therefore the technical solutions for TES may be quite simple, and for most cases the water is used as storage medium. Large volume sensible heat systems are promising technologies with low heat losses and attractive prices. Sensible heat based TES is based on the temperature change in the material and the unit storage capacity (J/kg) is equal to the medium heat capacitance times the temperature change. Phase-change based energy storage involves the material changes its phase at a certain temperature while heating the substance then heat is stored in the phase change. Reversing, heat is dissipated when at the phase change temperature it is cooled back. The storage capacity of the phase change materials is equal to the phase change enthalpy at the phase change temperature plus the sensible heat stored over the whole energy storage temperature range. The sorption or thermo-chemical reactions provide thermal storage capacity, based on the endothermic chemical reaction (when a chemical reaction absorbs more energy than it releases) principle:

$$\mathrm{AB} + \mathrm{Heat} \Leftrightarrow \mathrm{A} + \mathrm{B} \tag{12.21}$$

The heat used for a compound AB to be broken into components can be stored separately, and later by bringing the A and B compounds together to from again the AB compound the heat is released.

The energy storage capacity is the reaction heat or the reaction free energy. TESs based on chemical reactions have negligible losses whereas the sensible heat storage systems dissipate the stored heat to the environment and need to be isolated to perform efficiently. Materials are the key issues for TES. There are a large range of different materials that can be used for thermal storage as shown by Table 12.1. One of the most common storage medium is the water. TES in various solid and liquid media or materials is used for solar water heating, space heating, and cooling as well as high-temperature applications, such as solar-thermal power generation. Important parameters in a storage system include the storage duration, energy density (or specific energy), and the charging and discharging cycle (storage and retrieval) characteristics. However, the energy density is a critical factor for the size and application of any energy storage system. The rate of charging and discharging depends on thermo-physical properties, such as thermal conductivity and design of the TES systems. TES systems are dealing with the storage of energy by material

Table 12.1 Examples of materials suitable for thermal energy storage

Thermal storage methods	Material	Thermal storage methods	Material	Thermal storage methods	Material
Sensible heat	Water, ground rocks, ceramics	Phase change	Inorganic Salts Organic and Inorganic Compounds Paraffin	Thermo-chemical reactions	***Working fluid***: Water, ammonia, hydrogen, carbon dioxide, alcohols ***Sorption materials***: Hydroxides, hydrates, ammoniates, metal hydrides, carbonates, alcoholates

cooling, heating, melting, solidifying, or vaporizing, the thermal energy becomes available when the initial process is reversed. The materials that store heat are typically well insulated. Primary disadvantage of a TES system, similar to other energy storage technologies is the large initial investments required to build the energy storage infrastructure. However, it has two primary advantages: (1) the energy-system efficiency is improved with the implementation of a TES system (CHP has approximately 85%–90% efficient while conventional power plants are only 40% efficient or lower), and (2) these techniques have already been implemented with good results. On the negative side, TES does not improve flexibility within the transportation sector like the hydrogen energy storage system, this not being a critical issue. TES does have disadvantages but these are small comparing to its advantages. Due to the efficiency improvements and maturity of these systems, it is likely that they are becoming more prominent, enabling the utilization of intermittent renewable energy but also to maximize the fuel use within power plants. These systems already put into practice with promising results. Therefore, it is evident this technology can play a crucial role in future energy and power systems.

TES operation and system characteristics are based on the thermodynamics and heat transfer principles and laws. There are two major TES types for storing thermal energy, sensible heat storage, and latent heat storage. First consist of change the temperature of a liquid or solid, without changing its phase. Thermal energy quantities differ in temperature, and the energy required E to heat a volume V of a substance from a temperature T_1 to a temperature T_2 is expressed by the well-known relationship:

$$E = m\int_{T_1}^{T_2} CdT = mC(T_2 - T_1) = \rho VC(T_2 - T_1) \tag{12.22}$$

where C is the specific heat of the substance, m is the mass, and ρ is its density. The energy released by a material as its temperature is reduced, or absorbed by a material as its temperature is increased, is called the sensible heat. Second type of energy storage implies the phase change. The ability to store sensible heat for a

Table 12.2 Thermal capacities at 20 °C for some common TES materials

Material	Density (kg/m^3)	Specific heat (J/kg · K)
Aluminum	2,710	896
Brick	2,200	837
Clay	1,460	879
Concrete	2,000	880
Glass	2,710	837
Iron	7,900	452
Magnetite	5,177	752
Sandstone	2,200	712
Water	1,000	4,182
Wood	700	2,390

given material is strongly dependent of the value of ρC. For example, water has a high value, is unexpansive; however, being a liquid must be contained in a better and more expensive container than the one for a sold material. For high-temperature sensible heat TES (i.e., in the range of hundred Celsius degrees), iron and iron oxide have very good characteristics comparable to water, low oxidization in high-temperature liquid or air flow, with moderate costs. Rocks are unexpansive sensible heat TES materials; however, the volumetric thermal capacity is half of the water. Some common TES materials and their characteristics are listed in Table 12.2. Latent heat is associated with the changes of material state or phase changes, for example, from solid to liquid. The amount of energy stored (E) in this case depends upon the mass (m) and latent heat of fusion (λ) of the material:

$$E = m \cdot \lambda \tag{12.23}$$

The storage operates isothermally at the melting point of the material. If isothermal operation at the phase change temperature is difficult to achieve, the system operates over a range of temperatures T_1 to T_2 that includes the melting point. The sensible heat contributions have to be considered in the top of latent heat, and the amount of energy stored is given by:

$$E = m\left[\int_{T_1}^{T_{melt}} C_{Sd}dT + \lambda + \int_{T_{melt}}^{T_2} C_{Lq}dT\right] \tag{12.24}$$

Here, C_{Sd} and C_{Lq} represent the specific heats of the solid and liquid phases and T_{melt} is the melting point. It is relatively straightforward to determine the value of the sensible heat for solids and liquids, being more complicated for gases. If a gas restricted to a certain volume is heated, both the temperature and the pressure increases. Here, the specific heat is the specific heat at constant volume, C_v. If, instead the volume is allowed to vary and the pressure is fixed, the specific heat at constant pressure, C_p, is obtained. The ratio $\gamma = C_p/C_v$ and the fraction of the heat produced during compression can be saved, affecting the energy storage system efficiency. TES specific applications determine the used method. Considerations include among others: storage temperature range, storage capacity, having a significant effect on the system operation, storage heat losses, especially for long-term storage, charging and discharging rate, initial, and operation costs. Other considerations include the suitability of container materials, the means adopted for transferring the heat to and from the storage, and the power requirements for these purposes. A figure of merit that is used occasionally for describing the performance of a TES unit is its efficiency, which is defined by (12.25). The time period over which this ratio is calculated would depend upon the nature of the storage unit. For a short-term storage unit, the time period would be a few days, while for a long-term storage unit it could be a few months or even 1 year. For a well-designed short-term storage unit, the value of the efficiency should generally exceed 80%.

$$\eta = \frac{T_{\max} - T_{\min}}{T_{\text{charging}} - T_{\min}} \tag{12.25}$$

where $T_{\min}$, $T_{\max}$ are the maximum and minimum temperatures of the storage during discharging respectively, and T_{charging} is the maximum temperature at the end of the charging period. Heat losses to environment between the discharging end and the charging beginning periods, as well as during these processes are usually neglected. Two particular problems of TES systems are the heat exchanger design and in the case of phase change materials, the method of encapsulation. The heat exchanger should be designed to operate with as low a temperature difference as possible to avoid inefficiencies. In the case of sensible heat storage systems, energy is stored or extracted by heating or cooling a liquid or a solid, which does not change its phase during this process. A variety of substances have been used in such systems, such as (a) liquids (water, molten salt, liquid metals, or organic liquids) and (b) solids (metals, minerals, or ceramics). In the case of solids, the material is invariable in porous form and the heat is stored or extracted a gas or a liquid flowing through the pores or voids. For incompressible type of thermal storage, for example, the ones using heavy oils or rocks, the maximum work that can be produced is given in terms of specific heat capacity, C and mass:

$$\begin{aligned} W_{\max} &= m\left[C(T_{str} - T_{amb}) + CT_{amb}\ln\left(\frac{T_{str}}{T_{amb}}\right)\right] \\ &= \rho V\left[C(T_{str} - T_{amb}) + CT_{amb}\ln\left(\frac{T_{str}}{T_{amb}}\right)\right] \end{aligned} \tag{12.26}$$

Here, T_{str} and T_{amb} are the storage material temperature and ambient temperature (in Kelvin degrees), respectively and m is the storage material mass, ρ the storage material density, and V the volume. The storage materials (water, steam, molten salt, heavy oil, or solid rocks) are at temperatures significantly higher than the ambient one, so the heat is continuously lost from the thermal storage, regardless the insulation quality. Given enough time, the stored energy, if not used is dissipated. For this reason, TES are suitable for short-term or intermediate period applications rather than long-term ones. The total rate of heat transfer, q, from a TES reservoir depends on the overall heat transfer coefficient, C_{trsf}, the reservoir instant temperature, T, the ambient temperature and the reservoir total surface, A_{tot}, expressed as:

$$q = C_{trsf} \cdot A_{tot} \cdot (T - T_{amb}) \tag{12.27}$$

A number of TES applications are used to provide building or facility heating and cooling including *aquifer thermal storage* (ATS), and *duct thermal storage* (DTS). However, these are heat-generation techniques rather than energy storage techniques. An aquifer is a ground water reservoir, consisting of highly water permeable materials such as clay or rocks, having large volumes and high thermal storage capacities. When heat extraction and charging performances are good, high heating and cooling powers are achieved by such systems. The energy that can be stored in an aquifer depends on the local conditions (allowable temperature changes, thermal conductivity, and groundwater flows). An aquifer storage system is used for short to medium storage periods, daily, weekly, seasonal, or mixed cycles. In terms of storing energy, there are two primary TES options. One option is a technology used to supplement building air conditioning. The TES can also be used very effectively to increase the energy system capabilities to facilitate the penetrations of renewable energy sources can be increased. Unlike other energy storage systems which enabled interactions between the electricity, heat and transport sectors, TES only combines the electricity and heat sectors with one another. By introducing district heating into an energy system, then electricity and heat can be provided from the same facility to the energy system using CHP plants. This brings additional flexibility to the system, enabling larger renewable energy penetrations.

Example 12.6: A residence requires 72 kWh of heat on a winter day to maintain a constant indoor temperature of 21 °C. (a) How much solar collector surface area is needed for an all-solar heating system that has 20% efficiency? (b) How large does the storage tank have to be to provide this much energy? Assume the average solar energy per square meter and per day for the area is 6.0 kWh/m²/day.

Solution:

(a) Daily thermal energy per unit of area converted into thermal energy is:

$$\text{Thermal energy} = \frac{6.0 \times 0.20}{1.0} = 1.20\,\text{kWh/m}^2/\text{day}$$

The minimum converter area is then:

$$A_{Collector} = \frac{72}{1.2} = 60\,\text{m}^2$$

(b) If we are assuming the storage medium, water, the most common storage medium in residential applications, then the heat capacity of water is1 kcal/kg/ °C, and the temperature difference is between the hot fluid and the cold water going into the storage tank, about 40 °C. Therefore, the required mass of water for a day's worth of heat is:

$$\text{Tank mass} = \frac{72}{1.116 \times 10^{-3} \times 40} = 1612.9 \text{ or} \approx 1613 \text{ kg}$$

12.4 Microgrids and building integrated renewable energy systems

Microgrid concepts are not new, in fact the first power grids were more or less organized as microgrids. However, as new technologies are coming into place to employ renewable energy, energy storage, more efficient electricity production methods coupled with the flexibility of power electronics and control, new industry areas are developing these technologies and organize them into microgrids for extracting the maximum benefits for owners, consumers, and the power grid. Microgrids may be a prospective power system that addresses RES technologies accompanying necessary growing deployment of DERs, micro-CHP, and small-scale renewable energy sources. Microgrid is a new approach of power generation and delivery system that considers DG units and loads as small controllable power distribution subsystems, having characteristics such as the ability to operate in parallel or in isolation from the electrical grid, capabilities to improve service and power quality, reliability, and operational optimality. Microgrid operation, either in interconnected to or islanded from the grid depends on factors like planned disconnection, grid outages or economical convenience. A typical microgrid configuration consists of a group of radial feeders, a point of common coupling, critical and noncritical loads and micro-sources. This subsystem is capable to operate in dual modes, i.e., normal operation in which the microgrid is connected to the main grid and islanded operation in which it is separated from the main grid. These may include: micro-turbines, fuel cells, photovoltaic systems, solar thermal arrays, fuel cells, and wind turbines installations, also, the storage, load control, power and voltage regulation and heat recovery units need to be grouped together into microgrids. Microgrids can be described as a self-contained subset of indigenous generation, distribution system assets, protection and control capabilities, and end-user loads that may be operated in either a utility connected mode or in an isolated (islanded from the utility) mode. In addition to provide reliable electric power supply, microgrids are also capable of providing a wide array of ancillary services, such as voltage support, frequency regulation, harmonic cancellation, power factor correction, spinning, and nonspinning reserves. A microgrid may be intrinsically distributive in nature including several DGs—both renewable and conventional sourced energy storage elements, protection systems, end user loads, and other elements. In order to achieve a coordinated performance of a microgrid (or several microgrids) within the scope of a distribution company, it is required to perform distributed or cooperative control. Agent technology being one of the techniques for achieving the objectives of distributed control of microgrids. An agent is a software or hardware entity that exhibits characteristics of autonomy, self-organization, decentralization, and limited purview so as to progress the entire system toward a common goal, such as in cooperative distributed problem solving. Agents in microgrids are expected to perform sensory, communication, and actuation (control) tasks. In order to achieve reliable levels of distributed control of microgrids, it is imperative that the microgrid possesses a self-configurable sensor network that aids in communication among the constituent agents. Microgrids seem to be the quite suitable to address these challenges.

Microgrids may be a prospective framework that addresses technical concerns accompanying necessary growing deployment of RES by DG technologies to meet the increasing requirements for power quality and reliability. Microgrids are embedded in distribution systems, especially small-scale CHP, and within customer commercial or industrial facilities by incorporating modern controls that enable them to operate with a degree of autonomy from the traditional macrogrid. The growing requirements for the energy services provided by electricity will be met using a collage of approaches, technologies, and solutions. Examination of the electricity demand growth problem indicates that microgrids are well suited to play a significant role in the future evolution of energy service provision. DER systems, which may include DG and distributed storage (DS), are located near local loads and can provide a variety of benefits if they are properly operated in the electrical distribution system. Integration of various DG technologies with the utility power grid is an important pathway to a clean, reliable, secure, and efficient energy system for developed economies with established levels of quality and reliability of electrical service. Various studies have found that a large number of utilities as well as consumers that have installed DGs at their facilities realize benefits, such as local waste heat capture, improved reliability, and reduced cost. A microgrid is usually created by connecting a local group of small power generators using advanced censoring, controls, communications, and protection techniques. The development of the microgrid based on the DG together with the opportunities offered by renewables, could have a major effect on the way rural electrification is approached, both in developed and developing countries.

12.4.1 Microgrid concepts and architecture

A microgrid can be simply defined as an aggregation of electrical generation, storages, and loads. A microgrid may take the form of shopping center, industrial park, college campus, or an apartment building. To the utility, a microgrid is an electrical load that can be controlled, the load can be constant, can increase at night when electricity is cheaper, or can be held at zero during the grid stress. Concept of microgrid supersedes all the advantages of single source DG and hybrid power system. Moreover, it also includes all the advantages of a grid, at mini scale. A microgrid combined with power electronic interface is a completely self-sufficient network, preferably autonomous control, communication and protection.

It is capable of providing capacity support to the transmission grid while in grid-connected mode, and with capacity in excess of coincident peak demand. So, the microgrids comprise low voltage LV distribution systems with integration of DER, such as PV, wind energy, bio-mass, biofuel, and fuel cells together with distributed energy storage, such as flywheels, ultracapacitors for energy storage, and battery banks and controllable and smart loads that behave as a coordinated entity networked by employing advanced power electronic conversion and control capabilities. It has been proposed that one solution to the reliability and stability issues is to take advantage of microgrid technologies. The term *microgrid* is a popular topic within the power community but it still remains vaguely defined. A microgrid offers three major advantages over a traditional electricity supply involving central generation stations, long distance energy transmission over a network of high voltage lines, then distribution through medium voltage networks:

1. Application of CHP technology;
2. Opportunities to tailor the quality of power delivered to suit the requirements of end users; and
3. Create a favorable environment for energy efficiency and small-scale RES investments.

The use of waste heat through cogeneration or CHP implies an integrated energy system, delivering both electricity and useful heat. CHP processes can convert as much as 90% of its fuel into useable energy. To maximize the system efficiency, the sources need to be placed closer to the heat load rather than the electrical load since it is easier to transport electricity over longer distances. Small local sources can be sited optimally for heat utilization, so that a distributed system that is integrated with distributed generation is very pro-CHP. Distributed energy sources have the potential to increase system reliability and power quality due to the decentralization of supply. Increase in reliability levels can be obtained if distributed generation is allowed to operate autonomously in transient conditions, mainly if there is an outage or disturbances upstream in the electrical supply. The integration of RES units into the power system provides unique challenges to the power system designers and engineers. Due to the RES intermittent nature, central generation is required to provide the base power supply as well as provide backup power when RES are not generating power. Systems with intermittent sources can experience similar problems as systems with large, intermittent loads. Distributed generation can ease the burden of the high RES penetration by filling in when intermittent generation is low and by smoothing the transmission system loading. Several RES studies tie to the electrical system through inverters, which lack the mechanical inertia that help bring stability to the grid. A traditional architecture of a microgrid is shown in Figure 12.3 with a microgrid connected to a larger system and a disconnect switch that *islands* the DG units to protect sensitive loads. A major factor in microgrids is the disconnect switch enabling the microgrid to maintain compliance with current standards, needed to realize the high reliability and power quality that microgrids offer. Small generators have a lower inertia and are better at automatic load following and help avoid large standby charges that

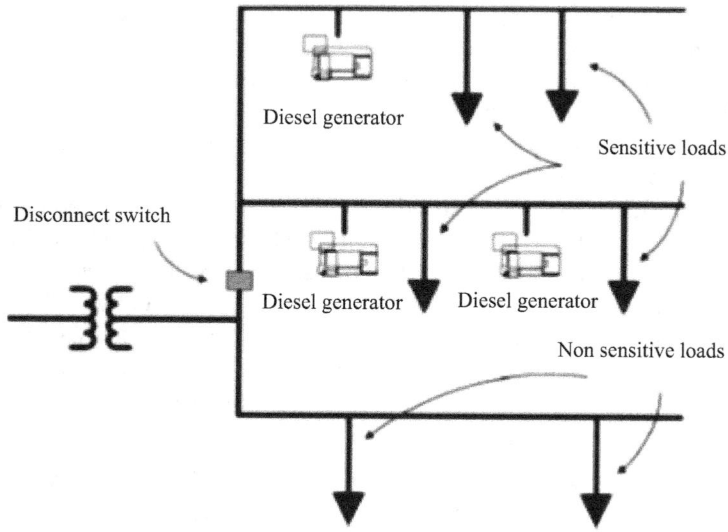

Figure 12.3 Microgrid architecture and structure

occur when there is only a single large generator. Having multiple distributed generators available makes the chance of an all-out failure less likely, especially if there are storage and backup generation capable of being quickly and easily connected to the system. This configuration of multiple independent generators creates a peer-to-peer network that insures that there is no master controller that is critical to the operation of the microgrid. A peer-to-peer system implies that the microgrid can continue to operate with the loss of any component or generator. With the loss of one source, the grid should regain all its original functionality with the addition of a new source, if one is available. This ability to interchange generators and create components with plug-and-play functionality is one requirement of microgrids. The concept can be extended to allowing generators to sit idly on the system when there is more electrical capacity than necessary. As the load on the system increases, additional generators would come online at a pre-determined set-point necessary to maintain the correct power balance. Intelligent devices could sense when there is extra generation capacity on the system and would disconnect and turn off generators to save fuel and increase machine efficiencies.

The microgrid of Figure 12.3 normally operates in a grid-connected mode through the substation transformer. However, it is also expected to provide sufficient generation capacity, controls, and operational strategies to supply at least a portion of the load after being disconnected from the distribution system at the point of common coupling and remain operational as an autonomous (islanded) entity. The existing power utility practice often does not permit accidental islanding and automatic resynchronization of a microgrid, primarily due to the human and equipment safety concerns. However, the high penetration of DER units potentially necessitates provisions for both islanded and grid-connected modes of operations

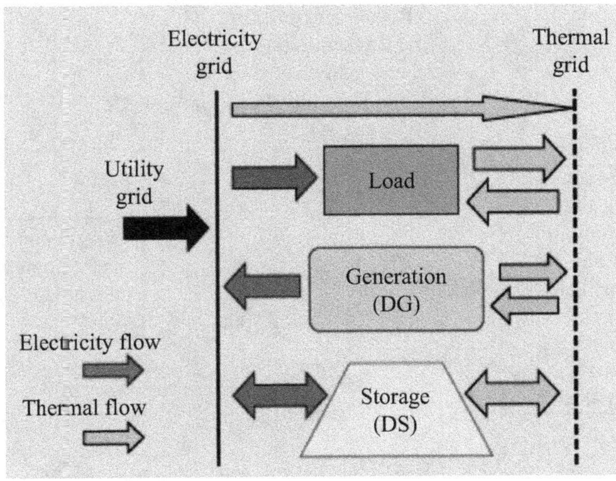

Figure 12.4 A general representation of the microgrid building blocks

and smooth transition between the two (i.e., islanding and synchronization transients) to enable the best use of the microgrid resources. DER units, in terms of their interface with a microgrid, are divided into two groups. The first group includes conventional or rotary units that are interfaced to the microgrid through rotating machines. The second group consists of electronically coupled units that utilize power electronic converters to provide the coupling media with the host system. The control concepts, strategies, and characteristics of power electronic converters, as the interface media for most types of DG and DS units, are significantly different than those of the conventional rotating machines. Therefore, the control strategies and dynamic behavior of a microgrid, particularly in an autonomous mode of operation, can be noticeably different than that of a conventional power system. Figure 12.4 shows a schematic diagram of the microgrid building blocks that includes load, generation/storage, electricity, and thermal grids. Figure 12.4 implies two levels of controls; i.e., component-level and system-level controls.

Both DG and DS units are usually connected at either medium- or low-voltage levels to the host microgrid. A DG unit usually comprises a primary energy source, an interface medium, and switchgear at the unit point of connection (PC). In a conventional DG unit (e.g., a synchronous generator driven by a reciprocating engine or an induction generator driven by a fixed-speed wind turbine) the rotating machine: (1) converts the power from the primary energy source to the electrical power, and (2) also acts as the interface medium between the source and the microgrid. For an electronically coupled DG unit, the coupling converter: (1) can provide another layer of conversion and/or control; e.g., voltage and/or frequency control, and (2) acts as the interface medium with the microgrid. The input power to the interface converter from the source side can be ac, at fixed or variable frequency or dc. The microgrid-side of the converter is at the frequency of either 50 or 60 Hz. Figure 12.5 also provides a high-level representation of a DS unit for which the "primary energy source" should be replaced by the "storage medium."

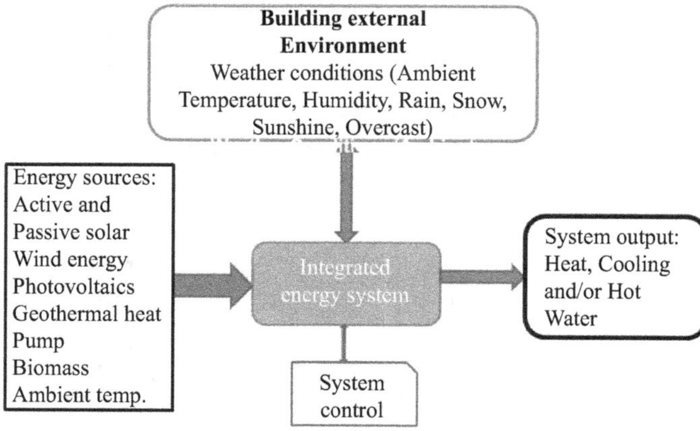

Figure 12.5 Building integrated energy system

Table 12.3 DER and DG units' interface and control

DER and DG type	Primary energy source	DER interface	Power flow control
Conventional DG	Reciprocating engines Small hydro Fixed-speed wind turbine	Synchronous or induction generators	Power flow and/or wind turbine control
Nonconventional DG	Microturbines, variable-speed wind turbines, solar PVs, fuel cells	Power electronics Converters	DG control, speed, output, frequency-voltage control
Energy storage	Batteries, ultracapacitors, flywheels, TESs	Power electronics Converters	DG control, state-of-charge, speed, output, frequency-voltage control

Table 12.3 outlines typical interface configurations and methods for power flow control of DG and DS units for the widely used primary energy sources and storage media, respectively. It should be noted that in addition to the two basic types of DG and DS units, a DER unit can be of a hybrid type; i.e., a unit that includes both "primary energy source" and "storage medium." A hybrid DER unit is often interfaced to the host microgrid through a converter system that includes bidirectional ac-dc and dc-dc converters. In terms of power flow control, a DG unit is either a dispatchable or a nondispatchable unit. The output power of a dispatchable DG unit can be controlled externally, through set points provided by a supervisory control system. A dispatchable DG unit is either a fast-acting or a slow-response unit. An example of a conventional dispatchable DG unit is the configuration shown in Figure 12.4, which utilizes a reciprocating engine as its primary energy source. A reciprocating-engine-based DG unit is normally equipped with a

governor for speed control and fuel in-flow adjustment. The automatic voltage regulator (AVR) controls the internal voltage of the synchronous generator. The governor and the AVR control real and reactive power outputs of the DG unit based on a dispatch strategy. In contrast, the output power of a nondispatchable DG unit is normally controlled based on the optimal operating condition of its primary energy source. For example, a nondispatchable wind unit is normally operated based on the maximum power tracking concept to extract the maximum possible power from the wind regime. Thus, the output power of the unit varies according to the wind conditions. The DG units that use renewable energy sources are often non-dispatchable units. To maximize output power of a renewable energy-based DG unit, normally a control strategy based upon maximum point of power tracking (MPPT) is used to deliver the maximum power under all viable conditions.

IEEE Standards Coordinating Committee 21 is supporting the development of *IEEE P1547.4 Draft Guide for Design, Operation, and Integration of Distributed Resource Island Systems with Electric Power Systems*. It will provide alternative approaches and good practices for the design, operation, and integration of the microgrid and cover the ability to separate from and reconnect to part of the Area EPS while providing power to the islanded local power systems. The guide covers the DR, interconnection systems, and participating microgrids. It is intended for use by designers, operators, system integrators, and equipment manufacturers. Its implementation will expand the benefits of DER by enabling improved EPS reliability and building on the requirements of *IEEE 1547-2003*. Technical challenges include the design, acceptance, and availability of low-cost technologies for microgrids.

12.4.2 Building thermal energy storage applications

A variety of TES techniques for space heating/cooling and domestic hot water have been developed over the past decades, including underground TES, building thermal mass, phase change materials for TES, and energy storage tanks. Seasonal thermal storage requires large inexpensive storage volumes and the most promising technologies were found underground. Underground TES (UTES) has been used to store large quantities of thermal energy to supply space cooling/heating, and ventilation air preheating. Energy sources include winter ambient air, heat-pump reject water, solar energy, process heat, etc. The most common UTES technologies are aquifer storage and borehole storage. It is not possible, for geological or geo-hydrological reasons, to construct these systems at any location. Borehole systems are the most generally applicable ones. The use of seasonal TES can substantially reduce the cost of providing solar energy systems that can supply 100% of buildings energy needs. The use of the ground as a seasonal thermal storage of solar energy has been used in a number of countries in conjunction with district heating systems. In seasonal geothermal energy storage, high temperatures usually result in the storage at the end of the summer period. Such high temperatures have two drawbacks. First, the return temperature to the solar collectors is quite high, leading to low solar collector efficiencies. Second, heat losses from the storage systems can be as high as 60% of the injected heat. Although different studies have shown potential for high/low temperature systems for space heating and cooling, the focus of innovative systems lies in the concept of ground source heat pumps (GSHP) for combined heating and cooling applications. In this case heat pump is employed to decrease or increase the storage temperature for cooling or heating. The GSHP technology can offer higher energy efficiency for air-conditioning compared to conventional systems. For UTES systems one of the most important factors is the required temperature level for the heating/cooling case involved. UTES systems are more efficient if the temperature requirement for space heating is low and for cooling is high. Using low-temperature heating/high-temperature cooling systems will result in high COP for the whole system. Seasonal storage of solar energy with commercial buildings and floor heating systems for residential buildings has proven successful in practice. Thermal mass is defined as the building mass that can be used to store thermal energy for heating/cooling purposes. In general the application has been found to be particularly suitable for climates with big diurnal temperature variations. Cooling by night-time ventilation can be used if night temperatures are low enough to release heat from the building's thermal mass. If a TES system is installed and charged using waste heat otherwise released to the environment, and if the energy is held and later used in place of added primary energy, overall energy consumption is reduced. To be economically feasible, the cost of the replaced primary energy should exceed the capitalization, maintenance, and operating costs of the TES system. The stored energy can in a sense be considered free, since it is otherwise lost. Useful waste or surplus thermal energy is available from many sources, hot or cold water drained to a sewer, hot flue gases, exhaust air streams, hot or cold gases or waste gases, heat collected from solar panels, ground source thermal energy, heat rejected from the condenser of refrigeration and air-conditioning equipment, and the cooling effect from the evaporator of a heat pump.

The principle of thermal storage is already being actively used in many buildings simply by utilizing the thermal mass, or thermal capacity of the building itself. One common example is to use the ventilation plant at full capacity during night in order to cool down the building. This can save on peak cooling costs during the day when the building gets gradually warmer. Night operation can in itself also be cheaper due to lower off-peak electricity cost. Using the building mass for thermal storage has the potential of finding much broader application, especially in relatively moderate climates. The thermal mass of buildings can be enhanced by conscious selection of building materials. The thermal capacity of concrete can for instance be increased by using aggregates with higher specific heat than common rock materials, for instance olivine rock or iron ore. Phase change materials can also be incorporated in the structure of the building, for instance inside floor/roof elements, or as micro capsules embedded in plaster wallboard, paints, and other surface coatings. A conceptual illustration of an integrated energy system based on new renewable energy sources with thermal storage is shown in Figure 12.5. Any application may incorporate one or more of the energy sources and system outputs. Buildings have a large mass and are reacting slowly to changes in heating/cooling demands. The figure therefore indicates the use of weather forecasts in order to optimize system efficiency and output by proactive rather than reactive control.

12.5 Micro-combined heat and power generation

The CHP concept or cogeneration has been known for a long time, being successfully implemented in industry. Cogeneration can provide heat and power in a more efficient way than the separate production of the heat and electricity. CHP, or cogeneration, refers to the simultaneous production of mechanical energy, which in most cases is used to produce electric energy, and useful heat. Whereas the conventional power stations release waste heat into the environment as a by-product of electricity generation, CHP units capture the by-product heat to supply process heat or to supply a district heating network with hot water. Large-scale CHP systems have been used for decades in district heating and industries, using gas and steam turbines and internal combustion (IC) engine systems from a few hundreds of kWe to MWe of power output. Recently, the interest in CHP has widened to encompass systems in the mini (up to a few kWe to 100 kWe) and micro (less than 10 kWe) scale providing heat and power to individual buildings or industrial processes. Micro-cogeneration (micro-CHP) is an emerging solution for the improvement of energy and environmental assessments of residential buildings. CHP paradigm is the simultaneous conversion of primary energy to useful heat and power at the point of consumption. Integrated micro-CHP system solutions represent an opportunity to address all of the following requirements at once: conservation of scarce energy resources, moderation of pollutant release into our environment, and assured comfort for home-owners. Micro-CHP has been defined, as the applying to systems with an electrical output of 16 A per phase or less, but this implies units up to 3.5 kWe for single phase and 12 kWe for three-phase units. In this paper, micro CHP refers only to individual units in individual homes and is, to all intents, limited to around 3 kWe. However, this definition may be somewhat misleading as other characteristics than the size alone are distinguishing, the *micro-CHP* from simply "very small" CHP. Micro-CHP promises significant economic and environmental benefits to energy suppliers and society at large. However, it also has the potential to substantially disrupt the established electricity supply industry both economically and technologically. It has a predicted capacity of similar order of magnitude to the existing nuclear generating capacity in the emerging liberalized energy markets in the US, EU, Japan, Canada, and most of developed and many developing countries. Although this concept might initially appear to be simply a scaled down version of large scale CHP, nothing could be further from the truth. The operational parameters and technical requirements of micro-CHP are fundamentally different from larger scale CHP and the market conditions which are adversely affecting CHP in general are less likely to impact on micro-CHP. The economic viability of a CHP system depends primarily on matching the thermal and electrical loads of a particular site with the heat and power outputs of the system, so that the CHP plant runs for a maximum number of hours per year. In many applications;

however, it is not possible to utilize all the available heat at certain periods of the year as the demand for space heating is greatly reduced, as the often coincides with periods of hot weather with a demand for air conditioning, the excess heat from the CHP unit can be used to operate a heat-driven cooling cycle.

12.5.1 Micro-combined heat and power system structure and configurations

The common solution for the buildings is to cover the electrical demand by the connection to the grid, while a boiler is generally used to meet the thermal energy demand which includes the domestic hot water and heating. If the users install a micro-CHP system, it is possible to couple it with an auxiliary boiler to cover the peak thermal demand, while also can generated power. Furthermore, the buildings can feed the grid when the electricity demand is minor than the electricity production and to cover the peak when the users' request is high. There are also the possibilities to use these systems in standalone applications. The types of micro CHP technologies are: reciprocating internal combustion engines, microturbines, micro Rankine cycle engines, Stirling engines, thermo-PV systems, and fuel cells. The opportunity for these technologies to become available on the market is due to factors connected with operational environment of energy generation, including political, economic, social and technological aspects. The availability of these systems on the market is impeded by other factors, with the most important the role of public administrations and energy market policy. There are several technologies being developed for micro-CHP generation. The most promising clean-fuel CHP technologies are Stirling engines and fuel cells. While fuel cells are the most attractive, due to their undisputed advantages as power generators they are still in the development stage, before are fully introduced to the market and for building energy applications. Residential Stirling engines in the meantime appear to be closest to commercialization and the developers have realized the significant opportunities they offer as the micro-CHP systems in residential market. Microgeneration CHP systems such as fuel cells, Stirling engine, internal combustion engine, etc., in size of 1 kWe to 10 kWe would be able to provide all or part of the power and heat load required by a typical household or residence. Most of the systems will be suitable for both off-grid and grid-connected applications. In either case, heat produced during the electricity generation can be recovered and used to satisfy space and water heating of the house in winter and to provide thermal cooling during the summer season. Due to small sizing, micro-CHP differs from larger CHP by the available conversion processes and by their efficiency. Nowadays, four main families of existing or emerging micro-CHP can be considered, differing by their conversion process: reciprocating engines, fuel cells, Stirling engines and steam engines (Rankine cycle). The Stirling engine micro-CHP systems seem to present several advantages for a use in residential buildings: low noise emission level, high heat recovery efficiency (60%–90%) and electricity production efficiency range from 10% to 25%. Moreover, the Stirling engine can operate with renewable fuels, such as liquid biofuels, biogas, or wood, which may improve the environmental performance of the system, and consequently, of the building.

Among the most widely used and most efficient prime movers are reciprocating (or internal combustion) engines. Electric efficiencies ranging from 25% to 50% make reciprocating engines an economic CHP option in many applications. Several types of reciprocating engines are commercially available; however, two designs are of most significance to stationary power applications and include four cycle-spark-ignited (Otto cycle) and compression-ignited (Diesel cycle) engines. They can range in size from small fractional portable gasoline engines to large units. Reciprocating engines have electric efficiencies of 25%–50% lower heat value (LHV) and are among the most efficient of any commercially available prime mover. Reciprocating engines have proven performance and reliability. With proper maintenance and a good preventative maintenance program, availability is over 95%. Improper maintenance can have major impacts on availability and reliability. Engine maintenance is comprised of routine inspections and adjustments and periodic replacement of engine oil, coolant, and spark plugs every 500–2,000 h. An oil analysis is an excellent method to determine the condition of engine wear. The time interval for overhauls is recommended by the manufacturer but is generally between 12,000–15,000 h of operation for a top-end overhaul and 24,000–30,000 for a major overhaul.

Microturbines were developed and customized for customer-site electric user applications by the industry through improvements in auxiliary power units originally designed for aircrafts and helicopters. Microturbines or turbo-generators are very small combustion turbines with outputs of 30–400 kW. Individual units can also be packaged together to serve larger loads. Several companies are developing systems with targeted product rollout within the next 2 years. Turbo-generator technology has evolved from automotive and truck turbochargers, auxiliary power units for airplanes, and small jet engines used for pilotless military aircraft. These are used for distributed electrical power generation or in combined cooling, heating, and power (CCHP) systems. Microturbines can use a large variety of fuels, such as: natural gas, gasoline, diesel fuel, kerosene, naphtha, alcohol, propane, methane, and digester gases. However, majority of commercial systems are using natural gas as primary fuel. Modern microturbines have dramatically progressed with advanced components, such as inverters, heat exchangers (recuperator units), power electronics, communications, and control systems. In most configurations, the turbine shaft spinning at up to 120,000 rpm drives a high-speed generator. The high-frequency output from the generator is first rectified and then converted to AC power. The systems are capable of producing power at around 25%–30% efficiency by employing a recuperator that transfers heat energy from the exhaust stream back into the incoming air stream. The systems are air-cooled, and some even use air bearings, eliminating water, and oil systems. Microturbines are appropriately sized for commercial buildings or light industrial markets for CHP or power-only applications. Recent development of these microturbines has been focused on this technology as the prime mover for hybrid electric vehicles and as a stationary power source for the DG market. In most configurations, the turbine shaft spinning at up to 100,000 rpm drives a high speed generator. This high frequency output is first rectified and then converted to 60 Hz or 50 Hz. Like larger turbines, these units are capable of operating on a variety of fuels.

Fuel cells offer the potential for clean, quiet, and very efficient power generation, benefits that have driven their development in the past two decades. Fuel cells offer the ability to operate at electrical efficiencies of 40%–60% (LHV) and up to 85% in CHP. Development of fuel cells for commercial use began in earnest in the 1970s for stationary power and transportation applications. Although several fuel cell designs are under development, only the phosphoric acid fuel cell (PAFC) is commercially available. The price of the most competitive PAFC is still around \$3,000/kW which is still too high for most industrial and commercial applications. The fuel cell requires continued research and development before it becomes a serious contender in the CHP market. Fuel cells, as discussed in previous chapter are similar to batteries in that they both produce a direct current (DC) through an electrochemical process without direct combustion of a fuel source. However, whereas a battery delivers power from a finite amount of stored energy, fuel cells can operate indefinitely provided that a fuel source is continuously supplied. Two electrodes (a cathode and anode) pass charged ions in an electrolyte to generate electricity and heat. A catalyst is used to enhance the process. Individual fuel cell elements are generating between 0.5 and 0.9 V, DC voltage. Fuel cells are combined into "stacks" like a battery to obtain usable voltage and power output. A fuel cell consists of several major components including a fuel reformer to generate hydrogen-rich gas, a power section where the electrochemical process occurs and a power conditioner to convert the direct current (DC) generated in the fuel cell into alternating current (AC). The five main types of fuel cells are defined by their electrolyte and include alkaline, proton exchange membrane (PEMFC), phosphoric acid (PAFC), molten carbonate (MCFC), and solid oxide (SOFC) fuel cells. Alkaline fuel cells which are very efficient and have been used successfully in the space applications are requiring very pure hydrogen that is expensive to produce and for this reason are not considered major contenders for the stationary power market. The PAFC represents the most mature technology and is commercially available today, having been installed in over 80 locations in the US, Europe and Japan. The electric efficiency of fuel cells is dramatically higher than combustion-based power plants. The current efficiency of PAFC is 40% with a target of 40-60% (LHV) estimated. With the recovery of the thermal byproduct, overall fuel utilization could approach 85%. Fuel cells retain their efficiency at part load. The capital cost of fuel cells is currently much higher than competing distributed resources. The commercial PAFC currently costs approximately \$3,000/kW. Fuel cell prices are expected to drop to \$500–\$1,500/kW in the next decade with further advancements and increased manufacturing volumes. Substantial cost reductions in the stationary power market are expected from advancements in fuel cells used for transportation.

Heat exchangers do not produce heat; they simply transfer heat from one system to another. Heat exchangers are designed from different materials depending upon the application. Stainless steel is expensive and not a high heat conductor but it holds up well against the corrosive effects of condensate from exhaust gases. Current heat exchangers are capable of capturing about 80% of the heat from exhaust gas and transfer it to the absorption chiller. The two major CHP technologies that provide cooling are absorption chillers and desiccant dehumidification systems. Absorption cooling is a way of using heat to drive a refrigeration cycle instead of the mechanical energy required to run a compressor, via electricity. Absorption cooling cycles take advantage of chemical processes using a refrigerant and an absorbent that combine at low pressure and low temperature to form a solution. Water, lithium bromide, and ammonia (NH_3-H_2O) are the most common refrigerant/absorbent combinations. Desiccant dehumidification and cooling is a method of removing latent heat load from the air as moisture. This reduces cooling loads and allows air conditioning systems to operate more efficiently. Conventional cooling systems remove latent heat by using a cooling coil to condense water from the air. This requires more energy than is cooling the air that has been dehumidified with a desiccant dehumidification system. By dehumidifying air prior to cooling, desiccant systems can reduce HVAC electricity use by 30%–60% and peak electricity demand by 65%–70%. Payback periods for desiccant systems typically ranged from 2 to 4 years.

12.5.2 Micro-CHP economics

Integrating a CHP technology with a specific application together as a system, requires an understanding of the engineering and site-specific criteria providing the most economical solution. The final design must address siting issues like noise abatement and footprint constraints. Engineering information for designing a technically and economically feasible system should include electric and thermal load profiles, capacity factor, fuel type, or performance characteristics of the prime mover. CHP implies the simultaneous generation of two or more energy products that function as a system. One of the first and most important elements in the analysis of CHP feasibility is the accurate representations of electric and thermal loads. This is particularly true for load following applications where the prime mover must adjust its electric output to match the demand of the end-user while maintaining zero output to the grid. Thermal load profiles can consist of hot water use, low and high pressure steam consumption, and cooling loads. The shape of the electric load profile and the spread between minimum and maximum values dictate the number, size and type of prime mover. It is recommended that electric and thermal loads be monitored if such information is not available. Capacity factor is a key indicator of how the capacity of the prime mover is utilized during operation. Capacity factor is a useful means of indicating the overall economics of the CHP system, indicating the facility's proximity to base-load operation and is defined as:

$$\text{Capacity factor} = \frac{\text{Actual energy consumption}}{\text{Prime mover peak capacity} \times 8{,}760\,\text{h}} \tag{12.28}$$

Economic operation of CHP systems requires the limitations of device power rating and peak demand (the investment and fixed operational costs), for covering electrical peaks grid connection provides the basis, while thermal peaks are smoothed out by thermal storage. Efficient operation depends on the system design for a particular building, energy and load management. Energy management means here, cost-efficient supply of electrical and thermal loads by intelligent and anticipatory operation of fully interacting system components. Load management means controlling the operation of certain devices, especially larger electro-thermal loads with significant power demands. Other items to be considered in TES economic analyses are space requirements, system reliability, and the interface to the delivery system for the application. An optimal energy storage application achieves a balance between maximizing the savings accrued in utility charges and minimizing the initial installation cost needed to achieve the savings. Consequently, the decision to install an energy storage unit must be based on the loads, load characteristics, and generating capacity for extended periods. Uncertainty about the future economic outlook, life-style changes, and the availability of low-cost energy charging the storage may lead to differing investment decisions if alternative technical solutions are feasible. Technical characteristics of alternative technologies for situations in which TES is potentially attractive may affect decisions. A low capacity factor is indicative of peaking applications, deriving economic benefits through the avoidance of high demand charges. A high capacity factor is desirable for most CHP applications to obtain the greatest economic benefits. A high capacity factor effectively reduces the fixed unit costs of the system ($/kWh) and to remain competitive with grid supplied power. Most commercial end-users have a varying electric load profile, i.e., high peak loads during the day and low loads after business hours. Thermal demand of a commercial or institutional end-user often consists of hot water or low pressure steam demand in the winter and a cooling demand in the summer. The thermal requirements of the end-user dictate the CHP system feasibility or system component selection.

12.6 Chapter summary and discussions

First part of the chapter provides a summary of introductory aspects of thermal engineering, heat transfer, and related fundamental definitions and physical quantities, in order to have a sufficient thermal sciences background for understanding TES systems, micro-CHP and applications, as well as their operations. Other sections and subsections are dedicated the microgrid concepts, configurations and architecture, the microgrids benefits and functions, as well as the building integrated renewable energy systems. CHP technologies produce electricity or mechanical power and recover waste heat for process use. Conventional centralized power systems average about 33% or less delivered efficiency for electricity in the US, while the CHP systems can deliver energy with efficiencies exceeding 90%, while significantly reducing emissions per delivered MWh. CHP systems can provide cost savings for industrial and commercial users and substantial overall emissions reductions. Selecting a CHP technology for a specific application depends on many factors, including the amount of power needed, the duty cycle, space constraints, thermal needs, emission regulations, fuel availability, utility prices, and interconnection issues. Although energy may be stored in many ways (e.g., in mechanical, kinetic, electrochemical, or chemical forms), since large part of the economy involves thermal energy, the storage of thermal energy warrants careful attention. TES deals with the storing of energy by cooling, heating, melting, solidifying, or vaporizing a material, the energy becoming available as heat when the process is reversed. TES is a temporary storage of high- or low-temperature energy for later use. Where thermal energy is available and thermal demands exist, but these are not necessarily coincident, TES, the most direct means of energy savings avoids the needs to convert energy from one form to another and the ensuing conversion losses. TES systems and their applications are examined from an energy savings perspective, while possible energy-saving technologies are discussed in details. The selection of TES system mainly depends on the storage period required, economic viability, operating conditions, and so on. Specific parameters that influence the viability of a TES include facility thermal loads, thermal load profiles availability of waste or excess thermal energy, electrical costs, and rate structures, type of thermal generating equipment, and building type, and occupancy. The economic justification for TES systems usually requires that annual capital and operating costs are less than the cost for conventional systems and equipment supplying the same service loads and periods. Well-designed systems can reduce initial and maintenance costs and energy use and demand, and improve indoor environmental quality. Although the various energy storage solutions have found many applications in practice, it is still not quite clear how they can best be integrated into ultra-low energy buildings, capable of being replicated in a variety of climates and technical capabilities. Detailed studies on the dynamic performance and control strategies of the energy storage systems for different building types, weather conditions, and user behavior are needed and must be performed. Advanced design strategies for building and on-site integrated TES solutions should be developed.

References and further readings

[1] J. Jensen and B. Sorensen, Fundamentals of Energy Storage, Wiley-Interscience, New York, 1984.

[2] B. Norton, Solar Energy Thermal Technology, Springer-Verlag, London, UK, 1992.

[3] F. Dinter, M. Geyer, and R. Tamme, Thermal Energy Storage for Commercial Applications, Springer-Verlag, Berlin, 1991.

[4] A. Ter-Gazarian, Energy Storage for Power Systems, Peter Peregrinus, Stevenage, 1994.

[5] I. Dincer and M. Rosen, Thermal Energy Storage: Systems and Applications, Wiley, Chichester, UK, 2002.

[6] F.P. Incropera, and D. P. DeWitt, Introduction to Heat Transfer, Wiley, New York, 2002.

[7] R. Siegel, and J. R. Howell, Thermal Radiation Heat Transfer (4th ed.), Taylor & Francis, New York, 2002.

[8] B. Sorensen, Renewable Energy—Its Physics, Engineering, Environmental Impacts, Economics & Planning (3rd ed.), Elsevier/Academic Press, London, 2004.
[9] F.A. Farret, and M.G. Simões, Integration of Alternative Sources of Energy. Wiley-Interscience, New York, 2006.
[10] I. Dincer and M. Rosen, Energy: Energy, Environment and Sustainable Development, Elsevier, New York, 2007.
[11] J. Andrews and N. Jelley, Energy Science, Principles, Technologies, and Impacts, Oxford University Press, Oxford, UK, 2007.
[12] M. Kaltschmitt, W. Streicher, and A. Wiese, Renewable Energy: Technology, Economics and Environment, Springer, New York, 2007.
[13] H.O. Paksoy (Ed.), Thermal Energy Storage for Sustainable Energy Consumption: Fundamentals, Case Studies and Design. Springer, Dordrecht, 2007.
[14] S. Vazquez, S.M. Lukic, E. Galvan, L.G. Franquelo, and J.M. Carrasco, 2010. Energy storage systems for transport and grid applications, IEEE Trans. Ind. Electron Vol. 57, pp. 3881–3895.
[15] R. Zito. Energy Storage: A New Approach. Wiley, New York, NY, 2010.
[16] N. Hatziargyriou (Ed.), Micrograds Architectures and Control, Wiley, Chichester, UK, 2014.
[17] R. Belu, Microgrid Concepts and Architectures, in Encyclopedia of Energy Engineering & Technology (Online) (Ed: Dr. Sohail Anwar, R. Belu et al.), CRC Press/Taylor and Francis, 2014.
[18] R. Belu, Microgrid Protection and Control, in Encyclopedia of Energy Engineering & Technology (Online) (Ed: Dr. Sohail Anwar, R. Belu et al.), CRC Press/Taylor and Francis, 2013.
[19] R.A. Higgins, Energy Storage (2nd ed.), Springer, USA, 2016.

Questions and problems

1. What are the benefits of the energy storage systems?
2. List the major applications of thermal energy storage.
3. What types of the energy storage options are available for solar and wind energy applications?
4. What are the differences between energy conservation and energy efficiency?
5. List the essential criteria for comparing energy storage methods.
6. What are the benefits and applications of TES systems? List the various TES processes and explain each.
7. What are the selection criteria for a TES system?
8. List the technical, economic, and environmental benefits of TES.
9. Describe possible energy conservation strategies with TES.
10. How TES can contribute to the increase of the overall system efficiency?
11. A 2-kg mass of ice initially at −10 °C is heated until the mass is melted. Determine the sensible heat and the latent heat transferred to the water.

The specific heat of ice at 0 °C is 2.11 kJ/kg °C, and the latent heat of fusion of water at 0 °C is 334.9 kJ/kg.

12. How much energy is needed to heat a volume of 10 m^3 of water from 20 °C to 81 °C? Take the density of water as 1,000 kg/m^3 and the specific heat as 4.185 kJ/kg · °C.
13. The walls of a house are filled with glass wool insulation. The wall is 5 in. thick. Assuming that the average temperature on the inside wall is 75 °F and outside is about 10 °F. The total walls of the house are 15 ft. high and 540 ft. wide, and the thermal conductivity is 0.04 W/m °C. What is the rate of heat transfer through this wall?
14. Estimate the coefficient of performance of a refrigerator that consumes 1,000 W of power to remove heat at a rate of 5.2 BTU/s.
15. If the electrical heating is used at a cost of 10 cents/kW h, what is the cost to heat the house of the previous problem for 1 day (just on the basis of heat loss through this one wall)?
16. Due to a severe weather event, the insulation blew off the wall of the house in the previous two problems, so now there is a 75 °F wall exposed directly to 10 °F air. What is the rate of heat transfer now assuming buoyant convection only (i.e., the severe weather event ended and now the wind is calm) with heat transfer coefficient 10 W/m^2 °C.
17. In a power plant, the steam from the boiler reaches the turbine at a temperature of 800 °C. The spent steam leaves the turbine at 100 °C. Calculate the maximum efficiency of the turbine. Assuming the boiler, steam turbine, and generator efficiencies of Example 12.7, what is the overall power plant efficiency?
18. Summarize the key advantages of using TES in buildings and in industrial processes
19. What are the modes of heat transfer? Briefly explain the physical mechanism of each mode
20. Calculate the required time for a heat storage temperature to decrease from 36 °C to 8 °C, if there is no load heat removal from, and the ambient temperature is 12 °C, the reservoir storage capacity is 2.0 m^3, the heat storage area is 30 m^2, and the overall heat transfer reservoir liquid-environment is 6.5 W/m^2 · °C.
21. Calculate the theoretical efficiency of a TES operating between 122 °C and 25 °C, assuming a charging temperature of 150 °C.
22. Assuming a cylindrical reservoir, made of bricks, with 10 m diameter and height of 12 m, compute the heat transfer rate, assuming the ambient temperature of 20 °C degrees and the storage material temperature of 105 °C.
23. Calculate the required storage medium volume for the following thermal energy storage systems in order to store 1 MWh of thermal energy for the temperature range from 350 °C to 20 °C, assuming the solid rocks and heavy oil are used as storage materials.

Chapter 13

Energy management, RES, and distributed generation economics

Objectives and abstract

Sustainable energy management, a paradigm and theory, having concepts, principles, and methods that are only recently fully accepted and employed is an important and comprehensive framework, part of the sustainable development, attempting to plan the energy use on the past experience and future needs. The energy management fundamental goals are to produce goods and to provide services with the minimum energy use and environmental impacts. The term energy management has different meaning to different people and in different areas. The objective of *Energy Management* is to achieve and maintain optimum energy procurements and uses, throughout the organization and to minimize energy costs and energy waste without affecting production levels and quality, while minimizing the environmental energy use effects. This rather broad definition covers many operations from the services, product, and equipment design through the product shipment and delivery. Waste minimization and disposal, important aspects of an energy management plan are also presenting several and important energy management opportunities and solutions. Energy savings and waste reductions constitute primary measures for the protection of the environment and, in addition, for the reduction of exchange effluxes, which are used to purchase the polluting fossil fuels, coal, oil, and natural gas. Noticing that in most process industries, energy costs are second only to raw materials. Very often entire department is devoted to optimizing raw material choices and product slates, by using planning models, energy management, supply strategies, and optimization approaches. This chapter provides guidelines and information how to set up an energy management program. Energy efficiency is about getting the same or better services using less energy. This energy management aspect is in contrast to the energy conservation, which involves doing less with less. The critical issues for energy efficiency and energy management are to identify the services that are needed and make sure that these are being provided cost-effectively, with minimum energy use with the least environmental impacts. Irrespective of the energy cost size, the continuous process nature or the types of equipment employed, energy efficiency is a must. Understandably, though, management gives the greatest amount of attention to the largest costs. The basic principles of energy management and energy efficiency are universal but different types of facilities require different types of energy management programs. Energy management is a long-term commitment, not just something that is conducted once and then is forgotten. The term *energy audit*, an important tool of energy management is widely used and may have different meaning depending on the energy service companies. Energy auditing of buildings can range from a short walk-through of the facility to a detailed analysis with hourly computer simulation.

13.1 Introduction, DG, and RES economical aspects

Electricity is an intermediate product, generated by using primary energy sources (oil, gas, coal, hydro-power, or nuclear energy) that are converted to electrical energy and transported, via power transmission and distribution networks to the consumers. Established electrical power systems, developed over past 100 years, are feeding the electrical energy from large generation units through step-up transformers to high-voltage interconnected electrical network, the transmission systems, substations, power subtransmission and distribution networks to the consumers or end users. Consumers are purchasing electricity as an intermediate step toward final nonelectric products. The large-scale electricity usage is due to various factors, such as easy to generate and transmit over long distances at very high efficiency, easy to distribute to consumers any time, any amount, and almost anywhere, and large variety of uses (heat, light, electrical motors, computers, radio, etc.). In order to meet the power operation criteria (safety, reliability, quality, economy, and supply security), within the power systems operation the following tasks are performed: maintain accurate load-generation balance, maintain the reactive power balance to control the voltage profile, maintain an optimum generation schedule to control generation cost the environmental impacts, and ensure the network security against credible contingencies, requiring network protection against reasonable outages, equipment failure, or unauthorized interventions.

Energy, one of the important company resources must be managed and controlled by systematic methods in collaboration with the management of other resources. Energy management is referring to the methods making out and facilitating an optimum company program of purchasing, generating, and using different energy types based on the company or facility overall short-term and long-term management programs, with due consideration of costs, availability, or other economic factors. Energy management is required because it influences a number of aspects of operation and activities regarding the energy costs, which directly affecting the company profitability, market competitiveness, and indirect country energy supply-demand balance, national trade and financial health, local and global environments, occupational safety, loss prevention and waste disposal reduction, productivity, and quality. Energy management in the form of implementing new energy efficiency technologies, materials and manufacturing processes and the use of new technologies in equipment and materials helping companies improve their productivity and increase their product or service quality. Often, the energy savings is not the main driving factor when companies decide to purchase new equipment, use new processes, or new high-tech materials. However, the combination of increased productivity, quality, reduced environmental emissions, and reduced energy costs provides a powerful incentive for companies and organizations to implement new technologies. Energy management is mainly about reducing the cost of energy used by an organization, now with the added spin of minimizing carbon emissions as well. Reducing energy costs has two facets: price and quantity. Improving energy efficiency is, in part, a technical pursuit with a scientific basis. Energy efficiency can be defined as utilizing minimum amounts of energy for heating, cooling, lighting, and the equipment that is required to maintain conducive conditions in a building or facility. An important factor impacting energy efficiency is not only the building energy envelope but also the management of energy within the premises. The energy consumed varies depending on the building or facility design, the available electrical systems and how these systems operate. The heating and cooling systems consume the most energy in a building or facility, the control system of the building energy management systems (EMS) can significantly reduce the energy use of these systems. There are large opportunities to improve energy use efficiency by eliminating waste through process optimization. Applying today's computing and control equipment and techniques is one of the most cost-effective and significant opportunities for larger energy users to reduce their energy costs and improve profits. An EMS is an important element of a comprehensive energy management program, providing relevant information to key individuals and departments that enable them to improve energy performance and uses.

13.2 EMS in manufacturing, industrial, and commercial sectors

EMS lies at the heart of all infrastructures from communications, economy, and transportation. This has made the system more complex and more interdependent, increasing number of disturbances occurring that has raised the priority of EMS infrastructure which has been improved with the aid of technology and investments. Modern industrial and commercial facilities operate complex and interrelated power systems. Energy conservation and facilities/equipment are only part of the approach to improve energy efficiency. Most energy efficiency in industry is achieved through changes in how energy is managed in a facility, rather than through installation of new technologies. Systematic management and the behavior approach have become the core efforts to improve energy efficiency today. An EMS provides methods and procedures for integrating energy efficiency into existing industrial or commercial management systems for continuous improvements in the energy usage. Energy management is defined as a system, methods and procedures for an effective and optimum energy use in industrial processes and in operation of residential, commercial and industrial facilities, maximizing the profits, enhancing the competitive positions through organizational measures and optimization of energy efficiency in the industrial and commercial processes.

Profit maximization is also achieved through the reduction in energy costs during each production and operational phase, while the three most important operational costs are those for materials, labor, and energy (fuels, electrical, and thermal). Moreover, the competitiveness improvement is not limited to the reduction of sensible costs, but can be achieved with an opportune energy cost management, increasing the flexibility and compliance to the changes of market and international environmental regulations. Energy management is a well-structured process, both technical and managerial in nature. In this chapter, we discuss the structure, methods, and techniques used in energy management, as well as new approaches and developments in the field. A rich, comprehensive and up-to-date literature is also included in the chapter for professionals, engineers, students, and interested readers in the energy management topics and problems.

In the past, most of the manufacturing, industrial, and large commercial organizations and facilities have lacked complex energy monitoring and control systems when compared to the ones frequently found in electrical generating plants. The reason is that the primary goal of any manufacturing organization is to cost effectively produce high quality products and thus the energy management often takes a somewhat secondary role to a production objectives. However, with increased energy prices and market competition, the companies are taking a closer look at energy saving opportunities and the potential for onsite electrical generation from alternative energy sources. Manufacturing firms are reexamining the energy management functional requirements and internal energy systems to increase efficiency and integration with the future power grid. Energy efficiency, conservation, and cost are top priorities all over the world, in particular, for heavy energy consumers. Electricity costs are about 50% or so of total energy in many manufacturing industries, motivating companies to employ various strategies to reduce the energy use and costs. Certainly many of our environmental problems today arise from the types of energy we use, and increased burning of fossil fuels will accelerate climate change. Energy conservation technology and facilities or equipment are only part of the approach to improve energy efficiency. Systematic management and the behavior approach have become the core efforts to improve energy efficiency today. Energy management represents a significant opportunity for organizations to reduce their energy use while maintaining or boosting productivity. Industrial and commercial sectors jointly account for approximately 60% of global energy use. Organizations in these sectors can reduce their energy use 10%–40% by effectively implementing an EMS. Fossil fuels are currently the major source of energy in the world. However, as the world is considering more economical and environmentally friendly alternative energy generation systems, the global energy mix is becoming more complex. Factors forcing these considerations are the increasing demand for electric power by both developed and developing countries, many of the developing countries lacking the resources to build power plants and distribution networks, some industrialized countries facing insufficient power generation and pollutant emission and climate change concerns. Renewable energy sources such as wind turbines, photovoltaic solar systems, solar-thermal power, biomass power plants, fuel cells, gas micro-turbines, hydropower turbines, combined heat and power (CHP), and hybrid power systems are now part of the power generation systems.

The modern manufacturing plant has a large set of IT equipment including EMS and manufacturing execution systems (MES) which are used to control processes and infrastructure. MES are single or even multiple software applications that perform such roles as equipment scheduling and monitoring, inventory and product tracking, quality monitoring, maintenance dispatching, and operating allocation and planning. These systems serving to manage aspects of manufacturing operations are not always well connected nor integrated despite their common efficiency objectives which include improving energy efficiency. With the increased availability of all types of real-time information from plant equipment, quality, reliability, and energy consumption monitoring, there is growing interests to take advantage of this valuable data to improve not only the productivity of the manufacturing operations but also simultaneously of the energy performances. However, due to the large amount of data and complex nature of most of manufacturing systems, extracting meaning and understanding from this rich source of data is often not an easy task. To complicate matters further, the manufacturing processes are highly dynamic environments where events like machine breakdowns, operator shortages, and quality problems are just a few examples. These events whether deterministic or stochastic, require effective management responses. Without a clear integrated view, it is quite difficult to take appropriate actions to ensure efficient operations from productivity and energy perspectives. The energy management key question is how to provide the best case for successful energy management within their organization, achieve the desired buy-in at top management level, and implement a successful EMS. The energy management purpose is to provide an organizational framework to integrate energy efficiency into management practices, including fine-tuning production processes and improving the system energy efficiency. It also seeks to apply to the energy use the same culture of continual improvement that is successfully used by companies to improve quality and safety practices. Its guidelines recommend that companies need to track energy consumption, benchmark, set goals, design action plans, evaluate progress and performances, and create energy awareness throughout the organization, enabling the energy efficiency and conservation integration into existing management system for continuous improvements and to reach the requirements for high efficient companies and ultimate to reduce cost and increase revenue. Among the high efficient company requirements are: efficiency is a core strategy, leadership and organizational support is real and sustained, company has energy efficiency goals, there is in place a robust tracking and measurement system, resources are put for energy efficiency, and energy efficiency strategy is working and results are communicated. An energy management standard is needed to influence how energy is managed in an industrial facility, realizing immediate energy use reduction through changes in operational practices, and creating a favorable environment and settings for the adoption of more capital-intensive energy-efficiency measures, practices and technologies. Efficient energy management requires the identification of where energy is used, where is wasted and where any energy saving measures have effects. The key feature of a successful EMS is that it is owned and fully integrated as an embedded management process within an organization or company, its implications are considered at all stages of the development process of new projects, and that these implications are part of any change control process. Standards should lead to reductions in energy cost, pollutant emissions, minimizing negative environmental impacts. A change in the organizational culture is needed in order to realize industrial energy efficiency potential. An EMS standard can provide a supportive organizational framework necessary to move beyond an energy saving project approach to an energy efficiency approach that routinely and methodically seeks out opportunities to increase energy efficiency, no matter how large or small.

Businesses, wasting energy are reducing their profitability, while causing avoidable pollution, primarily through increased carbon emissions. If comparisons are made between industries, corporations, or manufacturing plants, there are a wide variation in the structure of individual energy programs. Even with such variations in scope and implementation, there are good chances of successful energy management programs, while comparing the results of each company in energy management program variations and how the energy is consumed. A company, whose energy use is primarily in lighting and heating, develops a different program and strategy than a cement manufacturer, with hundreds of electric motors in addition to a significant thermal energy uses. Either a successful energy management program is implemented in different ways, there are common elements. A company with centralized management philosophy may dictate program structure across its plants and facilities, while one with decentralized management structure can set only broad program goals, allowing individual units to develop individual programs according to the local circumstances. Making businesses more energy efficient is still a largely untapped solution addressing environmental pollution, energy security and fossil fuel depletion. As pressures mount on businesses to become more energy efficient, managing resources effectively is proving more essential than ever. In addition, customers are increasingly asking for assurance from organizations that are environmentally responsible. Energy management is the means to control and reduce the energy consumption of the organization, which is important because reducing production costs, pollutant emissions, and the cost-related implications of the pollution, while reducing the risks, as more energy the organization consume, the greater are the risks that any energy price increases or supply shortages can affect the profitability, or even make impossible for the organization to continue. Energy management can reduce such risks by reducing the energy demands and controlling it to make it more predictable. On top of these reasons, it is advisable to have some rather aggressive energy usage reduction targets.

13.2.1 Identification of energy usage factors and parameters

With energy costs and the importance attributed to climate change having risen over the past several years, energy efficiency has become paramount. Identifying key energy performance indicators is vital for the planning process, as it provides managers with a clear picture of how their company uses energy and can highlight ways to manage resources better. The first step of energy management plan is to determine the energy consumption *structure* of the company, in other words, what energy resources are needed by an organization to run its operations (e.g., natural gas, coal, gasoline, or electricity). In order to maintain and evaluate the effectiveness of an energy management program, energy data must be collected and analyzed. Collecting credible utility data in the plant data historian provides tracking capability at a quite low cost. Ultimately, all energy use and energy-saving projects and actions must be collected, preferable in a database format. A management process is required to proactively assess, manage, and measure energy use and conservation. However, many companies and organizations have limited levels of expertise necessary to achieve these reductions and so need guidance, etc. Improvements in energy efficiency will require systems and processes necessary to improve energy performance. Companies and organizations need to manage the way in which the energy is used in order to reduce pollutant emissions and other environmental impacts, as well as reductions in energy costs and wastage. Energy management plan set the energy efficiency benchmarks that measure consumption and the energy consumption assets in operation and set the opportunities and ways to reduce the energy usage. Such benchmarks allow companies to compare their EMSs against best practices, and to compare the energy efficiency results against national standards or country's energy-saving technologies policy. Energy management is showing the company energy performance ratings, is helping to track the energy and water consumption, to set targets and indicators and as a result, to improve policies, identify areas that need improvements, and measure the investments required for facility upgrades. The EMSs aim at reduced energy use and costs, are representing a key element in any company energy management program. Energy management must be based on real time information obtained from process monitoring and control systems, and on production plans received from production planning systems. It is also useful to compare each unit and site with its own past performance to determine energy savings but the internal view is not very useful in determining how the process compares with its competitors. Often total and comprehensive solutions include planning and scheduling tools to optimize energy use and supply, energy balance management tools to support the real-time monitoring and control of the energy balance, and reporting tools to evaluate and report energy consumption, costs, efficiency, and other energy-related information. Opportunities for cost reduction are greatest when both electricity consumption and prices vary over time, which is common in process industries, and open electricity market environments.

Energy management concepts are built upon the plan-do-check cycle, being used as a basis for many management systems. Figure 13.1 illustrates how continued use of the management action process leads to continuous improvement. An EMS is a collection of procedures, methods, and tools designed to engage staff at all levels within an organization in managing energy use on an ongoing basis. It allows industrial plants, commercial, institutional, and governmental facilities and

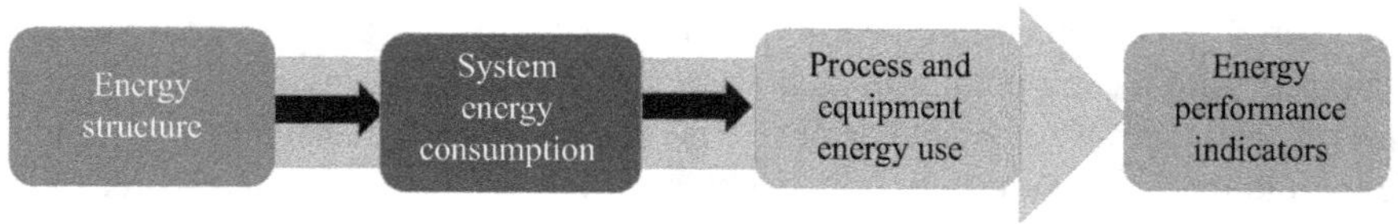

Figure 13.1 Energy management flow chart and structure

entire organizations to systematically track, analyze, and plan their energy use, enabling greater control of, and continual improvement in energy performance. Organizations may choose to pursue only certain components of energy management, based on their needs. While EMS implementation expertise is usually concentrated in technical and engineering positions within an organization, a variety of nontechnical personnel also exert significant influence on energy decision-making (e.g., executive staff, accountants, and financial managers). While these nontechnical staff members need not be knowledgeable in all aspects of energy management, are included in this analysis since they can be critical to the success of an EMS within their organization. Many long-standing professional training or credentialing programs cover some skills or knowledge areas relevant to EMS implementation but most falls short of providing the entire spectrum of skills and expertise needed to implement an EMS effectively. Energy is one of the management resources of a company, and shall be managed and controlled by a systematic method in harmony with the management of other resources. Energy management is managing all kinds of energy used in the company by making out an optimum program of purchasing, generating and consuming various types of energy based on the company's overall short-term and long-term management program, with due consideration of costs, availability, economic factors, and so on. Energy management is required because it influences a number of aspects of company operation and activities including the following aspects, energy costs, which affect the company profitability, energy costs which affect the competitiveness in the world market, national energy supply-demand balance, national trade and financial balance, local and global environments, occupational safety and health, loss prevention and waste disposal reduction, improved productivity, and quality. Energy management in the form of implementing new energy efficiency technologies, new materials, and new manufacturing processes and the use of new technologies in equipment and materials for business and industry is also helping companies improve their productivity and increase their product or service quality. Often, the energy savings is not the main driving factor when companies decide to purchase new equipment, use new processes, and use new high-tech materials. However, the combination of increased productivity, increased quality, reduced environmental emissions, and reduced energy costs provides a powerful incentive for companies and organizations to implement these new technologies. Energy management is all about reducing the cost of energy used by an organization, now with the added spin of minimizing carbon emissions as well. Reducing energy costs has two facets: price and quantity. Improving energy efficiency is, in part, a technical pursuit with a scientific basis.

The knowledge and skills of the energy management report impart guidance for development, generate opportunities for collaboration among developing or expanding training programs, facilitate greater consistency among existing professional programs, and increase awareness about the energy efficiency potentials. Steps which are needed during implementation of an EMS are:

1. *Initiating an energy management program*: Understanding basic concepts and requirements, getting organization higher management commitment, establishing an energy management team, and developing an energy policy and planning;
2. *Conducting an energy review*: Collecting energy data, analyzing energy consumption and costs, identifying major energy uses, conducting energy assessments, and identifying potential energy saving opportunities;
3. *Energy management planning*: Setting a baseline, determining performance metrics, evaluating energy saving opportunities and selecting best projects, and developing action plans;
4. *Implementing energy management*: Obtaining resource commitments, providing training and raising energy saving awareness, communicating to all stakeholders, and executing action plans;
5. *Measurement and verification*: Providing the knowledge and skills required to monitor, measure, verify, track, and document energy use and savings; and
6. *Management review*: Reviewing progress, modifying goals, objectives, and action plans as needed.

These steps are embedded in the *Plan-Do-Check-Act* cycle, which in addition to the common knowledge areas, includes the identified ancillary knowledge and skills that are enhancing the understanding of key energy management topics. To manage the energy a complete understanding of the company energy use, structure and demand is critical. This is based on a comprehensive energy review, consisting of the analysis of all energy use and consumption, determining the significant energy uses, and then identifying and prioritizing potential opportunities for the improvements and energy savings. An energy review is an essential element of energy management planning and must be performed by personnel with a broad range of knowledge and skills. An energy review requires the collection of energy consumption data from utility bills, energy meters, and other information sources. The data must be analyzed and interpreted within the context of the sites, facilities, processes, business units, and equipment. For example, in many of the smaller facilities, lighting and space heating and cooling are often the dominant energy users, while larger facility large energy parts are used in process or large equipment. These are just a tiny percentage of the cost at large petrochemical sites, where the dominant energy users are associated with moving, transforming, and separating feed and product materials. This analysis requires personnel that not only understand buildings, processes, energy using equipment, and other process or facility factors but also possess the knowledge and skills necessary to identify viable improvement opportunities. Some of the fundamental tasks involved in conducting an energy review include: data logging and collection, metering, monitoring, measurement, and verification, facilitating and managing the process for identifying the energy efficiency and conservation opportunities.

13.3 Energy management principles and methods

Business, industry, and government organizations are all under tremendous economic and environmental pressures in the last decades. Being economically competitive in the marketplace and meeting increasing environmental standards to reduce pollution are the major driving factors in most of the recent operational cost and capital cost investment decisions. Energy management is an important tool to help organizations meet these critical objectives for their short-term survival and long-term success. A common meaning of energy management, often that are signifying different things to different people is the efficient and effective energy use to maximize profits, while minimize the costs and to enhance competitiveness of the organization. Some goals and objectives of the energy management include the improving energy efficiency and reducing energy use, reducing costs and pollutant emissions, complaining to the various regulations, developing and maintaining effective monitoring, reporting, and management strategies for intelligent the energy use. Other goals are the finding better ways to increase returns from energy investments through research and development, developing interest in and dedication to the energy management program from all employees, and the deducing the impacts of curtailments or any interruption in energy supplies. Although energy conservation is an important energy management aspect, it is not the only consideration. In fact, for many organizations the energy management is one of the most promising revenue and profit improvements, and cost reduction programs available in today industries. Energy management programs are vital today. Commercial activity is very diverse, and this leads to varying energy intensities depending on the nature of the facility. Recording energy use in a building or a facility of any kind and providing a history of this use is necessary for the successful implementation of an energy management program. A time record of energy use allows analysis and comparison so that results of energy productivity programs can be determined and evaluated.

The most important organizational step affecting the success of an energy management program is the appointment of one person or a small group who have the full responsibility for its operation. Preferably that person reports directly to the top company management and has enough authority in directing technical and financial resources within the bounds set by the level of management commitment to implement the energy management plan. It is difficult to stress enough the importance of making the position of plant energy manager a full-time job. Any diversion of interest and attention to other aspects of the business may adversely affect the energy management program. One reason is that the greatest opportunity for energy cost control and energy-efficiency gains is in improved operational and maintenance practices. As the energy management program progresses to the energy audit and beyond, it is necessary to keep all employees informed as to its purposes, its goals, and how its operation are impacting plant or facility operations and employee routines, comfort, and job performances. The education can be delivered through several channels as best benefits the organizational structure. In addition to general education about energy conservation, it may prove worthwhile to offer specialized courses for boiler, mechanical and electrical equipment operators and other workers whose jobs can affect energy utilization in the plant. Staff training is the second important step and a key to staying on track for energy conservation and the success of any energy management program. It is the responsibility of the management to ensure that technical and operating personnel are trained to operate the equipment safely and in a proper manner. Effective training is not accomplished in a single session that once completed, may be quickly forgotten. Training must be through and continuous to help not only to inform but also to change attitudes. Training allows the staff to explore new ideas, interchange them with experts and with other trainee participants, and feel more comfortable with the role they must fulfill. In turn, trained technical and management staff should be encouraged to provide in-house training to operating and lower level technical staff. Staff training is the primary tool by which awareness is generated and knowledge is transmitted. As part of the energy management program, there are two major areas for employee training.

Frist one refers to the training in the development of new skills in technologies, and training to adopt new attitudes towards energy wastage and reduction of waste. The introduction of new technologies, process equipment, operating and maintenance procedures, and energy documentation methods requires training at many levels. There is a need to train new as well as experienced personnel in energy efficient operation of company facilities. The need for training in each should be reviewed periodically to assure that all new personnel are properly trained and to refresh the skills of existing personnel. Staff training is typically at three levels: management, engineering (technical level), and at supervisory level (operators). It is appropriate and even necessary to select an energy reduction goal for the first year of the energy management program or very early in the program. The purpose is to gain the advantage of the competitive spirit of those employees that can be aroused by a target goal.

13.4 Energy audit and energy conservation

Energy audits are the critical factors to a systematic approach for the decision-making in the energy management, attempting to balance the total energy inputs with its use, helping to identify all the energy streams in a facility or a building. An energy audit is periodic examination of an energy system to ensure that energy is used efficiently and to quantify the energy uses according to their functions. Industrial energy audit is an effective tool in defining and pursuing a comprehensive energy management program. Saving money on energy bills is attractive to businesses, industries, and individuals alike. In any industry, the three top operating expenses are the energy (electrical and thermal), labor, and materials. If one were to relate to the manageability of the cost or potential cost savings, the energy invariably emerges as a top ranker, and thus energy management constitutes a strategic area for cost reductions. Energy audits are helping to understand the ways in which the energy and fuel are used, and to identify areas where waste occur and where rooms for improvement exists. Energy audit gives a positive orientation to the energy cost reduction, preventive maintenance, and quality control programs, which are vital for production and utility activities. An energy audit program helps to keep focus on variations which occur in the energy costs, availability and reliability of the energy supply, to decide on appropriate energy mix, to identify energy conservation technologies or the retrofit for energy conservation equipment. In general, an energy audit is the translation of conservation ideas into realities, lending technically feasible solutions with economic and other organizational considerations within a specified time frame. The primary energy audit objective is to determine ways to reduce energy consumption per unit of product or to lower operating costs. Energy audit provides a *benchmark* (reference point) for managing energy in the organization and also provides the basis for planning a more effective use of energy throughout the organization.

Customers, for which a substantial fraction of their company's operating costs is due to the energy, have stronger motivations to initiate and continue energy cost-control programs. The energy audit is one of the first tasks to be performed in an effective energy cost control and reduction program. An energy audit consists of detailed examinations of how a facility uses energy, what the facility is spending for the energy, and finally, a recommended program for changes in operating practices or energy-consuming equipment that are cost-effectively to save on energy bills. The energy audit, sometimes called energy survey or energy analysis is a positive experience with significant benefits to the business or individual. Energy audits are performed by different groups. Electric and gas utilities are offering free residential energy audits, consisting of the analysis of the monthly energy bills, the inspection of the building construction and all of the energy-consuming appliances in the house. Ceiling and wall insulation is measured, ducts are inspected, and appliances, heaters, air conditioners, water heaters, refrigerators, freezers and the lighting system are examined and checked. An initial summary of the basic steps involved in conducting a successful energy audit is provided here, and these steps are explained more fully in the sections that follow. This audit description primarily addresses the steps in an industrial or large-scale commercial audit, and not all of the procedures described in this section are required for every type of audit. The audit process starts by collecting information about a facility's operation and about its past record of utility bills. These data are then analyzed to get a picture of how the facility uses, and possibly wastes the energy, as well as to help the auditor learn what areas to examine to reduce energy costs.

Energy audit is a periodic examination of an energy system to ensure that energy is used as efficient as possible. Energy audit is a systematic study or survey to identify how energy is being used in a building or plant, and identifies energy savings opportunities. An energy audit, also known as an energy survey, energy analysis, or energy evaluation, the energy audit examines the ways energy is currently used in that facility and identifies some alternatives for reducing energy costs. The energy audit goals are to clearly identify the types and costs of energy use, to understand how that energy is being used, and possibly wasted, to identify and analyze alternatives, such as improved operational techniques and/or new equipment that substantially are reducing the energy costs, and to perform an economic analysis on those alternatives and determine which ones are cost-effective for the business or industry involved. Using proper methods and equipment, an energy audit provides essential information on how much, where and how energy is used within an organization (facility or building), indicating the performance at the overall plant or process level. These performances can be compared against past and future levels for a proper energy management. The major parts of the energy audit report are energy savings proposals comprising of technical and economic analysis of projects. Looking at the final output, an energy audit can also be defined as a systematic search for energy conservation opportunities. In many situations major cost savings can be achieved through the implementation of no cost or low cost measures, such as:

(a) Changing energy tariff;
(b) Rescheduling production activities to take advantage of preferential tariffs;
(c) Adjusting existing controls so that plant operation matches the actual requirements of the building or manufacturing process;
(d) Implementing good housekeeping policies, in which staff are encouraged to avoid energy-wasteful practices; and
(e) Investing in small capital items such as thermostats and time switches.

Sometimes it is necessary to undertake more capital intense measures, investments and changes. Energy audit can be categorized into two types: (a) walk-through or preliminary audit, and (b) detail audit.

The energy audit process starts with the examination of facility historical and descriptive energy data. Specific data gathered in this preliminary phase include the energy bills for the past 12 months, descriptive information about the facility, such as a plant layout, and a list of the equipment that significantly affects the energy consumption. Often the most effective way to find and begin implementing energy improvements is to get all the interested parties into for a structured discussion of the plant. Careful preparation and actively involved stakeholders are critical to success. A good energy review leaves a site with its own plan, developed by its people and supported by its management. Walk-through or preliminary audit comprises: one day or half-day visit to a plant, where the output is a simple report based on observation and walk-through or preliminary audit historical data provided during the visit. The findings are general comments based on rule-of-thumbs, energy best practices or the manufacturer's data, seeking to establish the quantity and cost of each used energy form of energy. Quick overview of energy use patterns provides guidance for the energy accounting system and the personnel with perspectives of processes and equipment. The purpose is to identify energy intensive processes, system components and equipment, to identify the energy inefficiency at any phase of a detailed energy survey. In many situations major cost savings can be achieved through the implementation of no cost or low cost measures, such as changing energy tariff, rescheduling production activities to take advantage of preferential tariffs, adjusting existing controls so that plant operation matches the actual requirements of the building or manufacturing process, implementing good housekeeping policies, in which staff are encouraged to avoid energy-wasteful practices, or investing in small capital items, such as thermostats and time switches. However, sometimes are necessary to undertake more capital intense measures.

An energy balance is a set of relationships accounting for all the energy which is produced and consumed, and matches inputs and outputs, in a system over a given time period. The system can be anything from a whole country to an area to a process in a factory. An energy balance is usually made with reference to a year, though it can also be made for consecutive years to show variations over time. Energy balances provide overviews, and are basic energy planning tools for analyzing the current and projected energy situation. The overviews aid sustainable resource management, indicating options for energy saving, or for policies of energy pricing and redistribution, etc. There are a number of different types of energy balance which can be made, depending on the information you need; the energy commodity account, the energy balance, and the economic balance. The energy commodity account includes all flows of energy carriers, from the point of extraction through conversion to end-use, in terms of their original, physical units, such as kilotons of coal and kWh of electricity. The energy balance is similar to the energy commodity account, except for the fact that all physical units are converted into a single energy unit. This type of balance uses mean energy content values because of the inevitable variations in fuel composition. This means that there are inherent inaccuracies within the balance but as long as they are within accepted limits, this is common practice. Although this report does not analyze such policies, it acknowledges their vital role in improving workforce programs and work quality. In the economic balance, different forms of energy are accounted for in terms of their monetary value. An important part of preparing an energy balance is the construction of an energy chain to trace the flows of energy within an economy or system, starting from the primary source(s) of supply through the processes of conversion, transformation and transportation, to final/delivered energy and finishing with end use.

13.4.1 Types and structure of energy audits

The type of energy audit to be performed depends on the function and type of industry, depth to which final audit is needed, and the potential and magnitude of cost reduction desired. Thus, energy audit can be classified into the following two types: (i) *preliminary energy audit*, and (ii) *detailed energy audit*. Preliminary energy audit is a relatively quick exercise to establish energy consumption in the facility or organization, to estimate the scope for saving, to identify the most likely and the easiest areas for attention, to identify immediate, especially no- or low-cost improvements and/or energy savings, to set a *reference point*, to identify areas for more detailed study and measurement plans. Existing, historical, or easily obtained data are used often in a preliminary energy audit. This review can be led by knowledgeable internal resources (e.g., energy group members) or by an external consultant, including representatives from engineering, operations, and maintenance functions. The preliminary audit findings are a general comment based on rule-of-thumbs, energy best practices or the manufacturer's data. It aims to establish the quantity and cost of each form of energy used in a facility, plant or organization. A comprehensive audit provides a detailed energy project implementation plan for a facility, since it evaluates all major energy using systems. This type of audit offers the most accurate estimate of energy savings and cost. It considers the interactive effects of all projects, accounts for the energy use of all major equipment, and includes detailed energy cost saving calculations and project cost. In a comprehensive audit, one of the key elements is the energy balance. This is based on an inventory of energy using systems, assumptions of current operating conditions and calculations of energy use. This estimated use is then compared to utility bill charges. However, the audit structure and procedures are based on the industry-to-industry approaches, so the methodology of energy audits needs to be flexible.

Quick overview of energy use patterns are providing guidance for energy accounting system, personnel with perspectives of processes and equipment, identifying energy-intensive processes and equipment, energy inefficiency, if any, and can set the stage for eventually detailed energy surveys. A preliminary energy report structure consists of an introduction, overview of current systems in place, how much energy is being consumed, what type of energy is being consumed, the performance of the facility compared with other similar facilities, the characteristic performance of the facility, scope and goals of energy audit, recommendations and the associated costs and savings, and conclusions. Detailed energy audit is carried out for the energy savings proposal recommended in walk-through or preliminary audit.

It will provide detailed data on the energy inputs to, and energy flows within a facility and also technical solution options and economic analysis for the factory management to decide project implementation or priority. A feasibility study is required to determine the viability of each option. Detailed evaluations of the process and equipment energy use patterns, measurements of energy use parameters, and a review of operating characteristics are providing their efficiency evaluations, identifying the energy saving options and measures, and lead to the recommendations for the energy savings and conservation implementation.

13.4.2 Energy audit structure and phases

An energy audit begins with a detailed analysis of the energy bills for previous year or 2 years. His is an important task because: the energy bills show the use of each energy sources in the overall energy cost, where the energy is used and can identify the energy wastes, the total energy costs and the upper limit of the savings. A complete analysis of the facility or building energy bills requires detailed knowledge of the energy rate structures in the effect in order to determine the individual equipment and process energy costs, energy demand charges and the most accurate estimates of the energy savings for the energy management opportunities, such as: high-efficiency equipment, rescheduling of some of the on-peak electrical and/or thermal uses and processes. An initial site visit take one day and gives the Energy Auditor an opportunity to meet the personnel concerned, to familiarize with the site and to assess the procedures necessary to carry out the energy audit. During the initial site visit the Energy Auditor or Engineer caries out the following actions: discuss with the senior management the energy audit aims, as well as the economic guidelines associated with the recommendations of the audit, analyze the major energy consumption data with the relevant personnel, obtain site drawings where available, building layout, steam distribution, compressed air distribution, electricity distribution etc., and finally tour the site accompanied by engineering/production. The main aims of this visit are: structure and finalize the energy audit team, identify the main energy consuming areas or plant items to be surveyed during the audit, and any existing instrumentation/additional metering required, decide whether any meters have to be installed prior to the audit (kWh, steam, oil or gas meters), identify the instrumentation required for carrying out the energy audit, plan the audit time frame, collect macro data on plant energy resources, major energy consuming centers, and create awareness through meetings and program dissemination.

Depending on the site or facility nature and complexity, a comprehensive audit can take from several weeks to several months to be completed. Detailed studies to establish, and investigate, energy and material balances for specific plant departments or items of process equipment are carried out. Whenever possible, checks of plant operations are carried out over extended periods of time, at nights and at weekends as well as during normal daytime working hours, to ensure that nothing is overlooked.

The audit report includes a description of energy inputs and product outputs by major department or by major processing function, and will evaluate the efficiency of each step of the manufacturing process. Means of improving these efficiencies will be listed, and at least a preliminary assessment of the cost of the improvements will be made to indicate the expected payback on any capital investment needed. The audit report should conclude with specific recommendations for detailed engineering studies and feasibility analyses, which must then be performed to justify the implementation of those conservation measures that require investments. The information and data collected during the detailed energy audit includes: energy consumption by type, department, major process equipment, and by end-use. Part of this data and information collection are also: material balance data (raw materials, intermediate and final products, recycled materials, use of scrap or waste products, production of by-products for re-use in other industries, etc.), energy cost and tariff data, process and material flow diagrams, site power generation and distribution or other sit services (compressed air, steam), energy supply sources (e.g., electricity from the grid or self-generation), potential for fuel substitution, process modifications, and the use of cogeneration systems (combined heat and power generation), and last but not least energy management procedures and energy awareness training programs within the establishment.

Existing baseline information and reports are useful to get consumption pattern, production cost and productivity levels in terms of product output per raw material inputs. The audit team should collect the following baseline data: technology, processes used and equipment details, capacity utilization, amount and type of input materials used, water, fuels, steam, and electricity consumption, other inputs, such as compressed air, cooling water, quantity and type of wastes generated, percentage rejection or reprocessing, efficiencies, or yields. An overview of unit operations, important process steps, areas of material and energy use and sources of waste generation should be gathered and should be represented in a flowchart visually describing the unit material, energy fluxes, and the unit operation. Existing drawings, records, and shop floor walk through will help in making this flow chart. Simultaneously, the team should identify the various inputs and output streams at each process step. It is important to plan additional data gathering carefully. Here are some basic tips to avoid wasting time and effort: measurement systems must be easy to use and provide the information to the accuracy that is needed, not the accuracy that is technically possible, measurement equipment is desirable to be inexpensive (flow rates using a bucket and stopwatch), the quality of the data must be such that the correct conclusions are drawn (what grade of product is on, is the production normal etc.), define how frequent data collection should be to account for process variations., measurement exercises over abnormal workload periods (such as startup and shutdowns), design values can be taken where measurements are difficult (cooling water through heat exchanger).

The energy audit report begins with an executive summary that provides the audited organization or facility with a brief synopsis of the total savings available and the highlights of each energy conservation and savings opportunities. The report needs to describe the facility that has been audited, and provide information on the facility operations that are related to its energy costs. The structure and analysis of the energy bills is presented, in the most informative ways, such as tables and plots showing the costs and consumption. Following section of the energy cost analysis, the recommended measures and energy saving opportunities are presented, along with the calculations for the costs and benefits, and the cost-effectiveness criterion. Regardless of the audience for the audit report, it must be written in a clear, concise, and easy-to understand format and style. The executive summary is tailored to nontechnical personnel, and technical jargon must be minimized. A client who understands the report is more likely to implement the recommended measures. An outline for a complete energy audit report consist of: an executive summary, that includes a brief summary of the recommendations and cost savings, a table of contents, introduction, the purpose of the energy audit, emphasizing for the needs for a continuing energy cost control program, the facility description, consisting of the product or service descriptions, and materials flow, size, construction, facility layout, and hours of operation, equipment list, with specifications. In the report is included a comprehensive energy bill analysis and utility rate structures, tables and graphs of energy consumptions and costs, a discussion of energy costs and energy bills, as well as energy conservation opportunities and a listing of potential energy savings opportunities. In the report, cost and savings analysis, economic evaluation, an action plan, recommended measures and energy saving opportunities, an implementation schedule, the designation of an energy monitor and ongoing program, and a report summary and conclusions are also included. Additional comments not otherwise covered may be also included in more comprehensive reports or if needed.

13.5 Renewable energy economics

Renewable energy sources (RES) used in small-scale distributed generation systems are a promising alternative for additional energy supply toward smarter and more sustainable cities, facilities, and industrial structures. Although the resource of renewable energy systems, such as solar irradiation, wind energy, geothermal energy are free, the equipment required to collect, process and convert such energy types into useful energy forms (heat, thermal energy, or electricity) have a cost, lower or higher depending on each system. Many RES can be potentially implemented in smart buildings and industrial facilities. However, important factors, such as geo position characteristics, demand energy and load profiles, energy tariffs, local and national regulations need to be taken into account in order to select the optimal technology mix and development feasibility. An adequate and optimum selection of technologies and the optimal scaling of the implemented systems are crucial for the economic feasibility of DG projects. Integrating RES and DG into the power network implies changes in the way the energy is generated and scheduled, both utility and facility level. RES are not easy to control, and balancing generation and energy demand requires knowing almost instantaneously consumption and generation. On the other hand, the constant decrement of costs in RE systems, in addition to overall trends to higher energy costs, has derived in the profitability of RE in many applications; however, the adequate scaling of these systems, given the demand and geographical characteristics, is of prime importance in the economic feasibility analysis. Overall costs, including initial investments, operation, and maintenance costs are one of the important elements in all the distributed generation or renewable energy project evaluations.

Renewable energy conversion systems, such as solar or wind energy conversion systems are generally characterized by high initial or investment costs and quite low operating costs. To decide to employ any renewable energy system, the cost of system components, other required equipment, and conventional fuel required as backup must be lower than the cost of other conventional energy sources performing the same task. Thus, the economic analysis is to compare the initial known investments with the estimated future operating costs, including both the cost to run and maintain the renewable energy system and the auxiliary energy used as backup. Other factors that need to be considered include the interest paid on money borrowed, taxes if any, insurance cost if any, and resale of equipment at the end of its life.

However, the initial investment expenses alone are insufficient for complete analysis of RES and/or DG projects. Both reliability and power quality are also essential elements of the decision along with the type of DG selected. But cost is more often not fully taken into the primary considerations in most of the planning situations, since selecting the lowest cost option may not viable to achieve target reliability and power quality levels. Thus, evaluation of cost is usually the critical element in a DG planning evaluation. Inclusion of all costs is more important than the inclusion into the transmission and distribution planning, because each of the DG options often has significant differences among them. The operation and maintenance costs are rarely included in typical power distribution studies, because they usually do not vary significantly. However, operation and maintenance differs substantially among various DG options, and differs a great deal from the distribution operation and maintenance costs. In DG studies, energy cost (fuel) often varies a great deal, and such costs must be included in studies of distribution level generation. Various costs involved in the DG planning are: fixed costs (e.g., cost of the basic DG units, the annual costs of taxes, insurance, inspection, scheduled maintenance, testing, re-certification of a DG unit in service do not depend on run time), variable costs (DG fuel unit costs, if any some maintenance and operation costs) and continuing costs (facility in operation and include inspection and maintenance, fuel supplies, replacement parts, taxes, insurance, electrical losses, and perhaps other expenditures).

The decision-making process on renewable energy and distributed generation projects is similar to the decision-making process of any engineering projects and involves the following steps: the need for a project is identified and the project alternative solutions, and all the alternative projects are analyzed and evaluated, including a feasibility analysis and a detailed economic analysis of all the project alternatives, which are identified in previous decision-making step. In evaluating the alternative projects, all factors are examined and some of the proposed alternative solutions are often excluded for reasons other than the economic considerations. Finally, a decision is made on the best project alternative solution and the engineering project is identified and fully specified, technical, operational and cost. The final decision is specific enough to allow the planning and construction of the DG or RES facility but also allows sufficient flexibility for the next project stages. It is worth to have all the definitions of several concepts used in the fields of economics and management, which are helpful in the decision-making process for any energy projects. These concepts are briefly discussed here. The total of all fixed and variable costs calculated over a project time, divided by the total number of units produced, are representing the average costs. The average revenue represents the total revenue over a period of time, usually one year, divided by the total number of units produced, while the difference between the average revenue and the average cost is the average profit. All costs that are not affected by the level of business activity or production level, such as rents, insurance, property taxes, administrative salaries, and interest on borrowed capital. The sum of all costs, fixed and variable, associated with a project from its inception to its conclusion represents the life time cost. The life cycle costs include the planning costs, and any abandonment, disposal, or storage costs. Marginal or incremental cost is the cost associated with the production of one additional unit of output, while the marginal or incremental revenue is the one resulting from the production of one additional unit of output. Opportunity cost is the cost associate with the opportunity to use scarce resources, such as capital, to achieve monetary or financial advantage. Salvage values are the prices paid by a willing buyer for a plant or business after all operations have ceased. Sunk costs are all the costs associated with past activity that may not be recovered and do not affect any future costs or revenues. The horizon time is the one from the inception to the end of the project, including any disposal or storage of equipment and products. Variable costs are the costs associated with the level of business activity or output level, such as fuel cost, materials cost, labor cost, distribution cost, etc. The variable costs increase monotonically with the number of units produced. All these costs are associated with the DG or RES economic analysis. The decision making process for engineering projects follows a well-defined need and is accomplished by multistep, structured procedures, simulations, and modeling techniques. An economic analysis is made for a few carefully selected alternative solutions to the problem and the final decision takes into consideration the economic analysis as well as previous experience, environmental, and social factors as well as engineering input. The objective of the decision-making process is the evaluation or appraisal of several alternative solutions to the project needs. According to the decision making method, the initial investment on the project must be recovered from net receipts or income within a specified period of time, which ranges from 2 to 6 years. This is the payback period and is calculated by counting the number of years it takes for the cumulative cash flows to equal the initial investment. The payback period method makes use of the cash flows, which is the correct parameter for the calculation of the annual costs and benefits of the project. However, it does not differentiate between recent and distant cash flows or between capital expenditures that occur at the beginning or at another time during the life of the project.

13.6 Chapter summary

The concept of energy conservation, energy efficiency, or energy optimization became one of the major concerns of our times. Energy sustainability is the cornerstone to the health and competitiveness of industries that produce and manufacture in our global economy, being more than the process of being environmentally responsible and earning the right to operate as a business. It is the ability to utilize and optimize multiple sources of secure and affordable energy for the enterprise, and then continuously improve utilization through systems analysis and through an organizational drive for continuous improvement as a core principle. Management commitment to ensure the best energy efficiency management in existing process operations, as well as a dedicated pursuit of new system technologies and processes, is the only recipe for excellence. Diversification of energy resources and alternative methods for the energy generation becomes economic and environmental imperatives of today industry and society, being a major aspect of the energy sustainability paradigm. Renewable energy techniques can be used, both, as centralized and/or largely decentralized energy sources, while decentralized ones are particularly advantageous for rural and isolated consumers or special applications. Energy management is the philosophy of more efficient energy use, without compromising upon production levels, product quality, safety, and environmental standards. Energy accounting, monitoring, and control are the initial steps to be observed in any of the energy management program, being of great importance making any program of energy management a success. The effectiveness of energy utilization varies with specific industrial operations because of the diversity of the products and the processes required manufacturing goods and services. The organization of personnel and operations, involved also varies. Consequently, an effective energy management program should be tailored for each company and its plant operations. An energy audit is a systematic study or survey conducted to identify how the energy is being used in a building, facility or plant, and to identify energy savings opportunities, and indicating the performance at the overall plant, facility or process level. Using proper and well-designed energy audit methods and equipment, an energy audit provides the energy manager with essential information on how much, where, and how energy is used within an organization (factory, plant, or building). The main objectives and goals of the energy audit report are energy savings proposals comprising of technical and economic analysis of projects. Goals and objectives of an energy audit are clearly identifying the energy use types and cost, understanding the energy use and waste, analyzing, and identifying the ways to save or preserve energy, improved operational techniques and equipment, performing an economic analysis of alternatives and finally deterring the most cost effective one. In summary, energy management means lower the organization or company costs through eliminating unnecessary energy use and wastes, improving the efficiency of the needed energy, buying the energy at lowest cost possible, and adjusting the operations to allow purchasing the energy at lower prices.

References

[1] J.R. Cogdell, Foundations of Electric Power, Prentice-Hall, Upper-Saddle River, NJ, 1999.

[2] M.E. El-Hawary, Introduction to Power Systems, Wiley—IEEE Press, New York, 2008.

[3] Y. Goswami and F. Keith (Eds.), Energy Conversion, CRC Press, Boca Raton, FL, 2008

[4] M.A. El-Sharkawi, Electric Energy: An Introduction (2nd ed.), CRC Press, Boca Raton, FL, 2009.

[5] F.M. Vanek, L.D. Albright, and L.T. Angenent, Energy Systems Engineering, McGraw Hill, New York, 2012.

[6] W.C. Turner, Energy Management Handbook, John Wiley and Sons, New York, 2001.
[7] A.P. Rossiter and B.P. Jones (Eds.), Energy Management Efficiency for the Process Industries, Wiley, Hoboken, NJ, 2015.
[8] B.L. Capehart, Guide to Energy Management, Fairmont Press (7th ed.), Boca Raton, FL, 2011.
[9] R. Belu, and L.I. Cioca, Energy Management Systems, in Encyclopedia of Energy Engineering & Technology (Online) (Ed: Dr. Sohail Anwar, R. Belu et al.), Vol. 5, CRC Press/Taylor and Francis, 2015.

Questions and problems

1. List down the objective of energy management.
2. Explain the role of training and awareness in energy management program?
3. Define in energy audit in your words.
4. What are the steps or phases in the implementation of energy management in an organization or facility?
5. Explain briefly the difference between preliminary and detailed energy audits?
6. What is the significance of knowing the energy costs?
7. What are the benefits of benchmarking energy use?
8. List all the costs that are expected to have from the construction and operation of a geothermal power plant, which is to produce electrical energy for 40 years.
9. Do the same as in previous problem but for PV power plant, producing electrical energy for 25 years?
10. Do the same as in previous example but for a wind farm, producing energy for 30 years?
11. List the most used measurement instruments and equipment that are used in an energy audit.
12. What are the main steps in an energy audit?

Chapter 14
Post-face and pedagogical suggestions

14.1 Book overview

The book is covering three major disciples: basic and fundamental knowledge in power systems, such as power engineering basic and foundations, motors and transformers, power distribution basic, second the building and industrial power distribution, such as load characteristics and calculations, load and motor centers, building electrical systems and lighting, and motor protection and control, and the third fundamentals of the major renewable energy sources and energy storage, such as solar and wind energy, photovoltaics, geothermal energy and small hydro-power systems, major energy storage technologies, and a brief description of microgrids, distributed generation issues and energy management. Due to the fast pace of changes into the energy sectors and uses, more and more professional and the fresh graduates in the industrial and engineering fields, and not only are required to have better understanding of the energy industries, energy supply issues, energy conservation, alternative energy sources, or sustainability. The alternative energy sources, better and more efficient of energy uses, sustainability, energy use and conservation aspects are now intrinsic part of engineers, professional and managers. This book is trying to answer to some of these issues, providing foundations and basic concepts for students and professional. It originates from courses that the author taught during over 35 years academic career in the areas of energy and power engineering, renewable energy, industrial power distribution or building electrical systems. Each book chapter and/or topics included are self-sustained with almost no or very little connections with other book chapters or topics. The knowledge requirements for the readers and the students using this book are the basic physics and calculus engineering courses, making it accessible to a broad spectrum of users. Instructors can customize the topics order based on the course specifics, level and/or students who are taking the course. Book is primarily intended for engineering students or professionals working in the areas of industrial power and energy systems, building electrical systems or renewable energy. However, the book can also be used for major science courses interested in the alternative energy systems and industrial energy. No prior knowledge for the users is required with the exception, as we already mentioned of the general physics and calculus. Each book chapter contains several examples to support the student and reader topic understanding. In each chapter, several references and further reading suggestions are also included for readers who are interested in that specific topics. In the author's opinion, this book is a good education resource for anyone interested in the energy engineering, renewable energy, build electrical systems and industrial power distribution. It is also the hope of the author that this book will help new people to enter in the fascination filed of the energy engineering.

14.2 Pedagogical approaches and suggestions for instructors

Primarily the book is designated to sophomore and junior students in engineering programs. It can be used for a broad course spectrum, such as renewable energy technology, industrial power distribution, building electrical systems and electrical systems design, both as the required or recommended textbook. Over 450 problems and review questions are included in this textbook to help students, readers, and instructors to deepen the knowledge each of the covered topics. Each chapter contains several solved example helping the reader to fully understand the chapter topics and concepts. Comprehensive discussions of the system characteristics, performances, main applications advantages and disadvantages are also included and discussed extensively. The primary courses targeted by this book are introduction to renewable energy, industrial power distribution and electrical systems design (specifically building and industrial electrical systems). Each chapter is self-contained with almost no or very little connections with other book chapters, leaving the instructor full flexibility in customizing the topics and/or topics order. Book can be used regardless of the academic year structure, semester-based or term (quarter)-based. Tables 14.1 and 14.2 show the author suggestions for specific course content in the engineering areas targeted by this book, both for semester and term (quarter) academic year structure.

Table 14.1 Semester course and topics suggestions

RES	IPD	BES (ESD)
Chapter 1	Chapter 1	Chapter 1
Chapter 2 (2.1, 2.2, and 2.3)	Chapter 2	Chapter 2
Chapter 3 (3.1)	Chapter 3	Chapter 3 (3.1)
Chapter 9	Chapter 4	Chapter 4
Chapter 10	Chapter 5	Chapter 5 (5.1, 5.2, and 5.4)
Chapter 11	Chapter 6 (6.1 and 6.2)	Chapter 6
Chapter 12	Chapter 7	Chapter 7
Chapter 13 (13.1, 13.2, and 13.5)	Chapter 8	Chapter 8 (8.1)
	Chapter 13 (13.1 and 13.2)	Chapter 11 (11.1 and 11.2)
		Chapter 12 (12.1, 12.2, and 12.3)
		Chapter 13 (13.1 and 13.2)

Legend: RES: renewable energy systems; IPD: industrial power and energy distribution; and BES (ESD): (building) electrical systems design

Table 14.2 Term (quarter) course and topics suggestions

RES	IPD	BES (ESD)
Chapter 1 (1.1 and 1.5)	Chapter 1	Chapter 1
Chapter 2 (2.1 and 2.2)	Chapter 2	Chapter 2 (2.1, 2.2, and 2.3)
Chapter 9	Chapter 3	Chapter 3 (3.1)
Chapter 10	Chapter 4	Chapter 4
Chapter 11	Chapter 5 (5.1, and 5.4)	Chapter 5 (5.1, 5.2, and 5.4)
Chapter 12 (12.1, 12.2, 12.3, and 12.4)	Chapter 6 (6.1 and 6.2)	Chapter 6
	Chapter 7	Chapter 7
	Chapter 8 (8.1 through 8.6)	Chapter 11 (11.1 and 11.2)
		Chapter 12 (12.1 and 12.2)

Appendix A
Common parameters, units, and conversion factors

Table A1 Physical constants in SI units

Quantity	Symbol	Value
Avogadro constant	N	6.022169×10^{26} k/mol
Boltzmann	k	1.380622×10^{-23} J/K
First radiation constant	$C_1 = 2 \cdot \pi \cdot h \cdot c$	3.741844×10^{-16} Wm^2
Gas constant	R	8.31434×10^{3} J/kmol K
Planck constant	h	6.626196×10^{-34} Js
Second radiation constant	$C_2 = hc/k$	1.438833×10^{-2} mK
Speed of light in a vacuum	c	2.997925×10^{8} m/s
Stefan–Boltzmann constant	σ	5.66961×10^{-8} W/m^2 K^4
Speed of light	c	299,792.458 m/s
Elementary charge	e	1.602176×10^{-19} C

Table A2 Multiplication factors

Multiplication factor	Prefix	Symbol
10^{12}	Terra	T
10^{9}	Giga	G
10^{6}	Mega	M
10^{3}	Kilo	K
10^{2}	Hecto	H
10	Deka	da
1	N/A	—
0.1	Deci	d
0.01	Centi	c
10^{-3}	Mili	m
10^{-6}	Micro	μ
10^{-9}	Nano	n
10^{-12}	Pico	p
10^{-15}	Femto	f
10^{-18}	Atto	a

Table A3 System of units and conversion factors

US unit	Abbreviation	SI unit	Abbreviation	Conversion factor
Foot	ft	Meter	m	0.3048
Mile	mi	Kilometer	km	1.6093
Inch	in	Centimeter	cm	2.54
Square feet	ft^2	Square meter	m^2	0.0903
Acre	acre	Hectare	ha	0.405
Circular mil	cmil	μm^2	—	506.7
Cubic feet	ft^3	Cubic meter	m^3	0.02831
Gallon (US)	gal (US)	Liter	l	3.785
Gallon (UK)	gal (UK)	liter	l	4.445
Cubic feet	ft^3	Liter	l	28.3
Pound	lb	Kilogram	kg	0.45359
Ounces	oz	Gram	g	28.35
US ton	ton (US)	Metric ton	ton (metric)	0.907
Mile/hour	mi/h	Meter/second	m/s	0.447
Flow rate	ft^3/h	Flow rate	m^3/s	0.02831
Density	lb/ft^3	Density	kg/m^3	16.020
lb-force	lbf	Force	N	4.4482
Pressure	lb/in^2	Pressure	kPa	6.8948
Pressure	bar	Pressure	Pa	10^5
Torque	lb.force ft	Torque	Nm	1.3558
Power	ft.lb/s	Power	W	1.3558
Power (Horsepower)	HP	Power	W	745.7
Energy	ft.lb-force	Energy	J	1.3558
Energy (British thermal unit)	Btu	Energy	kWh	3412

Table A4 Summary of radiometric and photometric units

Quantity	Symbol	Unit
Wavelength	λ	m
Foot-candela	fc	$lm \cdot ft^{-2}$
Radiant (luminous) energy	Q	$W \cdot s$
Radiant (luminous) energy (Photometry)	Q_V	$lm \cdot s$
Radiant (luminous) Energy density	We	$W \cdot s \cdot m^{-2}$
Radiant (luminous) Energy density (photometry)	W_V	$lm \cdot s \cdot m^{-2}$
Radiant (luminous) Radiant energy density (per unit of volume)	$u_0 = \frac{dW_e}{dv}$	$J \cdot m^{-3}$
Energy flux (power)	Φ	W
Radiant (luminous) energy	Φ_V	lm

(Continues)

Table A4 (Continued)

Quantity	Symbol	Unit
Flux (power) (photometry)		
Radiant exitance (radiant flux per unit of source area)	M_0	$W \cdot m^{-2}$
Irradiance (radiant flux per unit of target area)	I_0	$W \cdot m^{-2}$
Irradiance and illuminance	E	$W \cdot m^{-2}$ (or $W \cdot cm^{-2}$)
Irradiance and illuminance (photometry)	E_V	lx (fc)
Radiance, intensity and luminance	L	$W \cdot m^{-2}$/sr (steradian)
Radiance, Intensity and luminance (photometry)	L_I	$lm \cdot m^{-2}$/sr (steradian)
	l	
Radiant and luminous intensity	l_V	W/sr
Radiant and luminous intensity (photometry)		cd (lm/sr)

Table A5 Common energy conversion factors

Energy unit	SI equivalent
1 electron volt (eV)	1.6021×10^{-19} J
1 erg (erg)	10^{-7} J
1 calorie (cal)	4.184 J
1 British thermal unit (Btu)	1055.6 J
1 Q (Q)	1018 Btu (exact)
1 quad (q)	1015 Btu (exact)
1 tons oil equivalent (toe)	4.19×10^{10} J
1 barrels oil equivalent (bbl)	5.74×10^{9} J
1 tons coal equivalent (tce)	2.93×10^{10} J
1 m^3 of natural gas	3.4×10^{7} J
1 liter of gasoline	3.2×10^{7} J
1 kWh	3.6×10^{6} J
1 ft^3 of natural gas (1000 Btu)	1055 kJ
1 gal. of gasoline (125,000 Btu)	131.8875 kJ

Table A6 Refractive index of some of the common materials (at 20 °C)

Material	Index of refraction	Material	Index of refraction
Diamond	2.419	Benzene	1.501
Fluorite	1.434	Carbon disulfide	1.628
Fused quartz	1.458	Carbon tetrachloride	1.461
Glass (crown)	1.520	Ethyl alcohol	1.361
Glass (flint)	1.660	Glycerin	1.473
Ice	1.309	Oil, turpentine	1.470
Polystyrene	1.590	Paraffin (liquid)	1.480
Salt ($NaCl_2$)	1.544	Water	1.333
Teflon	1.380	Air (0 °C, 1 atm)	1.000293
Zircon	1.923	Carbon dioxide (0 °C, 1 atm)	1.00045

Table A7 Properties of dry air

Temperature (K)	ρ (kg/m^3)	C_P (kJ/kg K)	k (W/m K)
293	1.2040	1.0056	0.02568
300	1.1774	1.0057	0.02624
350	0.9980	1.0090	0.03003
400	0.8826	1.0140	0.03365
450	0.7833	1.0207	0.03707
500	0.7048	1.0295	0.04038
600	0.5879	1.0551	0.04659
700	0.5030	1.0752	0.05230
800	0.4405	1.0978	0.05779
900	0.3925	1.1212	0.06279
1000	0.3525	1.1417	0.06752

Symbols: T, absolute temperature, degrees Kelvin; ρ, density; C_P, specific heat capacity; μ, viscosity; v, μ/ρ; and k, thermal conductivity. The values of ρ, μ, v, k, and C_P are not strongly pressure-dependent and may be used over a fairly wide range of pressures. Adapted from U.S. National Bureau Standards (U.S.), 1955.

Table A8 Properties of water

Temperature (°C)	ρ (kg/m^3)	C_p (kJ/kg K)	k (W/m K)
0.00	999.8	4.225	0.566
4.44	999.8	4.208	0.575
10.0	999.2	4.195	0.585
20.0	997.8	4.179	0.604
25.0	994.7	4.174	0.625
100	954.3	4.219	0.684

Table A9 Temperature conversion formulas

	Degree Celsius (°C)	DegreeFahrenheit (°F)	Kelvindegree (K)
Degree Celsius (°C)	—	$\frac{9}{5}(°C+32)$	K – 273.15
Degree Fahrenheit (°F)	$\frac{5}{9}(°F-32)$	-	1.8 × K – 459.67
Kelvin degree (K)	$°C+273.15$	$(459.67+°F)/1.8$	—

Table A10 Properties of metals used in cables and conductors

Property	**Unit**	**Copper**	**Silver**	**Aluminum**	**Lead**	**Steel**
Density	kg/m^3	8890	10520	2675	11370	7769
Thermal conductivity	W/m·°C	73	407	204	171	64
Thermal expansion coefficient	1/°C	$12 \cdot 10^{-6}$	$19.3 \cdot 10^{-6}$	$23 \cdot 10^{-6}$	$29 \cdot 10^{-6}$	$12 \cdot 10^{-6}$
Electrical resistivity (at 20 °C)	$\Omega m \cdot 10^{-8}$	1.724	1.626	2.803	21.40	13.80
Thermal coefficient of resistance	1/ °C	0.0039	0.0041	0.0040	0.0040	0.0045
Specific Heat	kJ/kg·°C	0.398	0.234	0.921	0.130	0.460

Appendix B

Design parameters, values, and data

Table B1 Terrain categories

Terrain category	Terrain description
0	Smooth, flat terrain, small water body
1	Open, flat, no trees, or obstructions or near/across water or desert, snow cover
2	Open, flat rural areas with no trees obstructions, few and low obstacles, for example, farmland, semiarid areas
3	Open, flat, or undulating with very few obstacles (open grass, farmland with few trees, hedgerows, other barriers), tundra, prairie
4	Build-up areas with some trees and buildings (suburbs, small towns, woodlands, shrubs, broken country with large trees, small fields with hedges)

Table B2-I Geometric characteristics of conduits (adapted from NEC)

Conduit trade size designator: English (metric)	Internal diameter in (mm)	Cross-sectional area in^2 (mm^2)
1/2 (16)	0.62 (15.7)	0.30 (195)
3/4 (21)	0.82 (20.9)	0.53 (345)
1 (27)	1.05 (26.6)	0.87 (559)
1 1/4 (35)	1.38 (35.1)	1.51 (973)
1 1/2 (41)	1.61 (40.9)	12.05 (1.322)
2 (53)	2.07 (52.5)	3.39 (2.177)
2 1/2 (63)	2.47 (62.7)	4.82 (3.106)
3 (78)	3.07 (77.9)	7.45 (4.794)
3 1/2 (91)	3.55 (90.1)	9.96 (6.413)
4 (103)	4.03 (102.3)	12.83 (8.268)
5 (129)	5.05 (128.2)	20.15 (12.984)
6 (155)	6.07 (154.1)	29.11 (18.760)

Table B2-II Maximum occupancy recommended for conduits (adapted from NEC)

Conduit trade size designator: English (metric)	1 cable = 53% fill in^2 (mm^2)	2 cables = 31% fill in^2 (mm^2)	3 + cables = 40% fill in^2 (mm^2)
1/2 (16)	0.16 (103)	0.09 (60)	0.120 (78)
3/4 (21)	0.28 (183)	0.16 (107)	0.21 (138)
1 (27)	0.46 (296)	0.27 (173)	0.35 (224)
1 1/4 (35)	0.80 (516)	0.47 (302)	0.60 (389)
1 1/2 (41)	1.09 (701)	0.64 (410)	0.82 (529)
2 (53)	1.80 (1.154)	1.05 (675)	1.36 (871)
2 1/2 (63)	2.56 (1.646)	1.49 (963)	1.93 (1.242)
3 (78)	3.95 (2.541)	2.31 (1.486)	2.98 (1.918)
3 1/2 (91)	5.28 (3.399)	3.09 (1.988)	3.98 (2.565)
4 (103)	6.80 (4.382)	3.98 (2.563)	5.13 (3.307)
5 (129)	10.68 (6.882)	6.25 (4.025)	8.06 (5.194)
6 (155)	15.43 (9.943)	9.02 (5.816)	11.64 (7.504)

Table B2-III Minimum radius of bends of conduits (adapted from NEC)

Conduit trade size designator: English (metric)	Layers of steel within sheath in (mm)	Other sheath in (mm)
1/2 (16)	6 (160)	4 (100)
3/4 (21)	8 (210)	5 (130)
1 (27)	11 (270)	6 (160)
1 1/4 (35)	14 (350)	8 (210)
1 1/2 (41)	16 (410)	10 (250)
2 (53)	21 (530)	12 (320)
2 1/2 (63)	25 (630)	25 (630)
3 (78)	31 (780)	31 (780)
3 1/2 (91)	36 (900)	36 (900)
4 (103)	40 (1.020)	40 (1.020)
5 (129)	50 (1.280)	50 (1.280)
6 (155)	60 (1.540)	60 (1.540)

Table B3-I Commonly used conductor cross-sectional area

Size AWG/ kcmil	Approximate cross-sectional area square inches/in^2					
	RHH/RHW with cover	**RHH/RHW without cover**	**TW or THW**	**THHN THWN**	**XHHW**	**BARE stranded conductors**
14	0.0293	0.0209	0.0139	0.0097	0.0139	0.004
12	0.0353	0.0260	0.0181	0.0133	0.0181	0.006
10	0.0437	0.0333	0.0243	0.0211	0.0243	0.011
8	0.0835	0.0556	0.0437	0.0366	0.0437	0.017
6	0.1041	0.0726	0.0726	0.0507	0.0590	0.027
4	0.1333	0.0973	0.0973	0.0824	0.0814	0.042
3	0.1521	0.1134	0.1134	0.0973	0.0962	0.053
2	0.1750	0.1333	0.1333	0.1158	0.1146	0.067
1	0.2660	0.1901	0.1901	0.1562	0.1534	0.087
0	0.3039	0.2223	0.2223	0.1855	0.1825	0.109
00	0.3505	0.2624	0.2624	0.2233	0.2190	0.137
000	0.4072	0.3117	0.3117	0.2679	0.2642	0.173
0000	0.4754	0.3718	0.3718	0.3237	0.3197	0.219

Table B3-II Resistivity and reactance per unit length (medium- and low-voltage)

Conductor Size	**Resistance (R) at 25 °C, 60 Hz (mΩ/conductor/1000 ft)**	**Resistance (R) at 25 °C, 60 Hz, 1 ft spacing (mΩ/conductor/1000 ft)**
	Aluminum Conductor Steel Reinforced (ACSR)	
1590.0	11.8	67.9
1272.0	14.7	70.4
954.0	19.4	7.38
556.5	32.1	78.6
477.0	33.8	80.2
336.4	48.0	84.3
4/0	76.2	109.9
2/0	121.3	121.2
2	243.5	121.5
4	386.7	124.0
6	614.7	127.3
	Copper	
1000.0	11.9	75.8
750.0	15.3	79.0
500.0	22.4	83.9
350.0	31.7	88.3
250.0	44.4	92.2
4/0	52.4	95.3
2/0	83.1	101.0
2	165.1	108.0
4	262.7	113.0
6	413.1	121.0

Table B4 Average luminance of the common light sources

Light source	Comment	Average (approximate) luminance (cd/m^2)
Sun (at Earth surface)	At meridian	1.60×10^9
Sun (at Earth surface)	Near horizon	6.0×10^4
Moon (at Earth surface)	Bright spot	2.50×10^3
Clear sky	Average luminance	8.0×10^3
Overcast sky	—	2.0×10^3
60 W incandescent lamp	—	1.20×10^6
Tungsten-halogen lamp (3000 K CCT)	—	1.30×10^7
Tungsten-halogen lamp (3400 K CCT)	—	3.90×10^7
CFL	36-W twin tube	3.0×10^4
T-5 fluorescent lamp	14–35 W	2.0×10^4
T-8 fluorescent lamp	36 W	1.0×10^4
T-12 fluorescent lamp	Cool white 800 mA	1.0×10^4
High-pressure mercury lamp	1,000 W	2.0×10^8
Xenon short-arc lamp	1,000 W	6.0×10^8

Table B5-I Percent effective ceiling (floor cavity reflectance) for various reflectance combinations (Part I)

% Ceiling or floor reflectance (ρ_{cc}, ρ_{fc})	90				80				70		
% Wall reflectance (ρ_w)	**90**	**70**	**50**	**30**	**90**	**70**	**50**	**30**	**70**	**50**	**30**
Room cavity ratio (RCR)											
0.2	89	88	86	85	78	78	77	76	68	67	66
0.4	88	86	84	81	77	76	74	72	67	65	63
0.6	87	84	80	77	76	75	71	68	65	63	59
0.8	87	82	77	73	75	73	69	65	64	60	56
1.0	86	80	75	69	74	72	67	62	62	58	53
1.2	85	78	72	66	73	70	64	58	61	57	50
1.4	85	77	69	62	72	68	62	55	60	55	47
1.6	84	75	67	59	71	67	60	53	59	53	45
1.8	83	73	64	56	70	66	58	50	58	51	42
2.0	83	72	62	53	69	64	56	48	56	49	40
2.2	82	70	59	50	68	63	54	45	55	48	38
2.4	82	69	58	48	67	61	52	43	54	46	37
2.6	81	67	56	46	66	60	50	41	54	45	35
2.8	81	66	54	44	65	59	48	39	53	43	33
3.0	80	64	52	42	65	58	47	37	52	42	32
3.2	79	63	50	40	65	57	45	35	51	40	31
3.4	79	62	48	38	64	56	44	34	50	39	29
3.6	78	61	47	36	63	54	43	32	49	38	28
3.8	78	60	45	35	62	53	41	31	49	37	27
4.0	77	58	44	33	61	53	40	30	48	36	26
4.2	77	57	43	32	60	52	39	29	47	35	25
4.4	76	56	42	31	60	51	38	28	46	34	24
4.6	76	55	40	30	59	50	37	27	45	33	24
4.8	75	54	39	28	58	49	36	26	45	32	23
5.0	75	53	38	28	58	48	35	25	44	31	22

Table B5-II Percent effective ceiling (floor cavity reflectance) for various reflectance combinations (Part II)

% Ceiling or floor reflectance (ρ_{cc}, ρ_{fc})	**50**				**30**			**10**		
% Wall reflectance (ρ_w)	**70**	**50**	**30**	**10**	**50**	**30**	**10**	**50**	**30**	**10**
Room cavity ratio (RCR)										
0.2	49	48	47	30	29	29	28	10	10	09
0.4	48	47	45	30	29	28	26	11	10	09
0.6	47	45	43	30	28	26	25	11	10	08
0.8	47	44	40	30	28	25	23	11	10	08
1.0	46	43	38	30	27	24	22	12	10	08
1.2	45	41	36	30	27	23	21	12	10	07
1.4	45	40	35	30	26	22	19	12	10	07
1.6	44	39	33	29	25	22	18	12	09	07
1.8	43	38	31	29	25	21	17	13	09	06
2.0	43	37	30	29	24	20	16	13	09	06
2.2	42	36	29	29	24	19	15	13	09	06
2.4	42	35	27	29	24	19	14	13	09	06
2.6	41	34	26	29	23	18	14	13	09	06
2.8	41	33	25	29	23	17	13	13	09	05
3.0	40	32	24	29	22	17	12	13	09	05
3.2	39	31	23	29	22	16	12	13	09	05
3.4	39	30	22	29	22	16	11	13	09	05
3.6	39	29	21	29	21	15	10	13	09	04
3.8	38	29	21	28	21	15	10	14	09	04
4.0	38	28	20	28	21	14	09	14	09	04
4.2	37	28	20	28	21	14	09	14	08	04
4.4	37	27	19	28	21	14	09	14	08	04
4.6	36	26	18	28	20	13	08	14	08	04
4.8	36	26	18	28	20	13	08	14	08	04
5.0	35	25	17	28	19	13	08	14	08	04

Table B5-III Adjustment coefficients for 30% and 10% floor cavity reflectance

% Ceiling or floor reflectance (ρ_{cc}, ρ_{fc})	**80**				**50**			**30**		
% Wall reflectance (ρ_w)	**70**	**50**	**30**	**10**	**50**	**30**	**10**	**50**	**30**	**10**
Room cavity ratio (RCR)	**Effective floor reflectance (ρ_{fc} = 30%) (20% = 1.00)**									
1	1.092	1.082	1.075	1.068	1.049	1.044	1.040	1.028	1.026	1.023
2	1.079	1.066	1.055	1.047	1.041	1.033	1.027	1.026	1.021	1.017
3	1.070	1.054	1 042	1.033	1.034	1.027	1.020	1.024	1.017	1.012
4	1.062	1.045	1.033	1.024	1.030	1.022	1.015	1.022	1.015	1.010
5	1.056	1 038	1.026	1.018	1.027	1.018	1.012	1.020	1.013	1.008
6	1.052	1.033	1.021	1.014	1.024	1.015	1.009	1.019	1.012	1.006
7	1.047	1.029	1.018	1.011	1.022	1.013	1.007	1.018	1.010	1.005
8	1.044	1.026	1.015	1.009	1.020	1.012	1.006	1.017	1.009	1.004
9	1.040	1.024	1.014	1.007	1.019	1.011	1.005	1.016	1.009	1.004
10	1.037	1.022	1.012	1.006	1.017	1.010	1.004	1.015	1.009	1.003
Room cavity ratio (RCR)	Effective floor reflectance (ρ_{fc} = 10%) (20% = 1.00)									
1	0.923	0.929	0.935	0.940	0.956	0.960	0.956	0.973	0.976	0.979
2	0.931	0.942	0.950	0.958	0.962	0.968	0.974	0.976	0.980	0.985
3	0.939	0.951	0.961	0.969	0.967	0.975	0.981	0.978	0.983	0.988
4	0.944	0.958	0.969	0.978	0.972	0.980	0.986	0.980	0.986	0.991
5	0.949	0.964	0.976	0.983	0.975	0.983	0.989	0.981	0.988	0.993
6	0.953	0.969	0.980	0.986	0.977	0.985	0.902	0.982	0.989	0.995
7	0.957	0.973	0.983	0.991	0.979	0.987	0.994	0.983	0.990	0.996
8	0.960	0.976	0.986	0.993	9.981	0.988	0.995	0.984	0.991	0.997
9	0.963	0.978	0.987	0.994	0.983	0.990	0.996	0.985	0.992	0.998
10	0.965	0.980	0.965	0.998	0.984	0.991	0.997	0.986	0.993	0.998

Table B5-IV Coefficient of utilization (CU)

Effective floor reflectance (ρ_{fc})	20	20	20	20	20	20	20	20	20
Effective ceiling reflectance (ρ_{cc})	**80**	**80**	**80**	**70**	**70**	**70**	**50**	**50**	**50**
Effective wall reflectance (ρ_{w})	**50**	**0**	**10**	**50**	**30**	**10**	**50**	**30**	**10**
Room cavity ratio (RCR)									
0	0.83	0.83	0.83	0.72	0.72	0.72	0.50	0.50	0.50
1	0.72	0.69	0.66	0.62	0.60	0.57	0.43	0.42	0.40
2	0.63	0.58	0.54	0.54	0.50	0.47	0.38	0.35	0.33
3	0.55	0.49	0.45	0.47	0.43	0.39	0.33	0.30	0.29
4	0.48	0.42	0.37	0.42	0.37	0.33	0.29	0.26	0.23
5	0.43	0.36	0.32	0.37	0.32	0.28	0.26	0.23	0.20
6	0.38	0.32	0.27	0.33	0.28	0.24	0.23	0.20	0.17
7	0.34	0.28	0.23	0.30	0.24	0.21	0.21	0.17	0.15
8	0.31	0.25	0.20	0.27	0.21	0.18	0.19	0.15	0.13
9	0.28	0.22	0.18	0.24	0.19	0.16	0.17	0.14	0.11
10	0.25	0.20	0.16	0.22	0.17	0.14	0.16	0.12	0.10

Table B6 Power and efficiency of the common light sources

Light source	Power (W)	Lamp efficiency (lm/W)
Incandescent lamp	100	17
Linear tungsten-halogen lamp	300	20
T-5 fluorescent lamp (4 ft)	28	100
T-8 fluorescent lamp (4 ft)	32	90
CFL	26	70
Mercury vapor lamp	175	45
Metal-halide, low-voltage	100	80
Metal-halide, high-voltage	400	90
High-pressure mercury lamp	1000	50
Xenon short-arc lamp	1000	30
High-pressure sodium, low-voltage	70	90
High-pressure sodium, high-voltage	250	100
Low-pressure sodium, U-type	180	180

Table B7 Color temperature

Color temperature	Kelvin degrees
Candle	1,800 K
Sun on the horizon	2,000 K
Sodium vapor lamp	2,200 K
Incandescent light bulb	2,400 K – 2,700 K
Warm white fluorescent tube	2,700 K – 3,000 K
Metallic halogen lamp	3,000 K – 4,200 K
Halogen lamp	3,000 K – 3,200 K
Neutral white fluorescent tube	3,900 K – 4,200 K
Midday sunshine (cloudless sky)	5,500 K – 5,800 K
Solar spectrum AM 0	5,900 K
Daylight fluorescent tube	5,400 K– 6,100 K
Electronic flash	5,000 K – 6,500 K
Cloudy sky	7,000 K – 9,000 K

Note: Color temperature value describes the apparent color of the light source, which varies from the orange red of a candle flame (1,800 K) to the bluish white of an electronic flash (between 5,000 K and 6,500 K, depending on the manufacturer).

Table B8 Lamp voltage factor (VF)

Lamp type	−5	−4	−3	−2	0	+2
	Rated lamp voltage deviation (%)					
Fluorescent (magnetic ballast)	95	96	97	90	100	102
Fluorescent (electronic ballast)	–	–	–	–	100	–
Mercury (ballast)	88	90	92	95	100	105
Halogen	80	84	88	92	100	108
Incandescent	83	86	89	94	100	106
Mercury (constant power)	97	98	99	99	100	102
Metal halide	91	93	95	97	100	–
High-pressure sodium	–	–	–	–	100	–

Table B9 Typical mounting height for standard luminaires

Luminaire type/wattage	Type of road	Preferred mounting height (m)	Minimum mounting height (m)
CFL32	Minor	7.5	5.5
2 × LF14	Minor	7.5	5.5
2 × LF24	Minor	7.5	5.5
S70	Minor	7.5	5.5
S100	Major	9.0	7.5
S150	Major	10.5	9.0
S250	Major	12.0	10.0
S400		15.0	12.0

Legend: CFL compact fluorescent; LF, linear fluorescent; S, high pressure sodium

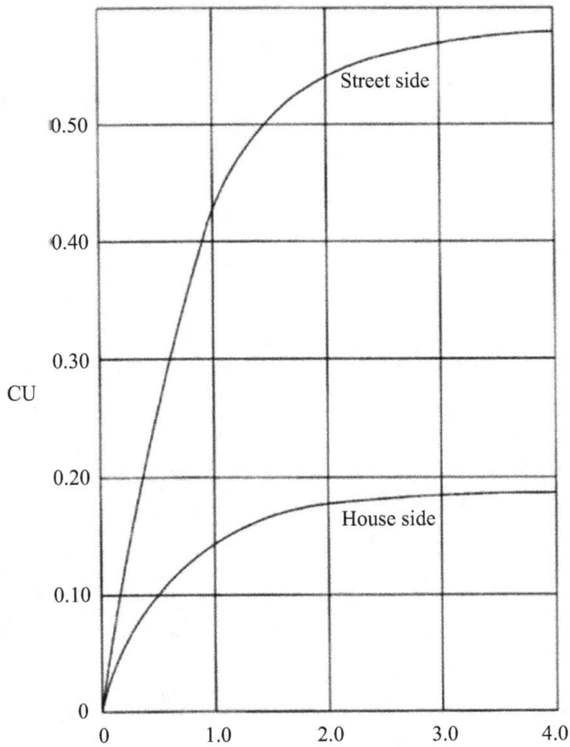

$$Ratio = \frac{\textit{Transverse distance (House side)}}{\textit{Luminaire moubnting height}}$$

Appendix C

Design parameters, conversion factors, and data for renewable energy conversion systems

Table C1 Solar collector important angles

Azimuth	Tilt	θ	Orientation
N/A	0°	90° − β	Horizontal (flat)
—	90°	Varies	Vertical wall
0°	90°	Varies	South facing vertical
−90°	90°	Varies	East facing wall
+90°	90°	Varies	West facing wall
α	90° − β	0°	Tracking system

Table C2 Standard test condition (solar energy)

Parameter	Value
Solar radiation	1,000 W/m^2
Solar spectrum	AM 1.5
Temperature	25 °C

Table C3 Monthly day number and recommended average days

Month	Day of month (*n*)	Date	Monthly average day (*n*)	Declination (δ-degrees)
January	i	17	17	−20.9
February	31 + i	15	47	−13.0
March	59 + i	16	75	−2.40
April	90 + i	15	105	9.40
May	120 + i	15	135	18.8
June	151 + i	10	162	23.1
July	181 + i	17	198	21.2
August	212 + i	16	228	13.5
September	243 + i	15	258	2.20
October	273 + i	15	288	−9.60
November	304 + i	14	318	−18.9
December	334 + i	10	344	−23.0

Table C4 Solar thermal collectors

Collector type	Temperature range (°C)	Concentration ratio
Flat-plate collector	30–80	1
Evacuated-tube collector	50–200	1
Compound parabolic collector	60–240	1–5
Fresnel lens collector	60–300	10–40
Parabolic trough collector	60–250	15–45
Cylindrical trough collector	60–300	10–50
Parabolic dish reflector	100–500	100–1,000
Heliostat field collector	150–2,000	100–1,500

Table C5 Mid-season monthly average daily extraterrestrial radiation (MJ/m^2)

Latitude	January	April	July	October
90° N	0.00	19.3	41.2	0.00
80° N	0.00	19.2	40.5	0.60
70° N	0.10	23.1	38.7	4.90
60° N	3.50	27.6	38.8	10.8
50° N	9.10	31.5	40.0	16.7
40° N	15.3	34.6	40.6	22.4
30° N	21.3	36.8	40.4	27.4
20° N	27.0	37.9	39.3	31.6
10° N	32.0	37.9	37.1	35.0
0°	36.2	36.7	34.0	37.3
10° S	39.5	34.5	29.9	38.5
20° S	41.8	31.3	25.2	38.6
30° S	43.0	27.2	19.9	37.6
40° S	43.1	23.3	14.2	35.5
50° S	42.3	16.8	8.40	32.4
60° S	41.0	10.9	3.10	28.4
70° S	40.8	5.00	0.00	24.0
80° S	42.7	0.60	0.00	20.6
90° S	43.3	0.00	0.00	204

Table C6 Thermal properties of some common materials

Material	**Density (kg/m³)**	**Heat capacity C_p (kJ/kg·K)**	**Temperature range (ΔT in °C)**
Water	1,000	4.190	0–100
Ethanol	780	2.460	−117 to 79
Glycerin	1,260	2.420	17–290
oil	910	1.800	−10 to 204
Synthetic oil	910	1.800	−10 to 400
Engine oil	883	1.880	0–160
Propanol	800	2.500	0–97
Brick	1,600	0.840	20–70
Concrete	2,240	1.130	20–70
Sandstone	2,200	0.712	20–70
Granite	2,650	0.900	20–70
Marble	2,500	0.880	20–70
Clay sheet			

Index

www.ingramcontent.com/pod-product-compliance
Lightning Source LLC
LaVergne TN
LVHW011854200926
848484LV00002B/453